# INSECT MOVEMENT
## MECHANISMS AND CONSEQUENCES

# Insect Movement Mechanisms and Consequences

## Proceedings of the Royal Entomological Society's 20th Symposium

*Edited by*

**I.P. Woiwod**
*Entomology and Nematology Department*
*IACR-Rothamsted*
*Harpenden*
*UK*

**D.R. Reynolds**
*Natural Resources Institute*
*University of Greenwich*
*Chatham Maritime*
*UK*

*and*

**C.D. Thomas**
*Centre for Biodiversity and Conservation*
*University of Leeds*
*Leeds*
*UK*

CABI *Publishing*

CABI *Publishing* is a division of CAB *International*

CABI Publishing
CAB International
Wallingford
Oxon OX10 8DE
UK

CABI Publishing
10 E 40th Street
Suite 3203
New York, NY 10016
USA

Tel: +44 (0)1491 832111
Fax: +44 (0)1491 833508
Email: cabi@cabi.org
Web site: http://www.cabi.org

Tel: +1 212 481 7018
Fax: +1 212 686 7993
Email: cabi-nao@cabi.org

© CAB *International* 2001. All rights reserved. No part of this publication may be reproduced in any form or by any means, electronically, mechanically, by photocopying, recording or otherwise, without the prior permission of the copyright owners.

A catalogue record for this book is available from the British Library, London, UK.

**Library of Congress Cataloguing-in-Publication Data**
Royal Entomological Society of London. Symposium (20th : 1999)
 Insect movement : mechanisms and consequences : proceedings of the
Royal Entomological Society's 20th Suymposium / edited by I. Woiwood,
D. R. Reynolds, and C. D. Thomas
   p. cm.
 Includes bibliographical references and index.
 ISBN 0-85199-456-3 (alk. paper)
   1. Insects--Flight--Congresses.  I. Woiwood, I. (Ian)  II. Reynolds, D. R.
(Donald Russell), 1947–   III. Thomas, C. D.  IV. Title

QL496.7.R68 1999
573.7'98157--dc21                                                            00-046778

ISBN 0 85199 456 3

Typeset by AMA DataSet Ltd, UK.
Printed and bound in the UK by Biddles Ltd, Guildford and King's Lynn.

# Contents

| | | |
|---|---|---|
| **Contributors** | | vii |
| 1 | Introduction and Overview<br>I.P. Woiwod, D.R. Reynolds and C.D. Thomas | 1 |
| 2 | The Biomechanics and Functional Diversity of Flight<br>R. Dudley | 19 |
| 3 | How Insect Wings Evolved<br>R.J. Wootton | 43 |
| 4 | Physiology and Endocrine Control of Flight<br>G. Goldsworthy and M. Joyce | 65 |
| 5 | Insect Behaviours Associated with Resource Finding<br>J. Hardie, G. Gibson and T.D. Wyatt | 87 |
| 6 | Host Location by Parasitoids<br>W. Powell and G. Poppy | 111 |
| 7 | Flight Trajectories of Foraging Insects: Observations Using Harmonic Radar<br>J.R. Riley and J.L. Osborne | 129 |

| 8 | The Evolution of Migratory Syndromes in Insects<br>*H. Dingle* | 159 |
|---|---|---|
| 9 | Orientation Mechanisms and Migration Strategies Within the Flight Boundary Layer<br>*R.B. Srygley and E.G. Oliveira* | 183 |
| 10 | Characterizing Insect Migration Systems in Inland Australia with Novel and Traditional Methodologies<br>*V.A. Drake, P.C. Gregg, I.T. Harman, H.-K. Wang, E.D. Deveson, D.M. Hunter and W.A. Rochester* | 207 |
| 11 | Significance of Habitat Persistence and Dimensionality in the Evolution of Insect Migration Strategies<br>*R.F. Denno, C. Gratton and G.A. Langellotto* | 235 |
| 12 | Predation and the Evolution of Dispersal<br>*W.W. Weisser* | 261 |
| 13 | Evolution of Mass Transit Systems in Ants: a Tale of Two Societies<br>*N.R. Franks* | 281 |
| 14 | Dispersal and Conservation in Heterogeneous Landscapes<br>*N.F. Britton, G.P. Boswell and N.R. Franks* | 299 |
| 15 | Scale, Dispersal and Population Structure<br>*C.D. Thomas* | 321 |
| 16 | Gene Flow<br>*J. Mallet* | 337 |
| 17 | Use of Genetic Diversity in Movement Studies of Flying Insects<br>*H.D. Loxdale and G. Lushai* | 361 |
| 18 | Coping with Modern Times? Insect Movement and Climate Change<br>*C. Parmesan* | 387 |
| 19 | Analysing and Modelling Range Changes in UK Butterflies<br>*J.K. Hill, C.D. Thomas, R. Fox, D. Moss and B. Huntley* | 415 |
| Index | | 443 |

# Contributors

**Graeme P. Boswell**, Centre for Mathematical Biology and Department of Mathematical Sciences, University of Bath, Bath BA2 7AY, UK.

**Nicholas F. Britton**, Centre for Mathematical Biology and Department of Mathematical Sciences, University of Bath, Bath BA2 7AY, UK.

**Robert F. Denno**, Department of Entomology, University of Maryland, College Park, MD 20742, USA.

**Edward D. Deveson**, Australian Plague Locust Commission, Agriculture Forestry and Fisheries Australia, GPO Box 858, Canberra, ACT 2601, Australia.

**Hugh Dingle**, Department of Entomology, University of California-Davis, One Shields Avenue, Davis, CA 95616, USA.

**V. Alistair Drake**, School of Physics, University College, The University of New South Wales, Australian Defence Force Academy, Canberra, ACT 2600, Australia.

**Robert Dudley**, Section of Integrative Biology, University of Texas, Austin, TX 78712, USA, and Smithsonian Tropical Research Institute, PO Box 2072, Balboa, Republic of Panama.

**Richard Fox**, Butterfly Conservation, Conservation Office, PO Box 444, Wareham, Dorset BH20 5YA, UK.

**Nigel R. Franks**, Centre for Mathematical Biology, Department of Biology and Biochemistry, University of Bath, Bath BA2 7AY, UK.
*Present address*: School of Biological Sciences, University of Bristol, Woodland Road, Bristol BS8 1UG, UK.

**Gabriella I. Gibson**, Pest Management Department, Natural Resources Institute, University of Greenwich, Central Avenue, Chatham Maritime, Kent ME4 4TB, UK.
**Graham Goldsworthy**, Department of Biology, Birkbeck College, University of London, Malet Street, London WC1E 7HX, UK.
**Claudio Gratton**, Department of Entomology, University of Maryland, College Park, MD 20742, USA.
**Peter C. Gregg**, School of Rural Science and Natural Resources, University of New England, Armidale, NSW 2351, Australia.
**Jim Hardie**, Aphid Biology Group, Department of Biology, Imperial College at Silwood Park, Ascot, Berkshire SL5 7PY, UK.
**Ian T. Harman**, School of Physics, University College, The University of New South Wales, Australian Defence Force Academy, Canberra, ACT 2600, Australia.
**Jane K. Hill**, Environmental Research Centre, Department of Biological Sciences, University of Durham, Durham DH1 3LE, UK.
*Present address*: Department of Biology, University of York, PO Box 373, York YO10 5YW, UK.
**David M. Hunter**, Australian Plague Locust Commission, Agriculture Forestry and Fisheries Australia, GPO Box 858, Canberra, ACT 2601, Australia.
**Brian Huntley**, Environmental Research Centre, Department of Biological Sciences, University of Durham, Durham DH1 3LE, UK.
**Michael Joyce**, Department of Biology, Birkbeck College, University of London, Malet Street, London WC1E 7HX, UK.
**Gail A. Langellotto**, Department of Entomology, University of Maryland, College Park, MD 20742, USA.
**Hugh D. Loxdale**, Entomology and Nematology Department, IACR-Rothamsted, Harpenden, Hertfordshire AL5 2JQ, UK.
**Gugs Lushai**, Biodiversity and Ecology Division, School of Biological Sciences, Basset Crescent East, Southampton University, Southampton SO16 7PX, UK.
*Present address*: Crop Protection Program, Eastern Cereal and Oilseed Research Centre, K.W. Neatby Building, C.E.F., Ottawa, Ontario, Canada, K1A 0C6.
**James Mallet**, Galton Laboratory, Department of Biology, University College London, 4 Stephenson Way, London NW1 2HE, UK.
**Dorian Moss**, Centre for Ecology and Hydrology – Monks Wood, Abbots Ripton, Huntingdon, Cambridgeshire PE17 2LS, UK.
**Evandro G. Oliveira**, Departamento de Biologia Geral, Instituto de Ciências Biológicas, Universidade Federale de Minas Gerais, 30161-970 Belo Horizonte, MG, Brasil, and Smithsonian Tropical Research Institute, Apdo. 2072, Balboa, Republic of Panama.
**Juliet L. Osborne**, Entomology and Nematology Department, IACR-Rothamsted, Harpenden, Hertfordshire AL5 2JQ, UK.

**Camille Parmesan**, Integrative Biology, Patterson Labs Building, University of Texas, Austin, TX 78712, USA.
**Guy Poppy**, Entomology and Nematology Department, IACR-Rothamsted, Harpenden, Hertfordshire AL5 2JQ, UK.
**Wilf Powell**, Entomology and Nematology Department, IACR-Rothamsted, Harpenden, Hertfordshire AL5 2JQ, UK.
**Don R. Reynolds**, Natural Resources Institute, University of Greenwich, Central Avenue, Chatham Maritime, Kent ME4 4TB, UK.
**Joseph R. Riley**, Radar Entomology Unit, Natural Resources Institute, University of Greenwich, Leigh Sinton Road, Malvern, Worcestershire WR14 1LL, UK.
**Wayne A. Rochester**, Department of Zoology and Entomology, University of Queensland, Brisbane, Queensland 4072, Australia.
**Robert B. Srygley**, Department of Zoology, University of Oxford, South Parks Road, Oxford OX1 3PS, UK, and Smithsonian Tropical, Research Institute, PO Box 2072, Balboa, Republic of Panama.
**Chris D. Thomas**, Centre for Biodiversity and Conservation, School of Biology, University of Leeds, Leeds LS2 9JT, UK.
**Hai-kou Wang**, School of Physics, University College, The University of New South Wales, Australian Defence Force Academy, Canberra, ACT 2601, Australia.
**Wolfgang W. Weisser**, Zoology Institute, University of Basel, Rheinsprung 9, 4051 Basel, Switzerland.
*Present address*: Institute of Ecology, Friedrich-Schiller-University, Dornburger Str. 159, 07743 Jena, Germany.
**Ian P. Woiwod**, Entomology and Nematology Department, IACR-Rothamsted, Harpenden, Hertfordshire AL5 2JQ, UK.
**Robin J. Wootton**, School of Biological Sciences, University of Exeter, Exeter EX4 4PS, UK.
**Tristram D. Wyatt**, Department of Zoology and Department of Continuing Education, University of Oxford, South Parks Road, Oxford OX1 3PS, UK.

# Introduction and Overview      1

Ian P. Woiwod,[1] Don R. Reynolds[2] and Chris D. Thomas[3]

[1]*Entomology and Nematology Department, IACR-Rothamsted, Harpenden, Hertfordshire AL5 2JQ, UK;* [2]*Natural Resources Institute, University of Greenwich, Central Avenue, Chatham Maritime, Kent ME4 4TB, UK;* [3]*Centre for Biodiversity and Conservation, School of Biology, University of Leeds, Leeds LS2 9JT, UK*

This book originated in the 20th Symposium of the Royal Entomological Society on 'Insect Movement', held in London during September 1999. It is now over a quarter of a century since a previous symposium of the Society on a similar theme, namely 'Insect Flight', which was the subject of the 7th Symposium in 1973 (Rainey, 1976). In this introductory chapter, we briefly review progress and identify some of the 'landmark' developments during the intervening period. The present book is broader in scope than the 1976 one, as it includes material on pedestrian movements as well as flight (for example, the movement of army ants, see Chapters 13 and 14). Moreover, as the title of this book suggests, we have not only included chapters on what might be called the *mechanisms* of movement (e.g. flight biomechanics, physiology and behaviour), but also some of its *consequences* for population structure and gene flow, and for geographical changes in species' ranges. Also, the more practically orientated chapters tend to be directed towards conservation goals (e.g. Chapters 14, 18 and 19), rather than towards pest management, as were some of the contributions to Rainey (1976) and several more recent books on insect movement (e.g. Goldsworthy and Wheeler, 1989; Drake and Gatehouse, 1995).

## Flight Biomechanics and Physiology

Among invertebrates, only the pterygote Insecta have evolved flight, and their great diversity and abundance in terrestrial habitats can be partly attributed to this ability (notwithstanding the secondarily apterous condition of many taxa). The origin of insect wings, and of powered flight, are fascinating subjects of wide general interest and are discussed in the chapters by Robert Dudley (Chapter 2) and Robin Wootton (Chapter 3). In this connection, we note that an eminent contribution to the 1973 symposium was V.B. Wigglesworth's revival of the 'gill-plate theory' of wing origin (Wigglesworth, 1976) – an unfashionable theory at the time, but one which has now become the majority view (Wootton, Chapter 3).

In the absence of direct evidence of Devonian 'missing link' fossils, there is much room for speculation on the early evolution of flight. Did the 'proto-wings' of ancestral insects arise from paranotal lobes which increased in size while being used for, say, display or thermoregulation purposes until they were large enough to allow parachuting/gliding from tall plants? Alternatively, did the original winglets evolve from gill covers, or some homologous structure, in aquatic insects in order to assist them in skimming across the water's surface? The jury is apparently still out over which of these two major hypotheses (or perhaps others) is the most likely (Dudley, Chapter 2; Wootton, Chapter 3; Grodnitsky, 1999). Dudley's chapter then goes on to explore several topics in the later evolution of insect flight in a review which provides a taste of his new book (Dudley, 2000). For example, he discusses the implications for flight of the miniaturization of body size which has occurred in many insect lineages, and how reciprocal interactions between insect flight capabilities and biological selective agents acting on them have led to escalating increases in flight performance.

Wootton (Chapter 3) reviews various evolutionary developments in the biomechanics and aerodynamics of insect flight, and shows how deductions from the structure of fossil and recent insects, combined with theoretical studies and laboratory experiments, can reveal how Palaeozoic insects actually flew. It has become progressively clearer that conventional aerodynamics, which deals mainly with steady airflows (e.g. Nachtigall, 1989), is inadequate to explain many aspects of insect flight. The first of the 'unsteady' mechanisms which insects can use to generate lift, the so-called 'clap and fling', was described by Weis-Fogh to the 1973 symposium (Weis-Fogh, 1976). Recently, however, elegant laboratory work, which has included the use of kinematics and mechanical models (Ellington *et al.*, 1996; Dickinson *et al.*, 1999), has revealed three other mechanisms (delayed stall, rotational circulation and wake capture) which may be important in explaining lift generation and manoeuvrability of a range of insect species (see also Wootton, 1999).

Flying is a very energetically expensive activity (metabolic rates can be 50–200 times greater than in a resting individual (Dudley, Chapter 2)) and it can make great physiological demands on the individual. In Chapter 4, Graham Goldsworthy and Michael Joyce review an important aspect of flight-related physiology – how energy-providing fuels are supplied to the flight muscles. They first discuss the metabolic pathways of carbohydrate and fat utilization, and the hormonal control mechanisms, in locusts (about which most is known) before going on to outline some recent results with several species of beetles where the amino acid proline is an important fuel both for flight and for pre-flight warm-up. Attention is drawn to the apparent advantages of proline as a flight fuel, compared with lipid or carbohydrate.

## Defining Types of Movement Behaviour

One of the most significant advances of recent years has been the improvement in the conceptual definition of various movement behaviours, particularly following Kennedy's (1985) classic paper, in which he removed the confusion caused by the confounding of behavioural and ecological definitions of migration. It is now widely (but by no means universally) accepted that migration is best defined as a *behavioural* process, and it is the consequences of migratory (and other) movements that can be explained in ecological terms (Gatehouse, 1987, 1997; Drake and Gatehouse, 1995; Dingle, 1996). Kennedy's (1985) definition (see p. 88, this volume) brings out the following characteristics of migration:

- Migratory movement may be accomplished by the animal's own locomotory exertions, *or* it may involve a transporting 'vehicle' such as wind, water, or phoresis on another organism, but in each case the migration is *actively* initiated and maintained through specialized behaviour patterns. Initiation may take a number of forms, ranging from the sustained climbing flight to high altitude (as observed in many radar studies, e.g. Schaefer, 1976) to the behaviour which increases drag on the body in minute scale-insect 'crawlers' before these are carried away by the wind (Washburn and Washburn, 1984).
- Migratory flight is often very persistent, and the insect's track is typically straightened out (by the wind, for example, or due to the lack of turning movements) so that the migrant covers new ground rather than remaining in a localized area (Kennedy, 1975).
- Responsiveness to stimuli which promote 'station-keeping' flight and settling are inhibited or depressed to a certain extent, so that escape from a 'vegetative' (see below) or diapause site can occur.
- As migration continues, there is a gradual increase in responsiveness to the stimuli that will eventually halt it. In some aphids, at least, interruption of settling responses can promote further and stronger migratory activity, so there appears to be a complex system of antagonistic inhibitory and excitatory interactions between migratory and 'vegetative' behaviours (see summary of Kennedy's work on *Aphis fabae* in Dingle, 1996, and in Chapter 8, this volume).

Some common features of insect migration, and a list of co-evolved traits that make up the 'migratory syndromes' are given in Drake *et al.* (1995; see also Chapter 8 of this volume).

'Vegetative' movements differ from migration in that they are directed towards the exploitation of resources, particularly those required for growth and reproduction (food, shelter, mates, oviposition sites, etc.), and these movements are readily interrupted by an encounter with the resource items (Kennedy, 1985; Dingle, 1996 and Chapter 8; Hardie *et al.*, Chapter 5). They often tend to be station-keeping and retain the animal within a locality which can be equated with the 'home-range' of vertebrates (Dingle, 1996). If the required resource is not available, or is not found within a given time, an individual may switch to a different mode of behaviour which may allow it to explore a wider area than its local habitat patch or home-range. These movements, sometimes called 'ranging', will still tend to cease if the resource is found – there is no prolonged inhibition of the vegetative responses as there is with migration.

Modern behavioural definitions have made the term 'dispersal' redundant as a synonym for migration, and it has been argued (Kennedy, 1985; Dingle, 1996) that its use in this sense should be discouraged because it confuses individual behaviour with the consequences for the population. Nevertheless, some authors (e.g. Weisser, Chapter 12) have preferred to retain the term 'dispersal', partly because it is not always possible to know the behavioural state of an individual insect during a particular movement. Dispersal is perhaps best reserved to describe the increase in separation between members of a population which may occur as a result of their migratory, vegetative or accidental movements. A more neutral term such as *population redistribution* may, however, be a better way of describing the ecological consequences of movement, because some so-called 'dispersals' may well result in *aggregation* of the population rather than spatial spread (for example, the bringing together of African armyworm moths and other airborne migrants by wind convergence – see Gatehouse, 1997). At present, there is no unanimity on the definitions of insect movement (or animal movement generally), but attempts to attain this should identify 'exceptions which prove the rule' and thus improve the conceptual framework of movement behaviour as Kennedy (1985) and Dingle (1996) have emphasized.

## 'Vegetative' or Resource-finding Movements

The next chapters are devoted to movement behaviours associated with resource finding, which include foraging and 'search' movements (Bell, 1991). Jim Hardie, Gabriella Gibson and Tristram Wyatt provide an overview of the subject (Chapter 5) while Wilf Powell and Guy Poppy (Chapter 6) concentrate on host location in parasitoids, and Joe Riley and Juliet Osborne (Chapter 7) outline some recent empirical studies of bee and moth foraging flights.

Hardie et al. start with the very simple 'building blocks'– the taxes and kineses which underlie some more elaborate movement behaviours – before going on to discuss more complex 'indirect' orientations (such as those used by insects flying to the source of odour plumes) and 'direct' visual orientation to, say, shape and colour of hosts. They then discuss the mechanisms used in three examples of resource-directed movements: mate location by male moths, and host location by haematophagous insects and by phytophagous insects. A common theme is that all insects are trying to find their resources against a background of environmental 'noise', and that the efficiency of sensory processes involves a combination of sensitivity and selectivity.

Recent studies have begun to show the complexity and subtlety of host finding behaviour in parasitoid species, particularly in their use of chemical cues (Powell and Poppy, Chapter 6). For example, it has now been shown that when parasitoids are host searching some utilize the plant volatiles induced by their hosts' herbivory whilst others make use of the pheromones produced for communication amongst their hosts. In the first case it is still not clear whether the plants are actually producing the chemicals to recruit parasitoids and predators of their unwelcome herbivores or the chemicals are just by-products of feeding damage. Whatever the reason, this response is clearly advantageous to plant and parasitoid alike. The intriguing insights described by Powell and Poppy are mainly derived from carefully controlled laboratory experiments and, as they point out, we still know relatively little about what actually happens under field conditions. This is even more true for long-range movements of parasitoids, particularly as *Aphidius*, *Praon* and *Diaeretiella* spp. have all been found in a preliminary examination of 12-m suction trap catches of the Rothamsted Insect Survey. Interestingly such catches are predominantly (> 80%) female (W. Powell, personal communication) and the implication that high level wind-borne migration of parasitoids may occur regularly poses important questions about the effect this might have on the population dynamics of these species.

Several field investigations of foraging behaviour using a newly developed harmonic scanning radar are described by Riley and Osborne (Chapter 7). One study provided direct evidence that foraging bumblebees (*Bombus* spp.) could compensate accurately for sideways drift whilst in flight, and maintain straight tracks towards their nests or forage sources in a variety of wind speeds and directions. Subtraction of the local wind vector from the bees' velocity showed that they were flying obliquely over the ground, but how did the bees know how far to turn off course to achieve the correct track to their destinations? The simplest explanation appears to be that the bees simply adjusted their headings so that the direction of ground image flow across their retinas was maintained at an angle *relative to the sun's azimuth* which corresponded to their intended track direction.

The foraging behaviour of army ant workers is spectacular, with vast numbers of workers moving out of a nest site in repeatedly branching columns which allows the ants to scour extensive areas and capture, cut up and carry back to the nest any small creature found in their path. Nigel Franks (Chapter 13) shows that although these raids are extremely efficient and flexible, their 'self-organization' can be accounted for by a few simple rules. He also shows that, notwithstanding the remarkable convergent evolution of the New World *Eciton* and Old World *Dorylus* army ants, differences between the two genera, for example in the polymorphism of the prey-carrying workers, and in the timing of raids (and colony migrations), can be explained as adaptations to differing rates of recovery of the respective prey populations.

## Migration

Hugh Dingle in the introduction to his chapter (8) briefly reviews the historical development of our ideas on what migration is, and how it can be separated from the other main class of movements, emphasizing the pioneering work of C.G. Johnson and J.S. Kennedy, before moving on to discussing *migration syndromes* in insects. Migratory behaviour is not an isolated trait, but evolves as part of a suite of characters influencing the behaviour, physiology and morphology of insects. Although this has long been suspected, or known in outline (e.g. the 'oogensis-flight' syndrome of Johnson (1969)), much of the detailed evidence has come from a very extensive series of studies by Dingle and his collaborators on the evolution and genetics of migratory syndromes in the milkweed bug, *Oncopeltus fasciatus* and, more recently, from work on the soapberry bug, *Jadera haematoloma*. Traits, which may include variation in flight duration, wing length, wing muscle development, fecundity and (in *Jadera*) length of the mouthpart stylets, often show a high degree of genetic correlation (indicating that the traits may share genes in common) and this has probably facilitated the evolution of adaptive syndromes in migratory forms of the two hemipterans. Dingle also discusses evidence that genes mediate behaviour and other manifestations of the syndromes through control of hormone, particularly juvenile hormone, production.

Some insects, particularly some butterflies, typically migrate within their *flight boundary layer* (Taylor, 1974), a layer of air (extending a variable distance up from the ground) where the wind speed is lower than insects' flight speeds, and where they can therefore control their displacement direction. Orientation mechanisms in these migrants are reviewed in Chapter 9 by Robert Srygley and Evandro Oliveira. Typical long-distance 'boundary-layer' migrants are usually relatively large day-flying insects, and their low-altitude flight is obviously easier to observe than that of most high-altitude windborne migrants. Despite this, the migrations have seldom been accurately quantified in terms of the migrants' velocities with respect to the air and to the ground. Srygley and Oliveira's chapter also describes some innovative experiments in which this was achieved by using boats to pace Lepidoptera and Odonata flying over the sea or a lake, and then measuring the individual's airspeed (or ground speed) and track with onboard navigational equipment. The experiment revealed that some species were able to compensate for fluctuating crosswinds and maintain a straight track over large bodies of water, although further work is required to identify exactly which mechanisms are used to achieve this compensation.

Most long-distance insect migrants (and even some butterflies when the wind direction coincides with their 'preferred' migration direction) utilize the wind to transport them faster, and usually further, than would be possible by self-powered flight (see Pedgley *et al.*, 1995; Johnson, 1995; Gatehouse, 1997 for recent reviews). In doing so, most species have evolved specialized behaviour which permits them to ascend out of their flight boundary layer and into fast-moving airstreams, often at heights of several hundred metres from the ground. These migrations are intrinsically difficult to study, and radar and ancillary technologies have had an important role to play (see Technologies section below). Current developments are indicated by Alistair Drake and his colleagues (Chapter 10) who describe a large multidisciplinary study (utilizing a new insect-monitoring radar, remote sensing, ground trapping and surveys) of the migration and population distribution of the Australian plague locust *Chortoicetes terminifera* and the native budworm *Helicoverpa punctigera* in inland Australia. The study aims to relate population movements to the changing distribution of resources, with the twin objectives of developing understanding of how the species' migratory adaptations function and of improving pest-forecasting capabilities. Findings are discussed within the framework of a holistic conceptual model of migration systems, particularly those of highly mobile organisms in rapidly changing environments (Drake *et al.*, 1995; Drake and Gatehouse, 1996).

As indicated above, insect groups showing distinctive polymorphisms in wing length or flight muscle development are good subjects for studies of the ecological and physiological constraints influencing the evolution of migration and other life-history traits. Robert Denno, Claudio Gratton and Gail Langellotto (Chapter 11) review evidence from a long-running study by Denno and his colleagues on the saltmarsh delphacid planthoppers *Prokelisia*, and from analyses of the literature on other wing-dimorphic delphacid species, which shows that habitat persistence and structure are important determinants in the evolution of insect migration strategies. They also discuss how selection for migratory capability may constrain associated reproductive traits in *both* sexes of delphacids – in other words, there is a trade-off between maintaining flight capacity and maximizing reproductive success in planthopper populations.

Wolfgang Weisser (Chapter 12) provides evidence that the dispersal movement of aphids and other prey are affected by predation, and discusses the evolutionary implications of this. Most of the literature about migration implicitly or explicitly characterizes migration as an evolutionary response to deteriorating resources locally and/or to improving resources elsewhere. Of course, the important issue is whether fitness is, on average, increased or decreased by moving, and escape from natural enemies could be a much more important determinant of migration than is generally appreciated.

## Spatial Ecology and Ecological Genetics

Insect movements obviously have profound implications for the ecology and genetics of insect populations. Within the discipline of ecology, the growing appreciation of the central importance of the spatial dimension in population dynamics is one of the most significant developments over the last quarter century. Before this, efforts were made to monitor the distribution and abundance of some highly migratory species, particularly locusts (Pedgley, 1981) and aphids (Tatchell, 1991), and some ecologists had attempted to include spatial and temporal dynamics in a single model, but generally abundance and distribution were treated separately (Turchin, 1998). Frequently, difficulties in measuring the movements of insects, and explicitly modelling their effects on populations, led to the topic being effectively ignored. With the advances in theory (e.g. metapopulation dynamics), the widespread availability of powerful personal computers and software able to cope with more realistic spatial models, and the development of new molecular techniques in ecology and population genetics, there has been an enormous increase in both theoretical and empirical work in this area (e.g. Hanski and Gilpin, 1997; Tilman and Kareiva, 1998; Turchin, 1998; Hanski, 1999; Hassell, 2000).

The integration of spatial processes into studies of population dynamics is exemplified by the development of metapopulation theory, in which networks of local populations are 'glued' together by movement (Hanski and Gilpin, 1997; Hanski, 1999). Each local population may be small, and prone to extinction, but the entire metapopulation survives because re-colonizations of empty habitat patches are sufficient to balance the extinctions that do take place. Without movement, these systems would become extinct, as the local populations die out, one by one. With movement, the metapopulation may survive indefinitely. Much of the conceptual development of metapopulation theory has been achieved by models that do not take any account of the precise geographic locations of each population and of each empty habitat patch. However, to test the theory empirically, the broad conceptual models must be converted into spatially explicit models that make real predictions about real landscapes. This has been achieved in various ways, mainly by logistic regression models, incidence function models (both of which usually assume that habitat patches are distinct), and by cellular automata (CA) models (which can be used with single cells representing patches, or with many cells per habitat patch; Hanski, 1999).

Critics of this approach have argued that few natural population systems fit a strict definition of a metapopulation, and that most so-called metapopulations could more appropriately be described in some other way (Harrison, 1991; Harrison and Taylor, 1997). The trouble is that empirical systems are extremely complex, and do not fit *any* strict definition. A way out of the quagmire of definitions is outlined in Chapter 15 (Chris Thomas), which redirects attention away from definitions of population types, back towards the fundamental processes: spatial patterns of birth and death, along with the movement of individuals. It is not necessary to decide whether metapopulations are common or rare, but to decide, for a particular system at the spatial scale of interest, which type of conceptual and modelling approach will provide the greatest insight (and predictive power). The cellular automata approach is adopted by Nick Britton, Graeme Boswell and Nigel Franks (Chapter 14) and applied to army ants, showing that within-habitat spatial dynamics may influence the ability of colonies to persist on habitat fragments of different total area. These predictions are at least in qualitative agreement with the distribution of army ants still surviving on forested islands that became isolated when the Panama Canal was formed. This shows nicely how spatial processes at one spatial scale (within islands) may lead to spatial patterns observed at another spatial scale (relationship between island area and whether they are occupied by army ants).

Concern over issues of habitat fragmentation (Britton *et al.*, Chapter 14; Thomas, Chapter 15) has focused attention on the role of movement in determining the distributions and persistence of non-migratory species at quite large spatial scales. Interest in the responses of insects to recent climate change is beginning to illustrate how these same processes are resulting in changes to entire geographic distributions (Camille Parmesan, Chapter 18; Jane Hill *et al.*, Chapter 19). Very many species are now shifting their distributions in directions that are consistent with these changes being caused by climate warming, and these shifts are taking place even in non-migratory species (Parmesan, Chapter 18). Interestingly, one does not need to invoke incredibly rare, long-distance movements for these species to achieve these changes in distribution. The 'normal' processes of local extinctions and distance-dependent colonization, year after year, can produce shifts to species ranges that have, in many cases, approximately kept up with the rate of climate warming. The serious concern is whether habitat specialists will be able to track the distribution of suitable thermal environments, across increasingly fragmented landscapes. There is some indication that even some relative generalists may be lagging behind the climate, because of habitat fragmentation (Hill *et al.*, Chapter 19).

These studies of range changes have been particularly useful in detecting the tail of the distribution of dispersal distances – the individuals that move the furthest and that are responsible for establishing new populations beyond the existing range margin. Such individuals are difficult or impossible to detect directly during mark–release–recapture programmes. One possible way out of this is to use genetic markers, estimating the level of movement between populations from the degree of genetic differentiation between them, based on the assumption that genetic drift and gene flow are at equilibrium. Hugh Loxdale and Gugs Lushai (Chapter 17) show how this approach may be adopted. Jim Mallet (Chapter 16) makes an important contribution to this field, arguing that estimates of migration rates are likely to be wrong, based on equilibrium assumptions. Gene flow and genetic drift in neutral genetic markers can take a very long time to come to equilibrium, in many cases longer than the present interglacial: the last 10,000 years. The whole field of research is dominated by work on patterns of neutral genetic variation. He argues, correctly, that selected traits come to equilibrium with gene flow far faster than does neutral variation. Provided we can measure the strength of selection in different locations, reliable estimates of gene flow may be possible.

## Technologies

Finally, we give a brief outline of some of the technological developments which have revolutionized insect movement research over the last quarter of a century. Further references to methodologies and techniques for quantifying insect movement and studying changes in spatial distribution can be found in Bell (1991), Dingle (1996), Reynolds *et al.* (1997), Wyatt (1997), Turchin (1998) and Southwood and Henderson (2000). (Note also forthcoming issue of *Computers and Electronics in Agriculture*, edited by V.A. Drake and J.R. Riley.)

The main over-arching technical advances have been the increasingly widespread availability of powerful micro- and mini-computers which, as in other branches of science, has transformed data handling and analysis, and more recently, the role of the Internet in facilitating data transfer, information dissemination and general communication between scientists. More specifically, the steadily increasing performance of desktop computers has allowed researchers to work more easily with complex models of spatial dynamics. Some of the types of simulation models used to analyse the spatial structure of populations and the causes and effects of insect movement have been mentioned above.

Sometimes it is important to retain the actual geographical coordinates of an insect population, particularly when one wants to look for relationships with other geographically referenced data such as those from environmental remote sensing (see below). Management of such spatial information is usually carried out within a geographical information system (GIS) which is a computer system for collecting, storing, manipulating and displaying spatial referenced data (Liebhold *et al.*, 1993). It is clear that the integration of movement modelling in population and landscape ecology will often require the integration of GIS and simulation technologies into a single environment with both database and simulation strengths. In entomology, GIS management systems have been developed for processing data on the habitats and populations of migrant pests in order to predict outbreaks and movements (Tappan *et al.*, 1991; Robinson, 1995; Williams and Liebhold, 1995; Healey *et al.*, 1996; Drake *et al.*, Chapter 10). GIS-based analysis has also been used for conservation-related purposes. For example, the mapping of range changes in butterflies and relating these to long-term changes in climate (Parmesan, Chapter 18; Hill *et al.*, Chapter 19).

Another obvious way in which computer modelling has influenced work on insect movement is through *operational* forecasting systems for migrant pests. These often consist of a data-management system combined with a set of 'decision tools' to aid surveillance and outbreak prediction (Knight *et al.*, 1992; Day and Knight, 1995; Day *et al.*, 1996; Tucker and Holt, 1999; Deveson and Hunter, 2000).

## Remote sensing

Remote sensing is the science of obtaining information about any object by measurement made at a distance from that object, and it thus includes radar, optical and acoustical methods of directly observing insects (Riley, 1989 – see below). Nowadays, however, it often refers to earth observation from airborne or satellite platforms (Sabins, 1997). A quarter of a century ago, the entomological application of satellite remote sensing was very much in its infancy (Pedgley, 1973). Today the products from several earth resources and meteorological satellites are used routinely to map the insect habitats and to monitor environmental variables (for example, vegetation type or condition, surface temperature, rainfall estimation, soil moisture or inundation) which may affect the distribution and movement of insect populations (e.g. Hay *et al.*, 1997; Tucker and Holt, 1999; Drake *et al.*, Chapter 10). Remote-sensing data are often placed in a GIS with other information so that the relationship between variables can be studied in a spatial context. New technology has also greatly facilitated the recording of the geographical locations of insect populations or individuals in the field. For example, the location of groups of desert locusts or the edge of an armyworm infestation can be determined to an accuracy of a few metres using cheap handheld global-positioning systems (GPS).

## Radar and radio-telemetry

The intrinsic difficulty of studying the movement of populations of many small invertebrates in the field continues to be a great impediment to progress, and there remains an urgent need for new technologies which allow monitoring of insect movements over all but the most restricted spatial scales (which can often be monitored by video recording). A seminal contribution in the 1973 symposium was G.W. Schaefer's (1976) detailed account of his use of X-band scanning radars to investigate the behaviour of high-flying insect migrants, and his description of the various striking phenomena caused by the interaction between atmospheric processes and the insects' flight behaviour. A relatively new form of entomological radar, which is much more suitable for continuous long-term automated operation than the earlier azimuthially scanning systems, is the vertical-looking, nutating-beam radar. Following an original idea from Schaefer, control and analysis algorithms had been developed by the early 1990s (Smith *et al.*, 1993), but it is only recently that these insect-monitoring radars have started to be used routinely for long-term research studies and operational tasks (Drake *et al.*, Chapter 10; Smith *et al.*, 2000).

Insect monitoring radars can categorize insect targets in terms of mass, body shape and wing-beat frequency, but if a definitive identification is required, this still entails trying to capture specimens at altitude. Various platforms have been used, e.g. aircraft, kites or tethered balloons (Reynolds *et al.*, 1997), but a cheap, convenient and practical method of sampling insects at various flight heights is still a difficult technical problem: recent experiments with nets attached to model airplanes (Shields and Testa, 1999) may have potential in this respect.

Radars generally do not work well for insects flying near the ground, because the strong reflections from ground features and vegetation ('clutter') tend to obscure those from the intended targets. This is a serious limitation, because practically all 'vegetative' flights, and some migratory ones, take place near the ground. The problem of ground reflections can be resolved by tagging insects with a small transponder (a combination of a diode and an antenna) which, unlike the ground features, will generate signals at harmonics of the illuminating frequency. The first step was the development of a harmonic direction-finder by Mascanzoni and Wallin (1986) which could be used to follow or locate (from a range of a few metres) the walking movements of insects, such as carabid beetles, which were large and robust enough to carry the diode fitted with a trailing wire antenna. Miniaturization of the transponder tag (0.4 mg) allowed a similar system to be used on flying insects (Roland *et al.*, 1996). A substantial advance in the application of the harmonic technique came when Riley *et al.* (1996) developed a true radar which provided both direction *and range* information, thus allowing precise and detailed descriptions of low-altitude flight trajectories over distances of several hundreds of metres, at least for larger-sized insects in open habitats. Riley and Osborne (Chapter 7) give a taste of the types of behavioural investigation (course control, orientation and navigation mechanisms, foraging strategies) which are now possible using this powerful new technique.

The harmonic transponders fitted to insects are passive devices which obtain their energy from the illuminating radar beam (and thus do not require batteries). It would obviously be useful to have an active radio device, like those fitted to vertebrates, to track insects over substantial distances. Until recently, radio-transmitters with a range of several hundred metres were still too heavy for flying insects, and have been restricted to large walking beetles (e.g. Riecken and Raths, 1996). Currently, however, Jonas Hedin and Thomas Ranius of Lund University (personal communication) are using 0.48 g radio-transmitters to record the movement (including flights) of the hermit beetle, *Osmoderma enemita* (Scopoli).

Kutsch (1999) has developed a radio system that transmits muscle potentials (EMGs) developed during the free flight of the locust, *Schistocerca gregaria*. The system was used in combination with high-speed video recording and showed how the electrical activity of certain muscles was associated with several flight parameters. The requirement for small, lightweight batteries means that systems of this sort have limited range, and they are at present more suitable for laboratory rather than field studies.

### Video observation of movement in the laboratory and field

For recording the movements of insects over short ranges (a few metres) but with very high resolution, optical methods rather than radar or radio-tracking are required, and methodology in this field has been revolutionized by the continual commercial development of video cameras and camcorders. The nocturnal flight of insects in the field can be recorded with low-light video cameras, usually supplemented with artificial near infra-red illumination (Riley, 1993). The use of two synchronized cameras with overlapping fields allows the reconstruction of accurate three-dimensional flight trajectories of flying insects (Riley, 1993; Hardie and Young, 1997). Commercial computerized video-tracking packages are now available (e.g. EthoVision®) which can be used for automated analysis of movement in laboratory arenas.

### Biochemical and molecular techniques

One of the most promising of recent developments is the application of new biochemical and molecular techniques to entomological research (Hoy, 1994; Loxdale *et al.*, 1996; Symondson and Hemingway, 1997; Loxdale and Lushai, 1998; Parker *et al.*, 1998). In particular, electrophoretic markers, either proteins (isoenzymes, allozymes) or DNA markers (where strands of DNA are examined directly) have found many uses in systemics, evolution, ecology and genetics. Loxdale and Lushai (Chapter 17) show how the genetic diversity of insect populations can be used to investigate movement and gene flow. As discussed by Mallet (Chapter 16; see above), this is not without its problems. However, continuing increases in the speed of collecting detailed genetic data, and the likelihood that we will soon be able to identify spatial variation at selected loci, will make genetic approaches increasingly attractive.

## Concluding Remarks

From the above comments, and more importantly the chapters that they are based on, it is clear that there have been some very significant developments in our understanding of insect movement over the 25 years since the previous Royal Entomological Society symposium on a similar topic (Rainey, 1976). Indeed, some of these advances have come from concepts and techniques, such as metapopulation dynamics and molecular biology, that hardly existed a quarter of a century ago. However, we should not ignore the real advances that have also accrued from much longer-established disciplines such as insect physiology and behaviour. As there are few research areas in entomology that do not have some bearing on insect movement, it has not been possible to cover all aspects in a volume of this size, but we hope that most biologists with an interest in the subject will find something new and interesting in the pages that follow.

## Acknowledgements

We would like to express our thanks to the Royal Entomological Society for asking us to convene the 20th Symposium, to the Registrar of the Society and his staff for shouldering much of the organizational burden, and to all who contributed to the symposium – the speakers, poster contributors, session chairmen and contributors to the discussions. We are also very grateful to the speakers and their co-authors for extending and developing their oral presentations into these chapters, to the colleagues who found time to review the chapters, and to our editors at CAB *International*, Rebecca Stubbs and Rachel Robinson, for seeing the book through to publication.

## References

Bell, W.J. (1991) *Searching Behaviour: the Behavioural Ecology of Finding Resources*. Chapman & Hall, London.

Day, R.K. and Knight, J.D. (1995) Operational aspects of forecasting migrant insect pests. In: Drake, V.A. and Gatehouse, A.G. (eds) *Insect Migration: Tracking Resources through Space and Time*. Cambridge University Press, Cambridge, pp. 323–334.

Day, R.K., Haggis, M.J., Odiyo, P.O., Mallya, G.A., Norton, G.A. and Mumford, J.D. (1996) WormBase: a data management and information system for forecasting *Spodoptera exempta* (Lepidoptera: Noctuidae) in eastern Africa. *Journal of Economic Entomology* 89, 1–10.

Deveson, E.D. and Hunter, D.M. (2000) Decision support for Australian locust management using wireless transfer of field survey data and automatic internet weather data collection. In: Laurini, R. and Tanzi, T. (eds) *Telegeo 2000 –*

Proceedings of 2nd International Symposium on Telegeoprocessing. Ecole des Mines de Paris, Sophia Antipolis, Nice, pp. 103–110.

Dickinson, M.H., Lehmann, F.O. and Sane, S.P. (1999) Wing rotation and the aerodynamic basis of insect flight. *Science* 284, 1954–1960.

Dingle, H. (1996) *Migration: the Biology of Life on the Move.* Oxford University Press, Oxford.

Drake, V.A. and Gatehouse, A.G. (eds) (1995) *Insect Migration: Tracking Resources through Space and Time.* Cambridge University Press, Cambridge.

Drake, V.A. and Gatehouse, A.G. (1996) Population trajectories through space and time: a holistic approach to insect migration. In: Floyd, R.B., Sheppard, A.W. and De Barro, P.J. (eds) *Frontiers of Population Biology.* CSIRO Publishing, Melbourne, pp. 399–408.

Drake, V.A., Gatehouse, A.G. and Farrow, R.A. (1995) Insect migration: a holistic conceptual model. In: Drake, V.A. and Gatehouse, A.G. (eds) *Insect Migration: Tracking Resources through Space and Time.* Cambridge University Press, Cambridge, pp. 427–457.

Dudley, R. (2000) *The Biomechanics of Insect Flight: Form, Function, Evolution.* Princeton University Press, Princeton, New Jersey.

Ellington, C.P., van den Berg, C., Willmott, A.P. and Thomas, A.L.R. (1996) Leading-edge vortices in insect flight. *Nature* 284, 626–630.

Gatehouse, A.G. (1987) Migration: a behavioural process with ecological consequences. *Antenna* 11, 10–12.

Gatehouse, A.G. (1997) Behavior and ecological genetics of wind-borne migration by insects. *Annual Review of Entomology* 42, 475–502.

Goldsworthy, G.J. and Wheeler, C.H. (eds) (1989) *Insect Flight.* CRC Press, Boca Raton, Florida.

Grodnitsky, D.L. (1999) *Form and Function of Insect Wings.* Johns Hopkins University Press, Baltimore, Maryland.

Hanski, I. (1999) *Metapopulation Ecology.* Oxford Series in Ecology and Evolution, Oxford University Press, Oxford.

Hanski, I.A and Gilpin, M.E. (1997) *Metapopulation Biology: Ecology, Genetics, and Evolution.* Academic Press, San Diego, California.

Hardie, J. and Young, S. (1997) Aphid flight-track analysis in three dimensions using video techniques. *Physiological Entomology* 22, 116–122.

Harrison, S. (1991) Local extinction in a metapopulation context: an empirical evaluation. *Biological Journal of the Linnaean Society* 42, 73–88.

Harrison, S. and Taylor, A. D. (1997) Empirical evidence for metapopulation dynamics. In: Hanski, I.A. and Gilpin, M.E. (eds) *Metapopulation Biology: Ecology, Genetics and Evolution.* Academic Press, San Diego, California, pp. 27–42.

Hassell, M. (2000) *The Spatial and Temporal Dynamics of Host–Parasitoid Interactions.* Oxford Series in Ecology and Evolution, Oxford University Press, Oxford.

Hay, S.I., Packer, M.J. and Rogers, D.J. (1997) The impact of remote sensing on the study and control of invertebrate intermediate hosts and vectors for disease. *International Journal of Remote Sensing* 18, 2899–2930.

Healey, R.G., Robertson, S.G., Magor, J.I., Pender, J. and Cressman, K. (1996) A GIS for desert locust forecasting and monitoring. *International Journal of Geographical Information Systems* 10, 117–136.

Hoy, M.A. (1994) *Insect Molecular Genetics.* Academic Press, London.

Johnson, C.G. (1969) *Migration and Dispersal of Insects by Flight*. Methuen, London.

Johnson, S. J. (1995) Insect migration in North America: synoptic-scale transport in a highly seasonal environment. In: Drake, V.A. and Gatehouse, A.G. (eds) *Insect Migration: Tracking Resources through Space and Time*. Cambridge University Press, Cambridge, pp. 31–66.

Kennedy, J.S. (1975) Insect dispersal. In: Pimentel, D. (ed.) *Insects, Science and Society*. Academic Press, New York, pp. 103–119.

Kennedy, J.S. (1985) Migration, behavioural and ecological. In: Rankin, M.A. (ed.) *Migration: Mechanisms and Adaptive Significance. Contributions in Marine Science* 27 (supplement), 5–26.

Knight, J.D., Tatchell, G.M., Norton, G.A. and Harrington, R. (1992) FLYPAST: an information management system for the Rothamsted Aphid Database to aid pest control advice and decision making. *Crop Protection* 11, 419–426.

Kutsch, W. (1999) Telemetry in insects: the "intact animal approach". *Theory in Biosciences* 118, 29–53.

Liebhold, A.M., Rossi, R.E. and Kemp, W.P. (1993) Geostatistics and geographical information systems in applied insect ecology. *Annual Review of Entomology* 38, 303–27.

Loxdale, H.D. and Lushai, G. (1998) Molecular markers in entomology. *Bulletin of Entomological Research* 88, 577–600.

Loxdale, H.D., Brookes, C.P. and De Barro, P.J. (1996) Application of novel molecular markers (DNA) in agricultural entomology. In: Symondson, W.O.C. and Liddell, J.E. (eds) *The Ecology of Agricultural Pests: Biochemical Approaches*, Systematics Association Special Volume 53. Chapman & Hall, London, pp. 149–198.

Mascanzoni, D. and Wallin, H. (1986) The harmonic radar: a new method of tracing insects in the field. *Ecological Entomology* 11, 387–390.

Nachtigall, W. (1989) Mechanics and aerodynamics of flight. In: Goldsworthy, G.J. and Wheeler, C.H. (eds) *Insect Flight*. CRC Press, Boca Raton, Florida, pp. 1–29.

Parker, P.G., Snow, A.A., Schug, M.D., Booton, G.C. and Fuerst, P.A. (1998) What molecules can tell us about populations: choosing and using a molecular marker. *Ecology* 79, 361–382.

Pedgley, D.E. (1973) *Testing the Feasibility of Detecting Locust Breeding Sites by Satellite*. Final report to NASA in ERTS-1 Experiment. Centre for Overseas Pest Research, London.

Pedgley, D.E. (ed.) (1981) *Desert Locust Forecasting Manual*, Vols 1 and 2. Centre for Overseas Pest Research, London.

Pedgley, D.E., Reynolds, D.R. and Tatchell, G.M. (1995) Long-range insect migration in relation to climate and weather: Africa and Europe. In: Drake, V.A. and Gatehouse, A.G. (eds) *Insect Migration: Tracking Resources through Space and Time*. Cambridge University Press, Cambridge, pp. 3–29.

Rainey R.C. (ed.) (1976) *Insect Flight*. Symposia of the Royal Entomological Society, No. 7. Blackwell, Oxford.

Reynolds, D.R., Riley, J.R., Armes, N.J., Cooter, R.J., Tucker, M.R. and Colvin, J. (1997) Techniques for quantifying insect migration. In: Dent, D.R. and Walton, M.P. (eds) *Methods in Ecological and Agricultural Entomology*. CAB International, Wallingford, UK, pp. 111–145.

Riecken, U. and Raths, U. (1996) Use of radio telemetry for studying dispersal and habitat use of *Carabus coriaceus* L. *Annales Zoologici Fennici* 33, 109–116.

Riley, J.R. (1989) Remote sensing in entomology. *Annual Review of Entomology* 34, 247–271.
Riley, J.R. (1993) Flying insects in the field. In: Wratten, S.D (ed.) *Video Techniques in Animal Ecology and Behaviour*. Chapman & Hall, London, pp. 1–15.
Riley, J.R., Smith, A.D., Reynolds, D.R., Edwards, A.S., Osborne, J.L., Williams, I.H., Carreck, N.L. and Poppy, G.M. (1996) Tracking bees with harmonic radar. *Nature* 379, 29–30.
Robinson, T.P. (1995) Geographic Information systems and remotely sensed data for determining the seasonal distribution of habitats of migrant insect pests. In: Drake, V.A. and Gatehouse, A.G. (eds) *Insect Migration: Tracking Resources through Space and Time*. Cambridge University Press, Cambridge, pp. 335–352.
Roland, J., McKinnon, G., Backhouse, C. and Taylor, P.D. (1996) Even smaller radar tags on insects. *Nature* 381, 120.
Sabins, F.F. (1997) *Remote Sensing: Principles and Interpretation*, 3rd edn. W.H. Freeman, New York.
Schaefer, G.W. (1976) Radar observations of insect flight. In: Rainey R.C. (ed.) *Insect Flight*. Symposia of the Royal Entomological Society, No. 7. Blackwell, Oxford, pp. 157–197.
Shields, E.J. and Testa, A.M. (1999) Fall migratory flight initiation of the potato leafhopper, *Empoasca fabae* (Homoptera: Cicadellidae): observations in the lower atmosphere using remotely piloted vehicles. *Agricultural and Forest Meteorology* 97, 317–330.
Smith, A.D., Riley, J.R. and Gregory, R.D. (1993) A method for routine monitoring of the aerial migration of insects by using a vertical-looking radar. *Philosophical Transactions of the Royal Society of London* B 340, 393–404.
Smith, A.D., Reynolds, D.R. and Riley, J.R. (2000) The use of vertical-looking radar to continuously monitor the insect fauna flying at altitude over southern England. *Bulletin of Entomological Research* 90, 265–277.
Southwood, T.R.E. and Henderson, P.A. (2000) *Ecological Methods*, 3rd edn. Blackwell Science, Oxford.
Symondson, W.O.C. and Hemingway, J. (1997) Biochemical and molecular techniques. In: Dent, D.R. and Walton, M.P. (eds) *Methods in Ecological and Agricultural Entomology*. CAB International, Wallingford, UK, pp. 293–340.
Tappan, G.G., Moore, D.G. and Knausenberger, W.I. (1991) Monitoring grasshopper and locust habitats in Sahelian Africa using GIS and remote sensing technology. *International Journal of Geographical Information Systems* 5, 123–135.
Tatchell, G.M. (1991) Monitoring and forecasting aphid problems. In: Peters, D.C., Webster, J.A. and Chlouber, C.S. (eds) *Aphid–Plant Interactions: Populations to Molecules*. Oklahoma Agricultural Experiment Station Miscellaneous Publication No. 132, 215–231.
Taylor, L.R. (1974) Insect migration, flight periodicity and the boundary layer. *Journal of Animal Ecology* 43, 225–238.
Tilman, D. and Kareiva, P. (eds) (1998) *Spatial Ecology: the Role of Space in Population Dynamics and Interspecific Interactions*. Monographs in Population Biology, No. 30. Princeton University Press, Princeton, New Jersey.
Tucker, M.R. and Holt, J. (1999) Decision tools for managing migrant insect pests. In: Grant, I.F. and Sear, C. (eds) *Decision Tools for Sustainable Development*. Natural Resources Institute, Chatham, UK, pp. 97–128.

Turchin, P. (1998) *Quantitative* Analysis of *Movement*. Sinauer Associates, Sunderland, Massachusetts.
Washburn, J.O. and Washburn, L. (1984) Active aerial dispersal of minute wingless arthropods: exploitation of boundary-layer velocity gradients. *Science* 223, 1088–1089.
Weis-Fogh, T. (1976) Energetics and aerodynamics of flapping flight: a synthesis. In: Rainey R.C. (ed.) *Insect Flight*. Symposia of the Royal Entomological Society, No. 7. Blackwell, Oxford, pp. 48–72.
Wigglesworth, V.B. (1976) The evolution of insect flight. In: Rainey R.C. (ed.) *Insect Flight*. Symposia of the Royal Entomological Society, No. 7. Blackwell, Oxford, pp. 255–269.
Williams, D.W. and Liebhold, A.M. (1995) Herbivorous insects and global change: potential changes in the spatial distribution of forest defoliator outbreaks. *Journal of Biogeography* 22, 665–671.
Wootton, R. (1999) How flies fly. *Nature* 400, 112–113.
Wyatt, T.D. (1997) Methods in studying insect behaviour. In: Dent, D.R. and Walton, M.P. (eds) *Methods in Ecological and Agricultural Entomology*. CAB International, Wallingford, UK, pp. 27–56.

# The Biomechanics and Functional Diversity of Flight

## Robert Dudley

*Section of Integrative Biology, University of Texas, Austin, TX 78712, USA, and Smithsonian Tropical Research Institute, PO Box 2072, Balboa, Republic of Panama*

## Introduction

Insect presence on the earth is ancient, persistent and ubiquitous. Following the initial evolution of wings either in the late Devonian or early Carboniferous, pterygote insects rapidly diversified to become, by the mid- to late Carboniferous, a predominant feature of the terrestrial biota. This role has been retained to the present time, and today pterygote insects can be found in essentially all terrestrial ecosystems and on all continental land masses. The contemporary taxonomic richness of insects is famously high, with the number of described pterygote species exceeding 1 million, and the number of as yet undescribed species ranging potentially as high as 10 million. Among the arthropods, perhaps only mite diversity rivals that of the winged insects (Hammond, 1992; Walter and Behan-Pelletier, 1999), and much of mite species richness may derive from symbiotic associations with insects. Pterygotes today are major consumers of plant productivity and also serve as a nutritional resource for diverse arthropod and vertebrate taxa. A rough indication of the trophic influence exerted by insects is suggested by an estimate for their aggregate biomass. Approximately $10^{18}$ insects may be alive worldwide at any given time (Williams, 1960). Assuming an average body mass of 1 mg, this number of individuals corresponds to an approximate biomass of $10^{12}$ kg, a value roughly comparable to the total mass of the contemporary human population (Dudley, 2000). To this day, winged insects thus compete with humans at the level of primary trophic consumption.

Much of the pterygote success story can be ascribed to the initial evolution of flight and subsequent exploitation of the aerial environment. Winged insects have clearly diversified and exploited a variety of terrestrial habitats far more effectively than have their sister taxon, the wingless apterygotes. Ample testimony to this enhanced diversification is provided by comparison of species richness among the two clades – pterygote insect species outnumber their apterygote counterparts by at least three orders of magnitude. Along with complete metamorphosis and lateral wing flexion, the ability to locomote in three dimensions is regularly cited in entomological textbooks as a causal factor underlying adaptive radiation in winged insects. Monophyly of wing origins (as well as of holometaboly and of the Neoptera) makes it difficult to test such diversification hypotheses directly, but the ability to fly obviously facilitates dispersal to and subsequent evolution within novel environments. Flight also figures prominently in diverse aspects of insect behaviour and ecology, including pollination, phytophagy, haematophagy, aerial predation, escape from predators and mating systems. Understanding of the evolution of functional diversity in flight mechanisms can therefore yield insight into major themes of insect biology.

In order to fly, animals must possess lift-producing structures or wings as well as particular configurations of dedicated muscles that effect wing flapping. Furthermore, a flight-control system based on both extero- and proprioceptive sensory mechanisms is necessary to permit regulation of aerodynamic force output and to enable manoeuvres. Flapping motions of the wings yield time-varying vortex flows over the wings and body that sustain the body mass against gravity and that can propel the body forwards, upwards or laterally. Many features of wing and body motions exhibit strong allometric variation, and in turn the aerodynamics of flight in insects is strongly scale-dependent. Body size itself has undergone substantial evolutionary change in insects, and much of present-day pterygote diversity derives from the process of miniaturization and the correlated acquisition of high wingbeat frequencies during flight. Modification of wings for non-aerodynamic purposes is another major trend underlying morphological diversification of insects. For example, tegminization and elytrization of the forewings for protective purposes is diagnostic of the Orthoptera and Coleoptera, respectively, but related modifications can be found in at least five other orders. Much of the glorious morphological diversity that we see today among the insects derives from transformations of what ancestrally were two pairs of aerodynamically functional wings. Evolutionary origins of such novel structures must therefore be examined if we are to understand their subsequent elaboration for both aerodynamic and non-aerodynamic purposes.

## The Origins and Elaboration of Flight

As with other volant animals, the origins and early evolution of winged insects have been the subject of abundant speculation based on a paucity of empirical observations. Pterygotes are particularly difficult in this regard given that an approximately 45-million-year gap separates the occurrence of the earliest known winged insects (325 Mya) from fossils of their apterygote ancestors (395–390 Mya; Whalley and Jarzembowski, 1981; Shear et al., 1984; Nelson and Tidwell, 1987; Labandeira et al., 1988; Brauckmann and Zessin, 1989; Jeram et al., 1990). Because pterygote insects appear abruptly in the fossil record with no obvious transitional forms, both the anatomical precursors to wings as well as the selective forces promoting their initial evolution remain unresolved. None the less, palaeobiological reconstruction can at least delineate possible scenarios of pterygote evolution. Most indirect evidence suggests that ancestral pterygotes (i.e. protopterygotes) were terrestrial animals (Messner, 1988; Pritchard et al., 1993; Dudley, 2000). Apterygote insects, the sister taxon of winged pterygotes, are almost exclusively terrestrial, and in fact the possession of a tracheal system by all hexapods predisposes these animals to life in air. Present-day aquatic habits in some insect taxa, and particularly in the larvae of the palaeopterous Ephemeroptera and Odonata, appear to be secondarily derived (Hinton, 1968; Hennig, 1981; Pritchard et al., 1993). Invasion of either fresh- or saltwater by pterygote ancestors would also have had to surmount the formidable ecological obstacles that confront all metazoan taxa transiting from terrestrial to aquatic habitats (Vermeij and Dudley, 2000). Terrestriality in protopterygotes therefore seems the most likely possibility, although only fossil evidence can empirically confirm this inference.

Hypotheses concerning wing origins, as elaborated elsewhere in this volume (Wootton, Chapter 3), distil to two major possibilities: initially fixed paranotal lobes that subsequently acquired mobility in flapping, or pre-existing mobile gills, gill covers or styli that then served aerodynamic purposes. This latter scenario generally presupposes ancestral aquatic habits in protopterygotes, together with initially hydrodynamic use for what ultimately became aerodynamic structures. Related hypotheses propose transitional stages of drifting or skimming across the surface of water bodies as a precursor to free flight (Marden and Kramer, 1994, 1995; Kramer and Marden, 1997). In extant taxa, such behaviours appear to be derived rather than retained ancestral characters (Will, 1995; Ruffieux et al., 1998; but see Thomas and Norberg, 1996), and no example exists of an aquatic insect that locomotes via projection of winglike structures across the water–air interface. Any suggestion of protowings operating either completely underwater or partially in air (and against the forces of surface tension) is also biomechanically implausible given the major physical differences between water and air and the correspondingly different flow regimes and patterns of force production in the two media (Denny, 1993; Dudley, 2000).

Independent of anatomical origins and ecological context, fossil as well as neontological evidence suggests that both larvae and adults of ancestral winged insects expressed winglets or winglike structures on thoracic as well as abdominal segments (Kukalová-Peck, 1978; Carroll et al., 1995). In addition to a possible aerodynamic role, a variety of other functional possibilities have been ascribed to these structures, including epigamic display during courtship (Alexander and Brown, 1963; Alexander, 1964) and thermoregulation (Whalley, 1979; Douglas, 1981; Kingsolver and Koehl, 1985, 1994). None of these roles are mutually exclusive. However, the most parsimonious hypothesis for the evolution of winglike structures is aerodynamic utility, possibly but not necessarily in concert with other functions. If terrestriality can be assumed for protopterygotes, then lift on the body as well as on winglets or protowings would have advantageously facilitated gliding performance (Ellington, 1991a). Escape from predators is the behavioural context classically presumed for jumping and gliding in protopterygotes, particularly given the ancestral nature of jumping and startle responses in hexapods, if not all arthropods (Edwards and Reddy, 1986; Kutsch and Breidbach, 1994; Edwards, 1997). Predatory pressures on protopterygotes within late Palaeozoic ecosystems were likely substantial, and the increasing spatial complexity of contemporaneous vegetation provided further advantages to escape strategies that facilitated movement in all three spatial directions (Dudley, 2000).

Once the aerodynamic utility of winglets was established, then winglet mobility (either acquired ancestrally or subsequently derived) would have facilitated greater aerial manoeuvrability and/or enhanced flight performance. Greater mobility in the air would obviously have been advantageous for purposes other than escape from predators, in particular for mate location and for resource acquisition from the rapidly diversifying terrestrial flora. Wings on the prothoracic segment, as well as on abdominal segments, appear to have been lost early in pterygote evolution, possibly because of deleterious aerodynamic interaction among winglets of adjacent segments (Grodnitsky, 1995). Instead, wings of the two posterior thoracic segments increased in size and became aerodynamically dominant. This locomotor dedication of the meso- and metathoracic segments would have been accompanied by hypertrophy in pterothoracic musculature, elaboration of the axillary apparatus, and systematic increases in aerodynamic force production mediated by changes in stroke amplitude and wingbeat frequency (Dudley, 2000). By the middle of the Carboniferous, pterygotes are impressively diversified into about 15 orders, some with seemingly modern morphologies (Wootton, 1990; Kukalová-Peck, 1991; Labandeira and Sepkoski, 1993). Major ordinal-level differentiation of pterygotes thus occurred in the Lower Carboniferous over a time period spanning 30–40 Mya.

Many Carboniferous insects possessed homonomous wings (i.e. wings of approximately equivalent size, shape and function) that were probably limited to low-amplitude flapping (e.g. Diaphanopterodea, Megasecoptera, many Palaeodictyoptera; Carpenter, 1992). However, morphological differentiation in the size of meso- and metathoracic wings is equally evident in this fauna. Differences in relative wing size in turn suggest varying aerodynamic roles of the fore- and hindwings. Schwanwitsch (1943, 1958) applied to pterygotes the terms of anteromotorism, posteromotorism and bimotorism to describe the conditions of enlarged forewings, enlarged hindwings or homonomous wings, respectively. Because force production in flapping flight depends not only on wing area but also on the particulars of wing motion, relative size alone does not necessarily indicate the extent of aerodynamic contribution. In the contemporary fauna, however, these two measures are broadly congruent (Dudley, 2000). Ipsilateral wing differentiation and expression of heteronomous wings were well underway by the Upper Carboniferous. For example, some Palaeodictyoptera possessed hindwings only half the area of the forewings, whereas expanded hindwings of the contemporaneous Protorthoptera are characteristic of posteromotorism. Hindwings were also much reduced or absent in the paleodictyopteroid lineage Permothemistida (Carpenter, 1992).

Among extant orders, antero- and posteromotorism occur at approximately equal frequencies, although posteromotorism is (with the exceptions of the Coleoptera and Strepsiptera) mostly confined to the exopterygotes (Fig. 2.1). The most parsimonious reconstruction of the ancestral locomotor mode in extant pterygotes suggests anteromotorism rather than homonomous bimotorism (Dudley, 2000; see also Fig. 2.1), although this result may in part derive from exclusion of extinct Palaeozoic taxa from the analysis. For example, Ephemeroptera of the Carboniferous were characterized by equally sized fore- and hindwings (Carpenter, 1992), in contrast to the anteromotoric condition of extant mayflies. If bimotorism is assumed for the Ephemeroptera, parsimonious reconstruction of the ancestral locomotor mode in pterygotes yields bimotorism rather than anteromotorism. Bimotorism persists to this day in at least seven orders (Fig. 2.1), and the salient example of dragonfly manoeuvrability indicates no necessary correlation between flight performance and this particular locomotor mode. Evolution of heteronomous wing pairs has been, however, a major theme in ordinal-level innovation. Subsequent to the great Carboniferous radiations, the species-rich and posteromotoric Coleoptera appeared with fully elytrized forewings by the mid- to late Permian. Similarly, the major orders Hymenoptera and Diptera originated in the Triassic, with the hindwings relegated to either a reduced or no direct aerodynamic role, respectively.

Parallel with locomotor dedication of either the fore- or hindwings, wing transformation for non-aerodynamic purposes has been widespread in pterygote evolution (Fig. 2.2). True diptery characterizes the Strepsiptera and the eponymous Diptera, with the fore- and hindwings, respectively, being miniaturized and used for purposes of flight control (Pix et al., 1993; Dickinson, 1999). Dipterous anteromotorism is, however, a rare outcome among pterygotes. Much more common has been the assumption of a protective role by the forewings. Forewings in many insects tend to be slightly thickened relative to the hindwings. Also, tegminization and elytrization have occurred at least three times at the ordinal level, and possibly many more times if tetraptery in the Isoptera is a retained ancestral trait (Fig. 2.2).

Coriaceous tegmina are characteristic of the Blattaria, Dermaptera, Mantodea, Orthoptera, Phasmatodea, and of some Homoptera. In some Phasmatodea, the remigium of the hindwing is also thickened and serves to protect the vannus (R. Wootton, personal communication). Elytra of the

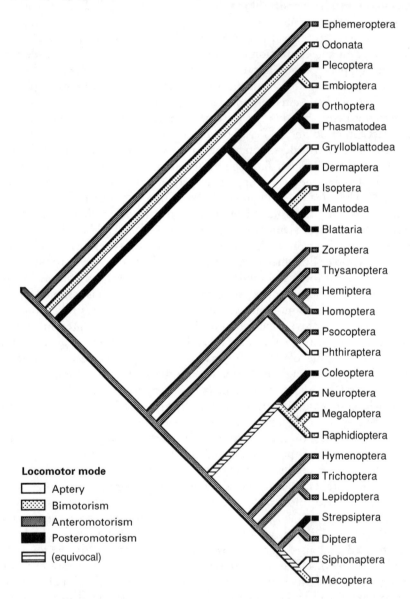

**Fig. 2.1.** Phylogenetic distribution of locomotor mode among extant pterygote orders (tree topology from Kristensen, 1991, 1997; Pashley et al., 1993; Whiting et al., 1997). MacClade 3.0 (Maddison and Maddison, 1992) was used to generate the most parsimonious reconstruction of ancestral character states.

Coleoptera have much reduced aerodynamic roles relative to the hindwings, and provide for greater mechanical resistance to crushing in conjunction with increased sclerotization of the body as a whole (Dudley, 2000). A similar functional role may be hypothesized for tegmina and for hemipteran

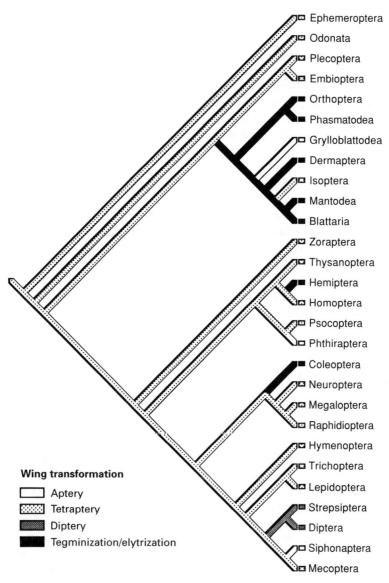

**Fig. 2.2.** Phylogenetic patterns of wing transformation among pterygote orders. These designations describe only the general trend within each order; further details as well as exceptions to ordinal-level characterizations can be found in Dudley (2000).

hemelytra; all such cases of mesothoracic specialization are necessarily associated with posteromotorism (Fig. 2.1).

By contrast, anteromotorism is typically characterized by the retention of aerodynamically functional hindwings, albeit with reduced surface area. Diptera are exceptional in this regard given their miniaturized halteres. In other anteromotoric orders, hindwings are only somewhat reduced in relative size and tend to operate in phase with the forewings. Such orders are functionally dipterous while retaining four wings, although some exceptions exist to this characterization (e.g. the miniaturized hamulohalteres of male Coccidae). In most anteromotoric groups, however, the hindwings are either mechanically coupled to or physically overlapped by the forewings. Physical overlap of ipsilateral wings is termed amplexiform coupling, and is characteristic of mayflies (Ephemeroptera) and butterflies. Alternatively, direct mechanical coupling between wings (e.g. hamuli of Hymenoptera, frenulum and jugum of moths) may be associated with the evolution of high wingbeat frequencies and with further reduction in relative hindwing area (Dudley, 2000). Evolutionary modifications of potentially ancestral wing homonomy have thus been functionally diverse. In addition to the roles of aerodynamic force production and possibly physical protection by the forewings, insect wings may also serve a variety of behavioural roles, including sound production and visual communication. None of these functions are mutually exclusive, although the role of aerodynamic force production must remain paramount for at least one wing pair. Having broadly delineated the range of pterothoracic diversification evident among extant orders, we now turn to the biomechanical specifics of aerodynamic force production through wing flapping.

## Kinematics and Aerodynamics

Diversity in wing shape and form is matched by functional divergence in wingbeat motions (i.e. kinematics) and in the underlying aerodynamics of flight. Two of the most important kinematic features of flight are the frequency of flapping motions and the speed of the entire body relative to the surrounding air (i.e. the airspeed). Much variation in these parameters is associated with differences in body mass ($m$), a quantity which in the extant fauna ranges over seven orders of magnitude. Such variation in body mass in turn yields major differences in wingbeat kinematics among different taxa. For example, large butterflies exhibit wingbeat frequencies as low as 5 Hz, whereas some small flies have frequencies as high as 1000 Hz. For the extant fauna, wingbeat frequencies span approximately three orders of magnitude and vary with $m^{-0.24}$ (Dudley, 2000). Insect flight speeds are also allometrically dependent, varying theoretically in proportion to $m^{0.17}$. Empirical results demonstrate that larger insects fly faster, and insect airspeeds in general range from 0.5 to 10 m s$^{-1}$. Steady forward flight requires the maintenance of a constant force balance to offset gravity and to create thrust that offsets drag on the body. Also, the transient generation of aerodynamic forces and rotational torques can yield near-instantaneous turns, accelerations, and directional changes that form the basis of aerial manoeuvrability in insects. Hovering, backward and even sideways flight are also common in many taxa.

Aerodynamically, the forces generated by flapping wings vary with wing morphology, with the kinematic details of motion, and with the density and viscosity of the surrounding air. For any object moving within a fluid, density of the fluid influences the inertial characteristics of the flow field around the object, whereas fluid viscosity determines the magnitude of shear forces exerted over the object's surface. The ratio of inertial to viscous forces thus varies with the ratio of fluid density and viscosity, and is also linearly proportional to the object's dimensions and to its velocity relative to the fluid. The four variables of density, viscosity, linear dimension and the relative velocity can be combined to yield a dimensionless parameter termed the Reynolds number ($Re$) that broadly characterizes the nature of forces within moving fluids (Vogel, 1994). At low $Re$, flow is highly viscous and shear forces within the fluid exert a predominant influence. Flow is usually laminar below $Re$ of $10^3$–$10^4$, whereas turbulence and inertially driven flows become more pronounced at higher $Re$. Values of the $Re$ for large insects typically range from $10^2$ to $10^4$ (Dudley, 2000). Most insects, however, are fairly small by anthropomorphic standards, and the $Re$ for the majority of these taxa is in the range of $10^0$–$10^2$ (Fig. 2.3). A quantitative comparison of the two allometries indicated in Fig. 2.3 suggests that approximately 85% of the fauna hovers with $Re$ for the wings below 1000, and that nearly one-half of the fauna (47%) hovers at $Re$ below 100. Flight at such low Reynolds numbers is aerodynamically challenging – viscosity exerts a predominant influence on moving appendages, and wing flapping is often analogized to swimming in molasses.

Given that most insects fly in fairly low $Re$ regimes, one basic question in biomechanical analysis concerns the magnitude of aerodynamic forces on the flapping wings, as well as the associated expenditure of mechanical power. In both hovering and forward flight, flapping of wings results in continuous changes in the speed and orientation of the wings relative to the surrounding fluid. These conditions violate many assumptions of traditional aerodynamic analysis, and the application of conventional aerodynamics to wing flapping by insects typically yields force balances inconsistent with those known to apply in free flight (Ellington, 1984; Dudley and Ellington, 1990; Dudley, 2000). Instead, unsteady aerodynamic mechanisms pertain that include such effects as wing acceleration, wing rotation and airflow along the length of the wing (Ellington, 1995). At $Re$ corresponding to the hovering flight of the hawkmoth *Manduca sexta* (L.) ($Re$~1000), direct visualization of flow around mechanically flapping wings together with computational modelling have revealed the presence of a leading-edge vortex attached dorsally to the

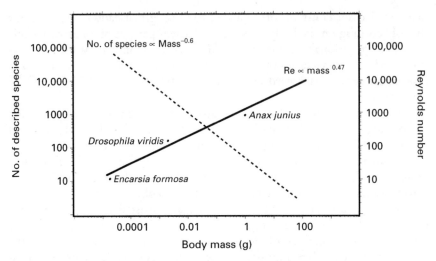

**Fig. 2.3.** Allometry of insect species diversity and of the Reynolds number (*Re*) during hovering flight. Values of body mass and the corresponding *Re* are indicated for three representative species. Species numbers are approximate worldwide values and do not reflect the reduced number of insect species at very small body masses. The *Re* refers to values for the wing chord in hovering flight; see Dudley (2000) for derivation of the associated allometry.

surface of the wing (Ellington *et al.*, 1996; van den Berg and Ellington, 1997a,b; Liu *et al.*, 1998). High-speed rotation of the leading-edge vortex creates a low-pressure zone above the wing, and transiently increases lift production above that feasible through linear translation at a constant velocity. At lower *Re* corresponding to the flight of smaller insects the size of *Drosophila* (*Re*~100), leading-edge vortices also predominate, but supplemental unsteady forces derive from wing rotation about the longitudinal axis at the ends of the half-stroke (Dickinson *et al.*, 1999). The high flapping velocities associated with small body size also yield high rotational velocities of the wings during pronation and supination at the ends of the up- and downstrokes, respectively. Thus, rotational circulation and associated force production are likely to be more pronounced in small insects. At such low *Re*, the presence of vortices shed previously from the flapping wings also appears to advantageously influence unsteady force production (Dickinson *et al.*, 1999). A diversity of unsteady aerodynamic mechanisms characterizes the flight of insects, and body size plays a major role in determining the relative importance of different kinds of aerodynamic forces.

Independent of the physical nature of these mechanisms, the predominant force produced by flying insects is directed vertically to offset gravity acting on the body mass. Regulation of forward airspeed is attained primarily by changes in the direction but not magnitude of the net aerodynamic forces on the wings. Except at very high forward speeds, thrust is small relative to vertical force production and the magnitude of the resultant aerodynamic vector remains approximately constant. Reorientation of the resultant force vector, however, is most simply attained by altering the angle of inclination between the plane of wing motions and the horizontal plane. Given the generally fixed dorsoventral nature of wing flapping motions, vertical and horizontal partitioning of total force output is effected through changes in body angle, a parameter which usually shows an inverse relationship with forward airspeed (Dudley, 2000). In turn, changes in more subtle kinematic features such as the mean positional angle of the wings and in rotational velocities during pronation and supination contribute to the regulation of body angle. Wing motions are bilaterally symmetric in forward and hovering flight. During manoeuvres, however, even small asymmetries in contralateral wing motions at different stages of the wingbeat have major implications for generation of roll, yaw and pitching moments. Relative body elongation, wing flexibility and the use of four rather than two wings (as characterizes the highly manoeuvrable dragonflies) also influence the extent of rotational and axial agility in the air (Dudley, 2000).

Application of aerodynamic models to known wingbeat kinematics allows calculation of the associated energetic costs of flapping flight. Insect flight represents an extreme of metabolic expenditure among animals, with rates of oxygen consumption during flight exceeding resting metabolic rates by a factor of 50–200 (Kammer and Heinrich, 1978; Casey, 1989). Also, flight in insects is powered exclusively through aerobic pathways. Metabolic power input and the actual expenditure of mechanical power for useful aerodynamic work (the mechanical power requirements) are related by the overall efficiency of the flight muscle. Because flight is so energetically expensive, selection has presumably acted to minimize mechanical power expenditure and to maximize flight muscle efficiency. Comparative studies of flapping energetics thus permit analysis of evolutionary optimization at the level of biomechanical performance. The variation in mechanical and metabolic power requirements with forward airspeed, the so-called power curve, is also of interest because of implications for optimal strategies of airspeed selection. Aerodynamic theories that evaluate the power curve quantitatively have been extensively applied to volant vertebrates (Norberg, 1990), but have received much less application to the flight of insects. Existing aerodynamic theories are also inadequate to incorporate the aforementioned unsteady aerodynamic effects now known to be pervasive in insect flight, but none the less provide a useful baseline estimate for the costs of flight.

Such aerodynamic modelling of bumblebees in forward flight suggests that mechanical power requirements are approximately constant over an airspeed of 0–4.5 m s$^{-1}$ (Dudley and Ellington, 1990). By contrast, two moth species as well as various odonates exhibit substantial increases in mechanical power expenditure with forward airspeed (Dudley and DeVries, 1990; Wakeling and Ellington, 1997; Willmott and Ellington, 1997). Power requirements at hovering and intermediate airspeeds are nominally similar, but then increase substantially at higher speeds, yielding a 'J'-shaped power curve (Ellington, 1991b). The precise shape of the power curve, however, derives from trends in individual components of power expenditure. The power required to overcome drag forces on the body (the parasite power) increases approximately with the cube of forward airspeed. Also, the power to overcome drag forces on the wing (the profile power) is strongly dependent on airspeed (Dudley, 2000), and a steep overall rise in mechanical power requirements is evident when insects fly faster. The power curve for most taxa is positively curvilinear under such circumstances, and the choice of airspeed during flight will accordingly have major energetic implications. Also, the relative costs of three-dimensional mobility may be reduced for small insects. Mass-specific power expenditure in forward flight increases with $m^{0.05-0.19}$, suggesting reduced relative costs of horizontal movement for smaller taxa (Dudley, 2000). Body size alone thus exerts a major influence on the kinematics, energetics and aerodynamics of flight in insects.

## Gigantism and Miniaturization

A major feature of pterygote evolution with important flight-related consequences has been historical change in body size. Although direct palaeontological evidence is not available, ancestral body lengths of pterygotes were probably in the range of 2–4 cm (Flower, 1964; Wootton, 1976; Labandeira et al., 1988). Following pterygote emergence and the diversification of the Palaeoptera and the exopterygote Neoptera, various endopterygote Neoptera appeared by the Upper Carboniferous and radiated extensively through the remainder of the Palaeozoic. Quantitative information of body-size distributions is not available for the late Palaeozoic fauna, but substantial increases in body length appear to have occurred by the mid-Carboniferous. Gigantism relative to today's forms was typical of many Carboniferous and Permian insects as well as of other contemporaneous arthropods (Briggs, 1985; Graham et al., 1995; Dudley, 1998). Giant forms were characteristic of at least three entire orders (Megasecoptera, Palaeodictyoptera, Protodonata), and also occurred among other pterygote as well as apterygote hexapods.

The most parsimonious explanation for late Palaeozoic gigantism is an increased oxygen concentration of the late Palaeozoic atmosphere, possibly to values as high as 35% relative to today's value of 20.9% (Berner and Canfield, 1989; Berner, 1997, 1998). For arthropods with tracheal respiratory systems, diffusional limits on oxygen supply probably constrain maximum body size. Greater partial pressures of oxygen, together with higher diffusion constants associated with an increased total pressure of the atmosphere, would relax these diffusional constraints and permit evolution of giant forms (Dudley, 1998). Furthermore, all giant arthropod taxa of the late Palaeozoic were extinct by the mid- to end-Permian (Carpenter, 1992; Graham et al.,

1995; Dudley, 1998), exactly as would be predicted by a decline in atmospheric oxygen concentration and concomitant asphyxiation of giant forms (Berner and Canfield, 1989; Graham *et al.*, 1995). Detailed analysis of phyletic size change within diffusion-limited taxa thus represents an important bioassay for effects of atmospheric hyperoxia in the late Palaeozoic, as well as for a secondary hyperoxic episode in the Cretaceous/ Tertiary (Dudley, 1998).

In sharp contrast to these Palaeozoic giants, the contemporary insect fauna is characterized by a remarkable diversity of miniaturized forms. For example, mean adult beetle body length lies between 4 and 5 mm (May, 1978; Crowson, 1981). Much of the wealth of dipteran and hymenopteran diversity is similarly associated with small body sizes, particularly among the parasitoid and hyperparasitoid taxa. No global estimate of insect body size distribution is presently available. However, existing data can be used at least to approximate body-length distributions for the North American fauna. Arnett (1985) provided species counts together with minimum and maximum body lengths (or in some cases, wing length) for each family present in the continental fauna. If the number of species within a family can be assumed to be inversely proportional to the square of body length (May, 1978), then mean body length within families and for the entire fauna can be calculated. This exercise suggests that the average interspecific body length for the North American insect fauna is about 7 mm. Note also that taxa composed of smaller insects are less likely to be well sampled than are those with more massive species. What are the historical origins and biomechanical correlates of such pronounced body miniaturization among pterygotes?

As mentioned previously, wingbeat frequencies are inversely proportional to body size. Insects less than 1 cm in body length typically fly with wingbeat frequencies and thus with contraction frequencies of the flight muscle in excess of 100 Hz (Dudley, 2000). Such rapid oscillations can only be attained using the specialized asynchronous flight muscle unique to particular lineages of pterygote insects. Asynchronous flight muscle contracts repeatedly in response to only one nervous impulse through a biophysical mechanism termed stretch-activation (Pringle, 1949, 1978). Contraction strains tend to be lower in asynchronous muscle relative to those of synchronous muscles characterized by only one contraction per activational nervous impulse. However, the higher oscillation frequencies of the former muscle type yield substantially higher power output during repetitive contraction. Energetic savings may also be associated with asynchronous flight muscle because of a much reduced need for calcium cycling that provides the molecular signal for myofibrillar activation (Josephson and Young, 1985).

Although phylogenetically basal pterygote orders possess synchronous flight muscles, asynchronous flight muscle has evolved repeatedly within and among more derived ptcrygote lineages (Fig. 2.4). The paraphyletic assemblage Homoptera is particularly characterized by a complex representation of synchronous and asynchronous muscle types (Cullen, 1974; Dudley,

2000). Asynchronous flight muscle is phylogenetically derived relative to synchronous muscle, although the exact nature of transitional forms remains unclear. Also, reversion from asynchronous to synchronous flight muscle is physiologically unlikely, but systematic phylogenetic analysis is not possible because muscle types are not well-resolved for key taxa (e.g. Zoraptera;

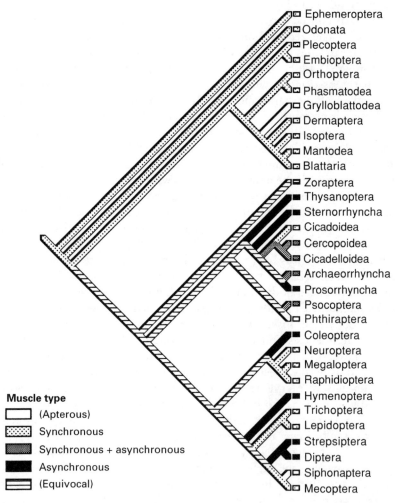

**Fig. 2.4.** Phylogenetic distribution of asynchronous flight muscle following Tiegs (1955), Boettiger (1960), Cullen (1974), Smith (1984), and Smith and Kathirithamby (1984). Note that the paraphyletic taxon Homoptera is here represented at the lower taxonomic levels of suborders and superfamilies. Relatedness of homopteran and hemipteran lineages follows Campbell *et al.* (1994, 1995), Sorensen *et al.* (1995), von Dohlen and Moran (1995), and Schaefer (1996). Equivocal branch designations indicate either an unknown (e.g. Zoraptera) or an unresolved character state.

Fig. 2.4). Most muscle type designations have also been made using morphological inference (i.e. reduced sarcoplasmic reticulum) rather than by direct physiological measurement. Furthermore, asynchronous flight muscle may be present in a number of insect taxa supplemental to those indicated in Fig. 2.4, most importantly in certain mayfly lineages (Dudley, 2000).

Because of the obligate association between elevated wingbeat frequencies and small body size, the taxonomic and numerical preponderance of small insects may have been facilitated by the acquisition of asynchronous flight muscle (Dudley, 1991, 2000). For example, three of the four largest orders (Coleoptera, Diptera and Hymenoptera) possess asynchronous flight muscle. Similarly, comparison of sister taxa that differ in flight muscle type suggests a decrease in body size and an increase in species number following acquisition of asynchronous flight muscle, at least for the North American fauna (Table 2.1). This association can be rigorously tested using an independent contrast analysis that controls for phylogenetic relatedness among taxa. Application of such a method to the character distribution portrayed in Fig. 2.4 demonstrates a statistically significant relationship between possession of asynchronous flight muscle and species richness (one-tailed sign test, $n = 6$, $P = 0.016$; Dudley, 2000). Higher wingbeat frequencies are furthermore associated with greater force production that enables a reduction in effective wing area relative to body mass.

Table 2.1. Number of species and mean body length for lineages in the North American fauna possessing only synchronous flight muscle, and for their sister clades characterized either partially or exclusively by possession of asynchronous flight muscle.

| Clade possessing synchronous flight muscle: | No. of species | Mean body length (mm) | Sister clade either partially or exclusively possessing asynchronous flight muscle: | No. of species | Mean body length (mm) |
|---|---|---|---|---|---|
| Cicadoidea | 166 | 34.5 | [Cercopoidea + Cicadelloidea] | 2,820 | 5.8 |
| Archaeorrhyncha (=Fulgoromorpha) | 530 | 6.7 | Prosorrhyncha (= Hemiptera) | 3,587 | 6.1 |
| [Neuroptera + Megaloptera+ Raphidioptera] | 349 | 17.9 | Coleoptera | 23,592 | 6.0 |
| [Mecoptera + Siphonaptera][a] | 393 | 3.3 | [Diptera + Strepsiptera] | 17,023 | 7.0 |

[a]The Mecoptera alone comprise 68 species with an estimated mean body length of 7.7 mm.
Species numbers are taken from Arnett (1985); calculations of mean body length are described in the text.

Correspondingly, one wing pair can be used in non-aerodynamic roles, as in the hemelytra of Hemiptera, the elytra of Coleoptera and the halteres of Diptera. Acquisition of asynchronous flight muscle has thus played a major role in ordinal-level insect diversification.

Another important consequence of miniaturization concerns a necessary reduction in flight speeds and a correspondingly increased influence of ambient winds on the flight trajectory. Most small insects cannot fly faster than typical wind speeds, and the majority of the extant fauna must accordingly fly close to the ground or within vegetation if intentionally directed flight is to be attained (Dudley, 1994). Conversely, long-distance dispersal in miniaturized taxa derives primarily if not exclusively from entrainment by ambient winds. The difficulty of measuring simultaneously the airspeed of small insects and their local wind speed precludes detailed analysis of endogenous contributions to wind-assisted displacement. However, even 'passive' drifting with winds must involve considerable energetic expenditure in order to stay aloft. Small insects being convectively dispersed must offset their body weight aerodynamically, either while maintaining a forward airspeed or while hovering within a moving air volume. If the goal of dispersal is to maximize the horizontal distance travelled, then flight at the minimum power speed maximizes time aloft and thus the extent of wind-assisted displacement. Given preceding arguments that power curves for the majority of insect taxa are strongly curvilinear, this speed is likely to be relatively low for 'J'-shaped power curves, and could be very close to the minimum power speed. A general rule for small insects maximizing long-distance displacement might then be to hover and simply to maintain vertical position in the moving air. Logistically, the measurement of airspeeds on minute insects flying under natural conditions hundreds of metres from the earth's surface remains a challenging problem. Systematic reduction in body size, however, must profoundly influence long-distance strategies of dispersal and resource location in insects.

## The Evolutionary Escalation of Flight Performance

Flight plays a central role in the lives of most pterygotes. A partial list of important insect behaviours mediated or facilitated by flight includes pollination, phytophagy, haematophagy, escape from predators and mate acquisition. In such various contexts, both natural and sexual selection have probably demanded ever-increasing flight performance through evolutionary time, whereas various agents of selection are often mutually reinforcing. Moreover, forces of both intra- and intersexual selection often act synergistically on manoeuvrability and flight capacity, thereby promoting rapid evolution in these traits. For example, dragonflies defend territories and chase out male conspecifics (intrasexual selection), pursue females (intersexual selection), capture prey items in the air (natural selection for aerial attack), and evade both aerial and terrestrial predators via flight (natural selection for aerial escape). Multiple modes of selection have thus probably acted to enhance flight performance in this famously manoeuvrable taxon.

Such an evolutionary outcome may be general for volant taxa. Flying insects are regularly the subject of attack by insectivorous birds and bats, escape from which requires increased manoeuvrability and/or greater flight speeds. Males of many insect species also attempt to capture females during chases, thus selecting for manoeuvrability in the context of mate choice (Thornhill and Alcock, 1983). Because performance traits that are under sexual selection can evolve rapidly (Andersson, 1994), synergistic interaction between reproductive behaviour and overall flight performance is likely to have been pronounced for those insects with aerial components to their mating systems. In contrast, sexual selection can potentially exaggerate secondary male characters at the expense of flight-related morphological investment (e.g. Kawano, 1997). For winged insects generally, however, intense selection on aerodynamic performance and the underlying biomechanics of flight is imposed in a variety of con- and heterospecific contexts. Such diverse evolutionary forces can, in general, result in reciprocal escalation between biotic selective agents and the target of selection (Vermeij, 1987), yielding ever-increasing levels of performance. Among all animal taxa, flying insects may best exemplify the extent of behavioural and physiological diversification that can arise as a consequence of such evolutionary escalation.

One of the major co-evolutionary interactions of the terrestrial biosphere concerns relationships between insects and plants. Phytophagy and pollination by insects are particularly influenced by three-dimensional aerial mobility, the capacity for which dramatically increases access to nutritional resources and suitable oviposition sites. The relative importance of phytophagy is difficult to overestimate for the present-day fauna, as approximately 85% of extant insect species feed on plants at some stage in their life cycle (Strong et al., 1984). Such interactions probably began at the very outset of flight in pterygotes, given that the fossil record demonstrates feeding on plants by insects in the Upper Carboniferous and early Permian (Labandeira and Phillips, 1996a,b; Rasnitsyn and Krassilov, 1996). Similarly, angiosperms are pollinated primarily by flying insects, in particular by the Coleoptera, Diptera, Hymenoptera, Lepidoptera and Thysanoptera (Kevan and Baker, 1983; Proctor et al., 1996). With the exception of the Lepidoptera, insects from these orders possess asynchronous flight muscle and are often miniaturized forms that hover at flowers either prior to or during pollination. Small insects can also act as wind-dispersed pollen vectors that become aerially entrained by prevailing winds, but also that exert some measure of behavioural choice once in the vicinity of nectar-bearing plants. In contexts both of phytophagy and pollination, selection for improved flight efficiency and capacity has probably been complementary to modes of selection for greater flight performance over much shorter time scales.

## Conclusions

The evolution of wings in the late Palaeozoic was a defining event for subsequent hexapod radiations on (and above) the surface of the earth. Equally important to the evolutionary diversification of insects have been repeated events of miniaturization, a process enabled by acquisition of asynchronous flight muscle and elevated wingbeat frequencies during flight. Ordinal-level patterns of wing transformation for non-aerodynamic purposes have similarly been influenced by high flapping frequencies, an effect most clearly evidenced by the elytra of Coleoptera and the halteres of Diptera. Most insects are fairly small and fly in Reynolds-number regimes strongly influenced by the viscosity of air. The combined effects of wing flapping and rotation at the ends of half-strokes yield strongly unsteady airflows, leading-edge vortices, and aerodynamic forces well in excess of those associated with steady-state wing translation. Flight has been an essential underpinning to diverse features of insect ecology and behaviour, including aerial mating systems, phytophagy and the pollination of angiosperms. Low flight speeds derived biomechanically from small body sizes, however, indicate that ambient flight trajectories are often dominated by ambient winds. Forces of both natural and sexual selection have contributed synergistically to the evolution of insect flight performance and manoeuvrability, yielding the wonderfully agile taxa that to this day compete with *Homo sapiens* for dominance of the terrestrial biosphere.

## Acknowledgements

I thank Carl Gans, Bob Srygley and Robin Wootton for useful comments, and the NSF (IBN-9817138) for research support.

## References

Alexander, R.D. (1964) The evolution of mating behaviour in arthropods. *Symposium of the Royal Entomological Society of London* 2, 78–94.
Alexander, R.D. and Brown, W.L. (1963) Mating behavior and the origin of insect wings. *Occasional Papers of the Museum of Zoology, University of Michigan* 628, 1–19.
Andersson, M. (1994) *Sexual Selection*. Princeton University Press, Princeton, New Jersey.
Arnett, R.H. (1985) *American Insects: a Handbook of the Insects of America North of Mexico*. Van Nostrand Reinhold, New York.
van den Berg, C. and Ellington, C.P. (1997a) The three-dimensional leading-edge vortex of a 'hovering' model hawkmoth. *Philosophical Transactions of the Royal Society of London B* 352, 329–340.

van den Berg, C. and Ellington, C.P. (1997b) The vortex wake of a 'hovering' model hawkmoth. *Philosophical Transactions of the Royal Society of London B* 352, 317–328.

Berner, R.A. (1997) The rise of plants and their effect on weathering and atmospheric $CO_2$. *Science* 276, 544–546.

Berner, R.A. (1998) The carbon cycle and $CO_2$ over Phanerozoic time: the role of land plants. *Philosophical Transactions of the Royal Society of London B* 353, 75–82.

Berner, R.A. and Canfield, D.E. (1989) A new model for atmospheric oxygen over Phanerozoic time. *American Journal of Science* 289, 333–361.

Boettiger, E.G. (1960) Insect flight muscles and their basic physiology. *Annual Review of Entomology* 5, 1–16.

Brauckmann, C. and Zessin, W. (1989) Neue Meganeuridae aus dem Namurian von Hagen Vorhalle (BRD) und die Phylogenie der Meganisoptera. *Deutsche Entomologische Zeitschrift, N.F.* 36, 177–215.

Briggs, D.E.G. (1985) Gigantism in Palaeozoic arthropods. *Special Papers in Paleontology* 33, 157.

Campbell, B.C., Steffen-Campbell, J.D. and Gill, R.J. (1994) Evolutionary origin of whiteflies (Hemiptera: Sternorrhyncha: Aleyrodidae) inferred from 18S rDNA sequences. *Insect Molecular Biology* 3, 73–88.

Campbell, B.C., Steffen-Campbell, J.D., Sorensen, J.T. and Gill, R.J. (1995) Paraphyly of Homoptera and Auchenorrhyncha inferred from 18S rDNA nucleotide sequences. *Systematic Entomology* 20, 175–194.

Carpenter, F.M. (1992) *Treatise on Invertebrate Paleontology*. Part R, Arthropoda 4, Vols 3 and 4 (Hexapoda). University of Kansas Press, Lawrence, Kansas.

Carroll, S.B., Weatherbee, S.D. and Langeland, J.A. (1995) Homeotic genes and the regulation and evolution of insect wing number. *Nature* 375, 58–61.

Casey, T.M. (1989) Oxygen consumption during flight. In: Goldsworthy, G.J. and Wheeler, C.H. (eds) *Insect Flight*. CRC Press, Boca Raton, Florida, pp. 257–272.

Crowson, R.A. (1981) *The Biology of the Coleoptera*. Academic Press, London.

Cullen, M.J. (1974) The distribution of asynchronous muscle in insects with special reference to the Hemiptera: an electron microscope study. *Journal of Entomology* 49A, 17–41.

Denny, M.W. (1993) *Air and Water: the Biology and Physics of Life's Media*. Princeton University Press, Princeton, New Jersey.

Dickinson, M.H. (1999) Haltere-mediated equilibrium reflexes of the fruit fly, *Drosophila melanogaster*. *Philosophical Transactions of the Royal Society of London B* 354, 903–916.

Dickinson, M.H., Lehmann, F.O. and Sane, S.P. (1999) Wing rotation and the aerodynamic basis of insect flight. *Science* 84, 1954–1960.

von Dohlen, C.D. and Moran, N.A. (1995) Molecular phylogeny of the Homoptera: a paraphyletic taxon. *Journal of Molecular Evolution* 40, 211–223.

Douglas, M.M. (1981) Thermoregulatory significance of thoracic lobes in the evolution of insect wings. *Science* 211, 84–86.

Dudley, R. (1991) Comparative biomechanics and the evolutionary diversification of flying insect morphology. In: Dudley, E.C. (ed.) *The Unity of Evolutionary Biology*. Dioscorides Press, Portland, Oregon, pp. 503–514.

Dudley, R. (1994) Aerodynamics of insect dispersal and the constraint of body size. In: Maarouf, A.R., Barthakur, N.N. and Haufe, W.O. (eds) *Proceedings of* the 13th International Biometeorology Congress, Part 2, Vol. 3. Environment Canada, Downsview, pp. 1035–1041.

Dudley, R. (1998) Atmospheric oxygen, giant Paleozoic insects and the evolution of aerial locomotor performance. *Journal of Experimental Biology* 201, 1043–1050.

Dudley, R. (2000) *The Biomechanics of Insect Flight: Form, Function, Evolution*. Princeton University Press, Princeton, New Jersey.

Dudley, R. and DeVries, P.J. (1990) Flight physiology of migrating *Urania fulgens* (Uraniidae) moths: kinematics and aerodynamics of natural free flight. *Journal of Comparative Physiology A* 167, 145–154.

Dudley, R. and Ellington, C.P. (1990) Mechanics of forward flight in bumblebees. II. Quasi-steady lift and power requirements. *Journal of Experimental Biology* 148, 53–88.

Edwards, J.S. (1997) The evolution of insect flight: implications for the evolution of the nervous system. *Brain, Behavior and Evolution* 50, 8–12.

Edwards, J.S. and Reddy, G.R. (1986) Mechanosensory appendages in the firebrat (*Thermobia domestica*, Thysanura): a prototype system for terrestrial predator invasion. *Journal of Comparative Neurology* 243, 535–546.

Ellington, C.P. (1984) The aerodynamics of hovering insect flight. VI. Lift and power requirements. *Philosophical Transactions of the Royal Society of London B* 305, 145–181.

Ellington, C.P. (1991a) Aerodynamics and the origin of insect flight. *Advances in Insect Physiology* 23, 171–210.

Ellington, C.P. (1991b) Limitations on animal flight performance. *Journal of Experimental Biology* 160, 71–91.

Ellington, C.P. (1995) Unsteady aerodynamics of insect flight. In: Ellington, C.P. and Pedley, T.J. (eds) *Biological Fluid Dynamics*. Company of Biologists Ltd, Cambridge, pp. 109–129.

Ellington, C.P., van den Berg, C., Willmott, A.P. and Thomas, A.L.R. (1996) Leading-edge vortices in insect flight. *Nature* 384, 626–630.

Flower, J.W. (1964) On the origin of flight in insects. *Journal of Insect Physiology* 10, 81–88.

Graham, J.B., Dudley, R., Aguilar, N. and Gans, C. (1995) Implications of the late Palaeozoic oxygen pulse for physiology and evolution. *Nature* 375, 117–120.

Grodnitsky, D.L. (1995) Evolution and classification of insect flight kinematics. *Evolution* 49, 1158–1162.

Hammond, P.M. (1992) Species inventory. In: Groombridge, B. (ed.) *Global Biodiversity, Status of the Earth's Living Resources*. Chapman & Hall, London, pp. 17–39.

Hennig, W. (1981) *Insect Phylogeny*. John Wiley & Sons, Chichester.

Hinton, H.E. (1968) Spiracular gills. *Advances in Insect Physiology* 5, 65–161.

Jeram, A.J., Selden, P.A. and Edwards, D. (1990) Land animals in the Silurian: arachnids and myriapods from Shropshire, England. *Science* 250, 658–661.

Josephson, R.K. and Young, D. (1985) A synchronous muscle with an operating frequency greater than 500 Hz. *Journal of Experimental Biology* 118, 185–208.

Kammer, A.E. and Heinrich, B. (1978) Insect flight metabolism. *Advances in Insect Physiology* 13, 133–228.

Kawano, K. (1997) Costs of evolving exaggerated mandibles in stag beetles (Coleoptera: Lucanidae). *Annals of the Entomological Society of America* 90, 453–461.

Kevan, P.G. and Baker, H.G. (1983) Insects as flower visitors and pollinators. *Annual Review of Entomology* 28, 407–453.

Kingsolver, J.G. and Koehl, M.A.R. (1985) Aerodynamics, thermoregulation, and the evolution of insect wings: differential scaling and evolutionary change. *Evolution* 39, 488–504.

Kingsolver, J.G. and Koehl, M.A.R. (1994) Selective factors in the evolution of insect wings. *Annual Review of Entomology* 39, 425–451.

Kramer, M.G. and Marden, J.H. (1997) Almost airborne. *Nature* 385, 403–404.

Kristensen, N.P. (1991) Phylogeny of extant hexapods. In: CSIRO (ed.) *The Insects of Australia*, 2nd edn, Vol. 1. Cornell University Press, Ithaca, New York, pp. 125–140.

Kristensen, N.P. (1997) The groundplan and basal diversification of the hexapods. In: Fortey, R.A. and Thomas, R.H. (eds) *Arthropod Relationships*. Chapman & Hall, London, pp. 281–293.

Kukalová-Peck, J. (1978) Origin and evolution of insect wings and their relation to metamorphosis, as documented by the fossil record. *Journal of Morphology* 156, 53–126.

Kukalová-Peck, J. (1991) Fossil history and the evolution of hexapod structures. In: CSIRO (ed.) *The Insects of Australia*, 2nd edn, Vol. 1. Cornell University Press, Ithaca, New York, pp. 141–179.

Kutsch, W. and Breidbach, O. (1994) Homologous structures in the nervous systems of Arthropoda. *Advances in Insect Physiology* 24, 1–113.

Labandeira, C.C. and Phillips, T.L. (1996a) A Carboniferous insect gall: insight into early ecologic history of the Holometabola. *Proceedings of the National Academy of Sciences USA* 93, 8470–8474.

Labandeira, C.C. and Phillips, T.L. (1996b) Insect fluid-feeding on Upper Pennsylvanian tree ferns (Palaeodictyoptera, Marattiales) and the early history of the piercing-and-sucking functional feeding group. *Annals of the Entomological Society of America* 89, 157–183.

Labandeira, C.C. and Sepkoski, J.J. (1993) Insect diversity in the fossil record. *Science* 261, 310–315.

Labandeira, C.C., Beall, B.S. and Hueber, F.M. (1988) Early insect diversification: evidence from a Lower Devonian bristletail from Québec. *Science* 242, 913–916.

Liu, H., Ellington, C.P., Kawachi, K., van den Berg, C. and Willmott, A. (1998) A computational fluid dynamic study of hawkmoth hovering. *Journal of Experimental Biology* 201, 461–477.

Maddison, W.P. and Maddison, D.R. (1992) *MacClade: Analysis of Phylogeny and Character Evolution*, Version 3. Sinauer Associates, Sunderland, Massachusetts.

Marden, J.H. and Kramer, M.G. (1994) Surface-skimming stoneflies: a possible intermediate stage in insect flight evolution. *Science* 266, 427–430.

Marden, J.H. and Kramer, M.G. (1995) Locomotor performance of insects with rudimentary wings. *Nature* 377, 332–334.

May, R.M. (1978) The dynamics and diversity of insect faunas. In: Mound, L.A. and Waloff, N. (eds) *Diversity of Insect Faunas*. Blackwell Scientific, Oxford, pp. 188–204.

Messner, B. (1988) Sind die Insekten primäre oder sekundäre Wasserbewohner? *Deutsche Entomologische Zeitschrift, N.F.* 35, 355–360.

Nelson, C.R. and Tidwell, W.D. (1987) *Brodioptera stricklani* n.sp. (Megasecoptera: Brodiopteridae), a new fossil insect from the Upper Manning Canyon Shale Formation, Utah (lowermost Namurian B). *Psyche* 94, 309–316.

Norberg, U.M. (1990) *Vertebrate Flight*. Springer-Verlag, Berlin.

Pashley, D.P., McPheron, B.A. and Zimmer, E.A. (1993) Systematics of holometabolous insect orders based on 18S ribosomal RNA. *Molecular Phylogenetics and Evolution* 2, 132–142.

Pix, W., Nalbach, G. and Zeil, J. (1993) Strepsipteran forewings are haltere-like organs of equilibrium. *Naturwissenschaften* 80, 371–374.

Pringle, J.W.S. (1949) The excitation and contraction of the flight muscles of insects. *Journal of Physiology* 108, 226–232.

Pringle, J.W.S. (1978) Stretch activation of muscle: function and mechanism. *Proceedings of the Royal Society of London B* 201, 107–130.

Pritchard, G., McKee, M.H., Pike, E.M., Scrimgeour, G.J. and Zloty, J. (1993) Did the first insects live in water or in air? *Biological Journal of the Linnaean Society* 49, 31–44.

Proctor, M., Yeo, P. and Lack, A. (1996) *The Natural History of Pollination*. Timber Press, Portland, Oregon.

Rasnitsyn, A.P. and Krassilov, V.A. (1996) First find of pollen grains in the gut of Permian insects. *Paleontological Journal* 30, 484–490.

Ruffieux, L., Elouard, J.-M. and Sartori, M. (1998) Flightlessness in mayflies and its relevance to hypotheses on the origin of insect flight. *Proceedings of the Royal Society of London B* 265, 2135–2140.

Schaefer, C.W. (ed.) (1996) *Studies on Hemipteran Phylogeny*. Entomological Society of America, Lanham, Maryland.

Schwanwitsch, B.N. (1943) Subdivision of Insecta Pterygota into subordinate groups. *Nature* 1943, 727–728.

Schwanwitsch, B.N. (1958) Alary musculature as a basis of the system of pterygote insects. *Proceedings of the 10th International Congress of Entomology*, Vol. 1, pp. 605–610.

Shear, W.A., Grierson, J.D., Rolfe, W.D.I., Smith, E.L. and Norton, R.A. (1984) Early land animals in North America: evidence from Devonian age arthropods from Gilboa, New York. *Science* 224, 492–494.

Smith, D.S. (1984) The structure of insect muscles. In: King, R.C. and Akai, H. (eds) *Insect Ultrastructure*, Vol. 2. Academic Press, New York, pp. 111–150.

Smith, D.S. and Kathirithamby, J. (1984) Atypical 'fibrillar' flight muscle in Strepsiptera. *Tissue & Cell* 16, 929–940.

Sorensen, J.T., Campbell, B.C., Gill, R.J. and Steffen-Campbell, J.D. (1995) Non-monophyly of Auchenorrhyncha ('Homoptera') based on 18S rDNA phylogeny: eco-evolutionary and cladistic implications within pre-Heteropterodea Hemiptera (S.L.) and a proposal for new monophyletic suborders. *Pan-Pacific Entomologist* 71, 31–60.

Strong, D.R., Lawton, J.H. and Southwood, R. (1984) *Insects on Plants*. Blackwell Scientific, Oxford.

Thomas, A.L.R. and Norberg, R.Å. (1996) Skimming the surface – the origin of flight in insects? *Trends in Ecology and Evolution* 11, 187–188.

Thornhill, R. and Alcock, J. (1983) *The Evolution of Insect Mating Systems*. Harvard University Press, Cambridge, Massachusetts.

Tiegs, O.W. (1955) The flight muscles of insects – their anatomy and histology; with some observations on the structure of striated muscle in general. *Philosophical Transactions of the Royal Society of London B* 238, 221–348.

Vermeij, G.J. (1987) *Evolution and Escalation*. Princeton University Press, Princeton, New Jersey.

Vermeij, G.J. and R. Dudley. (2000) Why are there so few evolutionary transitions between aquatic and terrestrial ecosystems? *Biological Journal of the Linnean Society*, 70, 541–554.

Vogel, S. (1994) *Life in Moving Fluids: the Physical Biology of Flow*. Princeton University Press, Princeton, New Jersey.

Wakeling, J.M. and Ellington, C.P. (1997) Dragonfly flight. III. Lift and power requirements. *Journal of Experimental Biology* 200, 583–600.

Walter, D.E. and Behan-Pelletier, V. (1999) Mites in forest canopies: filling the size distribution shortfall? *Annual Review of Entomology* 44, 1–19.

Whalley, P. and Jarzembowski, E.A. (1981) A new assessment of *Rhyniella*, the earliest known insect, from the Devonian of Rhynie, Scotland. *Nature* 291, 317.

Whalley, P.E.S. (1979) New species of Protorthoptera and Protodonata (Insecta) from the Upper Carboniferous of Britain, with a comment on the origin of wings. *Bulletin of the British Museum of Natural History (Geology)* 32, 85–90.

Whiting, M.F., Carpenter, J.M., Wheeler, Q.D. and Wheeler, W.C. (1997) The Strepsiptera problem: phylogeny of the holometabolous insect orders inferred from 18S and 28S ribosomal DNA sequences and morphology. *Systematic Biology* 46, 1–68.

Will, K.W. (1995) Plecopteran surface-skimming and insect flight evolution. *Science* 270, 1684–1685.

Williams, C.B. (1960) The range and pattern of insect abundance. *American Naturalist* 94, 137–151.

Willmott, A.P. and Ellington, C.P. (1997) The mechanics of flight in the hawkmoth *Manduca sexta*. II. Aerodynamic consequences of kinematic and morphological variation. *Journal of Experimental Biology* 200, 2723–2745.

Wootton, R.J. (1976) The fossil record and insect flight. In: Rainey, R.C. (ed.) *Insect Flight*. Blackwell Scientific, Oxford, pp. 235–254.

Wootton, R.J. (1990) Major insect radiations. In: Taylor, P.D. and Larwood, G.P. (eds) *Major Evolutionary Radiations*. Clarendon Press, Oxford, pp. 187–208.

# How Insect Wings Evolved 3

Robin J. Wootton

*School of Biological Sciences, University of Exeter, Exeter EX4 4PS, UK*

## Introduction

Insects were the first animal group to achieve powered flight, and are the only invertebrates to have done so. This has major implications on the evolution of their wings. The ancestors of flying vertebrates all developed wings by modifying forelimbs already equipped with a jointed endoskeleton, elaborate musculature, innervation and vascular supplies. Insects evolved wings from relatively far smaller structures whose precise design is unknown, but which must have been comparatively simple, with minimal or no musculature. Insect wings are mainly cuticular, with channels containing haemolymph, often tracheae and in places nerves, but scarcely any other tissues. There are no muscles beyond the extreme base.

Insect wings are none the less complex structures. This is necessary because flapping flight is a complex process. To develop the aerodynamic forces needed to support the insect's weight and to propel it appropriately, the wings must flap, twist and deform. These movements have to be finely adjustable in order to allow a range of speeds and manoeuvres. With no internal muscles, wing deformations must be remotely controlled, and in many cases automatic: elastic responses to the constantly changing aerodynamic and inertial forces which the wings are experiencing, and wholly dependent on their structure. The wings are in effect 'smart' aerofoils, with few if any parallels elsewhere (Wootton, 1981, 1992, 1999).

The evolution by insects of developed flight was therefore anything but a trivial process. Complex adaptations in wing design must have been accompanied by major modifications in the gross structure of the exoskeleton, musculature and associated systems of the thorax, together with the development of sophisticated neurosensory control. The whole required time, and sustained selection pressures operating at all stages in the evolutionary process, from the earliest experiments to the fully functional flight system.

Sadly we have no fossil record of these early stages. It seems probable that between 30 and 50 million years had elapsed between the beginning of flight and the first appearence of fossil Pterygota, in the Namurian division of the Carboniferous, c. 320 million years ago. By then they were already fully developed and varied fliers (Carpenter, 1992; Wootton et al., 1998; Wootton and Kukalová-Peck, 2000).

The homology and form of the precursors of wings, and the route by which flight evolved are extensively discussed in the literature, and will be reviewed only briefly here. The remainder of this chapter will be concerned with later stages. In particular, it will examine what structural characteristics would have been necessary components of the wings in a simple, but fully functional powered flight system, and what roles they would play. This differs in approach from the detailed analyses by Kukalová-Peck (1983, 1991), which are primarily concerned with details of wing and axillary homology, and are based on intensive comparison of living and fossil forms. Rather this chapter is an attempt at *a priori* reasoning based on our knowledge of flight processes. Inevitably hindsight is involved; after all, we know the results. However, as we increasingly understand the functional significance of wing characters it becomes possible to distinguish between those which would have been essential in the earliest stages of flight evolution and those which could safely have been left until later. Brodsky (1994) has made a similar attempt to reconstruct the transitional stages between the origin of flight and its developed forms, reasoning from the kinematics and the structure of the wakes of 'primitive' modern insects; a different approach leading to rather different conclusions.

The chapter will conclude with a brief review of the earliest known fliers: the wings of the Carboniferous Pterygota; with some interpretation of their operation in flight.

## Where Did Insect Wings Come From?

In 1973, at the last RES Symposium devoted to movement, Wigglesworth (1976) revived the long-outmoded view that insect wings began not as fixed lateral tergal outgrowths, but as mobile thoracic pleural structures homologous with the abdominal gill-plates of mayfly nymphs. Revolutionary then, this is now the majority view. Its most influential proponent, Kukalová-Peck (1978, 1983, 1991), has made a strong case for the origin of wings – and mayfly gills – from exites of basal leg segments long since incorporated into the body, and represented now by the pleural and laterotergal structures immediately surrounding the wing base, and by serially homologous but now obscure components of the abdominal wall. She and others have amassed a wealth of evidence from palaeontology, and from the embryology and morphology of modern insects, in support of this view. There are some differences in opinion about the precise homology of the wing precursors (e.g. Rasnitsyn, 1981; Kliuge, 1989; Trueman, 1990) but from a functional viewpoint this is relatively unimportant. The crucial development is the conclusion that wings probably evolved from structures which were already movable, with basal musculature and a nervous supply, *before* flight began, whereas previous theories assumed the origin of wings from fixed outgrowths, which only developed mobility at a later stage in the process.

If, as now appears probable, wings are serially homologous with the abdominal gills of mayfly nymphs, did they indeed evolve from plate-like thoracic gills, or did mayfly gill-plates and wings independently develop as flattened structures similarly adapted to their common role – that of accelerating a fluid? The former implies that the immediate ancestors of pterygotes were aquatic, at least as juveniles, and somehow made the transition to flight in air. This view has been supported by Riek (1971), Wigglesworth (1976), Kukalová-Peck (1978, 1983, 1991), Stys and Soldan (1980), Toms (1984) and others. If, however, gills and wings evolved in parallel from the same precursors, it is at least as probable that pterygote ancestors were fully terrestrial, as indeed are apterygote insects today.

## How Did Flight Begin?

Since the seminal papers of Wigglesworth (1976) and Kukalová-Peck (1978, 1983), there has been a marked renewal of interest in how flight arose. Usefully, there has been a move towards biomechanical experimentation and theoretical analysis. Kingsolver and Koehl (1985) used hypothetical model 'protopterygotes' in a wind tunnel to investigate the possible roles of incipient static wings, and concluded that they could have acted first in thermoregulation, and only later, when a critical size was reached, in stabilization and gliding. Their work prompted new reviews (Quartau, 1986; Wootton, 1986) and further experiments, this time with gliding models, appropriately scaled and with serial 'winglets' whose angle of attack could be altered (Wootton and Ellington, 1991). Ellington (1991) authoritatively discussed the aerodynamic implications of the origin of flight, and Kingsolver and Koehl (1994) published a further extensive review of the field.

The principal contribution of these biomechanical studies has been to clarify the bounds of physical possibility (Kingsolver and Koehl (1994) use the phrase 'bounded ignorance') within which the relative probability of available theories can be evaluated. This mature approach is greatly preferable to any dogged adherence to a particular scenario, when certainty is impossible. Biomechanics can also help to assess the probabilities of 'possible' hypotheses. Thus Wigglesworth's (1963) proposal that flight arose in insects passively drifting in air currents appears rather improbable because such a route would select for wings with high drag, rather than with high lift and low drag as required in true flight; and would also favour relatively small insects which would in due course need uncomfortably high flapping frequencies (Flower, 1964; Wootton, 1986; Ellington, 1991). Similarly any proposal that fast-running insects could use winglets to take off and plane over the ground has the disadvantage that they would lose airspeed as soon as their feet left the substrate. Using gravity to gain and maintain airspeed, by jumping, or by falling from a height, appears far more plausible, and Wootton and Ellington (1991) and Ellington (1991) have supported the traditionally popular 'flying squirrel' theory that flight arose in arboreal insects, via first stable parachuting, then gliding, and in due course flapping when selection for good gliding performance had led to wings large enough for flapping to be effective. They found (Wootton and Ellington, 1991) that winglets with suitably adjusted angles of attack stabilized and slowed the descent of their smaller models, and gave surprisingly shallow glide angles in larger models corresponding to insects about 70 mm long. Removal of abdominal winglets while enlarging those on the thorax destabilized the models, but stability was readily restored by attaching mayfly-like filaments to the abdomen; a convenient discovery in view of the presence of long caudal filaments in many Palaeozoic forms.

More recently, Marden and Kramer (1994) have put forward a wholly new theory, derived from observation of Plecoptera and other insects which use their wings, often reduced in size, to propel themselves across a water surface while maintaining foot-contact with the latter, which supports their weight. If the ancestors of pterygotes used their winglets similarly to gain thrust only, selection would favour their enlargement to a point where they could begin to support the weight of the insects and to lift them off the surface. This attractive theory is the first to give flapping a plausible function in the early stages of the process stage; in the parachuting/gliding hypothesis flapping would probably have only marginal value until the wings had reached a reasonable size (Ellington, 1991; Wootton and Ellington, 1991).

The surface-skimming scenario has attracted considerable attention (e.g. Thomas and Norberg, 1996) and appeals particularly to those who support the origin of pterygotes from aquatic insects, as it bridges the awkward gap between gills flapping in water and wings operating in air. It and the parachuting/gliding 'flying squirrel' theory now appear to be the two most probable hypotheses. At present we have insufficient evidence to decide between them.

## What Did Insects Need in Order to Fly at All?

Like all actively flying animals, insects use wing *lift* for both weight support and propulsion.

It is important to understand the true nature of fluid dynamic lift. It is defined not as a vertical force, but as a force perpendicular to the direction of relative movement of the lift-generating surface. An aircraft wing moving horizontally generates vertical lift (Fig. 3.1a); but the direction of the lift generated by an insect's flapping wings changes constantly with the wings', not the insect's, direction of movement relative to the air. When integrated over the entire stroke cycle, any net upward component of the resultant force

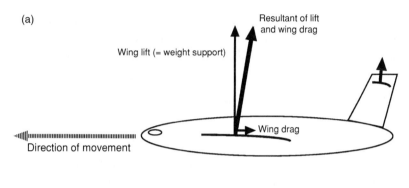

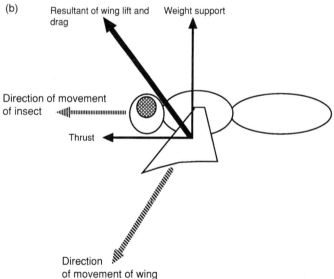

**Fig. 3.1.** The generation of thrust from lift. Broken arrows represent directions of movement relative to the air, continuous arrows represent directions of forces. (a) An aircraft in horizontal flight develops vertical lift, all of which is available to support the weight. (b) The flapping wings of an insect develop lift perpendicular to their own direction of movement; not to that of the body. At the instant illustrated, the resultant force of the lift and the drag of the wings has a vertically upward component, supporting the weight; and a component in the direction of movement of the insect, which is the thrust.

of the lift and the – much smaller – drag of the wings will tend to support the insect's weight. Any component in the direction of movement of the insect's *body* is the *thrust*, which will directly oppose the overall drag (Fig. 3.1b).

In fully developed flight, the insect needs to generate sufficient vertically upward force to support its entire weight, and enough thrust to overcome the drag which is an inevitable consequence of movement. However, in the early evolutionary stages, by whichever of the two most plausible routes, thrust may well have been the principal requirement. The weight of a skimming insect is supported by the water. A gliding insect would already be generating lift to support its weight, but would have only gravity to propel it. If the winglets were large enough, thrust from flapping would increase the insect's speed, and hence its overall lift and weight-supporting force, even if the vertical forces from the upstroke and downstroke themselves cancelled out, as would be the case if the wings simply flapped up and down without twisting or deformation (Fig. 3.2a). This might be a disadvantage, since an increase in speed could lead to dangerously high landing speeds. Indeed, Flower (1964). in a theoretical analysis of the 'flying squirrel' theory which was well ahead of its time, calculated that even unpowered steady gliding would be hazardous for protopterygotes significantly longer than 10 mm, and this was confirmed by Ellington's analysis (Ellington, 1991). However, these underestimate the insect's probable capacity to control its glide actively. There seems no reason why it should not utilize its mobile winglets to stall into a soft landing, as do many patagiate vertebrate gliders. Wootton and Ellington (1991) found the best glide performance in models of much larger insects. These were indeed travelling perilously fast, but minor adjustments to the 'winglets' altered the character of individual glides, and one at least stalled just before impact.

For basic thrust generation, the requirements would differ slightly in the two scenarios. Some needs would be common to both. In both cases the wings would need: (i) to be large enough to generate useful lift; (ii) to be rigid enough and strong enough to withstand both the aerodynamic forces and the significant inertial forces which would result from the flapping motion; and (iii) to achieve these needs with the lowest possible mass, so that these inertial

**Fig. 3.2** (opposite). The generation of net weight-support and thrust by flapping. The insects are viewed anterolaterally from above, and the wing-tip paths and forces are shown diagrammatically. In (a) the net aerodynamic force and its vertical and horizontal components are drawn. In (b)–(e) only the net force is shown. (a) Flapping without twisting develops net thrust, but the weight-supporting components of the downstroke and upstroke cancel out. (b) If the wings beat obliquely as the insect flies forward the downstroke is faster than the upstroke, and develops more lift. There is therefore a net upward force. (c, d) Net upward force can be achieved by twisting the wing so as to generate less (c) or no (d) lift on the upstroke. (e) The same effect may be achievable by cambering the wing on the downstroke, but not on the upstroke.

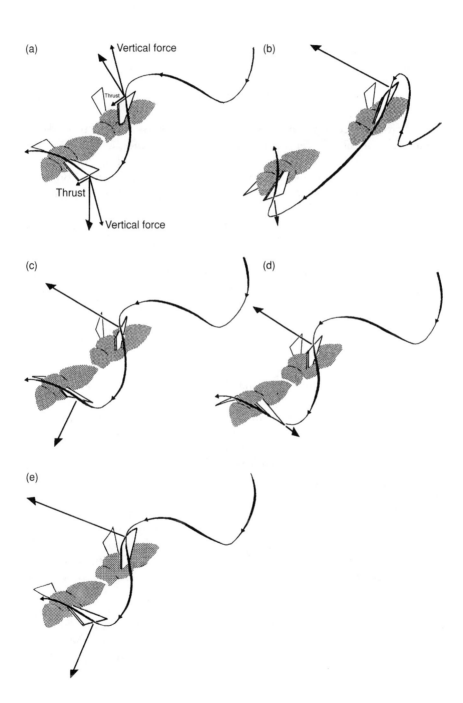

forces would be minimized. Insect cuticle has very high strength and stiffness for its weight, and it is very possible that an improvement in its ultrastructure and properties accompanied the origin of flight. Neville (1984) has suggested that cuticle of the kind in which the chitin microfibrillae in adjacent layers are arranged alternately helicoidally and unidirectionally evolved first in early Pterygota as an adaptation to provide both torsional and flexural rigidity to the veins of the wings. So far as is known, Apterygota have helicoidally arranged chitin only.

Rigidity, though, depends above all on macrostructure. Thick, plate-like wings would be rigid, but with high mass, and are today only found when adapted for protection. For rigidity with low mass, insects instead use frameworks of tubular veins supporting thin membranes, and relief in the form of corrugation and/or a cambered profile. Any or all of these would have been valuable in the earliest functional wings. Wing corrugation, with veins alternately occupying the ridges and troughs of the corrugations, appears to be a symplesiomorphy for the Pterygota as a whole (Kukalová-Peck, 1983, 1991). Despite the – conflicting – literature, we do not know the original number of branches to the various longitudinal veins, nor the primitive arrangement of cross-veins.

The principal difference in prerequisites for the two favoured scenarios relates to wing twisting. A gliding insect would already be moving fast, and its wings would already be meeting the oncoming air at a suitably low angle of attack for generating lift. If the winglets were large enough, simple flapping would produce some thrust, increase speed and hence overall weight-support, and so prolong the glide as needed. Thrust generation could be enhanced by some twisting of the wings between upstroke and downstroke, but this would not be essential. However, an insect stationary on the water surface would need to generate thrust from scratch. As there would be no pre-existing forward movement to provide lift, the wings would need to twist strongly between half-strokes in order to get underway, although the necessary angle of twist would diminish as the insect gathered speed.

This has implications on the nature of the wings' articulation with the thorax, and on the musculature. A strong articulation with the thorax would in any case be needed as the flight process advanced and basal stresses increased; but the combination of strength with controlled twisting is not easily achieved. Insect axillae are complex mechanisms. The problems involved in their evolution are beyond the scope of this review, but should not be underestimated. Mayfly gills have a single, simple articulation and simple musculature (Kliuge, 1989). Such a system would very soon become inadequate in the development of flight.

In due course – we cannot judge how soon – insects would begin to use flapping to gain weight support as well as thrust. For this it would be necessary to introduce asymmetry to the wing stroke, so that the vertical forces of the downstroke and upstroke did not cancel out. This could be achieved in several ways, separately or in combination:

1. The wings could be made to beat obliquely to the horizontal: anteroventrally in the downstroke, posterodorsally in the upstroke. If they were beating in simple harmonic motion, and the insect were flying forwards, the wings' airspeed and hence lift would be greater in the downstroke, when the vertical component would be upward, than in the upstroke, when it would be downward (Fig. 3.2b).
2. The wings could be twisted, so that they operated at a favourable angle of attack in the downstroke, and either 'feathered' for the upstroke or angled so as to limit the downward lift component (Fig. 3.2c,d).
3. Their area could be reduced between the downstroke and the upstroke.
4. The camber of the profile (cross-section) could be altered between downstroke and upstroke (Fig. 3.2e).
5. Force asymmetry could be introduced by employing a range of unsteady aerodynamic mechanisms, brought about by specific movements and deformations at particular phases of the stroke.

An oblique stroke-plane is usually achieved in modern insects by adjusting the angle of the body to the horizontal in flight, and this may indeed have been possible from the earliest stages of flight evolution. However, morphology can assist. The main wing articulation is frequently itself oblique to the longitudinal axis of the body: Odonata provide an extreme example.

Wing twisting is achieved in two ways: by active muscular twisting at the base, and by passive elastic torsion along the wing span, in response to the aerodynamic and inertial forces experienced (Wootton, 1981; Ennos, 1988, 1989). Controlled active twisting needs appropriate musculature and thoracic and axillary skeletal structure. The need for passive elastic torsion, however, introduces a new structural requirement in the wing: the concentration of the principal supporting veins towards the anterior edge. This would bring the axis of torsion, along which the wing naturally twists, to lie anteriorly to the centres of aerodynamic pressure and of mass. If this were the case, and the wing were fairly compliant to torsion, it would twist to some extent along its length as it passed through the air (Fig. 3.3a). This would be aerodynamically advantageous, since it would tend to reduce the angle at which the distal part of the wing met the oncoming air, and would help to prevent stalling at the tip. Furthermore, should the wing be inherently more resistant to pronatory ('nose-down') twisting than to supinatory ('nose-up') twisting it would help generate the required asymmetry between the up- and downstrokes. As we will see, this is often the case.

Locusts, and probably many other insects with specialized pleated hindwing fans, are capable of altering the effective area of their wings between downstroke and upstroke, and butterflies may be able to do so by altering the degree to which their apposed fore- and hindwings overlap. Neither effect, though, is likely to have been available in the early stages of wing evolution.

The ability to develop a cambered wing section is very important both structurally and aerodynamically. It is widespread, perhaps nearly universal,

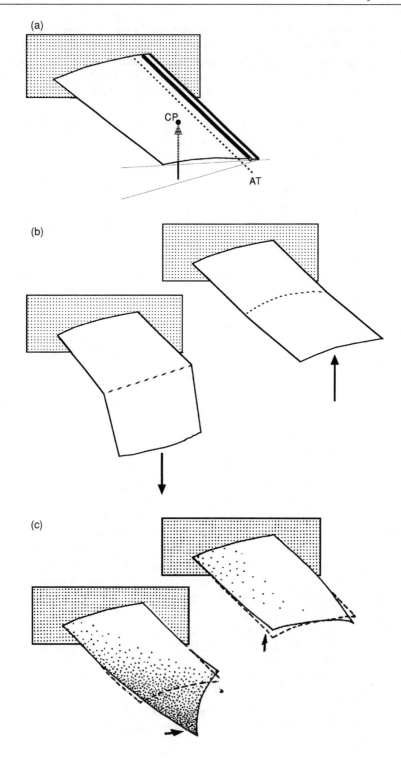

among flying insects today, and there seems every likelihood that it developed early. The wing base rests on the pleural wing process, which acts as the fulcrum for the stroke, and in most insects the basalar and subalar muscles act on either side of the process to depress the anterior and posterior regions of the wing respectively. As soon as this conformation had evolved, the possibility existed actively to camber the wing by contracting the muscles simultaneously. The wing, and the axilla, would need to be flexible enough to permit this. In particular, a sclerotized and thickened axilla would need some kind of longitudinally orientated flexible hinge-line.

Besides generally enhancing the wing's resistance to bending, the capability actively to generate and lose camber can contribute in several ways to the asymmetry of forces between the down- and upstrokes. First, a thin cambered plate of the size range relevant to insects, in a steady airflow at relevant velocities, generates higher lift than a flat one. The flow conditions around flapping insect wings are anything but steady, but it is probable that a wing cambered in the downstroke and flattened for the upstroke will generate a strong net upward force. This is known to be the case in the hindwings of fast forward fliers like locusts (Wootton, 1992; Wootton *et al*., 2000). It would be relatively easy to achieve at an early evolutionary stage.

Thin cambered plates are also asymmetrical in their mechanical properties. They are very rigid to loading from the concave side but bend far more easily when loaded from the convex side; the edges tend to buckle outwards, and the section to flatten (Wootton, 1981; Fig. 3.3b). A cambered wing would therefore be rigid and effective in the useful downstroke, but could flex ventrally and produce less force in the potentially disadvantageous upstroke.

Furthermore, and rather unexpectedly, cambered plates show asymmetry in resistance to twisting, exactly as discussed above. When a force is applied behind the torsional axis, as would happen to a wing in flight, the plate twists much more readily when loaded from the convex side than from the concave (Fig. 3.3c). Butterflies, at least, appear to make extensive use of this effect (Wootton, 1993), and it may well be important in Odonata (Wootton, 1991) and a wide range of other insects.

Insects are known to gain high values of lift by exploiting in various ways the *unsteady* airflows which result from flapping. Some of these effects would be available as soon as the wings reached a suitable size, and the wing movements became appropriate. The hawkmoth *Manduca*, for example, gains high lift by operating its wings at angles of attack which would cause

**Fig. 3.3** (opposite).   Elastic deformation under aerodynamic and inertial loads. (a) If the aerodynamic and inertial forces are centred behind the torsional axis, the wing will tend to twist nose-down along its length. CP: centre of pressure; AT: axis of torsion. (b) A cambered wing is far more resistant to a force from the concave side than the convex. (c) A cambered wing loaded behind its torsional axis is far more resistant to pronatory (nose-down) twisting than to supinatory (nose-up) twisting (modified after Wootton, 1993).

stalling, were it not for a vortex which is created above the leading edge and stabilized by moving out along the span as a result of the flapping motion (Ellington *et al.*, 1996), and which prevents the airflow from separating from the upper surface of the wing. This mechanism probably operates in most insects, at least those of moderate to large size, and would have been utilizable from a relatively early stage in flight evolution. Insects which beat their wings at high amplitudes, so that those on the left and right sides approach each other dorsally before moving apart into the downstroke, can gain extra lift from the mutual interference of the flow around the wings as they rotate at the top of the stroke (Ellington, 1984). This too would probably have been accessible to early insects, particularly if their wings were relatively broad at the base.

Summarizing, in the early stages of flapping flight one would expect development of the following wing adaptations, not necessarily in the order listed.

1. Progressive increase in wing area.
2. Effective support to the entire wing, minimizing mass while maximizing rigidity, particularly to dorsal bending, and providing adequate strength. This would be provided by tubular veins, by corrugation, or by a cambered section.
3. Establishment of a strong, efficient basal articulation, with basal sclerites linking the vein bases, hinging to the tergum at at least two points and resting on a pleural fulcrum.
4. Development of some torsional freedom in the axilla.
5. Concentration of main supporting veins towards the anterior part of the wing, giving scope for passive elastic torsion along the span.
6. Development of the capability to alter the wing's cross-section, with at least one longitudinal line of flexible cuticle crossing the axillary sclerotization, and adaptation for flexibility in the wing.
7. Some structural linking, by strong cross-veins or actual fusion, of the bases of the longitudinal veins into discrete groups, separated by the flexion-line(s). Kukalová-Peck (1997) has stressed the importance of these, both functionally and as early, stable characters of particular systematic value.

## What Did Insects Need in Order to Fly Well?

The characters so far listed are those which appear useful – indeed probably essential – in the initial evolution of wings fully capable of propelling and supporting the weight of an insect in air. With this equipment, together with associated thoracic skeletal, muscular, nervous and respiratory development, insects would be capable of basic flight – stable, relatively unmanoevrable and probably rather fast, since wing-loadings (weight carried by unit area of wing) would initially be high, and flight speeds tend to scale with (wing-loading)$^{0.5}$. Each of the characters would have clear selective value from an early stage, so that there are no difficulties in explaining the significance of intermediate conditions, such as those which we meet in discussing the initial origin of wings.

Thereafter, evolution would be concerned with improving and optimizing wing design for various kinds of flight behaviour, and also for functions other than flight. Foremost among the latter is the capacity to fold. The advantages of folding are obvious: the ability to penetrate small spaces and dense vegetation; to walk freely about without risk of being blown over or away; to utilize the forewings to protect the hindwings and abdomen. Traditionally, therefore, folding has been regarded as secondary, developing long after flight evolved. Rasnitsyn, however, believes that folding was already present at the proto-wing stage (Rasnitsyn, 1981) and that the inability to fold the wings in Odonata, Ephemeroptera and a wide range of extinct Palaeozoic forms is a derived condition. This view is supported by Grodnitsky (1999). Whichever is the case, folding would at first have been confined to the axilla and have little effect on the design of the main part of the wing; only later would longitudinal and transverse fold-lines become necessary in the wing itself.

In terms of flight adaptations we are now concerned with diversification: specialization for different kinds of flight performance. Brodsky (1994) and Grodnitsky (1999) have each discussed the evolution of flight techniques. The huge variety of wing designs is far beyond the scope of this chapter, but we can recognize several common trends, each of which has been followed independently several times.

**1.** More or less elliptical wings, firmly supported anteriorly and posteriorly by strong veins, often with high relief; with a deformable area between, mainly concentrated in the distal, most aerodynamically effective half of the wing, where the airspeeds are greatest. Wings of this kind are found in many neopterous orders. They are capable of only limited torsion, which tends to limit flight versatility, but this can partly be achieved by the development of transverse flexibility; see 3.ii, below.

**2.** Partial or total modification of the forewings for protection, by thickening the cuticle of the membrane and veins. This has usually been to some extent at the expense of aerodynamic function, and has been compensated by expansion of the hindwing, often requiring complex longitudinal and/or transverse folding mechanisms (Haas and Wootton, 1996). Flight in such forms is usually fairly unversatile, with a narrow range of speeds and little capacity for manoeuvre, but the expanded hindwings favour the development of unsteady lift at the beginning of the downstroke by the 'clap and peel' or 'near clap and partial peel' effects (Ellington, 1984). Examples today include Orthoptera, Dictyoptera, Dermaptera and Coleoptera.

**3.** Specialization for slow flight and hovering. This commonly involves adaptations for increased torsion, since strong wing twisting, combined with an oblique or horizontal stroke-plane, can recruit the upstroke as well as the downstroke for weight support (Fig. 3.4). Several trends are recognizable: (i) narrowing of the base, usually accompanied by vein fusion, reduction of posterior supporting veins, and their replacement by automatic mechanisms which lower the trailing edge and maintain an effective section in response to aerodynamic loading (Wootton, 1991, 1992). Today this is seen in zygopterous Odonata, in many Diptera, particularly Nematocera; and in some Neuroptera. (ii) Development of a transverse, often oblique flexion line partway along the wing, allowing the distal region to bend ventrally in the upstroke and twist into a more favourable angle of attack for the generation of upward force. Flexion of this kind is found both among insects with uncoupled wings – Plecoptera, Megaloptera, Mecoptera, Trichoptera – and among those with fore- and hindwings coupled in flight, see 4, below.

4.  Adaptations for flight over a wide speed range. These are to be found in anisopterous Odonata, which combine torsionally compliant wings with the broad bases usually associated with faster flight; and in a wide range of insect groups – many Hymenoptera, Hemiptera, Lepidoptera – in which the

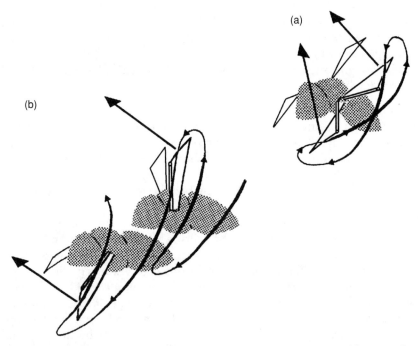

**Fig. 3.4.** Slow flight and hovering. Viewpoint as in Fig. 3.2. (a) An oblique stroke plane combined with strong wing twisting can allow the insect to gain weight support on both the downstroke and the upstroke, and to fly slowly. (b) If the stroke-plane is horizontal and the wings twist strongly, horizontal forces can cancel out, and the insect can hover or rise vertically. The wing is shown in both downstroke and upstroke positions.

hindwings are reduced and coupled to the longer forewings, forming a single effective aerofoil, with a broad base and a torsionally compliant distal region.

## How Did Known Palaeozoic Insects Fly?

So far we have been discussing probable events and evolutionary trends during a long period from which we have no direct information. When insects do eventually appear as fossils in the middle and later Carboniferous, we have, if we can interpret it, the first clear evidence of what has gone before: both a check on our predictions, and an impressive demonstration of how far flight had already diversified.

There is no space here to illustrate the diversity of Carboniferous insects, or to give any but the most superficial discussion. Carpenter's majestic review of fossil insects gives figures of many Palaeozoic forms (Carpenter, 1992), and Wootton and Kukalová-Peck (2000) provide functional interpretations of the wings of Palaeozoic Palaeoptera.

Understandably much use – in the early years rather too much – has been made of the Palaeozoic Pterygota in discussing the origin and early evolution of wings. It is entirely reasonable to search them for plesiomorphic characters, but one cannot too strongly emphasize that many features of Palaeozoic insects are already highly derived. The Pterygota had radiated extensively by the mid-Carboniferous (Wootton, 1990). Wing-folding and non-folding ('palaeopterous') types are present and highly varied in form and size, indicating a wide range of habitats and life styles, and many adaptations to specific modes of flight are already recognizable. These include examples of each of the categories described in the previous section, as well as some which appear to have no parallels today (Wootton *et al.*, 1998; Wootton and Kukalová-Peck, 2000).

A unique and much-discussed insight into the form of wing precursors is provided by the famous prothoracic winglets of some Carboniferous Palaeoptera, particularly in the Order Palaeodictyoptera (Fig. 3.5a), and in a few Permian Neoptera (F. Lemmatophoridae). They seem certain to be homologous with true wings (no such certainty exists in the case of similar structures in later insects), and look like genuine relics of the pre-flight wing condition. In their most fully developed form they are membranous, and supported by radiating veins which show no appreciable concentration towards the leading edge. Corrugation, if discernable, is never more than slight; the bending moments on them would be small. Their articulation with the thorax appears small and simple. It seems that they were movable, but there is no evidence that they could be flapped. In the insects where they are known they may have functioned as control surfaces, like the 'canard' aerofoils of some aircraft. They seem to be well developed only in insects adapted for reasonably rapid flight (Wootton and Kukalová-Peck, 2000) and indeed canards would be relatively useless in slow flight or hovering.

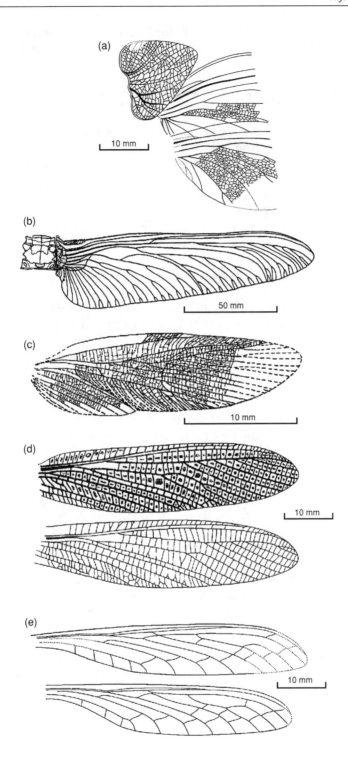

In the fully developed mesothoracic and metathoracic wings, separating the primitive from the derived is far from straightforward. There is a temptation to regard the wings with the most main veins and the richest branching as plesiomorphic, but this may be misleading. Many Carboniferous and Lower Permian insects are gigantic, and their rich venation almost certainly reflects the need to support their huge wings. However, the earliest pterygotes may not have been large. Gigantism seems to have been permitted by abnormally high atmospheric oxygen levels at that time, and while Graham et al. (1995) and Dudley (1998) have suggested that these may have played a significant role in the origin of flight, we do not actually know how big were the first winged insects and their immediate ancestors. The calculations of Flower (1964) and Ellington (1991) point to the advantages of relatively small size in gliding protopterygotes. The skimming hypothesis also favours small insects, since weight support from the water surface film scales approximately as (body mass)$^{0.33}$, and larger insects would sink.

Some of the wing characters predicted above as early requisites for flight are virtually ubiquitous. These include high relief through convex/concave alternation of the main veins, concentration of veins towards the anterior margin, linking of veins into functional zones by approximation, cross-veins and fusion. The apparent flexibility of the axillae, both in terms of camber generation and torsion, varies greatly. The wings of the giant palaeodictyopteran *Mazonopterum wolfforum* have broad, solid-seeming axillae (Fig. 3.5b), and their capacity for deformation may have been very restricted (Kukalová-Peck and Richardson, 1983), though the wings are long and slender enough to have allowed some passive spanwise torsion; others, like the slender-winged Megasecoptera, had exceptionally twisty-looking axillae, at least by the Lower Permian (Kukalová-Peck, 1974).

Evidence for active camber control is more variable. Longitudinal flexion lines in the wings have often gone unrecognized even in modern forms and would be hard to see in fossils, but can often be deduced from the absence of cross-veins between a pair of adjacent veins. In others, it is probable that camber was caused passively and automatically. Like a sail, a broad, flexible wing would automatically become cambered when generating lift, and would flatten when no longer doing so. In narrow-based wings in which the vein

**Fig. 3.5** (opposite). Some Palaeozoic insects. (a) Prothoracic winglets and fore- and hindwing bases of *Stenodictya pygmaea* Meunier (Palaeodictyoptera), U. Carboniferous, France. After Kukalová (1970). (b) Metanotum, axilla and hindwing of *Mazonopterum wolfforum* Kukalová-Peck and Richardson (Palaeodictyoptera). U. Carboniferous, Illinois, USA. After Kukalová-Peck and Richardson (1983). (c) Forewing of *Blattinopsis augustai* Kukalová ('Protorthoptera', F Blattinopsidae). L. Permian, Czech Republic. After Kukalová (1959). (d) Fore- and hindwings of *Caloneura dawsoni* Brongniart (Caloneurodea). U. Carboniferous, France. After Carpenter (1961). (e) Fore- and hindwings of *Sphecoptera brongniarti* Meunier (Megasecoptera). U. Carboniferous, France. After Carpenter (1951).

branches diverged pinnately from a leading edge spar – many Megasecoptera and early odonatoids – another automatic mechanism would operate: the generation of camber through passive torsion (Ennos, 1988; Wootton and Kukalová-Peck, 2000).

The results of later diversification are everywhere. Wing-folding was well established, and had apparently arisen twice independently: in the ancestors of Neoptera, and in the fluid-feeding palaeodictyopteroid Order Diaphanopterodea. Firmly supported wings of type 1, above, are particularly common among the neopterous types, and in Diaphanopterodea. Some, like the Blattinopsidae (Fig. 3.5c), show the transverse flexion lines mentioned earlier as adaptations to dorsoventral lift asymmetry and sometimes distal wing torsion, and there is evidence that some Palaeodictyoptera and most Diaphanopterodea were similarly adapted (Wootton and Kukalová-Peck, 2000). Early orthopteroids and dictyopteroids were present in the Carboniferous, and some show expanded hindwing fans like those of modern Orthoptera and Dictyoptera, and presumably had similarly straightforward flight techniques. In the Order Caloneurodea, by contrast, fore- and hindwings were long, slender and similar (Fig. 3.5d). They would probably have twisted extensively, in what was perhaps a rather termite-like flight, with the body steeply inclined. Carboniferous odonatoids were already dragonfly-like, and the wings of some of the earliest show 'smart' adaptations to versatile, manoeuvrable flight, paralleling extant Anisoptera, presumably in association with aerial predation (Wootton *et al.*, 1998). Carboniferous ephemeropteroids, however, were quite unlike modern mayflies, with the hindwings as large as or larger than the fore. The flight of the earliest forms must have been remarkably unsubtle, and forms like *Bojophlebia prokopi* (Kukalová-Peck, 1985), from the Westphalian division of the Carboniferous, may perhaps give us a glimpse of true archaism.

The greatest variety of wings, and probably of flight pattern, is to be found in the wholly extinct group of fluid and spore-feeding palaeopterous orders which includes the Palaeodictyoptera, Megasecoptera, Diaphanopterodea and Permothemistida. These are discussed at length by Wootton and Kukalová-Peck (2000) and the briefest summary will suffice here. Their linear size varied by a factor of perhaps 55, the largest rivalling the well-known giant dragonflies of the Carboniferous, the smallest the size of a large mosquito. Most had similar fore- and hindwings, but some Palaeodictyoptera showed Orthoptera-like hindwing expansion, while others had long forewings, and shorter, broader hindwings. Permothemistida, like some modern Ephemeroptera, reduced their hindwings to vestiges or lost them entirely. In some large Palaeodictyoptera, forewings and hindwings overlapped to a grotesque degree, and one can only conclude that they operated like flapping biplanes, capable only of fast, direct forward flight. In contrast, Megasecoptera (Fig. 3.5e) had slender, narrow-based fore- and hindwings, and appear highly adapted for slow flight and hovering, probably exploiting this skill to reach microsporangia at the flexible ends of branches.

## Conclusion

Flight was the key event in early insect evolution, and has proved one of the most far-reaching in the history of life on earth. The near absence of insect fossils from the long period within which flight must have originated and developed is one of the most frustrating in palaeontology. Eventually the gap will be filled; until then conjecture, informed by thorough, detailed comparative studies and a sound understanding of the biomechanics involved, remains our only available option.

## Summary

There is now majority support for the view that insect wings evolved from lateral segmental structures which were already mobile. The most plausible routes for the origin of flight appear to be either through parachuting and gliding, or through skimming on the surface of water. For the development of active flight, wings would initially need to enlarge, and to develop structural rigidity and a firm articulation to the thorax, then progressively to acquire structural adaptations for automatic, useful deformation when aerodynamically and inertially loaded. These would enable the insects to gain significant net lift for weight support and propulsion. Thereafter the way would be open for specialization into different modes of flight. By the first appearance of winged fossils, perhaps 30–50 million years after flight arose, a wide range of specializations is already evident in a rich, diverse fauna.

## References

Brodsky, A.K. (1994) *The Evolution of Insect Flight*. Oxford University Press, Oxford.

Carpenter, F.M. (1951) Studies on Carboniferous insects from Commentry, France: Part II. The Megasecoptera. *Journal of Paleontology* 25, 336–355.

Carpenter, F.M. (1961) Studies on Carboniferous insects from Commentry, France: Part III. The Caloneurodea. *Psyche, Cambridge* 68, 145–153.

Carpenter, F.M. (1992) *Treatise on Invertebrate Palaeontology. Part R, Arthropoda 4, Volume 3: Superclass Hexapoda*. Geological Society of America, Boulder, Colorado and University of Kansas, Lawrence, Kansas.

Dudley, R. (1998) Atmospheric oxygen, giant Paleozoic insects and the evolution of aerial locomotor performance. *Journal of Experimental Biology* 201, 1043–1050.

Ellington, C.P. (1984) The aerodynamics of hovering insect flight. IV. Aerodynamic mechanisms. *Philosophical Transactions of the Royal Society of London B* 305, 79–113.

Ellington, C.P. (1991) Aerodynamics and the origin of insect flight. *Advances in Insect Physiology* 23, 171–210.

Ellington, C.P., van den Berg, C., Willmott, A.P. and Thomas, A.L.R. (1996) Leading-edge vortices in insect flight. *Nature* 284, 626–630.

Ennos, A.R. (1988) The importance of torsion in the design of insect wings. *Journal of Experimental Biology* 140, 137–160.
Ennos, A.R. (1989) Inertial and aerodynamic torques on the wings of Diptera in flight. *Journal of Experimental Biology* 142, 87–95.
Flower, J.W. (1964) On the origin of flight in insects. *Journal of Insect Physiology* 10, 81–88.
Graham, J.B, Dudley, R., Aguilar, N. and Gans, C. (1995) Implications of the late Palaeozoic oxygen pulse for physiology and evolution. *Nature* 375, 117–120.
Grodnitsky, D.L. (1999) *Form and Function of Insect Wings*. Johns Hopkins University Press, Baltimore, Maryland.
Haas, F. and Wootton, R.J. (1996) Two basic mechanisms in insect wing folding. *Proceedings of the Royal Society of London B* 263, 1651–1658.
Kingsolver, J.G. and Koehl, M.A.R. (1985) Aerodynamics, thermoregulation, and the origin of insect wings. *Evolution* 39, 488–504.
Kingsolver, J.G. and Koehl, M.A.R. (1994) Selective factors in the evolution of insect wings. *Annual Review of Entomology* 39, 425–451.
Kliuge, N.Yu. (1989) The problem of the homology of branchial gills and paranotal lobes in mayfly nymphs to insect wings as related to the taxonomy and phylogeny of the mayfly order (Ephemeroptera). In: *Doklady na 41 Ezhegodnom Chtenii Pamiati N.A. Kholokhodskogo, April 1 1988*. Nauka, Leningrad, pp. 48–77.
Kukalová, J. (1959) On the Family Blattinopsidae Bolton, 1925. *Rozpravy Ceskoslovenske academie ved* 69, 3–27.
Kukalová, J. (1970) Study of the Order Palaeodictyoptera in the Upper Carboniferous shales of Commentry, France. Part III. *Psyche, Cambridge* 77, 1–44.
Kukalová-Peck, J. (1974) Pteralia of the Palaeozoic insect orders Palaeodictyoptera, Megasecoptera and Diaphanopterodea (Palaeoptera). *Psyche, Cambridge* 81, 416–430.
Kukalová-Peck, J. (1978) Origin and evolution of insect wings and their relation to metamorphosis, as documented by the fossil record. *Journal of Morphology* 156, 53–126.
Kukalová-Peck, J. (1983) Origin of the insect wing and wing articulation from the arthropod leg. *Canadian Journal of Zoology* 61, 1618–1669.
Kukalová-Peck, J. (1985) Ephemeroid wing venation based upon new gigantic Carboniferous mayflies and basic morphology, phylogeny, and metamorphosis of pterygote insects (Insecta, Ephemerida). *Canadian Journal of Zoology* 63, 993–995.
Kukalová-Peck, J. (1991) Fossil history and the evolution of hexapod structures. In: CSIRO (eds) *The Insects of Australia*, 2nd edn, Vol. 1. Melbourne University Press, Melbourne, pp. 141–179.
Kukalová-Peck, J. (1997) Arthropod phylogeny and 'basal' morphological structures. In: Fortey, R.A. and Thomas, R.H. (eds) *Arthropod Relationships*, Systematics Association Special Volume Series 55. Chapman & Hall, London, pp. 249–268.
Kukalová-Peck, J. and Richardson, E.S. Jr (1983) New Homoiopteridae (Insecta: Paleodictyoptera) with wing articulation from Upper Carboniferous strata of Mazon Creek, Illinois. *Canadian Journal of Zoology* 61, 1670–87.
Marden, J.H. and Kramer, M.G. (1994) Surface-skimming stoneflies: a possible intermediate stage in insect flight evolution. *Science* 266, 427–430.

Neville, C.P. (1984) Cuticle: organisation. In: Bereiter-Hahn, J., Matoltsy, A.S.G. and Richards, K.S. (eds) *Biology of the Integument*, Vol. 1. *Invertebrates*. Springer-Verlag, Berlin, pp. 611–625.

Quartau, J.A. (1986) An overview of the paranotal theory of insect wing origin. *Publicações do Instituto de Zoologia 'Dr Augusto Nobre', Faculdade de Ciencias do Porto* 194, 1–42.

Rasnitsyn, A.P. (1981) A modified paranotal theory of insect wing origin. *Journal of Morphology* 168, 331–338.

Riek, E.F. (1971) The origin of insects. *Proceedings of the 13th International Congress of Entomology* 1, 292–293.

Stys, P. and Soldan, T. (1980) Retention of tracheal gills in Ephemeroptera and other insects. *Acta Universitatis Carolinae, Biologica* 1978, 409–435.

Thomas, A.L.R. and Norberg, R.A. (1996) Skimming the surface – the origin of flight in insects? *Trends in Ecology and Evolution* 119, 187–188.

Toms, R.B. (1984) Were the first insects terrestrial or aquatic? *Suid-Afrikaanse Tydskrif vir Wetenskap* 80, 319–323.

Trueman, J.W.H. (1990) Comment – evolution of insect wings: a limb exite plus endite model. *Canadian Journal of Zoology* 68, 1333–1335.

Wigglesworth, V.B. (1963) The origin of insect flight. *Proceedings of the Royal Entomological Society of London* C 28, 23–24.

Wigglesworth, V.B. (1976) The evolution of insect flight. In: Rainey, R.C. (ed.) *Insect Flight*. Symposia of the Royal Entomological Society of London, No.7. Blackwell, Oxford, pp. 255–269.

Wootton, R.J. (1981) Support and deformability in insect wings. *Journal of Zoology, London* 193, 447–468.

Wootton, R.J. (1986) The origin of insect flight: where are we now? *Antenna* 10, 82–86.

Wootton, R.J. (1990) Major insect radiations. In: Taylor, P.D. and Larwood, G.P. (eds) *Major Evolutionary Radiations*. Systematics Association Special Volume 42. Clarendon Press, Oxford, pp. 187–208.

Wootton, R.J. (1991) The functional morphology of the wings of Odonata. *Advances in Odonatology* 5, 153–169.

Wootton, R.J. (1992) Functional morphology of insect wings. *Annual Review of Entomology* 37, 113–140.

Wootton, R.J. (1993) Leading edge section and asymmetric twisting in the wings of flying butterflies (Insecta, Papilionoidea). *Journal of Experimental Biology* 180, 105–119.

Wootton, R.J. (1999) Invertebrate paraxial locomotory appendages: design, deformation and control. *Journal of Experimental Biology* 202, 3333–3345.

Wootton, R.J. and Ellington, C.P. (1991) Biomechanics and the origin of insect flight. In: Rayner, J.M.V. and Wootton, R.J. (eds) *Biomechanics and Evolution*. Cambridge University Press, Cambridge, pp. 99–112.

Wootton, R.J. and Kukalová-Peck, J. (2000) Flight adaptations in Palaeozoic Palaeoptera (Insecta). *Biological Reviews* 75, 129–167.

Wootton, R.J., Kukalová-Peck, J., Newman, D.J.S. and Muzón, J. (1998) Smart engineering in the mid-Carboniferous: how well could Palaeozoic dragonflies fly? *Science* 282, 749–751.

Wootton, R.J., Evans, K.E., Herbert, R. and Smith, C.W. (2000) The hind wing of the desert locust (*Schistocerca gregaria* Forskål). 1. Functional morphology and mode of operation. *Journal of Experimental Biology* 203, 2921–2931.

# Physiology and Endocrine Control of Flight

## Graham Goldsworthy and Michael Joyce

*Department of Biology, Birkbeck College, University of London, Malet Street, London WC1E 7HX, UK*

## Introduction

Insects are able to utilize a range of fuels to supply the flight muscles with the energy required for flight: carbohydrates and/or fats are the most commonly used, but many Diptera and Coleoptera use the amino acid proline as a major fuel. In this account we will use the locust as a well-documented example of an insect using 'conventional' carbohydrate and fat fuels, and give a brief overview of the physiology and endocrine control of flight in locusts. Finally, we will discuss recent work on flight in beetles that suggests that proline may be the predominant fuel for flight in this order.

One common feature of insect metabolism is the occurrence of the major fuels in high concentrations in the haemolymph. This is related to the dynamics of the open circulatory system and the insect's dependence on relatively long diffusion pathways for making fuels available from the haemocoel to the flight muscles. Small amounts of fuel are usually available in the flight muscles themselves, and larger amounts of fuels are often stored in the fat body. However, the large amounts of carbohydrate that can be present in the haemolymph, even at the moment of take-off, can represent a reserve that is larger than that present in the fat body. It is therefore important to recognize the size of the pool of fuel in the haemolymph (Goldsworthy, 1983, 1990).

Changes in the concentrations of energy metabolites in the haemolymph during flight can often give useful information about what fuels are being oxidized. However, the concentration of a metabolite in the haemolymph at any time is influenced by both the input of that metabolite from any reserves and its output to tissues that take it up. Thus the interpretation of what these changes mean in terms of fuel utilization must be made with care. Nevertheless, simple analysis of the changes in levels of carbohydrate, fat and amino acids in locusts and some moths and beetles has provided extremely useful preliminary data concerning fuel utilization during flight in these insects.

## Why are hormones involved in insect flight?

In trivial flights, the flight muscles use their own reserves of glycogen and then take up and oxidize readily available fuels from the haemolymph. If flight activity is prolonged, hormones play an essential role in coordinating the further supply of fuels from reserves in the fat body: they may also influence the pattern of use of fuels by the flight muscles.

# Flight in Locusts

Locusts provide one of the best understood examples of the involvement of hormones in long-term flight and have been the subject of many reviews (for example Candy et al., 1997; Goldsworthy et al., 1997; van Marrewijk et al., 1998). Shortly after the onset of prolonged flight activity, octopamine levels in the haemolymph increase dramatically to reach a peak after about 10 min of flight (Goosey and Candy, 1980). Octopamine is a catecholamine closely related to vertebrate noradrenaline. The octopamine that appears in locust haemolymph during flight derives from cells in the metathoracic ganglia, and is released from neurohaemal sites on the surface of peripheral nerves (Braunig, 1995). Octopamine in the haemolymph may exert several effects: in the flight muscles it stimulates the synthesis of fructose 2,6-bisphosphate, favouring glycolysis, and it can modulate the power of muscle contraction (Wegener, 1996; Candy et al., 1997). It is also thought to induce some slight mobilization of energy reserves in the fat body, although it is not clear whether this is always a direct effect of the hormone on the fat body cells (Zeng et al., 1996).

Locust flight muscles oxidize carbohydrate predominantly during the first few minutes of flight, but switch over after about 15 min to the use of fats. The levels of trehalose in the haemolymph are high at the onset of flight, while those of fats are relatively low. It is the release of a second category of hormones, the adipokinetic hormones (AKHs), that causes a more pronounced mobilization of diacylglycerols from stores of triacylglycerol in the fat body. In the absence of octopamine, oxidation of fatty acids in the flight muscles inhibits glycolysis. Thus, the increased titres of octopamine in the haemolymph during the first few minutes of flight may be important in maintaining glycolytic flux at a time when the availability of fatty acids is increasing towards levels that would be adequate to support flight (Fig. 4.1). Within about 30 min of flight, the levels of diacylglycerol in the haemolymph

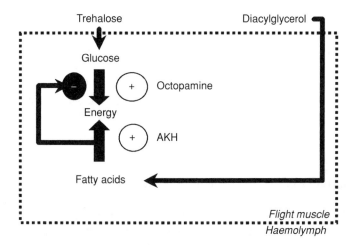

**Fig. 4.1.** Inhibition of glycolysis in the flight muscle, and the need to prevent stalling during flight. In the first few minutes of flight, octopamine stimulates glycolysis. This avoids stalling because at this time, adipokinetic hormone (AKH) is being released and beginning to encourage the utilization of lipid and discourage the use of carbohydrate in the flight muscles. Lipid mobilization takes some time to reach maximum capacity, so the interaction of octopamine and AKH ensures a smooth transition between carbohydrate and lipid-powered flight.

reach a steady state under the actions of the AKHs, and fatty acids become the predominant fuel for the flight muscles.

## Release and turnover of AKHs during flight

The stimulus mechanisms for the release of AKHs during flight are still not fully understood. The release of AKHs is certainly under the control of neurones from the brain. It has been suggested (Singh *et al.*, 1981; Pannabecker and Orchard, 1986) that octopaminergic neurones are involved in the release of AKH, but this is now thought unlikely: crustacean cardioactive peptide (Dircksen *et al.*, 1989; Passier *et al.*, 1997; Flanigan and Gäde, 1999) and locustatachykinin (Nassel *et al.*,1995; Passier *et al.*, 1997) are the prime candidates as neurotransmitters for bringing about the release of AKHs in locusts. Release of the hormones from the corpora cardiaca, where they are synthesized, is also modulated by high concentrations of trehalose in the haemolymph. At the onset of flight, the demand for energy substrates exceeds the amount of trehalose that can be mobilized from the fat body and consequently the trehalose concentration in the haemolymph decreases rapidly. It is tempting to imagine that this decrease in the concentration of trehalose in the haemolymph might in part be responsible for the release of AKHs. In effect, the decrease in concentration of trehalose releases the inhibition brought about by the normally high concentrations at rest.

However, this cannot be the whole explanation. Although sucrose is not effective *in vitro* (Passier *et al.*, 1997; Flanigan and Gäde, 1999), it can mimic the effect of injected trehalose in delaying the release of AKHs *in vivo* (Cheeseman *et al.*, 1976; van der Horst *et al.*, 1979). This is despite the fact that sucrose is not metabolized readily in the locust fat body. The precise links between wing-flapping, increased sensory information from mechanoreceptors registering wind speed, and metabolic changes in the haemolymph, remain to be determined.

## The importance of the lipoproteins in the haemolymph

The mobilization of fat stores in the fat body for energy utilization in the flight muscles requires the transport of the products of lipolysis in the haemolymph. It is likely that the need for high concentrations of fuels in the haemolymph makes the transport of free non-esterified amino acids for energy metabolism untenable: the high concentrations required would poison the tissues. The use of neutral lipids as a transport form avoids this problem but requires a mechanism to carry the water-insoluble diacylglycerols to the flight muscles. Insect haemolymph contains lipoproteins that can perform this task. These lipoproteins or lipophorins are macromolecular complexes, and Fig. 4.2 shows diagrammatically their general arrangement in a diacylglycerol-loaded state as they are during long-term flight in locusts. These complexes are called LDLp (low density lipophorin) particles and are easily visualized in the electron microscope (Wheeler *et al.*, 1984b). There are three apoproteins associated with the LDLp particles (Blacklock and Ryan, 1994): two non-exchangeable lipophorins (ApoLp-I and II), and an exchangeable protein, Apoprotein-III (Apo-III; see Figs 4.3 and 4.4).

The lipoproteins in the haemolymph of locusts show dramatic changes during the transition from rest to prolonged flight (Mayer and Candy, 1967; Mwangi and Goldsworthy, 1977a; Wheeler *et al.*, 1984b). In fact, they operate a continuous shuttle system: carrying diacylglycerol from the fat body to the flight muscle as LDLp and returning to the fat body as HDLp (high density lipophorin) to pick up more diacylglycerol (Fig. 4.3). On loading diacylglycerol from the fat body, the HDLp particles are transformed into LDLp and associate reversibly with Apo-III$^*$ (Wheeler and Goldsworthy, 1983). The shuttle system has clear advantages for the locust in comparison with the transport mechanisms for neutral lipids in vertebrates, where the lipoprotein complexes are not re-usable. The lipoprotein lipase in the fat body has been studied in detail, and several interesting similarities at the molecular level with vertebrate systems have been demonstrated (Dantuma *et al.*, 1997, 1998a,b, 1999). The localization and activity of the lipoprotein lipase in the flight muscles (van Antwerpen *et al.*, 1988, 1990; Wheeler, 1989; Wheeler *et al.*, 1984a), and the uptake of fatty acids into the flight muscles has also been studied (Haunerland, 1997). Figure 4.4 shows a speculative scheme

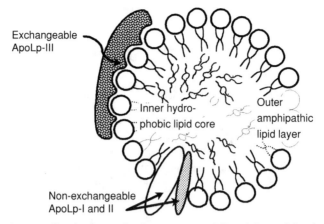

**Fig. 4.2.** A generalized scheme for the structure of diacylglycerol-loaded low density lipoprotein (LDLp) particles (lipophorins) in locust haemolymph during flight or after injection of adipokinetic hormone (AKH). For simplicity, only one molecule of the many molecules of each apoprotein is shown.

for the regulation of lipoprotein lipase in the flight muscles by Apo-III proteins, based on measurement of the enzyme activity (Wheeler *et al.*, 1986) and histochemical observation in preparations of membranes from flight muscle (Goldsworthy, 1990).

## Other actions for AKHs

One of the many questions that remain concerning the actions of AKHs in locusts concerns the fact that there are three separate AKHs in *Locusta migratoria* (L.) (Table 4.1). Each of these AKHs has a similar spectrum of biological activities. Why should a locust need three peptide hormones that appear to have identical activities? In searching for a possible answer to this, the possibility that these locust AKHs exert other actions than in mobilizing fat and carbohydrate from the fat body, and/or that they have differential potencies, has been investigated. It is clear that the AKHs are pleiotropic in locusts: they mobilize lipid and glycogen stores in the fat body, and inhibit the synthesis of fatty acids, RNA and proteins (Goldsworthy *et al.*, 1997). However, in all reported examples of AKH action in locusts but one, although the three peptides may differ in potency, they all elicit similar qualitative effects. The single exception appears to be the stimulation of the release of trehalose from locust fat body *in vitro*, where AKH-II appears to be active but AKH-I is not (Loughton and Orchard, 1981; Carlisle *et al.*, 1988). The activity of AKH-III in this system does not appear to have been investigated, nor has the physiological significance of this particular differential activity of AKH-I and II been determined.

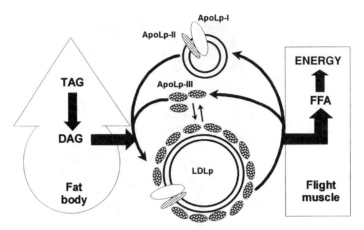

**Fig. 4.3.** The lipoprotein shuttle mechanism that operates during locust flight carrying diacylglycerol from the fat body to the flight muscle as LDLp (low density lipoprotein) and returning to the fat body as HDLp (high density lipoprotein) to pick up more diacylglycerol.

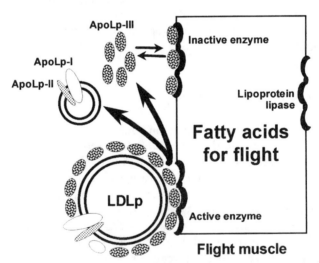

**Fig. 4.4.** A speculative model for the control of the membrane-bound lipoprotein lipase in locust flight muscle. The activity of the lipoprotein lipase in the flight muscles may be modulated by the concentration of 'free' Apo-III (i.e. not bound to low density lipoproteins (LDLp) particles). It is assumed that there are binding sites on the muscle membrane for Apo-III but these are of lower affinity than those on the LDLp (when this is present in the haemolymph). When these muscle binding sites are occupied, the lipoprotein lipase is inhibited. During long-term flight, when adipokinetic hormones (AKHs) are present, Apo-III will be mostly associated with LDLp, giving high lipase activity. During rest, 'free' Apo-III will be in high concentration, giving low lipoprotein lipase activity.

**Table 4.1.** AKH/RPCH peptides in *Locusta migratoria*. For compactness, the amino acid residues are given as the single letter codes (<Q, pyroglutamic acid; L, leucine; N, asparagine; F, phenylalanine; T, threonine; P, proline; W, tryptophan; G, glycine; S, serine; and A, alanine).

| Peptide | Sequence | Reference |
| --- | --- | --- |
| *Lom*-AKH-I | <QLNFTPNWGTamide | Stone *et al.* (1976) |
| *Lom*-AKH-II | <QLNFSAGWamide | Siegert *et al.* (1985) |
| *Lom*-AKH-III | <QLNFTPWWamide | Oudejans *et al.* (1991) |

Flight not only stimulates the release of AKHs from the corpora cardiaca, but also stimulates the synthesis of the hormones. Interestingly, the flight-induced increase in expression of the AKH-III gene is twice that for AKH-I and II (Bogerd *et al.*, 1995). Once released into the haemolymph, the locust AKHs are rather short-lived: the half-lives of AKH-I, II and III at rest are 51, 40 and 5 min, and 35, 37 and 3 min during flight (Oudejans *et al.*, 1996). These dramatic differences are difficult to interpret in terms of their possible significance to the physiology of the locust. It seems likely that the three AKHs do have distinct roles, although precisely what these might be remain to be identified.

## AKH and phase

An interesting hypothesis that has been explored recently concerns the possible influence of the phase polymorphism of locusts on the operation of AKHs. Because of the well-known migratory behaviour of gregarious phase locusts, it might seem likely that there could be differences in AKH-related flight physiology between solitary and crowded locusts. This has been studied extensively in *L. migratoria* and, indeed, there are more or less consistent phase-related differences in AKH-related physiology. The sensitivity of the lipid-mobilizing system to AKHs (Ayali and Pener, 1992, 1995; Ayali *et al.*, 1996); the content of fat stores (Ayali and Pener, 1995); and the levels of stored AKHs (Ayali *et al.*, 1996a) and their precursors (Ayali *et al.*, 1996b) in the corpora cardiaca in *L. migratoria* (Pener *et al.*, 1997; Pener and Yerushalmi, 1998) differ between solitary and crowded locusts. Similar phase-related differences have been reported in *Schistocerca gregaria* (Forskål) (Schneider and Dorn, 1994; Ogoyi *et al.*, 1995, 1996, 1998). The differences described are, however, small, and it is clear that solitary locusts are fully able to mobilise lipids during flight, and that this involves a functional AKH-response component. Perhaps this is not so surprising when the flight behaviour of solitary locusts is examined because, contrary to a misapprehension amongst many insect physiologists/endocrinologists, solitary locusts undertake migratory flights (Farrow, 1990). It seems therefore that, although it can be argued that gregarious locusts appear slightly better adapted for lipid mobilization in support of migratory flight, solitary locusts are not particularly physiologically or endocrinologically disadvantaged in this respect.

## AKHs in Beetles

The Coleoptera comprise > 350,000 known species and make up the largest of all the insect orders. However, it is only relatively recently that research has focused on the identification and possible role of AKHs in beetles. Early studies showed that injections of extracts of heads or corpora cardiaca from the mealworm *Tenebrio molitor* L. caused hyperlipaemia in *L. migratoria* (Goldsworthy *et al.*, 1972b; Gäde, 1989) and hypertrehalosaemia in *Periplaneta americana* (L.) (Gäde, 1989). Eight AKHs (Table 4.2) have been isolated and identified in species representing the families: Tenebrionidae, Meloidae, Chrysomelidae and the superfamily Scarabaeoidea (Gäde and Kellner, 1989; Gäde and Rosinski, 1990; Gäde, 1991, 1995, 1997b).

These peptides share the structural features common to the AKH/RPCH family as a whole. However, there are some features that are either unique or rare for this peptide family. *Ona*-CC and *Scd*-CC-I and II are the only known family members to contain three aromatic amino acid residues:

Table 4.2. AKH/RPCH peptides in Coleoptera. For compactness, the amino acid residues are given as the single letter codes (see Table 4.1; and V, valine; D, aspartate; Y, tyrosine).

| Peptide | Peptide sequence | Species peptide identified in | Reference |
| --- | --- | --- | --- |
| *Tem*-Hr-TH | QLNFSPNWa | *Tenebrio molitor, Zophobas rugipes, Onymacris rugatipennis, Decapatoma lunata, Onymacris plana* | Gäde and Rosinski (1990), Gäde (1994, 1995) |
| *Del*-CC | QLNFSPNWGNa | *Decapotoma lunata* | Gäde (1995) |
| *Pea*-CAH-I | QVNFSPNWa | (*Periplaneta americana*), *Leptinotarsa decemlineata* | Witten *et al.* (1984), Gäde and Kellner (1989) |
| *Pea*-CAH-II | QLTFTPNWa | (*Periplaneta americana*), *Leptinotarsa decemlineata* | Witten *et al.* (1984), Gäde and Kellner (1989) |
| *Mem*-CC | QLNYSPDWa | *Melolontha melolontha, Geotrupes stercorosus, Pachnoda marginata, Pachnoda sinuata* | Gäde (1991b), Gäde *et al.* (1992) |
| *Ona*-CC | QYNFSTGWa | *Onitis aygulus, Onitis pecuarius* | Gäde (1997a) |
| *Scd*-CC-I | QFNYSPDWa | *Scarabaeus deludens, Gareta nitens* | Gäde (1997b) |
| *Scd*-CC-II | QFNYSPVWa | *Scarabaeus deludens, Gareta nitens* | Gäde (1997b) |

in addition to those found normally in positions 4 and 8, there is a third aromatic amino residue in position 2. In *Ona*-CC, this is tyrosine and in *Scd*-CC-I and II, it is phenylalanine. Furthermore the presence of an aspartic acid residue at position 7 makes *Mem*-CC and *Scd*-CC-I negatively charged, again a relatively rare occurrence within this peptide family.

## AKHs and flight in beetles

A number of recent investigations have focused on the effects the AKHs of beetles have on the resting levels of haemolymph lipids, carbohydrates and amino acids, with a view to investigating their possible role in flight metabolism.

None of the AKHs identified in beetles induce hyperlipaemia in the beetles themselves. Furthermore flight experiments involving *Pachnoda sinuata flaviventris* (Gory & Percheron), *Decapotoma lunata* (Pallas), *Onitis pecuarius* Lanberge and *Scarabaeus rugosus* Hausmann all show no significant changes in lipid concentration of the haemolymph during flight times ranging from 1 to 30 min (Lopata and Gäde, 1994; Auerswald and Gäde, 1995; Gäde, 1997a). This does not exclude a possible increase in the turnover of lipid in the haemolymph during flight, but several lines of evidence argue against this. The low activity of β-hydroxy-CoA dehydrogenase (responsible for fatty acid oxidation) in flight muscles of *Pachnoda* and *Decapotoma* (Zebe and Gäde, 1993; Auerswald and Gäde, 1995), and the low respiratory rate of mitochondria isolated from flight muscles of *Lepinotarsa decemlineata* (Say) and supplied with palmitoyl carnitine as a substrate (Weeda *et al.*, 1980) indicate that lipid oxidation is unlikely to provide a significant contribution to energy production in these beetles. Only one beetle species, *Dendroctonus pseudotsugae* Hopkins has been reported to oxidize lipids directly during flight (Thompson and Bennet, 1971), although at present there are no published data regarding the presence and potential role of AKH peptides in this species.

Carbohydrate mobilization occurs in some species of beetle when they are injected with their own endogenous AKH peptide. For example, a hypertrehalosaemic response is obtained when adult *Pachnoda* are injected with *Mem*-CC (Lopata and Gäde, 1994). Flight experiments indicate that carbohydrates are oxidized during flight in this species (Zebe and Gäde, 1993), so during flight *Mem*-CC could regulate the release of carbohydrate from the fat body. A hypertrehalosaemic response is also observed in 1-day-old adult *Tenebrio* after injection of *Tem*-Hr-TH (Rosinski and Gäde, 1988). Further, Michalik *et al.* (1995) found that the same peptide stimulates the release *in vitro* of trehalose from the fat body of adult female *Tenebrio*. However, although this beetle uses flight for dispersal, to date there appear to be no published data concerning changes in potential fuels in its haemolymph and/or flight muscle during flight. Therefore the action of the peptide on resting haemolymph levels cannot be linked to any changes in haemolymph and flight muscle substrates during flight. Interestingly, *Tem*-Hr-TH is synthesized in the blister beetle *Decapotoma* (Gäde, 1995) and, according to Auerswald and Gäde (1995), carbohydrates are the major fuel for flight in this species. The latter authors reported that injections of *Tem*-Hr-TH or of *Del*-CC (also present in the blister beetle) failed to evoke a hypertrehalosaemic response in adult beetles. In a more recent study, however, Gäde and Auerswald (1999) conclude that in *Decapotoma* injections of *Tem*-Hr-TH or of *Del*-CC caused hyperprolinaemia along with a small but significant hypertrehalosaemia. It is worth noting, however, that such apparently contradictory responses to AKHs could be the result of using field-caught insects that are of uncertain age and physiological state, because there could be age-related changes in responsiveness. This is certainly the case in *Tenebrio* in which injection of up to 100 pmol of *Tem*-Hr-TH into adult beetles fails to elicit a hypertrehalosaemic response in adult beetles that are more than 1 day old (M. Joyce, unpublished observations).

## The use of proline during flight

A number of recent studies involving beetles from the superfamily Scarabaeoidea have focused on the use of proline in the flight metabolism of these beetles and the possible role of the endogenous AKH peptides present. In *Onitis aygulus* (Fab.), *O. pecuarius* and *S. rugosus*, Gäde (1997b) found that concentrations of proline in the haemolymph and flight muscles decrease during a flight of 1 min. At the same time, the concentrations of alanine in the haemolymph and flight muscles increase by equimolar amounts. Both amino acids return to their pre-flight concentrations within about 60 min. The AKH peptides isolated from these beetles, *Ona*-CC, *Scd*-CC-I and II (Table 4.2) all cause hyperprolinaemia in the resting beetles themselves, and this is accompanied by a significant decrease in the alanine concentration of the haemolymph. Hence these peptides may be responsible for the hormonal control of proline metabolism in these beetle species.

In *Leptinotarsa* (Weeda et al., 1979) and *Pachnoda* (Lopata and Gäde, 1994) proline is a major fuel for flight. Three enzymes involved in the metabolism of proline, malic enzyme, alanine aminotransferase and glutamate dehydrogenase, were found to be 3–8 times more active in the flight muscles of *Pachnoda* than in *Locusta* (Zebe and Gäde, 1993). Auerswald and Gäde (1999a) showed that injection of the endogenous AKH peptide, *Mem*-CC, causes hyperprolinaemia in *Pachnoda* (as well as hypertrehalosaemia). Interestingly in the same study the authors found that the charged peptide *Scd*-CC-I as well as a number of uncharged AKH peptides including *Del*-CC, *Tem*-Hr-TH and *Ona*-CC evoke a hyperprolinaemic response in this beetle.

The use of proline as a fuel for flight has been established in several insect species although the extent to which it is used varies from species to species.

At one end of the spectrum *Phormia regina* (Meigen) uses relatively small amounts of proline during the onset of flight to provide tricarboxylic inter mediates necessary for maximal oxidation of pyruvate (Sacktor and Wormser-Shavit, 1966; Sacktor and Childress, 1967). At the other extreme the tsetse fly, *Glossina morsitans* Westwood, uses proline almost exclusively as a substrate for flight (Bursell, 1981). With regard to those beetles studied to date, the species of onitine and scarabaeine dung beetles highlighted earlier have negligible stores of glycogen in their flight muscles and low levels of carbohydrate in their haemolymph. This suggests that they may use proline almost exclusively as a substrate for flight (Gäde, 1997b,c). Other species, including *Leptinotarsa*, *Pachnoda* and *Decapotoma* use a combination of proline and carbohydrates to power flight (Weeda et al., 1979; Zebe and Gäde, 1993; Auerswald and Gäde, 1995). Other studies (see, e.g. Crabtree and Newsholme, 1970; Pearson et al., 1979) similarly indicate that the use of proline as a fuel for flight may be a common occurrence in the Coleoptera.

The biochemical pathways of proline oxidation and its re-synthesis in the fat body have been elucidated fully in *Glossina* (Bursell, 1977, 1981; McCabe and Bursell, 1977) and to a lesser extent in *Leptinotarsa* (Khan and de Kort, 1978; Mordue and de Kort, 1978; Weeda et al., 1980). Proline is partially oxidized to alanine in the flight muscles. Alanine is then released into the haemolymph where it is taken up by the fat body. Proline is thus re-synthesized in the fat body where the additional $C_2$-units are derived from stored lipid as acetyl-CoA. Proline is then released into the haemolymph during and after flight (Fig. 4.5). The two overall reactions that describe the

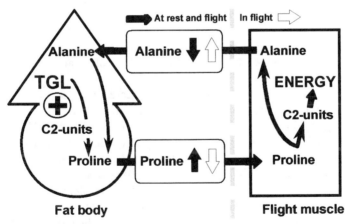

**Fig. 4.5.** The use of proline to transfer C2-units from the fat body to the flight muscles. A comparison is made between the effects on the concentrations of alanine and proline in the haemolymph of injecting adipokinetic hormones (AKHs), with those changes seen during flight. During flight, the fate of proline in the flight muscles reverses the direction of the AKH-induced changes in haemolymph metabolites compared with resting insects.

oxidation (1) and re-synthesis (2) of proline are:

$$Pro + 2.5O_2 + 14\ ADP + 14P_i \rightarrow Ala + H_2O + 2CO_2 + 14\ ATP \quad (1)$$

$$Ala + acetyl\text{-}CoA + NADH + H^+ \rightarrow Pro + H_2O + NAD^+ + CoAH \quad (2)$$

A possible hormonal control of the synthesis of proline in the fat body of beetles was suggested first by Weeda (1981) in *Leptinotarsa*. He showed that conspecific extracts of corpora cardiaca from these beetles, or synthetic *Lom*-AKH-I, bring about an increase in proline synthesis *in vitro*. It is now known that the corpora cardiaca of this beetle contain *Pea*-CAH-I and II (Gäde and Kellner, 1989) and it seems reasonable to conclude that these peptides are responsible for this effect. In the fruit beetle *Pachnoda*, Auerswald and Gäde (1999b) have shown that *Mem*-CC (Table 4.2) stimulates proline synthesis in the fat body, and they suggest that this operates via the production of intracellular cAMP (Auerswald and Gäde, 2000). The rate of synthesis of proline in the fat body is also subject to feedback inhibition. Feedback inhibition of proline upon its own synthesis was observed in *Leptinotarsa* (Weeda, 1981) and *Glossina* (Bursell, 1981) when fat body tissues of these insects were incubated *in vitro*. The disappearance of alanine from the incubation medium can be used as a measure of proline synthesis. In this way, Weeda (1981) demonstrated that the addition of increasing amounts of proline to the medium results in a significant decrease in the rate of its synthesis. Bursell (1981) showed that proline inhibits the transamination of alanine by alanine aminotransferase in the fat body of *Glossina*. Recently, such a feedback inhibition of the synthesis of proline has been described *in vivo*: Auerswald and Gäde (1999a) found that the maximum level of proline in the haemolymph of *Pachnoda* is always around 120 $\mu$mol ml$^{-1}$. Regardless of the amount of AKH peptide or alanine injected, the concentration of proline cannot be elevated beyond this (Auerswald and Gäde, 1999b).

Injection of AKH peptides into beetles and the process of flight itself have opposite effects on the concentrations of proline and alanine in the haemolymph (Fig. 4.5). AKHs initiate an increase in the rate of synthesis of proline, which is seen in resting beetles as an increase in the level of proline in the haemolymph and a decrease in that of alanine. If AKHs are released in beetles during flight (which remains to be shown directly), they will stimulate proline synthesis, but during the early stages of flight the rate of utilization of proline by the flight muscles far outstrips its rate of synthesis in the fat body. Consequently, until a new steady state is achieved, when the rates of synthesis and oxidation of proline are comparable, the concentration of proline in the haemolymph decreases while that of alanine increases.

It is noticeable that whereas the changes in the concentrations of proline and alanine in the haemolymph are equimolar during flight (Gäde and Auerswald, 1999), this is not the case when AKHs are injected into resting beetles. In resting beetles, the increase in proline is often up to four times that of the decrease in alanine. During flight, the decrease in concentration of proline in the haemolymph and the increase in that of alanine are almost direct reflections of the oxidation of proline in the flight muscles. In resting beetles, the decrease in the concentration of alanine in the haemolymph is insufficient to account for the observed increase in the concentration of proline. This suggests that a pool of alanine other than that in the haemolymph is being used for the synthesis of proline, or that other amino acids such as aspartate or leucine are being utilized. In *Dorcus parallelopipedus* (L.), analysis of the changes in the levels of amino acids in the haemolymph and tissues during AKH action support the former possibility (M. Joyce, unpublished observations).

## Fuels for pre-flight warm-up

To fly, many insects require their thoracic temperature to be above ambient, and need therefore to raise their body temperature prior to flight. This involves either basking in the sun or endothermic warm-up. Endothermic heat production can be achieved by muscle contraction with or without wingbeats (Heinrich, 1993) but requires the oxidation of suitable substrates. At present there is relatively little known about the fuels involved in these pre-flight warm-up processes, although it was demonstrated in *Manduca sexta* (L.) that carbohydrates are used to power pre-flight warm-up (Joos, 1987). In recent experiments, Auerswald *et al.* (1998a,b) found that the

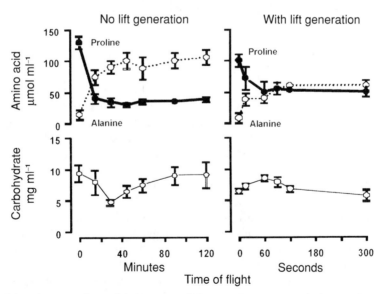

**Fig. 4.6.** A comparison of the pattern of change of energy metabolites in the haemolymph of *Pachnoda sinuata* during tethered flight with and without the opportunity to generate lift. Note that the changes are qualitatively similar for alanine and proline, but the time-scale and magnitude of the changes differ markedly. Redrawn from Auerswald *et al.* (1998b).

proline concentration in the flight muscles and haemolymph of *Pachnoda* decreased significantly during the warm-up period while the alanine concentration in the flight muscles and haemolymph increased. Because there is no significant change in the glycogen concentration of the flight muscles or haemolymph during the warm-up period, it seems likely that the energy required to raise the thoracic temperature (to about 35°C) is produced solely by the oxidation of proline.

## Questions for the Future

### How relevant is laboratory flight to 'natural' flight?

Many studies on the endocrine control of insect flight metabolism and physiology have been conducted using insects suspended on roundabouts. In such situations, the insect is required to produce varying amounts of forward thrust, depending on the apparatus used, but is not required to support its weight.

Recent studies on flight in fruit beetles have investigated whether the need for lift generation influences metabolic and physiological parameters. Auerswald *et al.* (1998a,b) investigated the pattern of metabolic changes during tethered flight with and without lift generation in *Pachnoda*. There are two distinct metabolic phases during lift-generating flight. During the first phase (from take-off until 30 s) there is a steep decrease in the concentrations of proline in the haemolymph and flight muscles, and a 'mirror-image' increase in the concentration of alanine in the haemolymph. The concentration of glycogen in the flight muscles is rapidly reduced while the concentration of carbohydrate in the haemolymph increases. Flight performance in this phase is high, as indicated by high rates of oxygen consumption. In the second phase metabolic levels appear to stabilise. Concentrations of proline and alanine in the haemolymph remain constant, while in the flight muscles they reach a plateau and then increase again at a lower rate. The glycogen levels in the flight muscles continue to fall, but at a lower rate while the concentration of carbohydrate (released into the haemolymph during the first phase) begins to fall as it is now oxidized by the flight muscles. As might be expected, flight performance is much lower during this second phase.

During flight without lift production, the metabolic changes are qualitatively similar to those observed when lift generation is possible, but they occur considerably more slowly. Moreover, the beetles fly 'to exhaustion' for approximately seven times longer than when lift generation is allowed. Tethered flight, with or without lift generation, does not equate with 'free flight', but it is clear from Auerswald *et al.*'s (1998) study that considerable care should be taken in extrapolating from laboratory to field situations.

A further potential problem relates to the use of insects that have been bred and maintained in the laboratory over many generations. It is likely that inbreeding and selection for non-escapees, together with likely qualitative and quantitative metabolic changes brought about by differences in diet between the laboratory and the field, will produce laboratory strains that behave/perform quite differently from insects in the field. On the other hand, animals caught in the field and used for laboratory experiments are usually of unknown age and physiological state. Many of the known responses to AKHs are age-related (Mwangi and Goldsworthy, 1976, 1977b; Lee and Goldsworthy, 1998).

## Why should proline be used as a major fuel to power flight or pre-flight warm-up in a number of beetles?

Early attempts to explain the use of proline in *Glossina* suggested it was an adaptation to the blood-sucking behaviour of the insect (Bursell, 1963; Bursell *et al.*, 1974). However, the discovery that phytophagous insects such as *Leptinotarsa* (Weeda *et al.*, 1979) and *Pachnoda* (Zebe and Gäde, 1993) also oxidize proline means that an alternative hypothesis is required. The high solubility of proline allows the accumulation of high concentrations of the compound in the haemolymph and in the flight muscles prior to flight (Bursell, 1981). This means that the high resting levels of proline in the haemolymph buffer the period between the onset of flight and the arrival of fresh fuels from the fat body. By contrast, the low solubility of diacylglycerols in the haemolymph means that the transportation of lipids requires a special transport mechanism (as described in locusts above). Because the mobilization of lipids from the fat body is relatively slow, there is a need to oxidize carbohydrates during the period between the onset of flight and the build-up of adequate amounts of lipid in the haemolymph. In addition, the energy yield from proline (0.52 mol ATP $g^{-1}$) is higher than that from glucose (0.18 mol ATP $g^{-1}$) and only slightly lower than that (0.65 ATP $g^{-1}$) derived from lipids (Bursell *et al.*, 1974). Hence these apparent advantages make proline an attractive alternative to lipids and carbohydrates as a fuel to power flight. In effect, in both beetles and locusts the ultimate fuel that is fed into the TCA cycle in the flight muscles is acetyl CoA. In both groups of insects these $C_2$ units derive initially from AKH-induced breakdown of triacylglycerol stored in the fat body, so the main difference is in the form in which the $C_2$ units arrive at the flight muscle. Perhaps we should ask the question as to why more insects do not use proline in preference to lipid as a transport fuel for flight.

## What about AKHs and locomotion in the broadest sense?

Actograph experiments show that a locust AKH stimulates locomotory (walking) activity in the firebug, *Pyrrhocoris apterus* (L.) (Socha *et al.*, 1999). There are at least three possible mechanisms by which this effect of injected

AKH could be mediated: a direct effect of the hormone on the CNS, a direct effect on the muscle, or an indirect response to changes in the levels of metabolites in the haemolymph. The dose–response and time–response relationships of this effect on behaviour correlate with that for the mobilization of lipid (Socha et al., 1999), supporting the latter possibility. Thus, changes in behaviour could be linked directly to the hormone-induced changes in metabolite levels in the haemolymph. Recently, the AKH from *Pyrrhocoris*, *Pya*-AKH, has been isolated and identified as des[Gly$^9$,Thr$^{10}$]-AKH-I (Kodrik et al., 2000), which could go some way towards explaining why the firebug responds to a locust AKH. It could be that endogenous AKHs or other neurohormones have a role to play in regulating locomotory behaviour in nymphs or larvae, or in adults of other flightless insects. These possibilities remain to be investigated.

## Summary

The utilization of lipid by insects to support long-term flight is a general phenomenon, but appears to be common in beetles even for relatively short flights. The endocrine system appears to be involved in initiating the mobilization of stores of triacylglycerol by releasing peptides called adipokinetic hormones. These neurohormones make available the two-carbon fuel units necessary for energy production in the TCA cycle, although in locusts (and moths) these fuels arrive at the flight muscle as diacylglycerol, whereas in beetles (and some other insects) they arrive in the form of proline. It is suggested that adipokinetic hormones may play a role in supplying energy substrates for other forms of locomotion.

## Note

\* Early literature used a different nomenclature from that presented here. The lipoprotein particles in resting haemolymph (HDLp) were called $A_{yellow}$; the lipoprotein particles in the haemolymph of flying locusts or ones injected with AKH (LDLp) were called A$^+$; and the exchangeable apoproteins (Apo-III) were called $C_L$-proteins (Goldsworthy, 1983; van Heusden et al., 1984).

## References

van Antwerpen, R., Linnemans, W.A.M., van der Horst, D.J. and Beenakkers, A.M.T. (1988) Immunocytochemical localization of lipoproteins in the flight muscles of the migratory locust. *Cell and Tissue Research* 252, 661–668.
van Antwerpen, R., Beekwilder, J., van Heusden, M.C., van der Horst, D.J. and Beenakkers, A.M.T. (1990) Interaction of lipophorin with the plasma membrane of locust flight muscles. *Biological Chemistry Hoppe-Seyler* 371, 159–165.

Auerswald, L. and Gäde, G. (1995) Energy substrates for flight in the blister beetle *Decapotoma lunata* (Meloidae). *Journal of Experimental Biology* 198, 1423–1431.

Auerswald, L. and Gäde, G. (1999a) Effects of metabolic neuropeptides from insect corpora cardiaca on proline metabolism of the African fruit beetle, *Pachnoda sinuata*. *Journal of Insect Physiology* 45, 535–543.

Auerswald, L. and Gäde, G. (1999b) The fate of proline in the African fruit beetle *Pachnoda sinuata*. *Insect Biochemistry and Molecular Biology* 29, 687–700.

Auerswald, L. and Gäde, G. (2000) Cyclic AMP mediates the elevation of proline by AKH peptides in the cetoniid beetle, *Pachnoda sinuata*. *Biochimica et Biophysica Acta–Molecular Cell Research* 1495, 78–89.

Auerswald, L., Schneider, P. and Gäde, G. (1998a) Utilisation of substrates during tethered flight with and without lift generation in the African fruit beetle *Pachnoda sinuata* (Cetoniinae). *Journal of Experimental Biology* 201, 2333–2342.

Auerswald, L., Schneider, P. and Gäde, G. (1998b) Proline powers pre-flight warm-up in the African fruit beetle *Pachnoda sinuata* (Cetoniinae). *Journal of Experimental Biology* 201, 1651–1657.

Ayali, A. and Pener, M.P. (1992) Density-dependent phase polymorphism affects response to adipokinetic hormone in *Locusta*. *Comparative Biochemistry and Physiology A–Physiology* 101, 549–552.

Ayali, A. and Pener, M.P. (1995) The relations of adipokinetic response and body lipid content in locusts (*Locusta migratoria migratorioides*) with special reference to phase polymorphism. *Journal of Insect Physiology* 41, 85–89.

Ayali, A., Golenser, E. and Pener, M.P. (1996a) Flight fuel related differences between solitary and gregarious locusts (*Locusta migratoria migratorioides*). *Physiological Entomology* 21, 1–6.

Ayali, A., Pener, M.P., Sowa, S.M. and Keeley, L.L. (1996b) Adipokinetic hormone content of the corpora cardiaca in gregarious and solitary migratory locusts. *Physiological Entomology* 21, 167–172.

Blacklock, B.J. and Ryan, R.O. (1994) Hemolymph lipid transport. *Insect Biochemistry and Molecular Biology* 24, 855–873.

Bogerd, J., Kooiman, F.P., Pijnenburg, M.A.P., Hekking, L.H.P., Oudejans, R.C.H.M. and van der Horst, D.J. (1995) Molecular-cloning of three distinct cDNAs, each encoding a different adipokinetic hormone precursor, of the migratory locust, *Locusta migratoria* – differential expression of the distinct adipokinetic hormone precursor genes during flight activity. *Journal of Biological Chemistry* 270, 23038–23043.

Braunig, P. (1995) Dorsal unpaired median (Dum) neurons with neurohemal functions in the locust, *Locusta migratoria*. *Acta Biologica Hungarica* 46(2–4), 471–479.

Bursell, E. (1963) Aspects of the metabolism of amino acids in the tsetse fly, *Glossina* (Diptera). *Journal of Insect Physiology* 9, 439–452.

Bursell, E. (1977) Synthesis of proline by fat body of the tsetse (*Glossina morsitans*): metabolic pathways. *Insect Biochemistry* 7, 427–434.

Bursell, E. (1981) The role of proline in energy metabolism. In: Downer, R.G.H. (ed.) *Energy Metabolism in Insects*. Plenum Press, New York, pp. 135–154.

Bursell, E., Billing, K.C., Hargrove, J.W. and McCabe, C.T. (1974) Metabolism of the blood meal in tsetse flies. *Acta Tropica* 31, 297–320.

Candy, D.J., Becker, A. and Wegener, G. (1997) Coordination and integration of metabolism in insect flight. *Comparative Biochemistry and Physiology B – Biochemistry and Molecular Biology* 117, 497–512.

Carlisle, J.A., Loughton, B.G. and Orchard, I. (1988) Hormone-mediated release of carbohydrate from locust fat body *in vitro*. *Canadian Journal of Zoology* 66, 191–194.

Cheeseman, P., Jutsum, A.R. and Goldsworthy, G.J. (1976) Quantitative studies on the release of locust adipokinetic hormone. *Physiological Entomology* 1, 115–121.

Crabtree, B. and Newsholme, E.A. (1970) The activities of proline dehydrogenase, glutamate dehydrogenase, aspartate-oxoglutarate aminotransferase and alanine-oxoglutarate aminotransferase in some insect flight muscles. *Biochemical Journal* 117, 1019–1021.

Dantuma, N.P., Pijnenburg, M.A.P., Diederen, J.H.B. and van der Horst, D.J. (1997) Developmental down-regulation of receptor-mediated endocytosis of an insect lipoprotein. *Journal of Lipid Research* 38(2), 254–265.

Dantuma, N.P., Pijnenburg, M.A.P., Diederen, J.H.B. and van der Horst, D.J. (1998a) Electron microscopic visualization of receptor-mediated endocytosis of DiI-labeled lipoproteins by diaminobenzidine photoconversion. *Journal of Histochemistry and Cytochemistry* 46(9), 1085–1089.

Dantuma, N.P., Pijnenburg, M.A.P., Diederen, J.H.B. and van der Horst, D.J. (1998b) Multiple interactions between insect lipoproteins and fat body cells: extracellular trapping and endocytic trafficking. *Journal of Lipid Research* 39(9), 1877–1888.

Dantuma, N. P., Potters, M., De Winther, M.P.J., Tensen, C.P., Kooiman, F.P., Bogerd, J. and van der Horst, D.J. (1999) An insect homolog of the vertebrate very low density lipoprotein receptor mediates endocytosis of lipophorins. *Journal of Lipid Research* 40(5), 973–978.

Dircksen, H., Müller, A. and Keller, R. (1989) Immunocytochemical demonstration of CCAP, the novel cardioactive peptide, in the nervous system of the crab, *Carcinus maenas*, and the insect, *Locusta migratoria*. *General and Comparative Endocrinology* 74, 241.

Farrow, R.A. (1990) Flight and migration in acridoids. In: Chapman, R.F. and Joern, A. (eds) *Biology of Grasshoppers*. John Wiley & Sons, New York, pp. 227–314.

Flanigan, J.E. and Gäde, G. (1999) On the release of the three locust (*Locusta migratoria*) adipokinetic hormones: effect of crustacean cardioactive peptide and inhibition by sugars. *Zeitschrift für Naturforschung C* 54, 110–118.

Gäde, G. (1988) Studies on the hypertrehalosaemic factor from the corpus cardiacum/corpus allatum complex of the beetle, *Tenebrio molitor*. *Comparative Biochemistry and Physiology A* 91, 333–338.

Gäde, G. (1989) Characterisation of neuropeptides of the AKH RPCH-family from corpora cardiaca of Coleoptera. *Journal of Comparative Physiology B* 159, 589–596.

Gäde, G. (1991) A unique charged tyrosine-containing member of the adipokinetic hormone/red-pigment concentrating hormone peptide family isolated and sequenced from two beetle species. *Biochemical Journal* 275, 671–677.

Gäde, G. (1994) Isolation and structure elucidation of a neuropeptide from three species of Namib Desert tenebrionid beetles. *South African Journal of Zoology-Suid-Afrikaanse Tydskrif Vir Dierkunde* 29(1), 11–18.

Gäde, G. (1995) Isolation and identification of AKH/RPCH family peptides in blister beetles (Meloidae). *Physiological Entomology* 20, 45–51.

Gäde, G. (1997a) Hyperprolinaemia caused by novel members of the adipokinetic hormone red pigment-concentrating hormone family of peptides isolated from corpora cardiaca of onitine beetles. *Biochemical Journal* 321, 201–206.

Gäde, G. (1997b) Distinct sequences of AKH/RPCH family members in beetle (*Scarabaeus*-species) corpus cardiacum contain three aromatic amino acid residues. *Biochemical and Biophysical Research Communications* 230, 16–21.

Gäde, G. (1997c) Metabolic neuropeptide from the corpus cardiacum of antlions (Neuroptera: Myrmeleontidae): purification and identification. *African Entomology* 5, 225–230.

Gäde, G. and Auerswald, L. (1999) Flight substrates in blister beetles (Coleoptera: Meloidae) and their regulation by neuropeptides of the AKH/RPCH family. *European Journal of Entomology* 96, 331–335.

Gäde, G. and Kellner, R. (1989) The metabolic neuropeptides of the corpus cardiacum from the potato beetle and the American cockroach are identical. *Peptides* 10, 1287–1289.

Gäde, G. and Rosinski, G. (1990) The primary structure of the hypertrehalosemic neuropeptide from tenebrionid beetles – a novel member of the AKH RPCH family. *Peptides* 11, 455–459.

Gäde, G., Lopata, A., Kellner, R. and Rinehart, K.L. (1992) Primary structures of neuropeptides isolated from the corpora cardiaca of various cetonid beetle species determined by pulsed-liquid phase sequencing and tandem mass spectrometry. *Biological Chemistry Hoppe-Seyler* 373, 133–142.

Goldsworthy, G. J. (1983) The endocrine control of flight metabolism in locusts. *Advances in Insect Physiology* 17, 149–204.

Goldsworthy, G. J. (1990) Hormonal control of flight metabolism in locusts. In: Chapman, R.F. and Joern, A. (eds) *Biology of Grasshoppers*. John Wiley & Sons, New York, pp. 205–225.

Goldsworthy, G.J., Johnson, R.A. and Mordue, W. (1972a) *In vivo* studies on the release of hormones from the corpora cardiaca of locusts. *Journal of Comparative Physiology* 79, 85–96.

Goldsworthy, G.J., Mordue, W. and Guthkelch, J. (1972b) Studies on insect adipokinetic hormones. *General and Comparative Endocrinology* 18, 545–551.

Goldsworthy, G.J., Lee, M.J., Luswata, R., Drake, A.F. and Hyde, D. (1997) Structures, assays and receptors for locust adipokinetic hormones. *Comparative Biochemistry and Physiology B – Biochemistry and Molecular Biology* 117, 483–496.

Goosey, M.W. and Candy, D.J. (1980) The D-octopamine content of the haemolymph of the locust *Schistocerca americana gregaria* and its elevation during flight. *Insect Biochemistry* 10, 393–397.

Haunerland, N.H. (1997) Transport and utilization in insect flight muscles. *Comparative Biochemistry and Physiology B – Biochemistry and Molecular Biology* 117, 475–482.

Heinrich, B. (1993) *The Hot Blooded Insects*. Springer Verlag, Berlin.

van der Horst, D.J., van Doorn, J.M. and Beenakkers, A.M.T. (1979) Effects of the adipokinetic hormone on the release and turnover of haemolymph diglycerides and on the formation of the diglyceride-transporting lipoprotein system during locust flight. *Insect Biochemistry* 9, 627–635.

van Huesden, M.C., van der Horst, D.J. and Beenakkers, A.M.T. (1984) *In vitro* studies on hormone-stimulated lipid mobilization from fat body and interconversion of haemolymph lipoproteins of *Locusta migratoria*. *Journal of Insect Physiology* 30, 685–693.

Joos, B. (1987) Carbohydrate use in the flight muscles of *Manduca sexta* during pre-flight warm-up. *Journal of Experimental Biology* 133, 317–327.

Khan, M.A. and de Kort, C.A.D. (1978) Further evidence for the significance of proline as a substrate for flight in the Colorado potato beetle, *Leptinotarsa decemlineata*. *Comparative Biochemistry and Physiology B* 60, 407–411.

Kodrík, D., Socha, R., Šimek, P., Zemek, R. and Goldsworthy, G.J. (2000) A new member of the AKH/RPCH family that stimulates locomotory activity in the firebug, *Pyrrhocoris apterus* (Heteroptera). *Insect Biochemistry and Molecular Biology* 30, 489–498

Lee, M.J. and Goldsworthy, G.J. (1998) New perspectives on the structures, assays and actions of locust adipokinetic hormones. In: Coast, G.M. and Webster, S.G. (eds) *Recent Advances in Arthropod Endocrinology*. Society for Experimental Biology Seminar Series, no. 65, Cambridge University Press, Cambridge, pp. 149–171.

Lopata, A.L. and Gäde, G. (1994) Physiological action of a neuropeptide from the corpora-cardiaca of the fruit beetle, *Pachnoda sinuata*, and its possible role in flight metabolism. *Journal of Insect Physiology* 40, 53–62.

Loughton, B.G. and Orchard, I. (1981) The nature of the hyperglycaemic factor from the glandular lobe of the corpus cardiacum of *Locusta migratoria*. *Journal of Insect Physiology* 27, 383–385.

van Marrewijk, W.J.A. and van der Horst, D.J. (1998) Signal transduction of adipokinetic hormone. In: Coast, G.M. and Webster, S.G. (eds) *Recent Advances in Arthropod Endocrinology*, Vol. 65. Society for Experimental Biology Seminar Series, Cambridge University Press, Cambridge, pp. 172–188.

Mayer, R.J. and Candy D.J. (1967) Changes in haemolymph lipoproteins during locust flight. *Nature* 215, 987.

McCabe, C.T. and Bursell, E. (1977) Interrelationships between amino acids and lipid metabolism in tsetse fly, *Glossina morsitans*. *Insect Biochemistry* 5, 781–789.

Michalik, J., Szolajska, E., Rosinski, G., Lombarska-Sliwinska, D. and Konopinska, D. (1995) Some physiological effects of AKH/RPCH peptides in *Tenebrio molitor* fat body *in vitro*. In: Konopinska, D., Goldsworthy, G., Nachman, R.J., Nawrot, J., Orchard, I., Rosinski, G. and Sobotka, W. (eds) *Insects: Chemical, Physiological and Environmental Aspects: 1994*. University of Wroclaw, Wroclaw.

Mordue, W. and de Kort, C.A.D. (1978) Energy substrates for flight in the Colorado beetle, *Leptinotarsa decemlineata* Say. *Journal of Insect Physiology* 24, 221–224.

Mwangi, R.W. and Goldsworthy G.J. (1976) Age related changes in the response to adipokinetic hormone in *Locusta*. *General and Comparative Endocrinology* 29, 291.

Mwangi, R.W. and Goldsworthy G.J. (1977a) Diglyceride-transporting lipoproteins in *Locusta*. *Journal of Comparative Physiology* 114, 177–190.

Mwangi, R.W. and Goldsworthy G.J. (1977b) Age-related changes in the response to adipokinetic hormone in *Locusta migratoria*. *Physiological Entomology* 2, 37–42.

Nassel, D.R., Passier, P., Elekes, K., Dircksen, H., Vullings, H.G.B. and Cantera, R. (1995) Evidence that locustatachykinin-I is involved in release of adipokinetic hormone from locust corpora-cardiaca. *Regulatory Peptides* 57, 297–310.

Ogoyi, D.O., Osir, E.O. and Olembo, N.K. (1995) Lipophorin and apolipophorin-III in solitary and gregarious phases of *Schistocerca gregaria*. *Comparative Biochemistry and Physiology B – Biochemistry amd Molecular Biology* 112, 441–449.

Ogoyi, D.O., Osir, E.O. and Olembo, N.K. (1996) Effect of phase status on responses to AKH I in the desert locust, *Schistocerca gregaria*. *Archives of Insect Biochemistry and Physiology* 32, 173–185.

Ogoyi, D.O., Osir, E.O. and Olembo, N.K. (1998) Fat body triacylglycerol lipase in solitary and gregarious phases of *Schistocerca gregaria* (Forskal) (Orthoptera: Acrididae). *Comparative Biochemistry and Physiology B – Biochemistry and Molecular Biology* 119, 163–167.

Oudejans, R.C., Kooiman, F.P., Heerma, W., Versluis, C., Slotboom, A.J. and Beenakkers, A.M.Th. (1991) Isolation and structure elucidation of a novel adopkinetic hormone (Lom-AKH-III) from the glandular lobes of the corpus cardiacum of the migratory locust, *Locusta migratoria*. *European Journal of Biochemistry* 195, 351–359.

Oudejans, R., Vroemen, S.F., Jansen, R.F.R. and van der Horst, D.J. (1996) Locust adipokinetic hormones: carrier-independent transport and differential inactivation at physiological concentrations during rest and flight. *Proceedings of the National Academy of Sciences USA* 93, 8654–8659.

Pannabecker, T. and Orchard, I. (1986) Octopamine and cyclic AMP mediate release of adipokinetic hormone I and II from isolated locust neuroendocrine tissue. *Molecular and Cellular Endocrinology* 48, 153–159.

Passier, P., Vullings, H.G.B., Diederen, J.H.B. and van der Horst, D.J. (1997) Trehalose inhibits the release of adipokinetic hormones from the corpus cardiacum in the African migratory locust, *Locusta migratoria*, at the level of the adipokinetic cells. *Journal of Endocrinology* 153, 299–305.

Pearson, D.J., Imbuga, M.O. and Hoek, J.B. (1979) Enzyme activities in flight and leg muscle of dung beetle in relation to proline metabolism. *Insect Biochemistry* 9, 461–466.

Pener, M.P. and Yerushalmi, Y. (1998) The physiology of locust phase polymorphism: an update. *Journal of Insect Physiology* 44, 365–377.

Pener, M.P., Ayali, A. and Golenser, E. (1997) Adipokinetic hormone and flight fuel related characteristics of density-dependent locust phase polymorphism: a review. *Comparative Biochemistry and Physiology B – Biochemistry and Molecular Biology* 117, 513–524.

Rosinsky, G. and Gäde, G. (1988) Hyperglycaemic and myoactive factors in the corpora cardiaca of the mealworm, *Tenebrio molitor*. *Journal of Insect Physiology* 34, 1035–1042.

Sacktor, B. and Childress, C.C. (1967) Metabolism of proline in insect flight muscle and its significance in stimulating the oxidation of pyruvate. *Archives of Biochemistry and Biophysics* 120, 583–588.

Sacktor, B. and Wormser-Shavit, E. (1966) Regulation of metabolism in working muscle *in vivo*. I. Concentrations of some glycolytic, tricarboxylic acid cycle and amino acid intermediates in insect flight muscle during flight. *Journal of Biological Chemistry* 241, 624–631.

Schneider, M. and Dorn, A. (1994) Lipid storage and mobilization by flight in relation to phase and age of *Schistocerca gregaria* females. *Insect Biochemistry and Molecular Biology* 24, 883–889.

Siegert, K., Morgan, P. and Mordue, W. (1985) Primary structures of locust adipokinetic hormones II. *Biological Chemistry Hoppe-Seyler* 366, 723–727.

Singh, G.J.P., Orchard, I. and Loughton, B.G. (1981) Octopamine-like action of formamidines on hormone release in the locust, *Locusta migratoria*. *Pesticide Biochemistry and Physiology* 16, 249–255.

Socha, R., Kodrík, D. and Zemek, R. (1999) Adipokinetic hormone stimulates insect locomotor activity. *Naturwissenschaften* 86, 85–86.

Stone, J.V., Mordue, W., Batley, K.E. and Morris, H.R. (1976) Structure of locust adipokinetic hormone, a neurohormone that regulates lipid utilisation during flight. *Nature* 263, 207–211.

Thompson, S.N. and Bennett, R.B. (1971) Oxidation of fat during flight of male Douglas-fir beetles, *Dendroctonus pseudotsugae*. *Journal of Insect Physiology* 17, 1555–1563.

Weeda, E. (1981) Hormonal regulation of proline synthesis and glucose release in the fat body of the Colorado potato beetle, *Leptinotarsa decemlineata*. *Journal of Insect Physiology* 27, 411–417.

Weeda, E., de Kort, C.A.D. and Beenakkers, A.M.T. (1979) Fuels for energy metabolism in the Colorado potato beetle, *Leptinotarsa decemlineata* Say. *Journal of Insect Physiology* 25, 951–955.

Weeda, E., de Kort, C.A.D. and Beenakkers, A.M.T. (1980) Oxidation of proline and pyruvate by flight muscle mitochondria of the Colorado beetle *Leptinotarsa decemlineata* Say. *Insect Biochemistry* 10, 305–311.

Wegener, G. (1996) Flying insects: model systems in exercise physiology. *Experientia* 52, 404–412.

Wheeler, C.H. (1989) Mobilization and transport of fuels to the flight muscles. In: Goldsworthy, G.J. and Wheeler, C.H. (eds) *Insect Flight*. CRC Press, Boca Raton, Florida, pp. 273–303.

Wheeler, C.H. and Goldsworthy, G.J. (1983) Protein-lipoprotein interactions in the haemolymph of *Locusta* during the action of adipokinetic hormone: the role of $C_L$-proteins. *Journal of Insect Physiology* 29, 349–354.

Wheeler, C.H., van der Horst, DJ. and Beenakkers, A.M.T. (1984a) Lipolytic activity in the flight muscles of *Locusta migratoria* measured with haemolymph lipoproteins as substrates. *Insect Biochemistry* 14, 261–266.

Wheeler, C.H., Mundy, J.E. and Goldsworthy, G.J. (1984b) Locust haemolymph lipoproteins visualised in the electron microscope. *Journal of Comparative Physiology* 154, 281–286.

Wheeler, C.H., Boothby, K.M. and Goldsworthy, G.J. (1986) $C_L$-proteins and the regulation of lipoprotein lipase activity in locust flight muscle. *Biological Chemistry Hoppe-Seyler* 367, 1127–1133.

Witten, J.L., Schaffer, M.H., O'Shea, M., Cook, J.C., Hemling, M.E. and Rinehart, K.L., Jr (1984) Structures of two cockroach neuropeptides assigned by fast atom bombardment mass spectrometry. *Biochemical and Biophysical Research Communications* 124, 350–358.

Zebe, E. and Gäde, G. (1993) Flight metabolism in the African fruit beetle, *Pachnoda sinuata*. *Journal of Comparative Physiology B – Biochemical Systemic and Environmental Physiology* 163, 107–112.

Zeng, H., Loughton, B.G. and Jennings, K.R. (1996) Tissue specific transduction systems for octopamine in the locust (*Locusta migratoria*). *Journal of Insect Physiology* 42, 765–769.

# Insect Behaviours Associated with Resource Finding

Jim Hardie,[1] Gabriella Gibson[2] and Tristram D. Wyatt[3]

[1]*Aphid Biology Group, Department of Biology, Imperial College at Silwood Park, Ascot, Berks SL5 7PY, UK;* [2]*Pest Management Department, Natural Resources Institute, University of Greenwich, Central Avenue, Chatham Maritime, Chatham, Kent ME4 4TB, UK;* [3]*Department of Zoology and Department for Continuing Education, Oxford University, South Parks Road, Oxford OX1 3PS, UK*

## Introduction

All insects move, but the degree of movement varies greatly between individuals and species. Thus scale insects and wingless aphids may spend their life-time on a small area of a single plant and move only centimetres, tsetse flies move hundreds of metres across the African savannah while desert locusts travel thousands of kilometres during their movements to feeding/breeding areas. For scale insects and aphids, the resources required, such as food and oviposition/larviposition sites, are to hand and determined by females of the previous generation. But, somewhere along the way, the resources required to sustain the life and reproductive potential of the individual as well as survival of the subsequent generations need to be located. The major resources required by insects are food, mates and oviposition sites but resting sites and diapause sites are also needed by some species/generations. All have to be found and the probability of success will depend not only on resource abundance and distribution but also on the efficiency of the sensory processes and behavioural mechanisms.

'Resource finding' is one of those anthropomorphic phrases that dog the study of animal behaviour. It has meaning in a list of the types of activities that insects engage in, but it is meaningless at the level of identifying the mechanisms that control behaviour. Insects do not set out to 'find' resources, rather they encounter them as they respond to external and internal stimuli. Evolution favours sensory systems which are sensitive and selective enough to detect appropriate resource cues, and it favours motor responses which optimize the likelihood of encountering suitable resources at the appropriate time, while minimizing costs (energy expenditure, risk of predation, etc.). This chapter aims to set out some of the constraints which affect 'resource-finding strategies', e.g. the type of resource cues or 'signals' available, the nature of irrelevant 'noise' from the environment and the range of possible motor responses which affect the chances of encountering resources.

The movements addressed in this chapter can be defined as 'station keeping'. That is, they do not take the insect out of its home range (Dingle, 1996) and terminate once a resource item has been located. 'Migratory' movements, on the other hand, tend to occur over longer distances and occur when movement is not distracted by resource items. Kennedy (1985) provided the most useful definition of behavioural migration:

> Migratory behaviour is persistent and straightened-out movement effected by the animal's own locomotory exertions or by its active embarkation on a vehicle. It depends upon some temporary inhibition of station-keeping responses but promotes their eventual disinhibition and recurrence.

Such a definition is pitched at the individual insect and specifically at the behaviour. It omits long-distance movement by accidental displacement that may play a role in population dispersal and 'migration' from an ecological viewpoint. Movements involved in migration are discussed in later chapters.

While the majority of insects may demonstrate an itinerant life style, some species are 'central place foragers' and demonstrate homing – movements to and from the hive/nest. These movements can be classified as resource finding but the mechanisms involved may be more akin to those used during directed migration. That is, the use of landmarks and compass-dependent cues from the sun or magnetic fields and these we shall exclude as they will be better covered under migratory movements (Riley and Osborne, Chapter 7; Srygley and Oliveira, Chapter 9). Learning can also play a major role in resource location and will undoubtedly reinforce and increase efficiency by interacting with basic behaviours (Powell and Poppy, Chapter 6). However, we shall concentrate on the basic behaviour and exclude learning.

Although many movements appear complex events, the building blocks can often be traced to relatively simple *kineses* and *taxes* (Fraenkel and Gunn, 1940; Kennedy, 1986; Dusenbury, 1992; Campan, 1997). *Kineses* are random movements governed by internal (idiothetic) motor programmes but are evoked by variable stimuli that have no inherent direction. The amount/degree of activity is a function of the intensity of the stimulus. Temperature and humidity are classical examples of such stimuli. The two basic kineses are 'orthokinesis' where the frequency or rate of turning changes and 'klinokinesis' where speed varies. Such behaviour can lead to aggregation or dispersal of insect groups in particular areas of favourable or unfavourable temperature, for example. When a stimulus has direction, e.g. sound, odour source, visual target, the movement directed towards (+) or away (−) from the stimulus is termed *taxis*. There are two basic taxes, *klinotaxis* where the environment is successively sampled by a single sense organ as the organism moves and any changes in stimulus strength are detected, and *tropotaxis*, where directed turns are based on simultaneous comparison of stimulus intensity with two sense organs, e.g. the paired antennae. The stimuli initiating taxes are often prefixed and defined in such terms as phototaxis (light), anemotaxis (wind) and chemotaxis (odour/taste).

The response to any stimulus depends upon the insect's sensory capacity. The major senses used by insects to locate resources are olfaction and vision although tactile/vibratory stimuli are used by some to detect, for example, leaf mining hosts by parasitoids (Meyhöfer and Casas, 1999) and mosquitoes use hearing (phonotaxis) to locate mates (e.g. Belton, 1994; Clements, 1999). The readiness of an insect to respond will also depend on its internal physiological condition (Harris and Foster, 1995). Thus there are restricted periods of activity with day-, night- and crepuscular-active species and the timing is often based upon endogenous circadian rhythms which are overtly entrained to 24 h, usually to the day–night light cycle. Not only does activity change throughout 24 h but there may also be circadian changes in the responsiveness of the sensory systems. The electrophysiological response of the antennal olfactory receptors of the fruit fly, *Drosophila melanogaster* Meigen, for example, shows a distinct circadian influence (Krishnan *et al.*, 1999).

The daily timing of activity and responsiveness to resources has evolved to be closely linked to environmental features. Thus, light intensity, wind, temperature and humidity vary during the day–night cycle and have associated benefits or handicaps (Gibson and Torr, 1999). Whatever the physiological condition or the stimulus, the behaviours involved in locating a resource begin at a time when the insect leaves a resting site and becomes active. Such activity is initiated internally, i.e. spontaneous activity, or more usually in the field by environmental stimuli, such as a particular light intensity. Activity in response to an environmental stimulus initially involves kinesis or it could arise from a response to a directional stimulus (i.e. taxis), such as host movement, e.g. tsetse fly (Vale, 1974; Torr, 1988, 1989). If no further resource cues are detected, activation is normally followed by 'ranging' behaviour, where movement is oriented to environmental cues such as wind direction or visual features of the landscape in such a way as to maximize the probability of detecting stimuli associated with resources (Dusenbury, 1992). We prefer the term 'ranging' to 'trivial', 'appetitive' movements or 'foraging' as it does not imply either a behaviour irrelevant to resource location nor a teleological 'goal'-seeking behaviour which can only be assessed subjectively (Jander, 1975; Kennedy, 1986). It may not be immediately apparent that ranging behaviour has a tactic element and it could be interpreted as a kinesis, where response to ambient temperature or light levels promote activity which is guided by idiothetic mechanisms. However, theoretical considerations of, for example, the best strategies for detection of a resource odour plume have indicated increased probabilities for detection with upwind or downwind flight in changing wind directions but across-wind flight in more stable conditions (Fig. 5.1; Sabelis and Schippers, 1984). In such cases, anemotactic behaviour is required. In the laboratory, Zanen *et al.* (1994) showed that *Drosophila* would fly with an across-wind preference in a stable wind but when the wind direction was varied, the flies tended to move parallel to the mean wind direction (Fig. 5.1). Examples from the field include tsetse flies which fly downwind in the absence of a host odour, as expected by predictions (Gibson *et al.*, 1991). Bidlingmayer classified mosquitoes into types depending on how they flew in relation to the vegetation: across fields, along the perimeter of fields or through forest with respect to their host preference (Bidlingmayer, 1974). Ranging behaviour may also follow an insect losing track of a resource stimulus.

The precise response to a given cue depends on many factors, such as strength and 'pattern' of presentation of the stimulus, and the presence of other resource cues and environmental conditions, such as wind speed and light intensity. Once a stimulus has been detected, oriented taxes come into play in response to a variety of cues. This phase of resource location has traditionally been divided, somewhat arbitrarily, into 'long-range' and 'short-range' responses. The behaviours elicited at the greatest distances from a resource generally bring the insect closer to the source as quickly as possible by, for example, increased airspeed and maximized upwind progress. This type of orientation is often described as *indirect* in that insects can use wind direction as an indication of where airborne molecules have come from. At 'short range' new stimuli bring insects even closer to the source by reducing airspeed and/or increasing turning frequency, until ultimately the cues

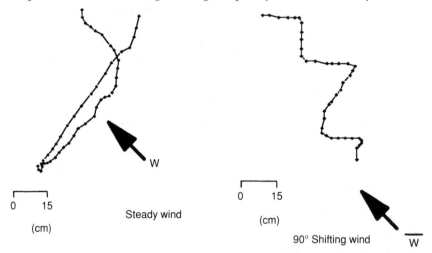

**Fig. 5.1.** Flight tracks taken from a video recording of a fruit fly, *Drosophila funebris*, in a steady wind (left) and a variable wind shifting through 90° (right). The camera was positioned vertically above the observation area. The insect flies across wind in the steady wind but moves parallel to the mean direction in wind with variable direction. From Zanen *et al.* (1994).

lead to arrestment and contact with the resource. Although this type of orientation can still be indirect, e.g. when the complex of odours indicates that the position of the resource is near (Warnes, 1990), it is often *direct*, in that the position of the resource is pinpointed and the insect can move directly to it, e.g. when visual or tactile contact is made.

In all cases, resource location takes place in a background of 'noise'. Noise can be both internal and external and comprises irrelevant signals which need to be successfully filtered for accurate detection and response (Dusenbury, 1992). Environmental cues used by insects for resource finding do not occur in isolation and it is the efficiency of the sensory processes that impacts greatly on the probability of success. With a low signal-to-noise ratio, resource items may well escape detection/location and this is utilized, for example, in crypsis. However, when the evolutionary advantage is for the resource item to be located (e.g. pollination, mate location), high signal-to-noise ratios have been selected.

Some of the best studied examples of invertebrate orientation behaviour towards resources come from insect groups important because of their impact on human wellbeing, for example, human health, crops, domesticated animals, housing or timber. Thus the orientation of moth males to female sex pheromones, tsetse flies to their vertebrate hosts, and phytophagous insects to their host plants have all been intensively investigated. The mechanisms revealed by studies of these economically important pest groups may form the basis of mechanisms used by all other species.

Common to all three examples is a long-distance response to odour plumes coming from the resource. Most insects appear to use similar mechanisms for locating the source of upwind odours (reviewed by Wyatt, 2001). After describing the common mechanisms for locating odour sources, we discuss the orientation behaviour of the three examples:

1. Mate location by male moths.
2. Host-animal location by haematophagous insects.
3. Host-plant location by phytophagous insects.

Each is affected by factors such as the time of day that searching is carried out and by the differences in scale between the sizes of the odour sources being searched for. For some systems, visual stimuli are dominant, for others, odours are most important. In most, if not all, cases it is a combination of cues that allows the insect to find the resource efficiently.

## Mechanisms of Odour-source Location

### Orienting to an odour plume in air

The majority of orientation to odours probably occurs outside the smooth boundary layers of still air, and quite beyond the physical scale at which diffusion dominates (Wyatt, 2001). Far from being a smooth concentration gradient from the source, odour plumes are turbulent and there is no evidence that flying insects use chemotaxis to find the source (see Kennedy (1992) for a history of research on this topic). Instead, the majority of signals have to be deciphered from a turbulent flow of molecules carried in wind currents. Nonetheless, insects not only receive signals but, despite the formidable difficulties of following the track of a fluctuating, meandering pheromone plume, are able to orient upwind to the odour source. Flying insects cope with even greater problems than insects in contact with the substrate as the flying insect cannot easily know if it is making progress upwind.

## Plume structure – what are the available signals?

Turbulent odour plumes form as air currents disperse odour molecules from their source. Odour plumes are, of course, normally invisible but swirling smoke clouds coming from a chimney provide a good visual analogy of the important features (Murlis *et al.*, 1992). The smoke forms a meandering cloud that drifts in snake-like curves downwind. If you are nearer the smoke, you can see the fine scale structure within the clouds, with filaments of high particle concentration interspersed with cleaner air.

As a cloud of odour molecules moves from the source, turbulence physically tears apart the cloud into elongated odour-containing filaments. The turbulent effects are greater than molecular diffusion which is comparatively slow (Murlis, 1997). Odour plumes could be imagined as cooked spaghetti pouring out of a chimney and as the plume widens, the strands of spaghetti become more widely separated. 'Flying spaghetti' is a good analogy, up to a point, as the concentration in 'odour-laden filaments' hundreds of metres from the source is similar to the concentration in filaments near the source. An important consequence of the relatively small effect of diffusion on the filaments is that a plume is far from being a uniform cloud of pheromone drifting downwind, getting more dilute as it diffuses; rather, it is composed of filaments which remain relatively concentrated. Thus the filaments will be above the response threshold much further downstream than a diffusion model would predict – but far downstream the filaments may become widely dispersed in a spreading plume. This fine filament structure is central to the responses evolved by orienting insects.

Two characteristics of odour plumes combine to give the intermittent signal which presents so many challenges to a receiver downwind. The first is the fine filament structure and small-scale eddies. The second is the changing wind direction, on a scale of seconds or minutes, which leads to the plume wandering across the landscape (made visible in field experiments with neutrally buoyant balloons or soap bubbles; Murlis *et al.*, 1992). The result is that a detector (or insect) at a point downwind receives a highly intermittent signal with two time-scales of variation: of minutes or seconds with or without any signal ('without' might be as much as 80% of the time) as the plume meanders over the detector and away again, and on a finer scale of milliseconds when the plume is over the detector, as individual odour filaments in the plume touch the detector.

## How do we know what happens in odour plumes and what are the signals that reach a searching insect like?

Detecting and describing the fine structure of odour plumes at the spatial and temporal resolution relevant to the organisms being studied has been a major challenge. The concentrations of pheromone detected by the animals are many orders of magnitude lower than the most sensitive human instruments and insect sensory cells respond at a millisecond frequency on a real time-scale. The solution has been to add an artificial marker that we can detect by physical or chemical instruments. In air this has often been ionization of air at the source and then detection with ionization detectors (Murlis, 1997). Recently, single cell recordings have been made from moth antennae in the field, downwind of a pheromone source. The sensory cells showed fluctuations in spiking, corresponding to contact with filaments of high pheromone concentration interspersed with clean air, at the fine time-scales previously predicted by ionized-air tracers and other techniques (van der Pers and Minks, 1997).

## Animals moving in contact with the substrate

Walking and crawling terrestrial insects respond to the plume coming from an odour source (Bell and Tobin, 1982). As the molecules are moving downwind, going upwind will bring an insect towards the source. Insects in contact with the substrate can potentially gain information on the wind flow from the way their mechanoreceptors are deflected by the current, and then respond with positive or negative anemotaxis. However, it is not always clear when animals are using the chemotactic response to odour, via klinotaxis or tropotaxis, as opposed to anemotaxis triggered by presence of odour, or a combination of these.

## Flying insects

Insects that fly to an odour source are offered the same intermittent odour stimuli in a turbulent plume as walking animals. However, they have the additional task of ensuring that they are making progress upwind in spite of being suspended in the medium – without an outside reference, the flow is not detectable to the animal immersed in it. The common analogy is that of swimming in a river: you cannot tell if you are making any progress against the current unless you can see landmarks on the bank. In a similar way, moths and other insects need visual feedback. For the moth, seeing the movement of the ground below tells whether it is making progress upwind (optomotor anemotaxis). If it is making forward progress, the images will be passing from front to back of its eye. If it is being blown backwards by a headwind, then the images will pass back to front. If being blown sideways (drift), then it will get a sideways movement of the image across the eye. To make upwind progress the moth flies to counter these movements, to counter the drift (David, 1986). The visual feedback can be exploited experimentally by using a moving floor in a windtunnel so moths can be kept flying on a 'visual treadmill' for long periods, without ever reaching the source (Millar and Roelofs, 1978). It is also recognized that over flat ground, even in wind with variable direction and a plume that 'snakes' from left to right (Fig. 5.2), air packets continue along nearly straight lines from an odour source. If an insect heads upwind immediately on odour detection, it will be heading towards the source (David et al., 1982).

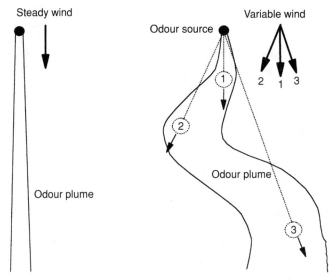

**Fig. 5.2.** Diagramatic representation of averaged odour plumes viewed from above in conditions of steady wind (left) and variable wind direction (right) over level ground without obstructions. Arrows on the right indicate wind directions. In variable wind, a snaking plume is formed and the dotted circles within the plume (1, 2 and 3) represent small packets of odour that moved away from the source at different times. The arrows leading from these packets indicate the instantaneous wind direction and the dashed lines show that back projections from these instantaneous wind directions lead directly to the odour source. Arrows to the right indicate wind direction at the odour source when odour packets 1, 2 and 3 were released. The turbulent changes in wind direction create the snaking plume but, counterintuitively, each odour packet has moved in a straight line from the source rather than along the curve of the odour plume as a whole.

Flying insects orienting to an odour source show similar characteristic zig-zag tracks and the detailed mechanisms will be now be discussed

## Mate Location by Male Moths

The best studied responses of flying insects to pheromones are undoubtedly male moths responding to female sex pheromones (Baker and Vickers, 1997; Cardé and Mafra-Neto, 1997). The turbulent, filamentous nature of real odour plumes plays a crucial role in upwind orientation: moths appear to respond with millisecond speed to the fine structure of odour plumes. Having evolved in response to turbulent plumes, moth orientation behaviour now depends on it (they are unable to progress upwind in a uniform pheromone cloud; Kennedy et al., 1980).

The current consensus is that upwind orientation by moths is the product of two programmes, both stimulated by pheromone: a brief surge of upwind flight when a pheromone filament is contacted and, second, self-steered counterturning (side-to-side flight or zig-zagging) when the filament is ended and the moth leaves the plume (Baker, 1990). The upwind flight is dependent on visual feedback for optomotor anemotaxis described above. The casting behaviour increases the chance of regaining contact with the shifting plume. The turn at the end of each zig and zag is self-steered, not chemotaxis at the edge of the pheromone plume (Kennedy, 1992).

Upwind flight is thus made up of repeated sequences of surge and cast as the moth encounters the filaments of the plume (Vickers and Baker, 1994). If the male encounters the next pheromone-laden filaments before he has switched to casting behaviour, a straighter flight will result. Different species have different times before switching to casting. Species that switch quickly, such as, for example, tobacco bud worm moth, *Heliothis virescens* (F.) (0.15 s latency; Vickers and Baker, 1994) will have zig-zagging flights whereas species with a longer period, such as the oriental fruit moth, *Grapholitha molesta* (Busck) (0.3 s; Baker and Haynes, 1987), will tend to have straighter flights because they are more likely to hit the next pheromone filament before switching.

Many lines of evidence from both field and wind-tunnel experiments support the cast–surge hypothesis (Baker and Vickers, 1997; Cardé and Mafra-Neto, 1997). Some of the most persuasive recent experiments have come from wind-tunnel work that mimics the turbulent plume by short puffs of pheromone at controlled rates. Working with almond moth males, *Cadra cautella* (Walker), Mafra-Neto and Cardé (1994) found that slowly pulsed plumes (*c.* 0.6 pulses $s^{-1}$) or a continuous, uninterrupted ribbon plume led to only very slow upwind progress with much zig-zagging. Males flew faster and with a straighter path in quickly pulsed (*c.* 5 pulses $s^{-1}$) or turbulent plumes (Fig. 5.3).

Independently, Vickers and Baker (1994) flew male *H. virescens* to puffs of pheromone. By plotting the track of the moths and pheromone puffs, Vickers and Baker could see the effect of each puff on moth behaviour. It

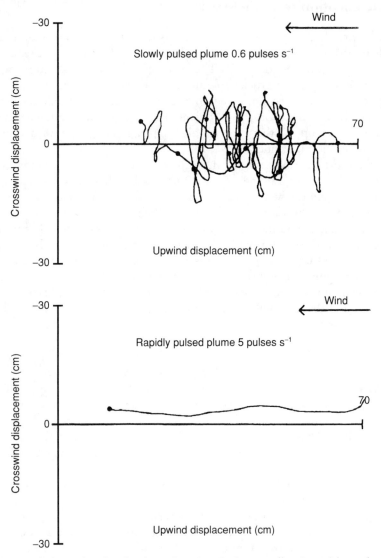

**Fig. 5.3.** Flight tracks of male almond moths, *Cadra cautella*, viewed from above and flying upwind in a sex pheromone plume comprising discrete short pulses of pheromone released at different rates. When the pheromone is pulsed slowly, 0.6 pulses $s^{-1}$ (top), the insect reaches the source slowly with much crosswind flight. When the pheromone is pulsed faster, 5 pulses $s^{-1}$ (bottom), the male reaches the source in 1 s and the flight is directly upwind. Markers represent 1-s intervals. From Mafra-Neto and Cardé (1994).

seemed to form a pattern, a template: each puff would start the male on a surge upwind, then, if he did not receive another puff in time, he would start casting side-to-side flight again (Vickers and Baker, 1994, 1996). In this species, the rather tortuous tracks in response to a plume of 4 pulses $s^{-1}$ were explained by a series of casts (clean air) and surges (pheromone pulse) matching their template. Moths flew nearly straight upwind tracks in response to 10 pheromone pulses $s^{-1}$.

Using an antenna (from a second male) attached across the head of the moth to give an in-flight electroantennogram (EAG) Vickers and Baker (1994) could relate behaviour to the pheromone filaments hitting the moth. The 'Cyclops' antenna responded with an EAG potential, read via fine wires, when pheromone puffs hit the flying moth and these could be matched, with a delay of 0.3 s, to surges in forward flight. When filament contacts ceased and the EAG recordings ceased for more than about 0.3 s, casting started.

The temporal and spatial resolution to which the moths are sensitive is very fine, presumably corresponding to the fine scale of the filaments. If pheromone components from females of sympatric species are given in the same puff (indicating to the male that the female is of the wrong species), males will turn back. If, however, these antagonist components are released in puffs separated by just 1 mm and at most 0.001 s from puffs of pheromone from his own species, male corn earworm moths, *Heliothis zea* (Boddie), continue to fly upwind (Baker *et al.*, 1998). Sensory cells for different components in a pheromone blend occur in the same sensory hair on the antenna so they will only be stimulated correctly if the odour filament, on this scale, simultaneously contains the correct components in the correct concentrations. This sophistication in the sensory system, together with the physical properties of pheromone plumes, means that the signal from a conspecific female can be recognized and responded to despite chemical noise where pheromone plumes from females of different species will be intermingling. The pheromone filaments from the two species remain separate, so by surging forward only to the filaments containing the correct blend without antagonists, the male can still make progress towards the conspecific female.

## Host-animal Location by Haematophagous Insects

Haematophagous insects are not unlike their phytophagous relations in the need to locate resting and oviposition sites and, in most cases, some nourishment from plants but they depend on vertebrate blood for reproduction. A blood-meal host is a unique resource in that it is both *not* in the host's interest to be found (unlike sex/aggregation pheromone-emitting insects) and the host can move (unlike a plant). 'Host-seekers' overcome these limitations by adopting a strategy of either locating the host while it is quiescent (at night usually), by moving faster than the host during its active phase or by remaining in one place and waiting for a host to appear close by (a 'sit and wait' strategy). As outlined in Table 5.1 (taken from Gibson and Torr, 1999), there are critical advantages and disadvantages to host location at particular times of day, associated with significant environmental conditions. Diurnal tsetse flies and nocturnal mosquitoes represent the extremes of the two main strategies which have evolved under these constraints. During the day many advantages, such as more reliable visual and directional cues (i.e. stronger signal-to-noise ratios) are offset by the disadvantages of increased host-defence behaviour and the greater risks of predation. Tsetse also take advantage of host mobility which increases their conspicuousness and can bring host and fly closer together. At night, host cues are generally weaker against background noise, e.g. light intensity is lower, background levels of carbon dioxide (the most common olfactory cue) are higher and the wind provides less reliable directional cues because wind speeds are lower. The host, however, is generally quiescent, allowing insects more time to locate a given host. For nocturnal mosquitoes, airspeeds during the crepuscular/night phase are generally low enough for them to fly against and the host's movements are minimal (see review by Gibson and Torr, 1999).

**Tsetse**

Tsetse are obligate blood-feeders, receiving no nourishment from plant sources. They have been likened to a turbo-charged engine (J. Brady,

**Table 5.1.** Opportunities and constraints for haematophagous Diptera feeding during the day or night.

|  | Day | Night |
|---|---|---|
| Disadvantage | 1. Greater risk of desiccation<br>2. Greater wind turbulence causing host-odour plumes to break up<br>3. Greater risk from predators<br>4. Host mobile (disadvantage for odour-responding flies?)<br>5. Greater risk from host defensive behaviour as host often active | 1. Poor visual cues (especially colour)<br>2. Low wind speed and so poor directional cues in host-odour plumes<br>3. Greater background noise of $CO_2$<br>4. Host less mobile, so 'sit and wait' strategies less feasible |
| Advantages | 1. Good visual cues<br>2. High winds providing good directional cues in host plumes<br>3. Reduced background levels of atmospheric $CO_2$<br>4. Host mobile making 'sit and wait' strategy feasible | 1. Less risk of desiccation<br>2. Host often quiescent with reduced risk from host defensive behaviour<br>3. Less risk from predators<br>4. Less atmospheric turbulence and thus odour plumes less broken up |

After Gibson and Torr (1999).

personal communication): the rich fuel derived from blood enables short bursts of fast, powerful flight. A female is capable of carrying her own weight again, for much of the time, in the form of the developing larva within her. This weight can double after a blood meal. Tsetse leave their resting sites by endogenously controlled spontaneous activity, by visual stimuli or when they are stimulated by airborne host odour during their active phase (Torr, 1988). In the absence of host odour, they generally fly downwind, thus maximizing their chances of encountering an odour plume near the source under conditions of variable wind direction (see above; Gibson et al., 1991). If host odour is present, tsetse turn upwind and follow host-odour plumes by visually guided, odour-mediated, positive anemotaxis, until the odour cue is lost. Flies losing odour make a large reverse turn or land. This behaviour increases the probability of the fly contacting odour again, whereupon it resumes flying upwind. When the host is within visual range of detection, or the odour profile indicates the host is near, tsetse reduce their flight speed, increase their turning rate and eventually, given the appropriate short-range cues, land on the host (Gibson and Torr, 1999).

As wind speed increases, turbulence also increases and breaks up the filament structure of odour plumes, thus reducing the strength of the signal (in terms of larger spaces between the odour filaments) as it moves downwind of the source. This probably means that tsetse need to sample an increasingly greater area to detect odour plumes as wind speed increases. With typical flight speeds of 6 m s$^{-1}$, a relatively large volume of air can be sampled in a short time. Tsetse eyes have evolved to maximize the available light; their relatively high acuity enables them to detect wind drift at high flight speeds (Gibson and Young, 1991; Land, 1997). Diurnal host location has also allowed tsetse to make full use of visual host cues, such as host movement, colour, shape and size, to detect them against background vegetation. Oddly, tsetse appear to be preferentially attracted to phthalogen blue, not a common colour in the animal kingdom. Possibly these wavelengths are not so much a positive identification for host animals, but rather a negative identification, i.e. 'not a plant' (see Fig. 5.4 and text below on phytophagous insects).

## Mosquitoes

Mosquitoes require blood only for egg production, and adults can survive on plant juices alone. Mosquitoes are lighter, weaker insects than tsetse, with mean flight speeds of less than 1 m s$^{-1}$, and hence their host location activity is limited to habitats and/or times of day when wind speeds are even lower than this. As wind speed decreases, so does turbulence, thus the filamentous structure of odour plumes, and hence the strength of the stimulus, remains intact further downwind (Griffiths and Brady, 1995). Wind direction, however, has a tendency to change direction more dramatically as wind speed decreases, hence the wind direction within an odour plume is a less reliable

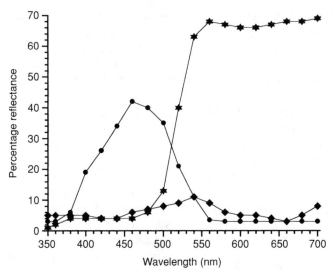

**Fig. 5.4.** Spectral reflectances of pthalogen blue which is visually attractive to tsetse flies (circles: redrawn from Green, 1988), the upper surface of a potato leaf (diamonds: redrawn from Vaishampayan *et al.*, 1975) and canary yellow (stars: redrawn from Hardie *et al.*, 1996).

indicator as to the source of the odour at lower wind speeds (Brady *et al.*, 1989).

Odour-plume following by nocturnal mosquitoes thus requires rather different specializations than for those of diurnal tsetse. For example, tsetse eyes are capable of detecting wind drift at high flight speeds, but they require ample light (Gibson and Young, 1991). Nocturnal mosquitoes, on the other hand, have sacrificed resolution to maximize the light-gathering capabilities of the eye to enable optomotor-guided anemotaxis with starlight (Gibson, 1995; Land *et al.*, 1997, 1999). These two types of eyes have evolved to operate maximally at light intensities six orders of magnitude apart.

These examples illustrate how important circadian rhythms are to efficient strategies for host location. If blood-feeding insects responded to airborne host chemicals at all times of day, they would risk a high failure rate by attempting to locate odour sources when their behaviour and physiology was not appropriate for the prevailing environmental conditions.

The degree of host preference amongst mosquito species varies significantly more than it does for tsetse. Although little is known about the mechanism which controls the response of mosquitoes to particular hosts, it is likely that generalist feeders respond most strongly to carbon dioxide and other volatiles common to many vertebrates. The strength of response of specialists is probably controlled by the presence of host-specific chemicals in addition to carbon dioxide. For example, *Anopheles gambiae* Giles *s.s.*, the malaria vector in sub-Sahara Africa, feeds almost exclusively on humans, and has evolved several behaviours which reflect this close association with the host. For example, Costantini *et al.* (1996) found that generalist species showed a reasonably linear dose–response to carbon dioxide-baited traps, whereas *An. gambiae* reached a plateau of response to carbon dioxide that was always less than the response to a human. What controls the differences in responsiveness between mosquito species to particular host cues is poorly understood, but probably involves differences in peripheral receptors as well as differences in central processing of sensory inputs.

Nocturnal, human-feeding mosquitoes provide a good example of how much the stimuli and responses can change with distance from the source. Often the presence of a human host must be detected initially by volatile chemicals carried downwind from dwellings. It is likely that some of the components can be detected further downwind than others, so the odour profile at a given distance from the source may give the mosquito information as to the proximity of the odour source. As the mosquito gets nearer the human's dwelling, visual responses to the size and shape of the house itself modify their flight path. Only after the mosquito has entered the dwelling can visual cues from the host be detected, and additional cues, such as convection currents created by the warm host, have an influence on behaviour. None of the odour cues detected at a distance have been lost, but in the presence of the new cues the mosquito's upwind progress is eventually arrested. For diurnal mosquitoes, visual stimuli of the host itself are probably the most significant short-range cues leading to arrestment at the source.

Similar patterns of host location are found amongst other haematophagous insects, but one case in particular merits special comment. Although only females of the sandfly *Lutzomyia longipalpis* (Lutz & Neiva) feed on blood, both males and females are attracted to host animals by various host cues, including odours (Dougherty *et al.*, 1999). In addition, males then form leks near the host and emit a sex pheromone, which also attracts females (Morton and Ward, 1989). The dynamics of this system are complex, and lead to a concentration of males and females on only a few of the potential hosts in a given area (Kelly and Dye, 1997).

## Host-plant Location by Phytophagous Insects

### Olfaction

Plants also produce volatile compounds which are released into the wind and can be utilized by phytophagous insects for host-plant location (Visser, 1986). 'Sit and wait' may be a strategy for insects requiring blood meals from mobile hosts but it is less useful for plant-eating insects. A change in wind direction might divert the host-plant odour to the insect olfactory system and it is suggested that cabbage root flies, *Delia radicum* (L.), do land after losing an aerial odour plume, wait to pick it up again and then fly upwind.

This behaviour can take them to the host-odour source without the need to detect wind direction in flight (Nottingham, 1988).

While some volatile breakdown products of secondary plant metabolites (e.g. the isothiocyanates from cruciferous plants) may offer specific odour cues, there are many volatile products of primary metabolism that are relatively ubiquitous. Phytophagous insects possess olfactory systems which detect plant volatiles and there is a widespread response to the green leaf volatiles (Visser, 1986). These are mainly 6-carbon compounds, e.g. ($E$)-2-hexenal and 1-hexanol, derived from unsaturated fatty acids. Such compounds are released from herbaceous plants and angiosperm trees but could be considered too universal a stimulus even for generalist herbivores. These compounds do, however, distinguish broad-leaf plants from conifers and have been shown to disrupt host-finding in pine and ambrosia beetles (Wilson et al., 1996; Deglow and Borden, 1998).

The volatile profile of plants also changes with age, physiological state, season and time of day (Visser, 1986; Agelopoulos et al., 1999) and seemingly offers an unreliable cue for host-specific behaviour. Insect attack can also change the volatile profile of a plant and these changed profiles can then influence the behaviour of parasites and predators (Powell and Poppy, Chapter 6). Nevertheless, the overall bouquet, or the ratio between a few key volatiles may be less variable and create a reliable signal for a phytophagous insect (Visser, 1986). Certainly insects are known to produce tactical responses to plant volatiles, e.g. Colorado potato beetle moves towards host odour (Visser and Avé, 1978), and olfactometer experiments with aphids show movement towards host and away from non-host odours (Nottingham et al., 1991). Aphids are also less responsive to a visually attractive target in the presence of a non-host odour, indicating an interaction between visual (see below) and olfactory responses (Nottingham and Hardie, 1993).

The most accurate information currently available on volatiles released from plants comes from air entrainment techniques (Blight, 1990) and these have been refined to access individual leaves of growing plants over short time periods (Agelopoulos et al., 1999). This provides for averaged concentrations over time but does not reveal the spatial composition. Given the studies indicating that male moth response to pheromones is crucially dependent upon odour *filament* structure and composition (see above), it appears likely that insect responses to plant odours will have similar requirements. Such an appreciation impacts on estimation of the signal-to-noise ratio. For sex pheromones, even the time-averaged background concentrations are extremely low and it is likely, for example, that only calling females of that species or a few closely related species produce the particular volatiles. Species can be distinguished and conspecific mating effected with mixtures of different pheromone components or different ratios. Plants, however, cover much of the earth's terrestrial surface and the average background concentrations of plant volatiles are high. There are two means by which insect herbivores might enhance the probability of locating the correct host by olfaction: one is to use a volatile specific to the host and the other is to utilize the ratio or mixture of key volatiles as described above. If the particular ratio is detected instantaneously in an odour filament by peripheral receptors on the antennae, then the time-averaged, apparent high background concentrations do not influence the insect's behavioural response. The latter is tuned specifically to the odour composition of the filaments which can easily be discriminated and there is some evidence that paired sensory cells in a single receptor hair can instantaneously detect specific volatile ratios (Tichy and Loftus, 1983; Blight *et al.*, 1995). Discrimination may also result from brain processing of the signals from different receptors (e.g. Laurent *et al.*, 1999).

When considering the detection of plant volatiles we need to appreciate that there is, in most cases, no point source, with a whole leaf/flower offering the smallest release site. Thus the odour plume will be larger initially than if released from a point source, but will divide into smaller elements with turbulence. Thus, if the phytophagous insect flies into an odour filament with the relevant volatile composition, an upwind, odour-modulated anemotaxis, as seen with male moths flying to female sex pheromone source, will provide the mechanism for source location that is not dependent upon the source dimensions. The fact that the overall size of the volatile source may be the size of a tree will not affect the efficiency of such a mechanism. Volatile profiles will differ between different parts (e.g. flower versus leaf) and allow for location of resources that are spatially separated on a single plant. How plant structures affect odour plumes is a largely unstudied area (Vogel, 1994). Some work on moths and beetles responding to pheromone sources on trees suggests that the tree has a significant effect on plume structure (Wyatt *et al.*, 1993, 1997; Willis *et al.*, 1994).

**Vision**

Vision is also important in host-plant location. Insects see the world differently to humans and most, or all, can detect wavelengths into the ultraviolet (350–400 nm) while we cannot detect wavelengths below 400 nm (blue). At the other end of the visible spectrum, insects and humans can detect wavelengths up to around 650 nm (red). There are exceptions to this with at least one forest beetle, *Melanophila acuminata* (De Geer), being reported to detect infrared wavelengths (>700 nm) which may assist them in locating recently burned forests with suitable oviposition sites (Schmitz *et al.*, 1997).

Visual stimuli provided by plants are colour and shape. Shape is highly variable throughout the year and will depend upon the position of the insect relative to the plant (Prokopy and Owens, 1983). Nevertheless, shapes are detected by insects and can provide cues as to plant suitability, and the shape of individual model leaves does influence landing and the choice of oviposition site by the carrot fly, *Psila rosae* (F.) (Degen and Städler, 1997). In this species, colour is also important for landing, as well as oviposition choice, and the texture of the model surface affected oviposition. Colour is relatively invariable for green leaves although age, senescence and pathogens can alter this, but it is well documented that many phytophagous insects land preferentially on yellow surfaces. Prokopy and Owens (1983) describe yellow as a super-normal foliage-type stimulus which reflects more energy in the wavelength region 500–580 nm characteristically reflected by green foliage (Fig. 5.4). For the main part, little UV is reflected by vegetation except for flowers but there are examples where leaves with different UV reflectances can assist the location of the correct host (Moericke, 1969).

Many insects can detect polarized light but this appears to be of little importance for host-plant location. Bees, for example, use polarized light for navigation but the receptors responsible are located in the dorsal region of the eye. Polarized sunlight reflected from plant leaves depends upon the angle of incidence and the cuticular wax covering. It would provide a highly variable and confusing signal (Land, 1993).

There are insects which specialize on plant parts which are not leaves and here visual cues of shape and colour may be crucial in resource location. Thus pollinating insects locate flowers by colour (including UV reflectance) and others oviposit in fruit and show distinct preference for shape and colour (e.g. the apple maggot fly, *Rhagoletis pomonella* (Walsh); Aluja and Prokopy, 1993). In these cases, wavelength reflectance is not used in isolation as both olfactory and visual cues are important (e.g. Reynolds and Prokopy, 1997).

The signal-to-noise ratio is also a crucial aspect of host-plant location for insects using visual cues. Thus catches of cabbage root flies, *D. radicum*, in water traps are dramatically affected by the background colour (Finch, 1995). Indeed, the observed reduction in pest numbers in brassica plants grown in bare soil compared to those undersown with clover may be a combination of odour-induced (by the correct host), but visually effected landing (possibly on the 'inappropriate' non-host). Thus, many cabbage root flies land on the clover and this reduces numbers observed on the crop plants (Finch and Keinegger, 1997).

## Concluding Remarks

This chapter highlights the principle role odours play in most resource-finding strategies. Although much has been learned from the male moth/sex pheromone system, it is not clear that this paradigm represents all resource-finding systems particularly well. Pheromones are released in discrete packets and involve a limited number of specific chemicals. Much needs to be elucidated about the filament structure of plant and host-animal odour plumes, especially from large sources such as fields of vegetation and herds of animals. There are many other opportunities for future research: the switch between ranging and orientated response, the control of response in specialists versus generalists and the interaction between the senses, to name but a few.

# References

Agelopoulos, N.G., Hooper, A.M., Maniar, S.P., Pickett, J.A. and Wadhams. L.J. (1999) A novel approach for isolation of volatile chemicals released by individual leaves of a plant *in situ*. *Journal of Chemical Ecology* 25, 1411–1425.

Aluja, M. and Prokopy, R.J. (1993) Host odor and visual stimulus interaction during intratree host finding behavior of *Rhagoletis pomonella* flies. *Journal of Chemical Ecology* 19, 2671–2696.

Baker, T.C. (1990) Upwind flight and casting flight: complementary phasic and tonic systems used for location of pheromone sources by male moths. In: Døving, K.B. (ed.) *Proceedings of the 10th International Symposium on Olfaction and Taste*. GCS A/S, Oslo, pp. 18–25.

Baker, T.C. and Haynes, K.F. (1987) Manoeuvres used by flying male oriental fruit moths to relocate a sex-pheromone plume in an experimentally shifted wind-field. *Physiological Entomology* 12, 263–279.

Baker, T.C. and Vickers, N.J. (1997) Pheromone-mediated flight in moths. In: Cardé, R.T. and Minks, A.K. (eds) *Insect Pheromone Research: New Directions*. Chapman & Hall, New York, pp. 248–264.

Baker, T.C., Fadamiro, H.Y. and Cosse, A.A. (1998) Moth uses fine tuning for odour resolution. *Nature* 393, 530.

Bell, W.J. and Tobin, T.R. (1982) Chemo-orientation. *Biological Reviews of the Cambridge Philosophical Society* 57, 219–260.

Belton, P. (1994) Attraction of male mosquitoes to sound. *Journal of the American Mosquito Control Association* 10, 297–301.

Bidlingmayer, W.L. (1974) How mosquitoes see traps: role of visual responses. *Journal of the American Mosquito Control Association* 10, 272–279.

Blight, M.M. (1990) Techniques for isolation and characterization of volatile semiochemicals of phytophagous insects. In: McCaffery, A.R. and Wilson, I.D. (eds) *Chromatography and Isolation of Insect Hormones and Pheromones*. Plenum Press, New York, pp. 281–288.

Blight, M.M., Pickett, J.A., Wadhams, L.J. and Woodcock, C.M. (1995) Antennal perception of oilseed rape, *Brassica napus* (Brassicaceae), by the cabbage seed weevil, *Ceutorhynchus assimilis* (Coleoptera, Curculonidae). *Journal of Chemical Ecology* 21, 1649–1664.

Brady, J., Gibson, G. and Packer, M.J. (1989) Odour movement, wind direction and the problem of host-finding by tsetse flies. *Physiological Entomology* 14, 369–380.

Campan, R. (1997) Tactic components in orientation. In: Lehrer. M. (ed.) *Orientation and Communication in Arthropods*. Birkhäuser Verlag, Basel, pp. 1–40.

Cardé, R.T. and Mafra-Neto, A. (1997) Mechanisms of flight of male moths to pheromone. In: Cardé, R.T. and Minks, A.K. (eds) *Insect Pheromone Research: New Directions*. Chapman & Hall, New York, pp. 275–290.

Clements, A.N. (1999) *The Biology of Mosquitoes*, Vol. 2. CAB International, Wallingford, UK.

Costantini, C., Gibson, G., Sagnon, N.F., Della Torre, A., Brady, J. and Coluzzi, M. (1996) Mosquito responses to carbon dioxide in a West African savanna village. *Medical and Veterinary Entomology* 10, 220–227.

David, C.T. (1986) Mechanisms of directional flight in wind. In: Payne, T.L., Birch, M.C. and Kennedy, C.E.J. (eds) *Mechanisms in Insect Olfaction*. Clarendon Press, Oxford.

David, C.T., Kennedy, J.S., Ludlow, A.R., Perry, J.N. and Wall, C. (1982) A reappraisal of insect flight towards a distant point source of wind-borne odor. *Journal of Chemical Ecology* 8, 1207–1215.

Degen, T. and Städler, E. (1997) Foliar form, colour and surface characteristics influence oviposition behaviour of the carrot fly. *Entomologia Experimentalis et Applicata* 83, 99–112.

Deglow, E.K. and Borden, J.H. (1998) Green leaf volatiles disrupt and enhance response to aggregation pheromones by the ambrosia beetle, *Gnathotrichus sulcatus*. *Recherche Forestiere* 28, 1696–1705.

Dingle, H. (1996) *Migration: the Biology of Life on the Move*. Oxford University Press, Oxford.

Dougherty, M.J., Guerin, P., Ward, R.D. and Hamilton, J.G.C. (1999) Behavioural and electrophysiological responses to the phlebotomine sandfly *Lutzomyia longipalpis* (Diptera: Psychodidae) when exposed to Canid host odour kairomones. *Physiological Entomology* 24, 251–262.

Dusenbury, D.B. (1992) *Sensory Ecology. How Organisms Acquire and Respond to Information*. W.H. Freeman, New York.

Finch, S. (1995) Effect of background on cabbage root fly landing and capture. *Entomologia Experimentalis et Applicata* 74, 201–208.

Finch, S. and Kienegger, M. (1997) A behavioural study to help clarify how undersowing with clover affects host–plant selection by pest insects of brassica crops. *Entomologia Experimentalis et Applicata* 84, 165–172.

Fraenkel, G.S. and Gunn, D.L. (1940) *The Orientation of Animals: Kineses, Taxes and Compass Reactions*. Oxford University Press, London.

Gibson, G. (1995) A behavioural test of the sensitivity of nocturnal mosquito *Anopheles gambiae* to dim white, red and infra-red light. *Physiological Entomology* 20, 224–228.

Gibson, G. and Torr, S.J. (1999) Visual and olfactory responses of haematophagous Diptera to host stimuli. *Medical and Veterinary Entomology* 13, 2–23.

Gibson, G. and Young, S. (1991) The optics of tsetse fly eyes in relation to their behaviour and ecology. *Physiological Entomology* 16, 273–282.

Gibson, G., Packer, M.J., Steullet, P. and Brady, J. (1991) Orientation of tsetse to wind, within and outside host odour plumes on the field. *Physiological Entomology* 16, 47–56.

Green, C.H. (1988) The effect of colour on trap- and screen-orientated responses in *Glossina palpalis palpalis* (Robineau-Desvoidy) (Diptera: Glossinidae). *Bulletin of Entomological Research* 78, 591–604.

Griffiths, N. and Brady, J. (1995) Wind structure in relation to odour plumes in tsetse fly habitats. *Physiological Entomology* 20, 286–292.

Hardie, J., Storer, J.R., Cook, F.J., Campbell, C.A.M., Wadhams, L.J., Lilley, R. and Peace, L. (1996) Sex pheromone and visual trap interactions in mate location strategies and aggregation by host-alternating aphids in the field. *Physiological Entomology* 21, 97–106.

Harris, M.O. and Foster, S.P. (1995) Behavior and integration. In: Cardé, R.T. and Bell, W.J. (eds) *Chemical Ecology of Insects 2*. Chapman & Hall, London, pp. 3–46.

Jander, R.J. (1975) Ecological aspects of spatial orientation. *Annual Review of Ecological Systems* 6, 171–188.
Kelly, D.W. and Dye, C. (1997) Pheromones, kairomones and the aggregation dynamics of the sandfly *Lutzomyia longipalpis*. *Animal Behaviour* 53, 721–731.
Kennedy, J.S. (1985) Migration, behavioural and ecological. In: Rankin, M.A. (ed.) *Migration: Mechanisms and Adaptive Significance. Contributions in Marine Science* 27 (supplement), 5–26.
Kennedy, J.S. (1986) Some current issues in orientation to odour sources. In: Payne, T.L., Birch, M.C. and Kennedy, C.E.J (eds) *Mechanisms in Insect Olfaction*. Clarendon Press, Oxford, pp. 11–25.
Kennedy, J.S. (1992) *The New Anthropomorphism*. Cambridge University Press, Cambridge.
Kennedy, J.S., Ludlow, A.R. and Sanders, C.J. (1980) Guidance system used in moth sex attraction. *Nature* 288, 475–477.
Krishnan, B., Dryer, S.E. and Hardin, P.E. (1999) Circadian rhythms in olfactory responses of *Drosophila melanogaster*. *Nature* 400, 375–378.
Land, M.F. (1993) Old twist in a new tale. *Nature* 363, 581–582.
Land, M.F. (1997) Visual acuity in insects. *Annual Review of Entomology* 42, 147–177.
Land, M.F., Gibson, G. and Horwood, J. (1997) Mosquito eye design: conical rhabdoms are matched to wide aperture lenses. *Proceedings of the Royal Society of London B* 264, 1183–1187.
Land, M.F., Gibson, G., Horwood, J. and Zeil, J. (1999) Fundamental differences in the optical structure of the ommatidia of nocturnal and diurnal mosquitoes. *Journal of Comparative Physiology A* 185, 91–103.
Laurent, G., MacLeod, K., Stopfer, M. and Wehr, M. (1999) Dynamic representation of odours by oscillating neural assemblies. *Entomologia Experimentalis et Applicata* 91, 7–18.
Mafra-Neto, A. and Cardé, R.T. (1994) Fine-scale structure of pheromone plumes modulates upwind orientation of flying moths. *Nature* 369, 142–144.
Meyhöfer, R. and Casas, J. (1999) Vibratory stimuli in host location by parasitic wasps. *Journal of Insect Physiology* 45, 967–971.
Millar, J.R. and Roelofs, W.L. (1978) Sustained-flight tunnel for measuring insect responses to windborne sex pheromones. *Journal of Chemical Ecology* 4, 187–198.
Moericke, V. (1969) Hostplant specific colour behaviour by *Hyalopterus pruni* (Aphididae). *Entomologia Experimentalis et Applicata* 12, 524–534.
Morton, I.E. and Ward, R.D. (1989) Laboratory response of female sandflies (*Lutzomyia longipalpis*) to a host and male pheromone source over distance. *Medical and Veterinary Entomology* 3, 219–223.
Murlis, J. (1997) Odor plumes and the signal they provide. In: Cardé, R.T. and Minks, A.K. (eds) *Insect Pheromone Research: New Directions*. Chapman & Hall, New York, pp. 221–231.
Murlis, J., Elkington, J.S. and Cardé, R.T. (1992) Odor plumes and how insects use them. *Annual Review of Entomology* 37, 505–532.
Nottingham, S.F. (1988) Host plant finding for oviposition by adult cabbage root fly, *Delia radicum*. *Journal of Insect Physiology* 34, 227–234.

Nottingham, S.F. and Hardie, J. (1993) Flight behaviour of the black bean aphid, *Aphis fabae*, and the cabbage aphid, *Brevicoryne brassicae*, in host and non-host plant odour. *Physiological Entomology* 18, 389–394.

Nottingham, S.F., Hardie, J., Dawson, G.W., Hick, A.J., Pickett, J.A., Wadhams, L.J. and Woodcock, C.M. (1991) Behavioural and electrophysiological responses of aphids to host and nonhost plant volatiles. *Journal of Chemical Ecology* 17, 1231–1242.

van der Pers, J.N.C. and Minks, A.K. (1997) Measuring pheromone dispersion in the field with the single sensillum recording technique. In: Cardé, R.T. and Minks, A.K. (eds) *Insect Pheromone Research: New Directions*. Chapman & Hall, New York, pp. 359–371.

Prokopy, R. and Owens, E.D. (1983) Visual detection of plants by herbivorous insects. *Annual Review of Entomology* 28, 337–364.

Reynolds, A.H. and Prokopy, R.J. (1997) Evaluation of odor lures for use with red sticky spheres to trap apple maggot (Diptera: Tephritidae). *Journal of Economic Entomology* 90, 1655–1660.

Sabelis, M.W. and Schippers, P. (1984) Variable wind directions and anemotactic strategies of searching for an odour plume. *Oecologia* 63, 225–228.

Schmitz, H., Mürtz, M. and Bleckmann, H. (1997) Infrared detection in a beetle. *Nature* 386, 773–774.

Tichy, H. and Loftus, R. (1983) Relative excitability of antennal olfactory receptors in the stick insect, *Carausius morosus* L.: in search of a simple, concentration-independent odor-coding parameter. *Journal of Comparative Physiology A* 152, 459–473.

Torr, S.J. (1988) The activation of resting tsetse flies (*Glossina*) in response to visual and olfactory stimuli in the field. *Physiological Entomology* 13, 315–325.

Torr, S.J. (1989) The host-oriented behaviour of tsetse flies (*Glossina*): the interaction of visual and olfactory stimuli. *Physiological Entomology* 14, 325–340.

Vaishampayan, S.M., Waldbauer, G.P. and Kogan, M. (1975) Visual and olfactory responses in orientation to plants by the greenhouse whitefly, *Trialeurodes vaporarium* (Homoptera: Aleyrodidae). *Entomologia Experimentalis et Applicata* 18, 412–422.

Vale, G.A. (1974) The responses of tsetse flies (Diptera: Glossinidae) to mobile and stationary baits. *Bulletin of Entomological Research* 64, 545–588.

Vickers, N.J. and Baker, T.C. (1994) Reiterative responses to single strands of odor promote sustained upwind flight and odor source location by moths. *Proceedings of the National Academy of Sciences USA* 91, 5756–5760.

Vickers, N.J. and Baker, T.C. (1996) Latencies of behavioural response of filaments of sex pheromone and clean air influence flight track shape in *Heliothis virescens* (F.) males. *Journal of Comparative Physiology A* 178, 831–847.

Visser, J.H. (1986) Host odor perception in phytophagous insects. *Annual Review of Entomology* 31, 121–144.

Visser, J.H. and Avé, D.A. (1978) General green leaf volatiles in the olfactory orientation of the Colarado beetle, *Leptinotarsa decemlineata*. *Entomologia Experimentalis et Applicata* 24, 538–549.

Vogel, S. (1994) *Life in Moving Fluids: the Physical Biology of Flow*, 2nd edn. Princeton University Press, Princeton, New Jersey.

Warnes, M.L. (1990) The effects of host odour and carbon dioxide on the flight of tsetse flies (*Glossina* spp.) in the laboratory. *Journal of Insect Physiology* 36, 607–611.

Willis, M.A., David, C.T., Murlis, J. and Cardé, R.T. (1994) Effects of pheromone plume structure and visual-stimuli on the pheromone-modulated upwind flight of male gypsy moths (*Lymantria-dispar*) in a forest (Lepidoptera, Lymantriidae). *Journal of Insect Behavior* 7, 385–409.

Wilson, I.M., Borden, J.H., Gries, R. and Gries, G. (1996) Green leaf volatiles as antiaggregants for the mountain pine beetle, *Dendroctonus ponderosae* Hopkins (Coleoptera: Scolytidae). *Journal of Chemical Ecology* 22, 1861–1875.

Wyatt, T.D. (2001) *Pheromones and Animal Behaviour: Communication by Smell and Taste*. Cambridge University Press, Cambridge.

Wyatt, T.D., Phillips, A.D.G. and Gregoire, J.C. (1993) Turbulence, trees and semiochemicals – wind-tunnel orientation of the predator, *Rhizophagus grandis*, to its barkbeetle prey, *Dendroctonus micans*. *Physiological Entomology* 18, 204–210.

Wyatt, T.D., Vastiau, K. and Birch, M.C. (1997) Orientation of flying male *Anobium punctatum* (Coleoptera: Anobiidae) to sex pheromone: separating effects of visual stimuli and physical barriers to wind. *Physiological Entomology* 22, 191–196.

Zanen, P.O., Sabelis, M.W., Buonaccorsi, J.P. and Cardé, R.T. (1994) Search strategies of fruit flies in steady and shifting winds in the absence of food odours. *Physiological Entomology* 19, 335–341.

# Host Location by Parasitoids 6

## Wilf Powell and Guy Poppy

*Entomology and Nematology Department, IACR-Rothamsted, Harpenden, Herts AL5 2JQ, UK*

## Introduction

Female parasitoids need to locate suitable hosts on or in which to lay their eggs. Foraging for hosts frequently requires extensive movements, especially by species which undergo solitary larval development and so have to locate a new host individual for each oviposition. Consequently, the majority of insect parasitoids are highly mobile, many of them being active fliers, although species with wingless and brachypterous females occur in some groups. The initial stages of the foraging process consist of host habitat and host location (Vinson, 1985) and these are the stages which often involve longer-range movements, usually by means of flight. Although short-range foraging movements, involving walking on host-infested substrates, occur once the parasitoid is in close proximity to its hosts, it is the foraging movements that bring female parasitoids into this close proximity, and the cues that govern these movements, that form the subject of this chapter.

At eclosion, the emerging female parasitoid will be confronted with one of several different situations with respect to host availability. If she is lucky, she will emerge directly into a host-occupied habitat and will not need to move very far to contact suitable hosts. Alternatively, she may emerge into an appropriate habitat but in an area locally devoid of hosts and so have to move considerable distances within that habitat to find them or, if she is very unlucky, she may find herself in the wrong habitat entirely. The chances of emerging in close proximity to appropriate hosts is influenced by a number of factors, including the biology of the parasitoid and the stability of the habitat in which its hosts occur. For example, species living in agricultural ecosystems and utilizing hosts on crop plants, particularly annual crops, are more likely to emerge into the wrong habitat or remote from host populations than are species operating in more stable natural and semi-natural habitats. Crops are harvested and often replaced by a completely different crop in the same field so that a parasitoid whose hosts feed on cruciferous plants, for example, could emerge into a cereal crop the following year. Even if emergence occurs in the right crop, the local host population may have disappeared or been drastically reduced by other agents, including chemical pesticides. Aphid parasitoids spin a cocoon inside the cuticle of their dead host to form a 'mummy', inside which they pupate, and this mummy case affords some protection from pesticides so that pupae which survive the treatment will emerge as adults into a habitat now devoid of hosts (Süss, 1983; Longley and Jepson, 1997; Longley, 1999). They will then rapidly disperse away from that habitat, the movement stimulated by a lack of available hosts and sometimes also by a repellent effect of pesticide residues on the foliage (Umoru et al., 1996).

The biology of the parasitoid itself is likely to have an effect on the situation in which it finds itself upon eclosion, and hence on the distance over which it may need to move to locate hosts. Multivoltine species such as aphid parasitoids, which can have generation times as short as 2–3 weeks, may have to adapt their foraging behaviour from generation to generation in response to changes in host and habitat availability. Most parasitoids of aphids are either oligophagous or polyphagous and may be forced to switch from one host species to another or one habitat to another (e.g. cereal crops to grass pasture and field margins) as the relative availability of different species and habitats changes through the course of the year. Many other parasitoids are univoltine, and some overwinter in pupal chambers in the soil, often as prepupae. In agricultural situations there is a significant chance of emerging into the wrong crop the following year, so host habitat location becomes very important for such species. Other species diapause as adults under bark or in grass tussocks (Gauld and Bolton, 1988), habitats that may be different from those likely to contain hosts the following spring, again necessitating considerable host location movements.

Parasitoid movements to locate hosts and host habitats are directed by their behavioural responses to environmental stimuli, which are predominantly olfactory during these early stages of foraging, although other stimuli such as visual cues can play a role at close range. To cope with changing conditions and so optimize their use of available resources, many parasitoids have evolved a remarkable degree of behavioural plasticity, modifying their responses to foraging cues through learning processes, as a result of experience (Vinson, 1984; Vet and Dicke, 1992; Tumlinson et al., 1993; Turlings et al., 1993; Vet et al., 1995). Studies of a wide range of host–parasitoid systems has revealed the importance of chemical information in mediating parasitoid foraging movements. It is not possible to review all this work here and so we will use two parasitoid systems as examples: parasitoids attacking aphids and parasitoids attacking lepidopteran hosts.

## Sources of Chemical Information Used for Host and Host Habitat Location

Female parasitoids use a variety of chemical cues whilst foraging for hosts, and this chemical information can originate from the host itself, from the host plant, in the case of herbivorous hosts or from sources associated with the host such as host excreta. The early stages of host location, which require long-range movement, are mediated by volatile chemicals, whereas close-range foraging often also involves contact chemical cues. The foraging parasitoid has to detect relevant chemical information amongst a vast array of volatile compounds present in the surrounding air. In so doing, it faces what has been called the 'reliability–detectability' problem (Vet and Dicke, 1992). Plants release a lot of volatile chemicals, providing an abundant supply of chemical information which is readily detectable, especially when the plant is a crop plant growing in monoculture. However, as this information does not emanate from the host itself, it is unreliable in indicating host presence even though it might lead the parasitoid to an appropriate habitat. Conversely, herbivorous arthropods generally exude much smaller amounts of volatiles, which are consequently more difficult to detect, but as they emanate from the host itself are much more reliable in indicating host presence. In the development of their theory, Vet and Dicke (1992) argue that in order to maximize searching efficiency and consequently fitness, parasitoids would most profitably respond to chemical information that is both reliable and easy to detect. Happily for the foraging female parasitoid, there are a number of ways in which this can be achieved:

**1.** By responding to more conspicuous chemical cues produced by stages different from the one under attack, e.g. Lepidopteran egg parasitoids utilizing the sex pheromone of adult females;
**2.** By responding to host-induced plant volatiles, generated by specific interactions between herbivores and their food plants, thus combining the detectability of plant cues with the reliability of host presence;
**3.** By learning to link detectable cues with reliable cues through associative learning processes.

All three of these mechanisms are discussed in this chapter, with emphasis on their role in parasitoid host location.

## Plant Volatiles and Herbivore-induced Plant Volatiles

The ability of parasitoids to locate their hosts by using volatile chemicals that are emitted by the plant in response to herbivore damage is well documented for parasitoids attacking a range of insect hosts, principally Lepidoptera (Turlings et al., 1990, 1995; Dicke, 1998) and Homoptera (Powell et al., 1998). This herbivore-induced volatile release has been termed plant signalling and is regarded as an indirect defence reaction by the plant (Chadwick and Goode, 1999). Foraging female parasitoids usually do not show significant flight responses to uninfested plants, especially when these are compared against host-infested plants in choice experiments. However, almost all of the research on the role of plant signalling in host location by parasitoids has been conducted in a laboratory environment. Therefore, the uninfested plants tested have never been exposed to any insects/pathogens and are kept under conditions of optimum nutrition and temperature, albeit with non-optimum lighting, which is obviously a rare phenomenon in the field. There is now an urgent need to conduct more field-based studies in order to elucidate further the functioning of plant signalling and its adaptive significance for parasitoid host location under more realistic ecological conditions.

### Are there control plants in the field?

To study parasitoid host location in the laboratory it is important to define closely and standardize plant treatments to ensure replicability. In the case of wind tunnel bioassays, this often means giving the parasitoid a choice between plants with a precise level of herbivore infestation and 'clean' uninfested plants which have never been exposed to any herbivory (Guerrieri et al., 1999). The age, growth stage and other biotic and abiotic conditions are standardized as far as possible as would be expected in any careful laboratory experiment investigating functional mechanisms. However, plants in the field are constantly being challenged by a range of biotic and abiotic factors and whether an undamaged 'control' plant, in the sense of laboratory standards, actually ever exists under natural conditions is questionable. This constant challenge could, in fact, make the task of host location easier for parasitoids. If plants are constantly adjusting to a range of stresses (e.g. buffeting by the wind), this may produce a uniform level of background chemical 'noise' against which a signal produced by a new challenge, namely herbivory, may be easier to detect. The concepts of signal/noise ratios and mechanisms of maximizing response are discussed in more detail by Hardie et al. (Chapter 5).

Plant pathologists have clearly demonstrated that plants attacked by pathogens develop a 'memory/sensitization', which leads to more effective induced resistance on subsequent pathogen challenges (Lyon and Newton, 1999). Whether or not plants develop a similar memory of herbivory, which could lead to dramatic changes in the dynamics of the signal and the behavioural response of parasitoids, has not yet been fully investigated. Although temporal patterns of signal production have been demonstrated in cotton and corn infested by lepidopteran larvae (Paré and Tumlinson, 1997, 1998) and in broad bean infested by aphids (Guerrieri et al., 1999), the effects of re-infestation of previously infested plants have not been studied, even in laboratory experiments. Plant 'memory/sensitization' may also increase the detectability of the signal produced by herbivore-infested plants in the field. There is also very little known about potential interactions between direct and indirect defence mechanisms at different stages of plant growth. It has been shown in the wild plant *Nicotiana* that direct defences (production of nitrogen-based nicotine) are switched off during flowering and indirect defences involving volatiles that attract parasitoids become more important (Baldwin, 1999). This balancing of more resource-expensive but reliable direct defences and the less resource-expensive and less reliable indirect defences of herbivore-induced volatiles is only just beginning to be investigated. Only by studying plants in more natural habitats will we begin to understand the trade-offs between direct and indirect defences in plants and thus the real functional significance of herbivore-induced volatiles.

In spite of concerns about whether responses to herbivore-induced volatiles observed in laboratory bioassays are directly pertinent to parasitoid foraging behaviour in the field, recent work has demonstrated that parasitoids can use induced plant volatiles to locate their hosts under field conditions and so these plant signals can be important in plant defence (de Moraes et al., 1998; Thaler 1999a,b). Recent work in our laboratory used a glasshouse arena to investigate host location by the aphid parasitoid *Aphidius ervi* Haliday in a more complex and realistic environment than that provided by wind-tunnel bioassays. A cohort of 50 aphids was placed on a broad bean plant that had been previously infested by several hundred aphids for several days, this previous infestation having been removed just prior to placing the new cohort on the plants. An identical cohort of aphids was placed on a previously uninfested plant. These two plants were placed in the glasshouse and positioned at random amongst a further 14 plants which remained uninfested. Female parasitoids were then released into the glasshouse for 24 h and any resulting parasitized aphids recorded. Newly infested plants, which had not been infested long enough to induce plant signalling, were never located by the foraging parasitoids, whereas the previously infested plants, which were assumed to be releasing aphid-induced volatiles, were frequently located (Poppy, unpublished results).

## Plant signal or plant by-product?

There is some current debate about whether herbivore-induced volatiles are 'distress signals', actively produced by the plant specifically to recruit parasitoids (Bruin et al., 1995), or whether they are simply a by-product of herbivorous feeding, to which parasitoids have developed a response, leading to their selection over evolutionary time (Poppy, 1999). It is clear that parasitoids exploit these signals very efficiently and that plants appear to benefit from producing such signals. The debate about the specificity of the signals, i.e. plants producing different signals in response to different herbivores (Dicke, 1999; Tumlinson et al., 1999; Vet, 1999), may suggest that the plant is producing context-dependent signals, although if the herbivores are differentially altering plant chemistry, for nutritional benefits or to combat direct defence mechanisms, then this specificity could have arisen as a by-product. There have been a number of recent demonstrations of such differential alteration of plant quality by aphids (Williams et al., 1998), and we are currently investigating whether this occurs in *Vicia faba*, which is known to produce a different signal in response to different aphid species (Du et al., 1996, 1998; Guerrieri et al., 1999).

Recently, a chemical elicitor, volicitin, has been isolated from the saliva of *Spodoptera exigua* Hübner caterpillars (Alborn et al., 1997), and it has been shown that this substance can induce the same volatile profile release from plants as does actual caterpillar attack (Alborn et al., 1997; Paré et al., 1998). Interestingly, volicitin does not appear to have a function for the caterpillar, yet part of the molecule is linolenic acid which originates in the plant. Thus, the plant could be generating the signal by producing a molecule within the caterpillar saliva, which then acts as an elicitor for the signal. Tumlinson's group have determined four biochemical pathways which are involved in volatile induction in the plant and, in a series of elegant experiments using cold-labelled $C^{13}O_2$, have shown in cotton that many of the induced volatiles are synthesized *de novo* after herbivore attack, whereas some other important compounds are either synthesized from non-labelled precursors or are released from storage (Tumlinson et al., 1999). The active involvement of photosynthesis and the increased output of volatiles with increased light intensity are further evidence of the active nature of signal generation in cotton and corn.

The situation with aphids is even less clear, although perhaps of more interest due to the minimal mechanical damage caused by aphids during host-plant feeding. In fact, some researchers compare aphids to pathogens and it is clear that many species have very intimate relationships with their host-plants. It has been suggested that aphids may actively be able to turn plant signals on and off (Schultz, quoted in Wadhams et al., 1999). The clonal nature of aphid populations, their telescopic generations, and the range of generalism–specialism found amongst both aphids and their parasitoids at different trophic levels, make the aphid parasitoid system the ideal model for investigating the proximate and adaptive function of herbivore-induced plant volatiles and their importance in parasitoid-host location. The relative importance of these cues in determining parasitoid movement patterns and the scale over which they operate remain to be elucidated in the field, but this will be a fruitful and exciting research avenue for the future.

## Host Pheromones

### Chemical espionage

Herbivores must defend themselves against attack by predators and parasitoids. One of the most effective means of defence is to remain as inconspicuous as possible and this tactic is most obvious in animals which employ physical camouflage to avoid visual detection. But, as many of their natural enemies use olfactory methods of prey location, herbivorous arthropods must also minimize the release of chemical information that would betray their presence and location. Nevertheless, most species still rely on olfactory means of intraspecific communication, in the form of pheromones. Pheromones have evolved as highly efficient mechanisms for communication over long distances and, therefore, provide an ideal opportunity for exploitation by parasitoids. Parasitoids are known to use a number of different host pheromones to aid host location, including sex pheromones, aggregation pheromones, alarm pheromones and spacing (epideictic) pheromones (Powell, 1999). This behaviour has been termed 'chemical espionage' or 'chemical eavesdropping' (Stowe et al., 1995). As yet, there are only a few known examples of parasitoids responding to host alarm pheromones or to epideictic (spacing) pheromones, but parasitoid responses to sex pheromones and aggregation pheromones are more regularly reported (Powell, 1999). Good evidence for the use of aggregation pheromones as host-location kairomones exists for parasitoids attacking bark beetles (Kennedy, 1979, 1984; Seybold et al., 1992), heteropteran bugs (Harris and Todd, 1980; Aldrich, 1995; Leal et al., 1995) and fruit flies (Wiskerke et al., 1993; Hedlund et al., 1996).

Sex pheromones are successfully exploited by parasitoids searching for two major host types: insect eggs and homopteran plant pests, such as scale insects and aphids, which often occur as aggregated colonies (Powell, 1999). Parasitoids that attack the larval and pupal stages of their hosts are unlikely to develop responses to chemical cues emanating from reproductive adult stages, unless their hosts occur as mixed-age populations, with broadly overlapping generations, living together in or on the same substrate.

### Aphid sex pheromones

It has been known for more than 25 years that aphelinid parasitoids of scale insects are attracted to the sex pheromones of their hosts (Sternlicht, 1973), but evidence for aphid parasitoids showing a similar response is more recent

(Hardie *et al.*, 1991, 1994; Powell *et al.*, 1993, 1998; Gabrys *et al.*, 1997; Glinwood *et al.*, 1999a,b). Although aphids exist as all-female, parthenogenic populations for most of the time, many species in cool, temperate regions are holocyclic, going through a sexual generation in the autumn to produce overwintering eggs. The sexual female morph of these aphids releases a sex pheromone to attract the winged males for mating, and the main chemical components of the sex pheromones of a range of species have been chemically identified as (+)-(4a$S$, 7$S$, 7a$R$)-nepetalactone and (−)−(1$R$, 4a$S$, 7$S$, 7a$R$)-nepetalactol or stereoisomers of the latter (Dawson *et al.*, 1987, 1990; Pickett *et al.*, 1992; Hardie *et al.*, 1999).

In initial field trials, using vials releasing aphid sex pheromone components attached to Petri-dish water traps, only parasitoid species of the genus *Praon* appeared to respond, the strongest response being to the nepetalactone component (Hardie *et al.*, 1991, 1994; Powell *et al.*, 1993). Pheromone water traps placed in winter wheat crops in autumn caught large numbers of *P. volucre*, a species which frequently attacks cereal aphids but, interestingly, the position of the trap in relation to the crop vegetation had a very significant effect on trap catches; traps placed immediately above the top of the young wheat plants caught 540 females whereas traps placed 1 m above the plants only caught 13 (Powell *et al.*, 1993). This suggests that aphid parasitoid movement at this stage of the foraging process occurs very close to the surface of the vegetation. In another study, direct observations of male *P. volucre* and *Aphidius rhopalosiphi* De Stefani–Perez, flying towards caged virgin females placed in cereal crops, revealed a movement pattern consisting of short flights close to the crop canopy alternating with brief landings at the top of plants, during which they appeared to test the air with their antennae (Decker *et al.*, 1993). Apparently, the parasitoids were assessing the presence and direction of the source of volatile chemical cues, in this case pheromones from the caged females, at frequent intervals.

Although only *Praon* females were caught in significant numbers by pheromone traps in these field trials, females from a range of other species gave obvious behavioural responses in laboratory wind tunnel and olfactometer bioassays (Hardie *et al.*, 1993; Glinwood *et al.*, 1999a,b). Electrophysiological responses to pheromone components, especially nepetalactone, were also obtained for a variety of other species (Hardie *et al.*, 1993). Responses to the pheromones by *Aphidius* species were confirmed in further field studies using aphid-infested potted plants, with and without pheromone-releasing vials attached, indicated by greatly increased parasitization rates in the presence of the pheromone (Glinwood *et al.*, 1999b; W. Powell, unpublished data). In these trials, nepetalactone significantly increased parasitization of the cereal aphid *Sitobion avenae* (Fab.) by both the generalist *P. volucre* and the cereal aphid specialist *A. rhopalosiphi*. These results raised the question of why *Aphidius* females were never caught by pheromone water traps in the same way as *Praon* females. This could be explained if the pheromone affects the foraging movements of *Praon* and *Aphidius* in different ways. The evidence suggests that the pheromone acts principally as an attractant for *Praon*, the parasitoids orienting to the point source of the pheromone before commencing the next stage of the foraging process, which involves walking on the substrate. In contrast, there appears to be more of an arrestant effect of the pheromone on *Aphidius*, the parasitoids landing on the substrate in the general vicinity of the pheromone source without flying directly to the point source, and so seldom encountering the water trap. If this hypothesis is correct, the use of sex pheromones to manipulate the foraging movements of aphid parasitoids (Powell *et al.*, 1998) may have a better chance of success with species showing the *Aphidius*-type responses.

## Lepidopteran sex pheromones

Sex pheromones are most frequently produced by female insects and so are potentially valuable host-location cues for egg parasitoids because female hosts often mate and oviposit in the same location (Noldus and van Lenteren, 1985). Egg parasitoids, particularly *Trichogramma* species, have been widely used in biological and integrated control strategies against lepidopteran pests on a range of arable, orchard and forest crops (Li, 1994) and host sex pheromones offer opportunities for their manipulation. However, as in the case of aphid parasitoids, we still have little idea about the distances over which lepidopteran sex pheromones can affect egg parasitoid foraging behaviour. Detailed laboratory studies have revealed parasitoid responses to host sex pheromones, such as increased walking times and path lengths as well as reduced walking speeds, increased turning and longer residence times, which are consistent with arrestment behaviour (Noldus, 1988; Noldus *et al.*, 1990, 1991a), but these studies focused mainly on short-range responses. Results of some field studies, using pheromone traps, have suggested that an attraction effect, consistent with longer-range movements, is also involved (Battisti, 1989; Arakaki *et al.*, 1996). In their primary role of intraspecific communication, most sex pheromones have evolved as long-range attractants and so it would not be surprising for them to function in the same way in their role as kairomones. It is likely that host pheromones, and indeed some other semiochemical foraging cues, could act both as attractants and arrestants, the type of behavioural response shown by the parasitoid depending upon the stage that it has reached in the host location process. More work on the movements of parasitoids in the field, especially in response to olfactory cues, is urgently needed to elucidate response patterns, so that these cues can be deployed more effectively in pest control strategies.

The value of female sex pheromones as host-location cues for egg parasitoids is, of course, drastically reduced if the host species does not oviposit at the mating site, as is the case for the pine processionary moth *Thaumetopoea pityocampa* Denis and Schiffermüller (Demolin, 1969). Some parasitoids have circumvented this problem by adopting the habit of phoresy; they locate virgin female hosts by means of their sex pheromones and then move to the oviposition site by attaching themselves to the body of the host. The encyrtid egg parasitoid *Ooencyrtus pityocampae* Mercet is phoretic and was attracted to pheromone traps releasing pityolure, the synthetic sex pheromone of *T. pityocampa*, whereas the non-phoretic eulophid *Tetrastichus servadeii* Domenichini, which also parasitizes *T. pityocampa* eggs, was never caught in host pheromone traps (Battisti, 1989). Similarly, the scelionid *Telenomus euproctidis* Wilcox is phoretic on *Euproctis*-species moths, the eggs of which it parasitizes, and was attracted to virgin female moths and to traps releasing synthetic components of host sex pheromones when these were placed in the field (Arakaki *et al.*, 1996, 1997).

Another constraint to the use of moth sex pheromones as host-location kairomones by parasitoids is temporal association; many moths release their sex pheromones at night whereas most parasitoids forage during the day. One hypothesis to explain this incongruity is that the pheromones are adsorbed onto the substrate, usually leaf surfaces, adjacent to the calling female moth, from which they gradually re-volatilize during the following day (Noldus and van Lenteren, 1985). In laboratory experiments, female *Trichogramma evanescens* Westwood showed arrestment responses when walking on Brussels sprout leaves that had been exposed to air immediately after it had passed over calling female host moths, an effect which lasted for several hours (Noldus *et al.*, 1991b). When the main component of the sex pheromone of the blackheaded fireworm *Rhopobota naevana* (Hübner) was adsorbed on to cranberry leaves by passive diffusion, the egg parasitoid *Trichogramma sibericum* Sorokina increased its searching and residence times on the treated foliage (McGregor and Henderson, 1998), further supporting the hypothesis.

## Factors Determining Parasitoid Responses to Semiochemicals

Parasitoid foraging behaviour is known to be influenced by interactions of genetic, physiological, environmental and experiential factors. The comparative importance of genetics, developmental conditioning and adult learning in determining foraging responses to chemical cues, and thus host location, by parasitoids is currently receiving considerable attention (van Emden *et al.*, 1996; Poppy *et al.*, 1997). In this section, we will describe examples of how each of these factors can influence the parasitoid response.

### Genetics

Theoretically, innate responses to host-derived cues should be most prevalent amongst specialist parasitoids with limited host ranges (Vet and

Dicke, 1992). Most of the research on the host-location behaviour of lepidopteran parasitoids has involved parasitoids that have been given an oviposition experience before their use in bioassays (Potting *et al.*, 1999) because naïve parasitoids often show low responses (T.C.J. Turlings, personal communication). The few studies which have suggested the existence of an innate response are rarely able to differentiate between a genetically determined response and a response conditioned during development or adult eclosion (see next section). However, one of the few responses that are known to be genetically fixed is the response of aphid parasitoids to aphid sex pheromone components (Poppy *et al.*, 1997; Glinwood, 1998; Glinwood *et al.*, 1999a,b). Behavioural bioassays demonstrating this response used parasitoids that were collected from and subsequently reared in asexual aphids. Therefore, these parasitoids had never been exposed to the aphid sex pheromone either during development or as emerging adults, and so were demonstrating a true innate response to the pheromones. Why should parasitoids respond to aphid sex pheromones when this chemical information is only available for a short period in the autumn? The most likely explanation is the co-location of parasitoids with their hosts before the onset of winter, although the response could be an evolutionary hangover from before the development of parthenogenesis as the dominant mode of reproduction in aphids.

## Conditioning

The term conditioning has been used in several different ways in insect behavioural studies. In this chapter, it is used to denote parasitoid behavioural responses to chemical cues that are acquired during larval/pupal development or at adult eclosion, but not responses which are learnt by the adult parasitoid during active foraging. Hopkins (1917) proposed that the chemical experience of the larva of an endopterygote insect can be transferred through the pupa to affect the response to the chemical by the adult. It is rather appropriate that Hopkins first proposed his theory during a discussion at a conference. 'Hopkins' host-selection principle' has subsequently been the topic of much discussion but there have been few conclusive demonstrations of its validity. In fact, Corbet (1985) proposed the 'Chemical Legacy Hypothesis', which suggested that the most likely explanation for apparent 'larval learning' is contact with plant compounds on the cocoon or pupal case during adult eclosion, and pointed out that most studies have failed to demonstrate that appropriate neural changes occur during development, which is viewed as essential to uphold Hopkins' principle (Jaenicke, 1982). Van Emden *et al.* (1996) conducted a number of clever experiments to see whether the host-location behaviour of aphid parasitoids could provide evidence for Hopkins' principle. They concluded that adult host-location behaviour was influenced by chemical information obtained from the mummy case during emergence of the adult and that Hopkins' host-selection principle was therefore unsupported.

It is important to remember that a parasitoid emerges from a host/host-plant situation that was determined by its mother at the time of oviposition, and this in turn determines its initial exposure to chemical information, encountered on the pupal/mummy case at eclosion. Therefore, the 'naïve' parasitoid has had some chemical experience even before it encounters either hosts or host plants. How important this initial experience is compared to subsequent adult learning experiences is currently under investigation. We regard this influence on adult parasitoid behaviour of exposure to chemical information during eclosion as a kind of conditioning, which we refer to as 'emergence conditioning', as distinct from true learning. These 'conditioned' responses, if they prove unrewarding, can be rapidly overridden during subsequent adult foraging experiences, emphasizing the behavioural plasticity of parasitoids.

**Adult learning**

The role of learning in insect behaviour is well recognized (Papaj and Lewis, 1993) and there have been a number of important reviews on learning in parasitoids (e.g. Turlings *et al.*, 1993; Vet *et al.*, 1995). Learning can strongly influence host location and thus affect parasitoid foraging movements. Learning and its role in host location has been most extensively studied in parasitoids attacking lepidopteran and dipteran hosts but recent work has highlighted the importance of learning for aphid parasitoids (Sheehan and Shelton, 1989; Grasswitz and Paine, 1993; Du *et al.*, 1997; Guerrieri *et al.*, 1997; Powell *et al.*, 1998). Rather than become engulfed in the terminology surrounding learning, we will describe aspects of learning which may influence the movement of parasitoids and thus ensure that they locate and parasitize suitable hosts.

Variation in the foraging behaviour of individuals can have an important influence on parasitoid population dynamics (Vet, 1995). Most studies on learning have concentrated on olfactory learning and its influence on host-location behaviour, and have shown that the behavioural preferences for plant/host combinations can be enhanced or even altered by learning. Returning to the scenario of an aphid parasitoid emerging from its mummy case, this parasitoid will potentially be influenced by its genotype and by any chemical cues present on the mummy case. This 'naïve' parasitoid will then utilize any innate or conditioned cues to try and locate a host. If it successfully finds a host, then learning has the potential to reinforce these cues and thus possibly lead to increased and faster responses (Du *et al.*, 1997). However, if it does not find these hosts or encounters a different host/plant complex, then this new experience may be positive or negative according to the host/plant range of the parasitoid. If this new host/plant combination is suitable, then the parasitoid can exploit this new host source and learning the associated chemical cues will enhance its ability to locate and attack this new resource. Such behavioural plasticity is a significant benefit to the parasitoid.

The interaction between genetics, 'conditioning' and learning offers the parasitoid considerable flexibility in its foraging movements and the ability to exploit hosts as and when they become available. In all but the most specialized parasitoids, plasticity of the behavioural response allows rapid exploitation of alternative hosts and/or host-plants, when the original hosts or host-plants become unrewarding foraging resources. Rapid learning abilities by parasitoids have been convincingly demonstrated in the laboratory but the real challenge is to demonstrate their importance in the field. If we can understand how these mechanisms act and interact in the field, then the door is open for manipulating these mechanisms to our advantage and ensuring that the balance between parasitoids and pests is tipped towards parasitoids and thus control of the pests.

## Conclusions

Movement plays a very important role in the biology of parasitoids, the females of which spend a large proportion of their adult lives foraging for hosts, and the pattern of this movement is largely governed by behavioural responses to environmental cues, particularly semiochemical information. Therefore, the behavioural responses of individuals to these cues are likely to have a major influence both on the spatial and temporal distribution patterns of the parasitoid population and on the genetic structure of populations through gene flow amongst host- or habitat-associated subpopulations. Regrettably, we still know very little about parasitoid foraging movements in the field, in either agricultural or natural ecosystems, or about the effects of movement on parasitoid population biology. Nevertheless, an understanding of these processes is important if we are to develop more effective strategies for both the conservation of parasitoid communities and their use as biological control agents of insect pests in field crops.

## References

Alborn, H.T., Turlings, T.C.J., Jones, T.H., Stenhagen, G., Loughrin, J.H. and Tumlinson, J.H. (1997) An elicitor of plant volatiles from beet armyworm oral secretion. *Science* 276, 945–949.

Aldrich, J.R. (1995) Chemical communication in the true bugs and parasitoid exploitation. In: Cardé, R.T. and Bell, W.J. (eds) *Chemical Ecology of Insects 2*. Chapman & Hall, New York, pp. 318–363.

Arakaki, N., Wakamura, S. and Yasuda T. (1996) Phoretic egg parasitoid, *Telenomus euproctidis* (Hymenoptera: Scelionidae), uses sex pheromone of Tussock moth

*Euproctis taiwana* (Lepidoptera: Lymantriidae) as a kairomone. *Journal of Chemical Ecology* 22, 1079–1085.

Arakaki, N., Wakamura, S., Yasuda, T. and Yamagishi, K. (1997) Two regional strains of a phoretic egg parasitoid, *Telenomus euproctidis* (Hymenoptera: Scelionidae), that use different sex pheromones of two allopatric Tussock moth species as kairomones. *Journal of Chemical Ecology* 23, 153–161.

Baldwin, I.T. (1999) Functional interactions in the use of direct and indirect defences in native *Nicotiana* plants. In: Chadwick, D.J. and Goode, J.A. (eds) *Insect–Plant Interactions and Induced Plant Defence*. John Wiley & Sons, Chichester, pp. 74–94.

Battisti, A. (1989) Field studies on the behaviour of two egg parasitoids of the pine processionary moth *Thaumetopoea pityocampa*. *Entomophaga* 34, 29–38.

Bruin, J., Sabelis, M.W. and Dicke, M. (1995) Do plants tap SOS signals from their infested neighbours? *Trends in Ecology and Evolution* 10, 167–170.

Chadwick, D.J. and Goode, J.A. (eds) (1999) *Insect–Plant Interactions and Induced Plant Defence*. John Wiley & Sons, Chichester.

Corbet, S.A. (1985) Insect chemosensory responses: a chemical legacy hypothesis. *Ecological Entomology* 10, 977–980.

Dawson, G.W., Griffiths, D.C., Janes, N.F., Mudd, A., Pickett, J.A., Wadhams, L.J. and Woodcock, C.M. (1987) Identification of an aphid sex pheromone. *Nature* 325, 614–616.

Dawson, G.W., Griffiths, D.C., Merritt, L.A., Mudd, A., Pickett, J.A., Wadhams, L.J. and Woodcock, C.M. (1990) Aphid semiochemicals – a review, and recent advances on the sex pheromone. *Journal of Chemical Ecology* 16, 3019–3030.

Decker, U.M., Powell, W. and Clark, S.J. (1993) Sex pheromones in the cereal aphid parasitoids *Praon volucre* and *Aphidius rhopalosiphi*. *Entomologia Experimentalis et Applicata* 69, 33–39.

Demolin, G. (1969) Comportement des adultes de *Thaumetopoea pityocampa* Schiff. Dispersion spatiale, importance écologique. *Annales des Sciences Forestières* 26, 81–102.

Dicke, M. (1998) Evolution of indirect defence of plants. In: Harvell, C.D. and Tollren, R. (eds) *The Ecology and Evolution of Inducible Defences*. Princeton University Press, Princeton, New Jersey, pp. 62–88

Dicke, M. (1999) Specificity of herbivore-induced plant defences. In: Chadwick, D.J. and Goode, J.A. (eds) *Insect–Plant Interactions and Induced Plant Defence*. John Wiley & Sons, Chichester, pp. 43–59.

Du, Y.-J., Poppy, G.M. and Powell, W. (1996) Relative importance of semiochemicals from first and second trophic levels in host foraging behavior of *Aphidius ervi*. *Journal of Chemical Ecology* 22, 1591–1605.

Du, Y.-J., Poppy, G.M., Powell, W. and Wadhams, L.J. (1997) Chemically mediated associative learning in the host foraging behavior of the aphid parasitoid *Aphidius ervi* (Hymenoptera: Braconidae). *Journal of Insect Behavior* 10, 509–522.

Du, Y.-J., Poppy, G.M., Powell, W., Pickett, J.A., Wadhams, L.J. and Woodcock, C.M. (1998) Identification of semiochemicals released during aphid feeding that attract parasitoid *Aphidius ervi*. *Journal of Chemical Ecology* 24, 1355–1368.

van Emden, H.F., Sponagl, B., Wagner, E., Baker, T., Ganguly, S. and Douloumpaka, S. (1996) Hopkin's 'host selection principle', another nail in its coffin. *Physiological Entomology* 21, 325–328.

Gabrys, B.J., Gadomski, H.J., Klukowski, Z., Pickett, J.A., Sabota, G.T., Wadhams, L.J. and Woodcock, C.M. (1997) Sex pheromone of cabbage aphid *Brevicoryne brassicae*: identification and field trapping of male aphids and parasitoids. *Journal of Chemical Ecology* 23, 1881–1890.

Gauld, I. and Bolton, B. (1988) *The Hymenoptera*. Oxford University Press, New York.

Glinwood, R.T. (1998) Responses of aphid parasitoids to aphid sex pheromones: laboratory and field studies. Ph.D. Thesis, University of Nottingham.

Glinwood, R.T., Du, Y.-J., Smiley, D.W.M. and Powell, W. (1999a) Comparative responses of parasitoids to synthetic and plant-extracted nepetalactone component of aphid sex pheromones. *Journal of Chemical Ecology* 25, 1481–1488.

Glinwood, R.T., Du, Y.-J. and Powell, W. (1999b) Responses to aphid sex pheromones by the pea aphid parasitoids *Aphidius ervi* and *Aphidius eadyi*. *Entomologia Experimentalis et Applicata* 92, 227–232.

Grasswitz, T.R. and Paine, T.D. (1993) Effect of experience on in-flight orientation to host-associated cues in the generalist parasitoid *Lysiphlebus testaceipes*. *Entomologia Experimentalis et Applicata* 68, 219–229.

Guerrieri, E., Pennacchio, F. and Tremblay, E. (1997) Effect of adult experience on in-flight orientation to plant and plant-host complex volatiles in *Aphidius ervi* Haliday (Hymenoptera, Braconidae). *Biological Control* 10, 159–165.

Guerrieri, E., Poppy, G.M., Powell, W., Tremblay, E. and Pennacchio, F. (1999) Induction and systemic release of herbivore-induced plant volatiles mediating in-flight orientation of *Aphidius ervi*. *Journal of Chemical Ecology* 25, 1247–1261.

Hardie, J., Nottingham, S.F., Powell, W. and Wadhams, L.J. (1991) Synthetic aphid sex pheromone lures female parasitoids. *Entomologia Experimentalis et Applicata* 61, 97–99.

Hardie, J., Isaacs, R., Nazzi, F., Powell, W., Wadhams, L.J. and Woodcock, C.M. (1993) Electroantennogram and olfactometer responses of aphid parasitoids to nepetalactone, a component of aphid sex pheromones. *Abstracts of the IOBC Global Working Group on Aphidophaga Symposium on Ecology of Aphidophaga*, La Colle sur Loup, France, September, p. 29.

Hardie, J., Hick, A.J., Höller, C., Mann, J., Merritt, L., Nottingham, S.F., Powell, W., Wadhams, L.J., Witthinrich, J. and Wright, A.F. (1994) The responses of *Praon* spp. parasitoids to aphid sex pheromone components in the field. *Entomologia Experimentalis et Applicata* 71, 95–99.

Hardie, J., Pickett, J.A., Pow, E.M. and Smiley, D.W.M. (1999) Aphids. In: Hardie, J. and Minks, A.K. (eds) *Pheromones of Non-Lepidopteran Insects Associated with Agricultural Plants*. CAB International, Wallingford, UK, pp. 227–250.

Harris, V.E. and Todd, J.W. (1980) Male-mediated aggregation of male, female and 5th-instar southern green stink bugs and concomitant attraction of a tachinid parasite, *Trichopoda pennipes*. *Entomologia Experimentalis et Applicata* 27, 117–126.

Hedlund, K., Vet, L.E.M. and Dicke, M. (1996) Generalist and specialist parasitoid strategies of using odours of adult drosophilid flies when searching for larval hosts. *Oikos* 77, 390–398.

Hopkins, A.D. (1917) (Contribution to discussion). *Journal of Economic Entomology* 10, 92–93.

Jaenicke, J. (1982) Environmental modification of oviposition behaviour in *Drosophila*. *American Naturalist* 119, 784–802.

Kennedy, B.H. (1979) The effect of multilure on parasites of the European elm bark beetle, *Scolytus multistriatus*. *Bulletin of the Entomological Society of America* 25, 116–118.

Kennedy, B.H. (1984) Effect of multilure and its components on parasites of *Scolytus multistriatus* (Coleoptera: Scolytidae). *Journal of Chemical Ecology* 10, 373–385.

Leal, W.S., Higuchi, H., Mizutani, N., Nakamori, H., Kadosawa, T. and Ono, M. (1995) Multifunctional communication in *Riptortus clavatus* (Heteroptera: Alydidae): conspecific nymphs and egg parasitoid *Ooencyrtus nezarae* use the same adult attractant pheromone as chemical cue. *Journal of Chemical Ecology* 21, 973–985.

Li, L.-Y. (1994) Worldwide use of *Trichogramma* for biological control on different crops: a survey. In: Wajnberg, E. and Hassan, S.A. (eds) *Biological Control with Egg Parasitoids*. CAB International, Wallingford, UK, pp. 37–53.

Longley, M. (1999) A review of pesticide effects upon immature aphid parasitoids within mummified hosts. *International Journal of Pest Management* 45, 139–145.

Longley, M. and Jepson, P. (1997) Effects of life stage, substrate and crop position on the exposure and susceptibility of *Aphidius rhopalosiphi* De Stefani–Perez (Hymenoptera: Braconidae) to deltamethrin. *Environmental Toxicology and Chemistry* 16, 1034–1041.

Lyon, G.D. and Newton, A.C. (1999) Implementation of elicitor mediated induced resistance in agriculture. In: Agrawal, A.A., Tuzun, S. and Bent, E. (eds) *Induced Plant Defences Against Pathogens and Herbivores. Biochemistry, Ecology and Agriculture*. APS Press, St Paul, Minnesota, pp. 299–318.

McGregor, R. and Henderson, D. (1998) The influence of oviposition experience on response to host pheromone in *Trichogramma sibericum* (Hymenoptera: Trichogrammatidae). *Journal of Insect Behavior* 11, 621–632.

de Moraes, C.M., Lewis, W.J., Paré, P.W., Alborn, H.T and Tumlinson, J.H. (1998) Herbivore-infested plants selectively attract parasitoids. *Nature* 393, 570–573.

Noldus, L.P.J.J. (1988) Response of the egg parasitoid *Trichogramma pretiosum* to the sex pheromone of its host *Heliothis zea*. *Entomologia Experimentalis et Applicata* 48, 293–300.

Noldus, L.P.J.J. and van Lenteren, J.C. (1985) Kairomones for the egg parasite *Trichogramma evanescens* Westwood. I. Effect of volatile substances released by two of its hosts, *Pieris brassicae* L. and *Mamestra brassicae* L. *Journal of Chemical Ecology* 11, 781–791.

Noldus, L.P.J.J., Lewis, W.J. and Tumlinson, J.H. (1990) Beneficial arthropod behavior mediated by airborne semiochemicals. IX. Differential response of *Trichogramma pretiosum*, an egg parasitoid of *Heliothis zea*, to various olfactory cues. *Journal of Chemical Ecology* 16, 3531–3544.

Noldus, L.P.J.J., van Lenteren, J.C. and Lewis, W.J. (1991a) How *Trichogramma* parasitoids use moth sex pheromones as kairomones: orientation behaviour in a wind tunnel. *Physiological Entomology* 16, 313–327.

Noldus, L.P.J.J., Potting, R.P.J. and Barendregt, H.E. (1991b) Moth sex pheromone adsorption to leaf surface: bridge in time for chemical spies. *Physiological Entomology* 16, 329–344.

Papaj, R.D. and Lewis, A.C. (eds) (1993) *Insect Learning. Ecological and Evolutionary Perspectives*. Chapman & Hall, New York.

Paré, P.W. and Tumlinson, J.H. (1997) Induced synthesis of plant volatiles. *Nature* 385, 30–31.

Paré, P.W. and Tumlinson, J.H. (1998) Cotton volatiles synthesized and released distal to the site of insect damage. *Phytochemistry* 47, 521–526.

Paré, P.W., Alborn, H.T. and Tumlinson, J.H. (1998) Concerted biosynthesis of an insect elicitor of plant volatiles. *Proceedings of the National Academy of Sciences USA* 95, 13971–13975.

Pickett, J.A., Wadhams, L.J., Woodcock, C.M. and Hardie, J. (1992) The chemical ecology of aphids. *Annual Review of Entomology* 37, 67–90.

Poppy, G.M. (1999) The raison d'être of secondary plant chemicals? *Trends in Plant Science* 4, 82–83.

Poppy, G.M., Powell, W. and Pennacchio, F. (1997) Aphid parasitoid responses to semiochemicals – genetic, conditioned or learnt? *Entomophaga* 42, 193–199.

Potting, R.P.J., Poppy, G.M. and Schuler, T.H. (1999) The role of volatiles from cruciferous plants and pre-flight experience in the foraging behaviour of the specialist parasitoid *Cotesia plutellae*. *Entomologia Experimentalis et Applicata* 93, 87–95.

Powell, W. (1999) Parasitoid hosts. In: Hardie, J. and Minks, A.K. (eds) *Pheromones of Non-Lepidopteran Insects Associated with Agricultural Plants*. CAB International, Wallingford, UK, pp. 405–427.

Powell, W., Hardie, J., Hick, A.J., Höller, C., Mann, J., Merritt, L., Nottingham, S.F., Wadhams, L.J., Witthinrich, J. and Wright, A.F. (1993) Responses of the parasitoid *Praon volucre* (Hymenoptera: Braconidae) to aphid sex pheromone lures in cereal fields in autumn: implications for parasitoid manipulation. *European Journal of Entomology* 90, 435–438.

Powell, W., Pennacchio, F., Poppy, G.M. and Tremblay, E. (1998) Strategies involved in the location of hosts by the parasitoid *Aphidius ervi* Haliday (Hymenoptera: Braconidae: Aphidiinae). *Biological Control* 11, 104–112.

Seybold, S.J., Teale, S.A., Wood, D.L., Zhang, A., Webster, F.X., Lindahl, K.Q. and Kubo, I. (1992) The role of lanierone in the chemical ecology of *Ips pini* (Coleoptera: Scolytidae) in California. *Journal of Chemical Ecology* 18, 2305–2329.

Sheehan, W. and Shelton, A.M. (1989) The role of experience in plant foraging by the aphid parasitoid *Diaeretiella rapae* (Hymenoptera: Aphidiidae). *Journal of Insect Behavior* 2, 743–759.

Sternlicht, M. (1973) Parasitic wasps attracted by the sex pheromone of their coccid host. *Entomophaga* 18, 339–342.

Stowe, M.K., Turlings, T.C.J., Loughrin, J.H., Lewis, W.J. and Tumlinson, J.H. (1995) The chemistry of eavesdropping, alarm, and deceit. *Proceedings of the National Academy of Sciences USA* 92, 23–28.

Süss, L. (1983) Survival of pupal stage *Aphidius ervi* Hal. in mummified *Sitobion avenae* F. to pesticide treatment. In: Cavalloro, R. (ed.) *Aphid Antagonists. Proceedings EC Experts Group Meeting, Portici*. A.A. Balkema, Rotterdam, pp. 129–134.

Thaler, J.S. (1999a) Induction and plant resistance to herbivores and effects on yield of field grown tomato plants. *Environmental Entomology* 28, 35–57.

Thaler, J.S. (1999b) Jasmonate-inducible plant defences cause increased parasitism of herbivores. *Nature* 399, 686–688.

Tumlinson, J.H., Lewis, W.J. and Vet, L.E.M. (1993) How parasitic wasps find their hosts. *Scientific American* 268, 100–106.

Tumlinson, J.H., Paré, P.W. and Lewis, W.J. (1999) Plant production of volatile semiochemicals in response to insect-derived elicitors. In: Chadwick, D.J. and

Goode, J.A. (eds) *Insect–Plant Interactions and Induced Plant Defence*. John Wiley & Sons, Chichester, pp. 95–109.

Turlings, T.C.J., Tumlinson, J.H. and Lewis, W.J. (1990) Exploitation of herbivore-induced plant odors by host-seeking wasps. *Science* 250, 1251–1253.

Turlings, T.C.J., Wäckers, F.L., Vet, L.E.M., Lewis, W.J. and Tumlinson, J.H. (1993) Learning of host-finding cues by Hymenopterous parasitoids. In: Papaj, D.R. and Lewis, A.C. (eds) *Insect Learning. Ecological and Evolutionary Aspects*. Chapman & Hall, New York, pp. 51–78.

Turlings, T.C.J., Loughrin, J.H., McColl, P.J., Röse, U., Lewis, W.J and Tumlinson, J.H. (1995) How caterpillar-damaged plants protect themselves by attracting parasitic wasps. *Proceedings of the National Academy of Sciences USA* 92, 4169–4174.

Umoru, P.A., Powell, W. and Clark, S.J. (1996) Effect of pirimicarb on the foraging behaviour of *Diaeretiella rapae* (Hymenoptera: Braconidae) on host-free and infested oilseed rape plants. *Bulletin of Entomological Research* 86, 193–201.

Vet, L.E.M. (1995) Parasitoid foraging: the importance of variation in individual behaviour for population dynamics. In: Floyd, R.B., Sheppard, A.W. and Debarro, P.J. (eds) *Frontiers of Population Biology*. CSIRO Publications, Melbourne, pp. 245–256.

Vet, L.E.M. (1999) Evolutionary aspects of plant-carnivore interactions. In: Chadwick, D.J. and Goode, J.A. (eds) *Insect–Plant Interactions and Induced Plant Defence*. John Wiley & Sons, Chichester, pp. 3–20.

Vet, L.E.M. and Dicke, M. (1992) Ecology of infochemical use by natural enemies in a tritrophic context. *Annual Review of Entomology* 37, 141–172.

Vet, L.E.M., Lewis, W.J. and Cardé, R.T. (1995) Parasitoid foraging and learning. In: Cardé, R.T. and Bell, W. (eds) *Chemical Ecology of Insects 2*. Chapman & Hall, New York, pp. 65–101.

Vinson, S.B. (1984) Parasitoid–host relationships. In: Bell, W.J. and Cardé, R.T. (eds) *Chemical Ecology of Insects*. Chapman & Hall, London, pp. 205–233.

Vinson, S.B. (1985) The behavior of parasitoids. In: Kerkut, G.A. and Gilbert, L.I. (eds) *Comprehensive Insect Physiology, Biochemistry and Pharmacology*, Vol. 8. Pergamon Press, Oxford, pp. 417–469.

Wadhams, L.J., Birkett, M.A., Powell, W. and Woodcock, C.M. (1999) Aphids, predators and parasitoids. In: Chadwick, D.J. and Goode, J.A. (eds) *Insect–Plant Interactions and Induced Plant Defence*. John Wiley & Sons, Chichester, pp. 60–73.

Williams, I.S., Dewar, A.M. and Dixon, A.F.G. (1998) The influence of size and duration of aphid infestation, and its effect on sugarbeet yellowing virus epidemiology. *Entomologia Experimentalis et Applicata* 89, 25–33.

Wiskerke, J.S.C., Dicke, M. and Vet, L.E.M. (1993) Larval parasitoid uses aggregation pheromone of adult hosts in foraging behaviour: a solution to the reliability–detectability problem. *Oecologia* 93, 145–148.

# 7 Flight Trajectories of Foraging Insects: Observations Using Harmonic Radar

Joseph R. Riley[1] and Juliet L. Osborne[2]

[1]Radar Entomology Unit, Natural Resources Institute, University of Greenwich, Leigh Sinton Road, Malvern, Worcestershire WR14 1LL, UK; [2]Entomology and Nematology Department, IACR-Rothamsted, Harpenden, Herts AL5 2JQ, UK

## Introduction

At a meeting of the Royal Entomological Society on 3 December 1969, the late Professor Glen Schaefer described how he had used a modified 3-cm marine radar to investigate the migratory flight of locusts, moths and butterflies in the Sahara (Schaefer, 1969). His pioneering experiment set the stage for the development of radar entomology into a discipline that found worldwide applications, and it opened a completely new window on to the behaviour of high-flying insects (Drake and Farrow, 1988; Reynolds, 1988; Riley, 1989; Reynolds and Riley, 1997).

We hope to show in this chapter that the advent of harmonic radar (Riley et al., 1996) has similarly marked a new era in the study of insects engaged in *low altitude* flight. Years of conventional observations have, of course, already provided many intriguing insights into the behaviour of low-flying insects, but harmonic radar not only extends the range of detection far beyond that of human vision, it also provides geometrically accurate maps of the insects' flight trajectories. Thus, quantitative investigations of course control and of navigational performance over many hundreds of metres have become possible for the first time. This chapter outlines four studies of this type. The first is an investigation of the flight of foraging bumblebees, *Bombus terrestris* L. In this study, the capabilities of the technique were still being assessed and our experiment was correspondingly exploratory. This latitude allowed us to examine how the bees coped with the wind, and thus wind compensation by bumblebees became the focus of the second study. By contrast, the third was specifically designed to investigate the development, or ontogeny, of flight patterns in young honeybees, *Apis mellifera* L., and the fourth to record the flight paths taken by turnip moths, *Agrotis segetum* (Denis & Schiffermüller), in the presence of synthetic sex pheromone.

## Flights Made by Foraging Bumblebees

### Background

In 1996 the generally held assumption that bumblebees forage close to their colonies when food is locally abundant (Heinrich, 1976; Teräs, 1976; Bowers, 1985) was questioned by Dramstad (1996), who cited the lack of convincing evidence about foraging range. Our harmonic radar had become ready for full-scale experiments at about that time, and so we decided to use it to try to resolve this question. A secondary objective was to assess the degree to which bumblebees are constant to established forage sites and follow the same routes on successive foraging trips. A knowledge of this fidelity is required for testing theoretical models of foraging strategy. Records of flight tracks were expected to demonstrate immediately the degree of route fidelity, and also offered an effective new method of investigating site fidelity which could be combined with traditional pollen analysis and mark-and-recapture techniques.

### Methods

Two field studies were undertaken in 1996, one in June and the other in August (Osborne *et al.*, 1999). Seven days before each study, a colony of *B. terrestris* (supplied by Koppert UK) was placed in the experimental arena – a relatively flat area (40 ha) of arable farmland with flowering crops and some hedges. Each colony consisted of a queen, approximately 100 workers and their comb with brood. The nest box was fitted with a transparent perspex tunnel to allow the identification of individual foragers (tagged with coloured and numbered discs), as they left and returned to the nest. The main sources of pollen and nectar in the arena were recorded so that we could interpret the bees' flight tracks and destinations in relation to the spatial distribution of available forage (Fig. 7.1).

We set up an azimuthally scanning harmonic radar (Riley *et al.*, 1996) at the southern edge of the arena (Fig. 7.1), and bees identified as regular foragers were caught on departure from the colony, fitted with a radar transponder and released at the nest-box entrance. These bees took off readily, and flew away quite normally, showing no signs of being perturbed by the transponders. The radar recorded their position once every 3 s, except when they flew beyond and below the local radar horizon formed by tall crops and

**Fig. 7.1.** The flight arena and outward tracks from the colony. Tracks ($n = 35$) of bees flying away from the colony (♦) and out over the experimental arena, in June. North is upwards; the range rings are at 200 and 400 m from the radar (○); the four anemometry stations are denoted by the symbol (⊗), and the central anemometry mast by (●). Field boundaries and hedges are shown respectively as thin and thick lines; flowering crops by diagonal hatching, and wooded areas by solid in-filling. Radar visibility was limited beyond buildings (cross-hatching to the south of the radar) and beyond hedges. Flowering crops included rape, spring field beans, winter field beans and lupins. Gardens beyond the hedges to the NE contained unknown forage. Tracks are shown as continuous lines unless intervals between radar fixes exceeded 60 s. Each symbol denotes an individual bee, and their tracks are identified as follows: G30 (×), Y35 (O), R8 (+), R33 (★), G22 (□), G20 (◊), Y18 (Δ), Y7 (∇), Y34 (∗). Bee Y18 (Δ) was notable in that it initially foraged in the area to the NE of the radar, but then, on an outward flight towards the NE, suddenly turned through 90° and went to a forage area to the NNW. Bee R8 (normally marked +) flew out on one trip, stopped for 5 min 54 s and then continued. The continuation is plotted as -------∞, and is counted as one track. (Printed with permission from *Journal of Applied Ecology*, Blackwell Science Ltd.)

hedges, or when they landed in vegetation. The resulting tracks described the length, direction, speed and straightness of flights to and from forage patches, but not the bees' activity whilst they were actually foraging. To avoid potential confusion between criss-crossing or truncated tracks, a maximum of three bees with transponders were allowed to fly at any one time.

Wind speed and direction were recorded every 10 s at a height of 2.7 m, by four anemometry stations set up in the experimental arena (Fig. 7.1). Wind speed was also recorded near the midpoint of the stations by four anemometers mounted on a mast, at heights of 1.9, 3.9, 5.9 and 8.8 m above ground level. Direction was recorded by the mast's wind vane, at a height of 1.9 m (Fig. 7.1).

## Results

The results of the experiment are more fully described in Osborne et al. (1999), and we present only a summary here. Most outward and return flights were impressively straight (Fig. 7.1), with ground speeds ranging from 3.0 to 15.7 m s$^{-1}$ ($n = 100$), depending on the wind. The bees' incoming flights ended at the nest, but their outward destinations could often not be fixed precisely, because tracks were truncated when a bee descended into a forage area or flew over the local horizon formed by hedgerows or tall crops. Within this constraint, the mean length of outgoing journeys was found to be 275.3 ± 18.5 m (65 tracks made by 21 bees). A result of direct relevance to our first experimental objective was that outward journeys were frequently to destinations lying well beyond the nearest available forage (Fig. 7.1). For example, in the June experiments there was flowering oilseed rape available 42 m from the nest, but 28 out of 35 outward tracks were to destinations more than 250 m away. This pattern was maintained in August, when 15 out of 30 tracks ended over 200 m from the nest despite a patch of flowering lupins at only 108 m.

Most bees tracked for more than one trip ($n = 14$) kept to their directions of outward travel (and the same destinations) over successive trips. In nine cases, this individual directionality was maintained to within ± 13° for all the observed journeys; four bees were directionally constant on all but one of their journeys, and the remaining bee switched between forage patches. She made four outward trips to the NW but then flew from this area in an easterly direction (Fig. 7.2), perhaps searching for new forage in the region to the NE of the nest. Her last three tracked trips from the nest went directly to this area.

## Discussion

It seems safe to assume that flight performance was not significantly perturbed by the transponders because bees fitted with them were seen to fly normally, and collected nectar weighing approximately 50% of their body-weight, both before and after transponder attachment. The transponder seems likely to have impeded the bees' access to certain flowers, and this might explain why bees with transponders tended to spend longer on trips than those without (Osborne *et al.*, 1999).

The straightness and continuity of the flight tracks demonstrated that the bumblebees were not searching for food en route, but were flying directly to known destinations. These 'bee-lines' seem to be characteristic of *in transit* flight (Dramstad, 1996), and are very different from the tortuous *foraging flights* between flowers and plants (Zimmerman, 1982; Rasmussen and Brødsgaard, 1992). We had only a few records of bees which had been tracked on sequential days, but like Menzel (1999) and Chittka *et al.* (1999) we found little evidence that the bees' memories of destination and of flight route became impaired overnight.

It has been postulated that bumblebees should work the forage patches closest to the nest (Heinrich, 1976; Bowers, 1985), thus minimizing the time and/or energy expended in commuting, and optimizing either net energy gain and/or overall foraging efficiency (Pyke, 1979; Hodges, 1985; Schmid-Hempel *et al.*, 1985). Our experiment examined this prediction in an environment with patchy, but relatively abundant, resources. While we did

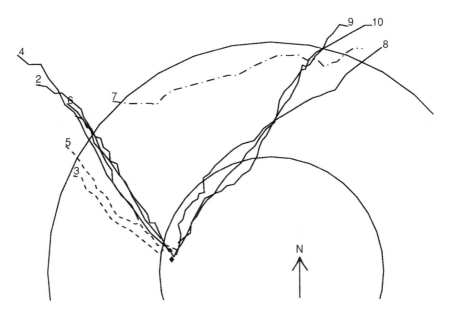

**Fig. 7.2.** Example of a bee changing her foraging area. A series of tracks from bee R33 from 12–16 June. Outward (——) and return tracks (-----) are numbered chronologically. The bee originally worked a forage area to the NW of the radar, but was then seen flying eastwards (track 7) towards a new forage area to the NE. Her subsequent outward flights from the nest went directly to this forage.

not attempt to quantify and compare the net energy gains of the individual bees in the study, the simple observation that bees tended to fly further than the nearest available nectar and pollen sources provides convincing qualitative evidence that bees do not necessarily forage close to the nest when local resources are abundant. There are other reasons for thinking that bumblebees might select more distant forage patches. For example, by travelling further, they may gain access to a wider variety of resources (Dukas and Edelstein-Keshet, 1998), and numerical models of the energy budgets of foraging flight have shown that long commuting flights need not necessarily be disadvantageous. This is especially the case when reward levels in the landscape are meagre (Cresswell et al., 2000). The models demonstrate that rewards at distant foraging sites do not have to be much greater than those nearby for long-range commuting to become the more profitable option. Another advantage of foraging further afield is that it probably reduces the risk of attracting predators and parasitoids into the proximity of the nest (Dramstad, 1996).

Our second objective was to assess the degree to which bees kept to the same routes, and the results showed that within the limited time span of our experiments, a high degree of route fidelity (length and direction) was maintained by most of the bees. Although route fidelity had previously been assumed for bees, it had never before been demonstrated on anything like the scale or with the accuracy made possible with the radar. This route constancy reflects the patterns found by Collett et al. (1992) in desert ants (*Cataglyphis* spp.) which do not lay pheromone trails but rely on visual landmarks to navigate.

Because we were working with experienced foragers who were already familiar with the local environment, we could not investigate how they found forage patches in the first place. However, on two occasions foragers changed destinations in a way that suggested they may have been looking for a new food source. The bees flew out along previously learned bee-lines and then changed direction, either before reaching the original destination (Fig. 7.1), or afterwards (Fig. 7.2). Further experiments are needed to show whether 'route branching' is a feature of searching strategy, and indeed, to investigate how individual bumblebees normally discover their forage sources. Intriguingly, Dornhaus and Chittka (1999) have recently shown that whilst in the nest, bumblebees can alert their nest mates to the discovery of a good source, but do not, apparently, indicate its location.

## Wind Compensation by Bumble Bees

### Background

Airborne insects very often need to travel in some particular direction. In the case of migrants, it may be sufficient to maintain an overall displacement towards some general geographic region. For central place foragers like bees, much more specific, goal-directed flight must be achieved, but in both cases the insects are liable to be drifted off course by any winds not parallel with the intended direction of travel. The question of whether and how this problem is solved by different species has been a topic of interest for many years (Williams *et al.*, 1942).

## Migrant butterflies

For migrants, information is limited mainly to butterflies, and here there seem to be two views about how compensation might be achieved. Thus Schmidt-Koenig (1993) observed monarch butterflies (*Danaus plexippus* L.) being passively drifted off course by 'direct wind action' in crosswinds, and by 'indirect wind action' causing the butterflies to fly along the lee side of shelters not lying in the intended track direction. He concluded, however, that in order to reach their overwintering grounds in Mexico they must keep track of, and ultimately compensate for, these off-course displacements. He describes a possible example of this process in which southwesterly bound monarchs experienced southerly drift whilst crossing a runway exposed to a northwesterly wind. Once they reached more sheltered areas, the monarchs adopted tracks more towards the west, appearing retrospectively to make up for the period of southwards drift.

In distinct contrast to this delayed 'drift and then compensate later' process, Williams (1958) quotes an anecdotal report that unidentified, eastbound skipper butterflies actively turned their headings towards a southerly crosswind to avoid being blown to the north. This evidence of 'immediate' compensation is supported by Walker's (1986) report that *Urbanus proteus* L. appear to achieve full compensation whilst in crosswind flight. This conclusion was based on his observation that the mean vanishing bearings of these butterflies remained close to their overall direction of autumn migration (141°) in Florida, in spite of crosswinds from either side.

Johnson (1969, p.157) has pointed out that 'the observer of butterflies is at a disadvantage . . . for he does not know to which place the migrant is going. He cannot tell therefore whether it steers a preordained track in spite of the wind, . . . or whether the wind is merely deflecting the insect from a track that it would have kept in calm weather'. In an ingenious experiment to overcome this problem, Srygley *et al.* (1996) showed that two species of butterflies (*Aphrissa statira* (Cramer) and *Marpesia chiron* (Fabricius)) in flight over a lake adjusted their headings in response to *changes* in the cross-track component of the wind. They interpreted this to mean that butterflies could compensate, at least in part, for systematic drift in sustained crosswinds.

Overall, there seems to be every reason to believe that some migrant species of butterfly can compensate, at least to some degree, for the drift induced by crosswinds. It seems probable, however, that the compensation mechanisms used by species with rather precisely defined destinations (like the monarch) will differ from those of butterflies where achievement of a generalized direction of migration is all that is required. This difference may perhaps be reflected in the contrast between the 'later' and 'immediate' compensation responses described above, but questions of this type seem unlikely to be resolved until quantitative measurements of butterfly flight paths are made in well-described windfields. Such experiments should be possible in the future with harmonic radar.

### Central place foragers

In the case of central place foragers, the only work on wind drift compensation prior to our own on bumblebees seems to be that of von Frisch and Lindauer (1955) and Heran and Lindauer (1963) on honeybees. Von Frisch and Lindauer used binoculars and a series of sighting poles in an attempt to determine the flight tracks of bees flying to and fro between hive and a feeder. Their surprising conclusion was that the insects seemed to anticipate wind drift by heading partially into the wind and flying in shallow curves on the upwind side of the hive-to-feeder line.

## Methods

We used the same flight tracks that were recorded in the course of the foraging study described above, but set out to examine them in the context of the winds in which the bees were flying. The procedure was to calculate each bee's average heading and its airspeed by subtracting the wind vector from its track vector. The results have been outlined elsewhere (Riley *et al.*, 1999), but are described in more detail here.

### Estimation of the mean wind vector experienced by the bees

In earlier compensation studies, data on wind speed and direction have been rather limited, so we made a particular effort to obtain good estimates of the winds experienced by our bees. The wind vector recorded at each of our four anemometer stations was averaged over the period of each flight track, and the results were interpolated to describe the mean wind field over the flight arena for the duration of the flight (Fig. 7.3). Near the ground, wind speed tends to increase as a logarithmic function of height (McCartney, 1997) so the track-averaged wind speeds recorded by the four anemometers on our mast were plotted against the logarithm of their height above ground. The positions of the points showed that speed was often better described by a quadratic rather than linear fit to log height (Fig. 7.4), so a quadratic equation was used, with new coefficients being found for each flight. Entering a bee's height of flight (see below) into this equation, and combining the result with the wind field interpolation data, allowed us to estimate the wind vector at its track centre, averaged over the period of its flight.

*Estimation of height of flight*

Because we were studying flight in the first few metres above ground level, we operated the radar with its beam axis close to the horizontal. Thus the only direct information about the height of a bee detected on the radar was that it lay somewhere below the top edge of the radar beam which lay typically from 4 to 8 m above ground level, depending on the range (Fig. 7.5). However, we had noticed that the bees tended to fly higher when on fast downwind trajectories and lower when moving slowly upwind. It was as though they were adjusting their heights of flight so as to stabilize the angular rate at which images of ground features passed over their retinae (the optical flow rate; Esch and Burns, 1996) at some 'preferred' value. Subsequent radar observations showed that in calm conditions, the bees typically flew with ground speeds of ~7 m s$^{-1}$, and we visually estimated their height of flight to be about 2 m. Thus the preferred rate (for the visual field pointing directly downwards) appeared to be ~7 m s$^{-1}$/2 m = 3.5 rad s$^{-1}$. We assumed that the bees would attempt to maintain this rate by changing their height of flight whenever their ground speeds were altered by favourable or contrary winds. To the extent that this assumption remained true, their height in metres could be found simply from the equation: $h = V/3.5$ m, where $V$ is the bee's ground speed in m s$^{-1}$ as measured by the radar.

## Results

*Compensation for crosswind drift*

Most bees maintained straight outward and return routes between their nest and forage areas, and they did so even in winds that had a strong cross-track component. Crosswind flight paths directly towards a known destination demonstrate in themselves that active wind compensation for drift is occurring, and we show two clear examples of this in Fig. 7.3. The bees approached their nest in winds that were almost at right angles to their flight paths so there is no doubt that they were laying off their headings into the wind and were moving obliquely over the ground. Subtraction of the wind vector from bees' displacement vectors allowed calculation of their headings and airspeeds, and when this was done for all the homeward tracks ($n = 53$), it was found that some degree of crosswind compensation occurred in most

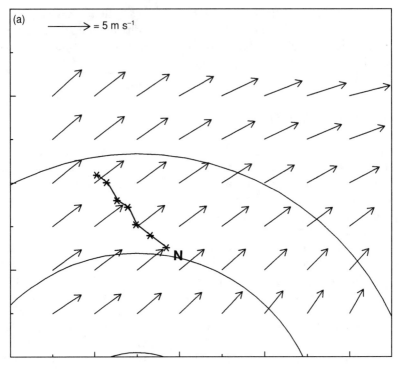

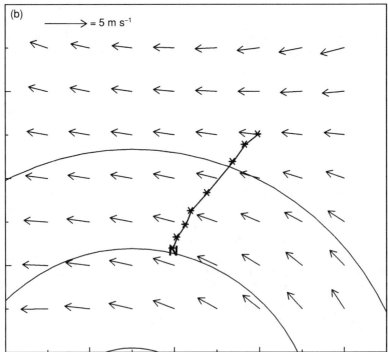

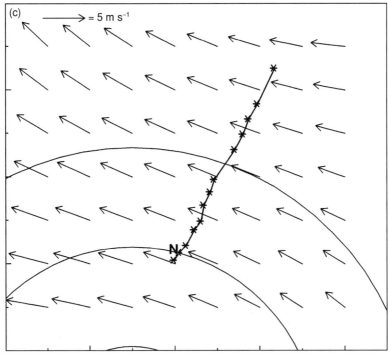

**Fig. 7.3** (a and b, opposite). Three examples of straight, compensated crosswind flight by bumblebees. The area represented in the figure is 450 m east to west, and 400 m south to north. The radar range rings here are at intervals of 115 m, and the bee's colony is indicated by N. The speed of the wind at a height of 2.7 m above ground level, and the direction towards which it is blowing, are shown by the grids of arrows. The grids are derived from data recorded by the four anemometry stations (shown in Fig. 7.1), for the duration of each flight. Bee flight trajectories are denoted by the asterisks, which represent sequential position fixes recorded by the radar. In (a), a bee is approaching its nest directly from the NW at 6.7 m s$^{-1}$, in a ~3.8 m s$^{-1}$ crosswind blowing from the SW. In (b), the crosswind is from the ESE at ~2.4 m s$^{-1}$, and the bee is travelling directly from the NE towards its nest, at 6.9 m s$^{-1}$. (c) shows an outward flight at 5.8 m s$^{-1}$ towards the NNE, in a cross-wind of ~3.5 m s$^{-1}$ from the ESE. In this case the bee's precise destination was not known, but it seems certain to have been in the forage areas to the NE and NNE that were regularly visited by the bees in our experiment (see Fig. 7.1).

of them (Fig. 7.6). The outward tracks from the nest towards forage areas produced very similar results.

*Airspeeds*

When the bees' airspeeds were examined, we found that the overall average for our June experiments was 7.3 ± 1.2 m s$^{-1}$ (SD) ($n = 74$), and for

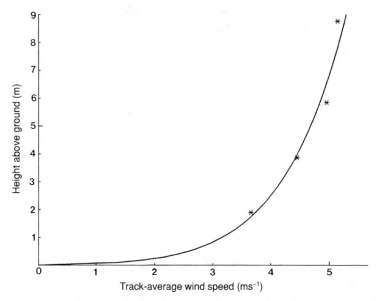

**Fig. 7.4.** A typical profile of wind speed as a function of height above ground. Wind speed was recorded at four heights once every 10 s, at the mast positioned near the centre of the flight arena (Fig. 7.1). The speeds are averaged over the period of a flight track (here 100 s), and are shown by the asterisks. The line is derived by fitting a second-order polynomial to these points and to a supplementary point of (assumed) zero wind speed at a height of 1 cm, when the points were plotted as a function of log (height).

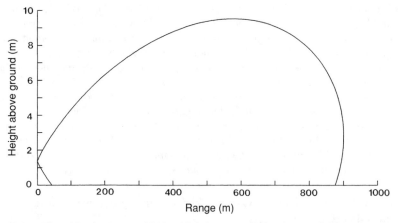

**Fig. 7.5.** The altitude range within which insects can be detected. The curve shows the envelope in the vertical plane within which the transponders can normally be detected by the harmonic radar. In this example the beam axis is tilted upwards at 0.1° to the horizontal. Note that for clarity, the vertical scale is shown 100× larger than the horizontal one. Over smooth and wet terrain, interference between direct and reflected signals (the Lloyds mirror effect) may substantially modify the envelope.

July/August 6.2 ± 1.2 m s$^{-1}$ ($n$ = 63), the difference between the means being highly significant ($P \ll 0.001$). There were also significant ($P < 0.005$) differences between the mean airspeeds found for upwind and downwind flights (Table 7.1), upwind flight being about 1 m s$^{-1}$ faster than downwind in both study periods. This result is consistent with the data in Fig. 7.7 that show a weak negative correlation of airspeed with the wind component along the

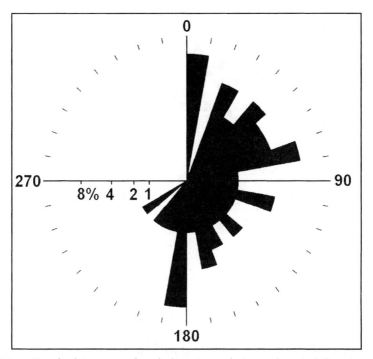

**Fig. 7.6.** Circular histogram of track directions relative to the wind direction. Distribution of θ = |Track direction − wind direction|, for bees flying on straight tracks directly towards the nest (end-to-end distance/along track distance >0.85), $n$ = 53. Radial scale shows the percentage of tracks falling within each 10° bin, and θ = 0 corresponds to bees flying directly downwind. In all cases other than 0 and 180°, the bees were exposed to crosswinds, and had compensated fully for them.

**Table 7.1.** Radar observations of the airspeed of bumblebees.

|  | June | | | Jul / Aug | | |
| --- | --- | --- | --- | --- | --- | --- |
|  | All | Upwind | Downwind | All | Upwind | Downwind |
| Mean (m s$^{-1}$) | 7.3 | 7.7 | 6.7 | 6.2 | 7.1 | 5.7 |
| SD (m s$^{-1}$) | 1.2 | 0.9 | 1.3 | 1.2 | 0.9 | 1.2 |
| $n$ | 74 | 28 | 14 | 63 | 14 | 16 |
| $P$ (up/down) |  | 0.0048 |  |  | 0.0012 |  |

body axis. Figure 7.7 also illustrates how regulating height of flight using an optical flow rate mechanism is equivalent to reducing altitude in headwinds and climbing in tailwinds, which is what we had observed the bumblebees to do.

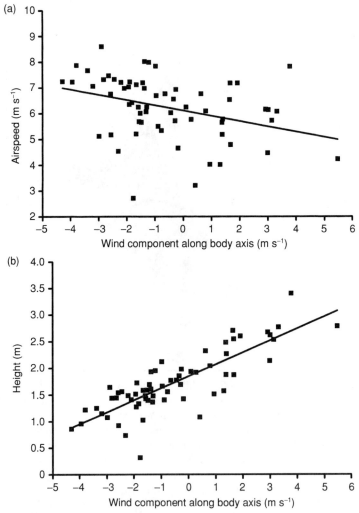

**Fig. 7.7.** Airspeed and height as a function of the wind. The component of the wind along the bees' body axis is plotted against (a) their airspeeds, found by subtraction of the wind vector from their track vectors, and (b) their flight altitudes which were calculated by dividing their ground speeds by 3.5. Negative values correspond to head winds, and positive values to tail winds. The data suggest that there is a weak negative correlation ($R^2 = -0.16$) between airspeed and wind component, and clear positive correlation ($R^2 = 0.52$) between height and wind component ($n = 53$).

## Discussion

*Compensation for wind drift*

Our observations of bumblebees flying directly towards their nests in sustained crosswinds provide definitive and unequivocal evidence that they were capable of accurate, in-flight compensation for wind drift. This rather remarkable phenomenon was observed in all the tracks that had a crosswind component, and it was clear that it is a routine feature of bumblebee flight. This raised the obvious question: how did the bees know how far to turn off course to achieve the correct track to their destinations in winds of different strengths and from different directions? In an attempt to answer it, we looked at earlier work to see what mechanisms had been proposed to explain the compensation observed in honeybees.

*Honeybees*

Von Frisch (1967, p.186) deduced that a honeybee would be able to 'recognise lateral [wind] drift by the oblique course of the ground beneath [her] and compensate for it in such a way that once again she moves towards her goal. The ground now runs again obliquely past the long axis of her body. Hence ... she must include her oblique stance in the calculation. Its amount she can read from the solar angle'. It is clear from these statements that he believed that bees observed how their longitudinal body axes (headings) differed from the direction of ground movement beneath them, and from the solar angle, and that they somehow used these angles to compute the correct heading to compensate for drift. As his co-worker Lindauer (1976) put it in an earlier RES Symposium, 'the angular direction of the optomotor ground-pattern, with respect to the body axis, is of great value for cross-wind compensation'.

The fact that honeybees lose the ability to compensate for wind drift when flying out over a featureless expanse of water (Heran and Lindauer, 1963) shows that the direction of ground image movement is, as one would indeed expect, a vital component of the compensation mechanism. However, it seems to us that rather than measuring image movement *relative to their body axes*, honeybees more probably compare it *directly* with the solar angle. This simple mechanism has the great advantage that it would not require bees to assess wind speed and direction and then calculate the heading and airspeed to stay on course. Instead, they would need only to rotate their body axes until the direction of optical flow across their retinae occurred at the angle relative to the sun's azimuth that corresponded to their intended tracks. Furthermore, the direct comparison of two optical directional cues, neither subject to parallax, would seem to be a relatively undemanding neurological task. Some support for our hypothesis comes from von Frisch's own observation that honeybees returning from crosswind trips indicated in their waggle dance the true sun compass direction of the nectar or pollen source, and not the headings they must have flown. He writes (von Frisch, 1967, p.194) that '. . . the accomplishment gains in significance from the fact that the strength and direction of the wind may vary in different segments of the flight course, and the extent and duration of these variations would have to be taken into account by the dancer in reaching her final goal'. This would indeed be a most remarkable feat of computation and integration, but our hypothesis suggests, by contrast, that the returning bee is simply communicating the (constant) angle that she maintained between optical flow direction and the sun's azimuth.

## Bumblebees

Although it has not been demonstrated that bumblebees are able to use a sun compass, two facts suggest that they can. Firstly, like other Hymenoptera known to use a sun compass (Wehner, 1984), bumblebees are sensitive to the direction of polarization in light (Wellington, 1974), and secondly, the straightness and accuracy of their outward and return flights to forage areas are very similar to those of honeybees. This leads us to conclude that the optical flow–sun compass hypothesis that we propose above for honeybees also provides the simplest explanation of the wind compensation we have observed in bumblebees.

## Regulation of height of flight and airspeed

We were unable to find any publications describing measurements of bumblebee heights of flight or airspeeds, and so examined our results in the context of the more extensive literature on honeybee flight.

## Height of flight

There is general agreement that honeybees fly lower on slow upwind flights than they do when moving fast downwind, although there are substantial differences in the heights which different authors suggest (Table 7.2). These differences are qualitatively understandable in terms of the 'preferred optical flow rate' hypothesis suggested above for bumblebees, because the speed of the wind (or more precisely, its component along the flight track direction) partly determines the bees' ground speed and should therefore influence their height. Thus, unless experiments happen to be performed in the same wind conditions, they would be expected to yield different heights of flight.

Heran and Lindauer (1963) derived equations relating height to the (ground) speed of honeybees, on the assumption that bees try to maintain a

**Table 7.2.** Observations of flight altitude of honeybees.

| Altitude | Reference |
|---|---|
| 2–3 m in normal flight conditions | Heran and Lindauer |
| 2.5 m at 8 m s$^{-1}$ airspeed | (1963) |
| 1.5 m over water | |
| ≈ 7–8 m in normal flight, i.e. calm or low wind speeds | Wenner (1963) |
| At wind speeds greater than 4 m s$^{-1}$, bees flew 'just above the ground' | |
| 2.5 m in calm air | Lindauer (1976) |
| 1.5 m in winds of 2 m s$^{-1}$ | |
| 0.5 m in winds of 4 m s$^{-1}$ | |

constant back-to-front velocity of image flow over the retina. This is closely analogous to our height model for bumblebees, although their equation coefficients were based on the optomotor responses of tethered bees flying in a wind tunnel, rather than on field observations; and they considered the geometrical effects of slightly forward fields of view. Esch and Burns (1996) used the formulae to derive their Fig. 5, and we note that this figure implies a preferred optical flow rate of 3.5 rad s$^{-1}$, which is exactly what we had found for bumblebees.

Overall, it seems that at least during their 'in-transit' flights, bumblebees vary their altitude in response to wind conditions in very much the same way as honeybees, and both insects probably use the same 'preferred optical flow rate' mechanism as a means of regulating height.

*Airspeeds*

Several field estimates of the airspeed of honeybees have been reported in the literature (Table 7.3). The estimates were obtained by timing flights between hives and feeding stations to obtain the bees' ground speeds, and then usually subtracting the wind speed. It was assumed that the flight paths were uninterrupted, straight 'bee-lines' – an assumption which may not always have been valid, especially on outward flights (Park, 1923). In all of the experiments, the equipment available for recording the wind vector was rather rudimentary; the vertical gradient of wind speed was not measured, and heights of flight were estimated visually, so overall it is perhaps not surprising that a rather wide range of speeds was reported. In spite of these differences, there was general agreement that honeybees tend to increase airspeed when flying upwind, and to lower it on downwind flights. Our finding that bumblebees also regulate their airspeed in this way suggests that the phenomenon may be a common feature in goal-directed flying insects.

**Table 7.3.** Observations of the airspeed of honeybees.

| Airspeed range (m s$^{-1}$) | Mean airspeed (m s$^{-1}$) | Factors affecting airspeed | Reference |
|---|---|---|---|
| 3.0–11.4 | 6.4 | Outward or return flight<br>Wind speed<br>Headwind, crosswind or tailwind | Park (1923, 1928) |
|  | 6.3 | Wind speed<br>Headwind, crosswind or tailwind | Ribbands (1953) quoting earlier work by Lundie |
| 5.5–9.4 | 8.2<br>7.7[a] | Outward or return flight<br>Quality of food source<br>Wind speed<br>Headwind, crosswind or tailwind | von Frisch and Lindauer (1955); von Frisch (1967) |
| 6.5–7.8 |  | Wind speed<br>Headwind, crosswind or tailwind | Wenner (1963) |
| 3.1–9.1 | 8.6[b] | Quality of food source<br>Wind speed<br>Headwind, crosswind or tailwind<br>Terrain (land, water, water with markers) | Heran and Lindauer (1963); Heran (1964) |
| 4.4–7.0 |  | Load carried by bee<br>Temperature (of thorax) | Coelho (1989) |

[a]Value quoted from earlier work by Boch and Schifferer. [b]Mean airspeed quoted from Heran's earlier work.

## The energy costs of flights to and from food sources

Although much attention has been paid to height of flight and airspeed of honeybees, these factors do not seem to have been considered in the context of the energy costs of in-transit flight. We therefore examined our data to see how the behaviour of bumblebees affected the amount of energy they used on journeys between their nests and forage patches. Kinematic measurements of bumblebee wing action have shown that the power output required for flight is a relatively insensitive function of airspeed in the range 0–4.5 m s$^{-1}$ (Dudley, 1995), but starts to increase rapidly at higher speeds (Fig. 7.8). If $P(a)$ is the mass specific power output in W kg$^{-1}$ of flight muscle needed to fly at an airspeed $a$ m s$^{-1}$, then the energy cost $E$ per metre of journey is given by:

$$E = P(a)/(a \times \cos(\alpha) + W \times \cos(\beta)) \text{ J kg}^{-1} \text{ m}^{-1} \tag{1}$$

where $W$ is the wind speed in m s$^{-1}$, $\alpha$ and $\beta$ are respectively the angles between the insect body axis and the track, and between the wind direction and the track. For directly upwind flight $\beta = 180°$, for downwind $\beta = 0°$, and in both cases $\alpha = 0°$. Basal metabolic rates are ignored in the expression, because they are small compared with the very high rates developed during

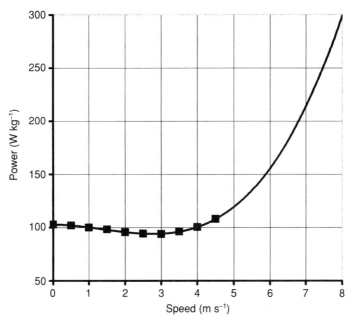

**Fig. 7.8.** Required power as a function of airspeed. The squares show how the power output per kg of flight muscle ($P(a)$) of a flying bumblebee worker varies with its airspeed ($a$). The data are reproduced from Cooper (1993), and were derived from kinematic analyses of the wing action of a bee with body-mass 206 mg, flying freely in a large wind tunnel. The solid line represents our third-order polynomial fit to Cooper's data ($P(a) = 0.8019 \times a^3 - 3.3347 \times a^2 - 0.0676 \times a + 102.81$), and extrapolation above 4.5 m s$^{-1}$ must clearly be treated with caution. However, dissipation of power by parasitic drag on the bee's body increases as the third power of its airspeed, so the trend of the extrapolated line probably provides a very reasonable estimate of power requirements at the higher speeds.

flight (Kammer and Heinrich, 1978). Figure 7.9 shows how $E$ varies with airspeed for different head- and tail-winds, and it can be seen that in calm conditions ($W = 0$), the airspeed at which the energy cost of travelling is minimized is ~5 m s$^{-1}$. It can also be seen, as one would expect (Pennycuick, 1969), that the airspeed required to minimize energy demand rises as tailwinds decrease and headwinds increase, and this may explain the tendency for bees to fly slightly faster in headwinds than in tailwinds. However, the average speed (7.1 m s$^{-1}$) that we observed in still air is well above that predicted by the minima in energy cost shown in Fig. 7.9, and so we conclude that reducing transit flight time was much more important to the bumblebees than minimizing energy costs. Faster transit flights would give proportionately more time for pollen and nectar collection, and thus could arguably maximize the overall energy intake per foraging sortie, as has been proposed for birds (Hedenström and Alerstam, 1995).

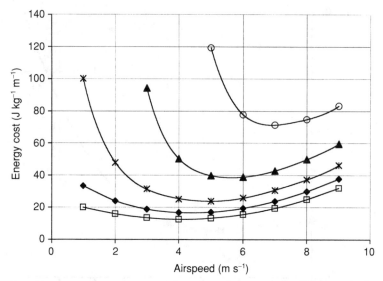

**Fig. 7.9.** Energy used per metre as a function of airspeed, for different wind speeds. The energy cost per metre of distance covered is calculated using the polynomial shown in Fig. 7.8 and equation (1). □ tail wind 4 m s$^{-1}$; ♦ tail wind of 2 m s$^{-1}$; * no wind; ▲ head wind of 2 m s$^{-1}$; ○ headwind of 4 m s$^{-1}$.

Flying lower in headwinds and higher in tailwinds has the dual advantage that it reduces both energy cost *and* travelling time. Thus, if a bee maintained its 'still air' altitude of 2 m and flew directly upwind with an airspeed of 7.1 m s$^{-1}$ in the wind conditions shown in Fig. 7.4, its ground speed would be 3.4 m s$^{-1}$ and the energy cost 64.7 J m$^{-1}$ (per kg of flight muscle). By reducing altitude to 1.1 m to maintain an optical flow rate of 3.5 rad s$^{-1}$, the bee increases its ground speed to 3.9 m s$^{-1}$ (+15%) and reduces energy cost to 56.4 J kg$^{-1}$ m$^{-1}$ (−13 %). Similarly, changing altitude when flying downwind would result in flight at 3.3 m, a ground speed of 11.3 ms$^{-1}$ (+5%) and an energy cost of 19.5 J kg$^{-1}$ m$^{-1}$ (−4%). In practice one would expect upwind journey speeds and energy costs to be a little higher, and downwind ones a little lower, because of the tendency for upwind airspeeds to be slightly faster than downwind ones, but the very substantial advantages of height regulation would, of course, remain.

Extrapolation of Alison Cooper's (1993) data in Fig. 7.8 to the average airspeed of 7.1 m s$^{-1}$ that we found for bumblebee workers implies that the bees deliver an average mass-specific muscle power output of over 200 W kg$^{-1}$. This is greater than the *maximum* value (180 W kg$^{-1}$) currently attributed to bee flight muscle (Cooper, 1993), and together with our observations of airspeeds as high as 9.5 m s$^{-1}$, suggests that the figure for maximum power output needs to be revised upwards. This conclusion is supported by a recent laboratory investigation of the flight muscle of the beetle *Cotinus mutabilis*

Gory & Perch., which showed a maximum output of 200 W kg$^{-1}$ (Josephson et al., 2000).

## Learning or Orientation Flights by Honeybees

### Background

It was well known from visual observations that honeybees make a number of short, 'orientation' or 'play' flights before becoming foragers, and also that the bees initially hover and turn to look back towards the hive (von Frisch, 1967; Lehrer, 1991). It seemed probable that the function of the orientation flights was to allow the bees to develop their aerial navigational skills, but because most of these flights took place beyond the range of human vision, very little was known about them. Harmonic radar offered a means to extend the range of observation to hundreds of metres, and an opportunity arose in 1998 to work with colleagues from the University of Illinois who were particularly interested in honeybee learning mechanisms. This study has been described elsewhere (Capaldi et al., 2000), and we present only an outline here. (Note: the orientation flights are regarded here as a sort of 'pre-foraging' which will prevent the bee becoming lost when true foraging starts. These flights can thus be grouped, along with foraging itself, under 'station-keeping movements' – see Dingle, 1996, p. 10.)

### Methods

Our colleagues began the investigation with an intensive visual study designed to record each exit, flight activity and re-entry of worker bees that had previously been tagged with numbered discs and introduced into the hive as 1-day-old adults. Artificial rain produced from a sprinkler device at the hive entrance prevented bees from leaving the hive before observers arrived at noon each day. Thereafter, observations were maintained until 21:00 h every day or until all the marked bees had signalled a transition to foraging flight by returning to the hive with pollen and/or nectar (mean ± SD, 14 ± 0.18 days). Records were obtained for 125 bees. For the subsequent radar observations, we used the same equipment and experimental arena as in our bumblebee studies. During this 5-week study, 625 marked 1-day-old bees were introduced into a colony of approximately 10,000 bees, and intensive visual observations were again made of all the flight activities of all the marked bees until they began to return with pollen. In this way we were able to select bees for transponder attachment and radar tracking which were still in their pre-foraging phase, and of known age and flight experience.

## Results

*Visual observations*

All the marked bees made orientation flights, but the number they made before becoming foragers was highly variable, ranging from 1 to 18, with a mean of 5.6. The age at which the bees began to fly varied from 3 to 14 days with a mean of 6.2. Bees were frequently seen to hover facing the hive entrance, before departing on their flights.

*Radar observations*

Bees fitted with transponders appeared to behave normally, at least within visual range, and performed the hovering flights characteristic of orienting bees. Each bee was tracked by radar only once, and we obtained 29 examples of complete out-and-back flight paths from bees in the age range 3–27 days, and with different degrees of experience (1st to 17th orientation flight). Figure 7.10 shows three examples of these flight paths, and for comparison, the flight of a forager.

Flight performance (ground speed, round trip distance and maximum range from the hive) tended to increase with flight experience, but flight duration did not, i.e. as the bees gained experience, they covered more ground by travelling faster. We found no significant correlation in the data set between age and the measures of flight performance. This lack of correlation is what one might expect given that the age at which orientation flights began was highly variable.

Examination of the spatial distribution of the radar tracks showed that only six of the 29 flights entered all four quadrants of azimuth round the hive, and in 28 flights, at least 50% of the flight path fell within a single quadrant.

## Discussion

Our experiment demonstrated that honeybees on orientation flights explore progressively larger areas of landscape, and acquire views of the hive from different ranges and from different perspectives. These novel findings strongly imply that the bees were improving their navigational competency with each flight, presumably because they were steadily accumulating more information about their environment.

The fact that some bees needed fewer orientation flights than others may mean that they assimilated information more efficiently than their hive mates, but it could also mean that the *quantity* of stored information required to trigger foraging behaviour was lower in these bees. Similar questions are raised by differences in the range of azimuthal exploration undertaken by different bees, but answers to these questions seem to require a radar study

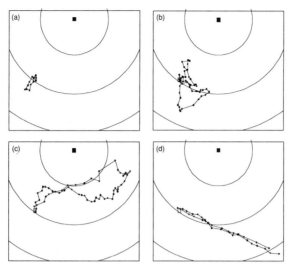

**Fig. 7.10.** Orientation and foraging tracks from honeybees. Three orientation and one foraging track from field experiments in June/July 1997. The position of the hive and the harmonic radar are shown by open and filled squares, respectively. The range rings are spaced at 116 m. (a) The first orientation flight of a 4-day-old bee, duration 228 s; (b) the 2nd orientation flight of a 6-day-old bee, duration 486 s; (c) the 14th orientation flight of a 6-day-old bee, duration 1011 s; (d) a foraging flight of a bee of unknown age, which returned to the hive with a pollen load, duration 1215 s. This track is not complete, as the bee left the area of radar coverage, re-entering on its homeward journey. Comparisons between orientation and foraging flights suggest that both ground speed and flight distance continue to increase as bees mature into foragers. Foraging bees had a significantly faster average ground speed than orienting bees ($5.6 \pm 1.0$ m s$^{-1}$ vs. $3.6 \pm 1.4$ m s$^{-1}$), and also flew further and straighter, in agreement with previous reports that honeybees forage over distances of hundreds and even thousands of metres (Ribbands, 1953).

specifically designed to assess and compare the navigational competencies of new foragers with different orientation flight histories. Such a study, for example, might record the first homing flights of new foragers with a known orientation history that had been captured at a feeder and then released at different points around the hive.

## The Flight Paths of Male Moths in the Presence of Pheromones

### Background

There has been sustained interest over recent years in the behaviour of airborne male moths exposed to volatile sex pheromones emitted by females.

This is mainly because the use of synthetic pheromones to disrupt mate-finding offers an environmentally sound means of pest control, but also because the mechanisms by which moths are able to find the source of airborne plumes are of intrinsic interest in themselves (Cardé and Minks, 1997). Much information about pheromone-following flight has been gained from video studies in wind tunnels (Cardé and Mafra-Neto, 1997) and in the field (Willis et al., 1991; Baker and Haynes, 1996), but because of their very short working range (Riley, 1993), video techniques are useful for recording flight only within a few metres of the pheromone source. The flight behaviour of male moths when they are initially searching for a pheromone plume at some distance from its source ('ranging'; Dusenbery, 1989), or when they are immersed in a blanket of artificial pheromone during field studies of mating disruption, has thus remained a matter for conjecture (Sanders, 1997). Harmonic radar seemed to be well suited to tracking moths in these circumstances, and we therefore took an opportunity to use our system to observe the flight of male turnip moths, *A. segetum*, with colleagues from the University of Lund. These joint investigations took place in 1997 and 1998, and are summarized briefly below; more complete descriptions are available elsewhere (Riley et al., 1998; G.P. Svensson et al., unpublished observations).

## Methods

We used the same flight arena and radar equipment as in the bee research, but the field experiments were conducted at night. The 1997 experiments were designed primarily to assess the feasibility of using harmonic radar for moth tracking, and began with visual observations of male moths flying in a wind tunnel towards an artificial pheromone bait whilst carrying a transponder. In the subsequent field study, male moths were fitted with transponders in the laboratory and then transported to the experimental site where they were released 30 m downwind of a line of pheromone traps. Their flight tracks were then recorded by the radar.

In 1998, our objective was to compare the flight behaviour of male moths in plots with and without disruptive treatments of sex pheromone. The two circular plots were each delineated by a 100-m-diameter ring of eight standard pheromone traps, and the (downwind) treatment zone had in addition a 160-m-diameter concentric ring of 16 devices dispensing artificial pheromone. The centres of the treatment and control plots were 436 m apart. Male moths fitted with transponders were placed in open glass cylinders at the centre of each experimental areas at least 2 h before being released and tracked. Radar observations were made between 20:00 h and 03:00 h for ten nights.

## Results

*The feasibility study*

Moths tagged with transponders and placed in the wind tunnel were seen to take off and fly upwind towards the pheromone bait in the same way that untagged moths did, and their flight behaviour appeared to be quite normal (Riley *et al.*, 1998; G.P. Svensson *et al.*, unpublished observations). In the field, radar flight tracks were up to 250 m long, though frequently interrupted when moths flew behind crops or hedges, or settled. We obtained examples of fast, linear crosswind displacements (which may have corresponded to ranging flights), and zig-zag flights, progressing both up- and downwind (which may have been plume-following, and plume-recovery flights, respectively; Riley *et al.*, 1998).

*The mating disruption experiment*

Over 4300 radar position fixes were acquired on transponder-carrying moths in the course of the experiments. However, because the moths often flew very low, behind hedges and tall crops, and also settled for highly variable periods, many of the fixes could not be used to generate reliable tracks attributable to specific moths. Most of the initial flight trajectories of males upon release were either downwind or crosswind, which indicates that the majority in both plots did not engage in optomotor anemotaxis at the time of their release. Some time later, however, there were brief periods of upwind flight towards the pheromone traps, and significantly more of these upwind-oriented movements occurred in the control plot. Surprisingly, only one tagged moth from each release site actually entered a trap. This result was in strong contrast to the finding that 16 out of 80 untagged males released in the control plot were later found in the control traps, but only one out of 84 released in the treatment zone was caught in its traps – convincing evidence that disruption was occurring.

## Discussion

The feasibility experiments clearly demonstrated the potential of harmonic radar to reveal information about the pheromone-mediated flight behaviour of male turnip moths on a scale of hundreds of metres. The more elaborate mating disruption study showed, however, that because moths flew so low and often landed (sometimes for hours), both the experiment itself, and the interpretation of the recorded flight tracks were much less straightforward than in the bee observations. It thus seems evident that harmonic radar observations of turnip moths require a flatter and much less obstructed experimental arena than was needed for the bee studies. The experiment nevertheless provided evidence that the disruptant effect of the pheromone treatment was caused by a general insensitivity in the male moths (probably due to habituation), rather than to 'false trail following' (G.P. Svensson et al., unpublished observations). Wind-tunnel observations later showed that the presence of the tag could impede moths from entering the traps, and this explains why tagged moths were not caught as frequently as expected in the control plot.

## Conclusion

The four experiments described in this paper serve to demonstrate some of the ways in which harmonic radar has been used to reveal previously inaccessible aspects of the flight behaviour of low flying insects. Many more applications can be easily envisaged – perhaps most obviously studies of the navigational performance of honeybees. Here, the uncertainties intrinsic to traditional 'vanishing bearings', and times-of-flight methods, will be replaced by detailed and geometrically accurate descriptions of complete flight tracks, and commensurate new insights into navigational mechanisms can be expected. Similar advances can be expected in studies of the foraging behaviour of a wide variety of insects, and the investigation of long-range, pheromone-following flight by moths is clearly now a practicable proposition.

Given the technical resources required to deploy scanning harmonic radars, it would be unrealistic to suppose they will proliferate throughout the entomological community. It would be equally unrealistic to ignore the limitations of the technique, principally the requirement for a clear field of view that largely precludes its use for forest-dwelling insects and in undulating terrain. Nevertheless, given the results already achieved, and outlined in this paper, it will be surprising indeed if harmonic radar does not turn out to have a profound impact on our understanding of the behaviour, and particularly of the navigational abilities, of insects engaged in low-altitude flight.

## References

Baker, T.C. and Haynes, K.F. (1996) Pheromone-mediated optomotor anemotaxis and altitude control exhibited by male oriental fruit moths in the field. *Physiological Entomology* 21, 20–32.

Bowers, M.A. (1985) Bumblebee colonization, extinction, and reproduction in subalpine meadows in Northeastern Utah. *Ecology* 66, 914–927.

Capaldi, E.A., Smith, A.D., Osborne, J.L., Fahrbach, S.E., Farris, S.M., Reynolds, D.R., Edwards, A.S., Martin, A., Robinson, G., Poppy, G.M. and Riley, J.R.

(2000) Ontogeny of orientation flight in the honeybee revealed by harmonic radar. *Nature* 403, 537–540.

Cardé, R.T. and Mafra-Neto, A. (1997) Mechanisms of flight of male moths to pheromone. In: Cardé, R.T. and Minks, A.K. (eds) *Insect Pheromone Research: New Directions*. Chapman & Hall, New York, pp. 275–290.

Cardé, R.T. and Minks, A.K. (eds) (1997) *Insect Pheromone Research: New Directions*. Chapman & Hall, New York.

Chittka, L., Thomson, J.D. and Waser, N.M. (1999) Flower constancy, insect psychology, and plant evolution. *Naturwissenschaften* 86, 361–377.

Coelho, J.R. (1989) The effect of thorax temperature and body size on flight speed in honey bee drones. In: *Proceedings of the American Bee Research Conference. American Bee Journal* 129, 811–812.

Collett, T.S., Dillman, E., Giger, A. and Wehner, R. (1992) Visual landmarks and route following in desert ants. *Journal of Comparative Physiology A* 170, 435–442.

Cooper, A.J. (1993) Limitations of bumblebee flight performance. PhD thesis, University of Cambridge.

Cresswell, J.E., Osborne, J.L. and Goulson, D. (2000) An economic model of the limits to foraging range in central place foragers with numerical solutions for bumblebees. *Ecological Entomology* 25, 249–255.

Dingle, H. (1996) *Migration: the Biology of Life on the Move*. Oxford University Press, Oxford.

Dornhaus, A. and Chittka, L. (1999) Evolutionary origins of bee dances. *Nature* 401, 38.

Drake, V.A. and Farrow, R.A. (1988) The influence of atmospheric structure and motions on insect migration. *Annual Review of Entomology* 33, 183–210.

Dramstad, W.E. (1996) Do bumblebees (Hymenoptera, Apidae) really forage close to their nests? *Journal of Insect Behavior* 9, 163–182.

Dudley, R. (1995) Aerodynamics, energetics and reproductive constraints of migratory flight in insects. In: Drake, V.A. and Gatehouse, A.G. (eds) *Insect Migration, Tracking Resources through Space and Time*. Cambridge University Press, Cambridge, pp. 303–319.

Dukas, R. and Edelstein-Keshet, L. (1998) The spatial distribution of colonial food provisioners. *Journal of Theoretical Biology* 190, 121–134.

Dusenbery, D.B. (1989) Optimal search direction for an animal flying or swimming in a wind or current. *Journal of Chemical Ecology* 15, 2511–2519.

Esch, H.E. and Burns, J.E. (1996) Distance estimation by foraging honeybees. *Journal of Experimental Biology* 199, 155–162.

von Frisch, K. (1967) *The Dance Language and Orientation of Bees* (translated by Leigh E. Chadwick). Oxford University Press, Oxford.

von Frisch, K. and Lindauer, M. (1955) Über die Fluggeschwindigkeit der Bienen und über ihre Richtungsweisung bei Seitenwind. *Naturwissenschaften* 42, 377–385.

Hedenström, A. and Alerstam, T. (1995) Optimal flight speed of birds. *Philosophical Transactions of the Royal Society of London B* 348, 471–487.

Heinrich, B. (1976) The foraging specializations of individual bumblebees. *Ecological Monographs* 46, 105–128.

Heran, H. (1964) Wie überwacht die Biene ihren Flug? *Umschau* 10, 299–303.

Heran, H. and Lindauer, M. (1963) Windkompensation und Seitenwindkorrektur der Bienen beim Flug über Wasser. *Zeitschrift für vergleichende Physiologie* 47, 39–55.

Hodges, C.M. (1985) Bumble bee foraging: the threshold departure rule. *Ecology* 66, 179–187.

Johnson, C.G. (1969) *Migration and Dispersal of Insects by Flight*. Methuen, London.

Josephson, R.K., Malamud, J.G. and Stokes, D.R. (2000) Power output by an asynchronous flight muscle from a beetle. *Journal of Experimental Biology* 207(17), 2667–2689.

Kammer, A.E. and Heinrich, B. (1978) Insect flight metabolism. *Advances in Insect Physiology* 13, 123–128.

Lehrer, M. (1991) Bees which turn back and look. *Naturwissenshaften* 78, 274–276.

Lindauer, M. (1976) Foraging and homing flight of the honey-bee: some general problems of orientation. In: Rainey, R.C. (ed.) *Insect Flight*. Symposia of the Royal Entomological Society, No. 7. Blackwell, Oxford, pp. 199–216.

McCartney, H. A. (1997) Physical factors in the dispersal of aerobiological particles. In: Agashe, S.N. (ed.) *Aerobiology*. Oxford & IBH Publishing Co., New Delhi, pp. 439–450.

Menzel, R. (1999) Memory dynamics in the honeybee. *Journal of Comparative Physiology A* 185, 323–340.

Osborne, J.L., Clark, S.J., Morris, R.J., Williams, I.H., Riley, J.R., Smith, A.D., Reynolds, D.R. and Edwards, A.S. (1999) A landscape scale study of bumble bee foraging range and constancy, using harmonic radar. *Journal of Applied Ecology* 36, 519–533.

Park, O.W. (1923) Flight studies of the honeybee. *American Bee Journal* 71, 71.

Park, O.W. (1928) Time factors in relation to the acquisition of food by the honeybee. *Research Bulletin of the Iowa Agricultural Research Station* 108, 185–225.

Pennycuick, C.J. (1969) The mechanics of bird migration. *Ibis* 111, 525–556.

Pyke, G.H. (1979) Optimal foraging in bumblebees: rule of movement between flowers within inflorescences. *Animal Behaviour* 27, 1167–1181.

Rasmussen, I.R. and Brødsgaard, B. (1992) Gene flow inferred from seed dispersal and pollinator behaviour compared to DNA analysis of restriction site variation in a patchy population of *Lotus corniculatus* L. *Oecologia (Berl.)* 89, 277–283.

Reynolds, D.R. (1988) Twenty years of radar entomology. *Antenna* 12, 44–49.

Reynolds, D.R. and Riley, J.R. (1997) Flight behaviour and migration of insect pests: radar studies in developing countries. *NRI Bulletin* 71. Natural Resources Institute, Chatham.

Ribbands, C.R. (1953) *The Behaviour and Social Life of Honeybees*. Bee Research Association, London.

Riley, J.R. (1989) Remote sensing in entomology. *Annual Review of Entomology* 34, 247–271.

Riley, J.R. (1993) Flying insects in the field. In: Wratten, S.D. (ed.) *Video Techniques in Animal Ecology and Behaviour*. Chapman & Hall, London, pp. 1–15.

Riley, J.R., Smith, A.D., Reynolds, D.R., Edwards, A.S., Osborne, J.L., Williams, I.H., Carreck, N.L. and Poppy, G.M. (1996) Tracking bees with harmonic radar. *Nature* 379, 29–30.

Riley, J.R., Valeur, P., Smith, A.D., Reynolds, D.R., Poppy, G. and Löfstedt, C. (1998) Harmonic radar as a means of tracking the pheromone-finding and pheromone-following flight of male moths. *Journal of Insect Behavior* 11, 287–296.

Riley, J.R., Reynolds, D.R., Smith, A.D., Edwards, A.S., Osborne, J.L., Williams, I.H and McCartney, H.A. (1999) Compensation for wind drift by bumble bees. *Nature* 400, 126.

Sanders, C.J. (1997) Mechanisms of mating disruption in moths. In: Cardé, R.T. and Minks, A.K. (eds) *Insect Pheromone Research: New Directions*. Chapman & Hall, New York, pp. 333–346.

Schaefer, G.W. (1969) Radar studies of locust, moth and butterfly migration in the Sahara. *Proceedings of the Royal Entomological Society of London C* 34, 33 and 39–40.

Schmid-Hempel, P., Kacelnik, A. and Houston, A.I. (1985) Honeybees maximize efficiency by not filling their crop. *Behavioral Ecology and Sociobiology* 17, 61–66.

Schmidt-Koenig, K. (1993) Orientation of autumn migration in the Monarch butterfly. *Natural History Museum of Los Angeles County, Sciences Series*, no. 38, 275–283.

Srygley, R.B., Oliveira, E.G. and Dudley, R. (1996) Wind drift compensation, flyways, and conservation of diurnal, migrant neotropical Lepidoptera. *Proceeding of the Royal Society of London B* 263, 1351–1357.

Teräs, I. (1976) Flower visits of bumblebees, *Bombus* Latr. (Hymenoptera, Apidae) during one summer. *Annales Zoologici Fennici* 13, 200–232.

Walker, T.J. (1986) Butterfly migration in the boundary layer. In: Rankin, M.A. (ed.) *Migration: Mechanisms and Adaptive Significance. Contributions in Marine Science* 27 (suppl.), 704–723.

Wehner, R. (1984) Astronavigation in insects. *Annual Review of Entomology* 29, 277–298.

Wellington, W.G. (1974) Bumblebee ocelli and navigation at dusk. *Science* 183, 550–551.

Wenner, A.M. (1963) The flight speed of honeybees: a quantitative approach. *Journal of Apicultural Research* 2, 25–32.

Williams, C.B. (1958) *Insect Migration*. Collins, London.

Williams, C.B., Cockbill, G.F., Gibbs, M.E. and Downes, J.A. (1942) Studies in the migration of Lepidoptera. *Transactions of the Royal Entomological Society London* 92, 236–280.

Willis, M.A., Murlis, J. and Cardé, R.T. (1991) Pheromone-mediated upwind flight of male gypsy moths, *Lymantria dispar*, in a forest. *Physiological Entomology* 16, 507–521.

Zimmerman, M. (1982) Optimal foraging: random movement by pollen collection bumblebees. *Oecologia* 53, 394–398.

# The Evolution of Migratory Syndromes in Insects

## Hugh Dingle

*Department of Entomology, University of California-Davis, One Shields Avenue, Davis, CA 95616, USA*

### Introduction: Migration and Migration Syndromes

The way we think about insect migration is primarily due to the work of four pioneers in the analysis of insect movements: C.G. Johnson, J.S. Kennedy, T.R.E. Southwood and L.R. Taylor. Johnson and Kennedy stressed that insect movements vary with respect to physiology and function, and their ideas revamped our notions about the behavioural and life history aspects of movements, especially insect migration (Johnson, 1960, 1963; Kennedy, 1961, 1966). In parallel studies, Southwood (1962) showed that the type of habitat determined the likelihood of migration among insect species, and Taylor (1961) noted the importance of movement in the distribution of populations. In this contribution on migration syndromes, I shall be mostly following on the work of Johnson and Kennedy, but the ecological consequences of movement are of equal significance and, indeed, are capably addressed by other contributors to this symposium.

To understand migration we need to view it in the context of the different sorts of movements that characterize organisms. These movements can be roughly divided between two general categories (Kennedy, 1985; Dingle, 1996). The first category consists of what can be described as 'station-keeping' responses that generally keep the organism within its home range, the area within which it carries out most of its life functions and spends most of its time. Included within station-keeping movements are kineses, many taxes, foraging, territorial behaviour and commuting, which is a periodic, usually although not always daily, round trip for resources. Foraging may be for any resource including food, shelter or mates, and commuting, which can also be considered extended foraging, may cover considerable distances, 3000 km in some albatrosses, for example (Jouventin and Weimerskirch, 1990). A characteristic of all these movements is that they cease when a resource is encountered, as when a foraging predator stops and eats when it makes its kill or a female butterfly stops searching when she finds a suitable host plant on which to oviposit.

Movements of the second major category take an organism beyond its home range (Dingle, 1996). One of these is ranging, the movement over an area to explore it and locate a new area of residence (Jander, 1975). Typical examples of ranging include movements to new territories (or home ranges) by young mammals and birds (often called 'natal dispersal'). Like station-keeping responses, ranging is a facultative response and ceases as soon as an appropriate resource is encountered, in this case a previously unclaimed living space. The second type of behaviour that takes an organism beyond its current home range is migration. It differs from all the above movements, including ranging, because inputs from resources or other stimuli that would usually stop the movement fail to do so. A characteristic of migration is that the organism undertaking it is undistracted by food or mates, those 'vegetative stimuli', in Kennedy's terminology, that are otherwise so necessary a part of its life functions. Furthermore, migration is usually triggered by cues such as photoperiod that act as surrogates for habitat change, rather than directly by a change in the current state of resources. Note that I have not used the term 'dispersal' to describe any individual movements because it is one of the population outcomes of movement, one of the ecological consequences that should be distinguished from the behaviour of individuals (Kennedy, 1985; Gatehouse, 1987; Dingle, 1996). Dispersal is a process that increases the distances among individuals in a population (Southwood, 1981). Other population consequences of individual movement include congregation, brought about by mutual attraction, and aggregation, where individuals collect in a favourable habitat.

The distinct nature of migratory behaviour was precisely delineated by Kennedy in his studies of the flight of the summer parthenogenetic virginoparae of *Aphis fabae* Scop. (Kennedy, 1958; Kennedy and Booth, 1963a,b, 1964; Kennedy and Ludlow, 1974). He used a flight chamber that allowed an aphid to fly freely in response to downward directed light and wind. The aphid would remain in the light beam, and its rate of climb could be controlled by varying the wind speed. Landings and take-offs could be regulated by presenting a leaf or other landing surface at the end of a lever arm. Several important characteristics of migratory flight were revealed by the flight chamber experiments. First, the migratory aphids would not initiate larviposition until they had undertaken some migration. Second, landing responses were primed by the migratory flight: the longer the flight the lower the threshold for landing. Finally, there was reciprocal interaction between flight and settling because settling responses involved in probing a leaf to test its suitability and subsequent moving to the underside of a leaf to larviposit could prime further flight if they failed to go to completion (attaching, larvipositing). This flight after incomplete settling could actually be stronger than the take-off and initiation of migration. Kennedy likened the flight/settling interaction to Sherringtonian reflexes because the stronger flight after settling (and vice versa) so resembled the rebound of a reflex after inhibition. Suffice to say here, it was apparent that migration was qualitatively different from other movement behaviours because station-keeping responses, landing, probing, etc., were inhibited by flight, but flight also primed them to promote their later recurrence. It was further evident that because of the interaction between flight and landing/reproduction, migration involves a syndrome of traits and not just movement alone. Subsequent studies (David and Hardie, 1988) have revealed that migration is stronger and the migratory syndrome is even more evident in the autumn sexual gynoparae of *A. fabae* and other aphids flying back to their winter woody hosts.

The existence of migratory syndromes was clearly recognized by Johnson who coined the term 'oogenesis-flight syndrome' in his book, *Migration and Dispersal of Insects by Flight* (1969). Johnson focused on the observation that in a high proportion of migratory insects, flight was limited to individuals with immature reproductive systems, and that this characteristic of migrants was especially likely to be true for females. He stressed that 'the adaptive process of migration by female insects is based on the relation of the development of ovaries to that of the flight apparatus' (Johnson, 1969, p. 9). Implicit in this notion was the assumption that migration and reproduction were alternative physiological states with trade-offs in the mobilization of energy and materials. This put the oogenesis-flight syndrome explicitly within the context of life histories. Johnson also postulated that the syndrome would be mediated by juvenile hormone, a postulate that has now been amply demonstrated (Rankin *et al.*, 1986; Rankin, 1991; Dingle, 1996).

The notion that the 'adaptive process of migration' led to a syndrome of traits indicated that natural selection was acting not on movement alone but rather on the complexes of traits that made up migratory syndromes. That being the case, it seemed likely that the traits making up the syndromes would share genes in common, or in other words, there would be genetic correlations among the traits, and the existence and strength of these correlations could themselves be the objects of selection (Bradshaw, 1986). In my laboratory we have examined the evolution and genetics of migratory syndromes in the context of life histories with comparative studies on two species of seed-feeding bug, the milkweed bug, *Oncopeltus fasciatus* (Dallas) (Lygaeidae), and the soapberry bug, *Jadera haematoloma* (Herrich-Schaefer) (Rhopalidae). I shall discuss some of our results with each of these species in turn. In the case of the soapberry bug, the syndrome involves a wing polymorphism, and I precede discussion of this species with a brief outline of polymorphisms in other species, focusing on crickets.

## Migration Syndromes in *Oncopeltus*

In our studies of *Oncopeltus* we took advantage of the fact that both migratory and non-migratory populations occur within the range of the species, and we were able to address the issue of whether similar or different flight/life history syndromes existed in the two types of bug. For a migratory population we used bugs collected in Iowa in the upper midwest of North America. This population invades milkweed (*Asclepias syriaca*) patches in the late spring and early summer at the latitude of Iowa, produces one and sometimes a partial second generation of offspring, and under the influence of short autumn days enters an adult reproductive diapause and migrates to more southerly regions to overwinter, probably mostly along the coast of the Gulf of Mexico (Dingle, 1981). Flight tests in the laboratory, using a tethered flight technique whereby a bug is mounted on a thin stick glued to the prothorax and its flight reflex activated by lifting it free of the substrate (Dingle, 1965), demonstrated that these migratory bugs frequently flew for periods of several hours. Furthermore, as with *A. fabae*, flight primes reproduction as females flying for long periods begin ovipositing earlier than unflown females (Slansky, 1980). The non-migratory population came from the Caribbean island of Puerto Rico. It is subject to a relatively benign environment throughout the year with continuous breeding and no selection for migration. When tested using tethered flight, very few individuals flew for more than a few seconds, and even the longest flights failed to match durations common to bugs from Iowa. Summaries of the ecology, genetics and general flight characteristics of these bugs can be found in Dingle (1981, 1996).

We used artificial selection in the laboratory to assess syndromes in the migratory and non-migratory populations (Palmer and Dingle, 1986, 1989; Dingle and Evans, 1987; Dingle *et al.*, 1988; Dingle, 1994). We first selected on wing length because this is a morphological character known to be positively associated with migration in many organisms. In both populations we created replicated selected lines, with directional selection for both long and short wings, and unselected controls. Both populations responded to selection on wing length so that short- and long-winged lines were created readily. Also, in both populations there was a correlated response in body size, with long-winged bugs larger and short-winged smaller than their respective controls. This result is *de facto* evidence for a positive genetic correlation between wing length and size, indicating that the traits share genes.

In other ways, however, the Iowa and Puerto Rico bugs responded quite differently to selection on wing length. Two correlated responses, in addition to size, were evident in the migratory Iowa bugs; these were fecundity and tethered flight duration. Our fecundity measure was the number of eggs oviposited by a female in her first 5 days of reproductive life. Early reproduction is what usually contributes most to population growth rate ($r$ or the 'intrinsic rate of increase') and is likely to be especially important for migrant insects invading new habitats. In the migratory population of milkweed bugs this fecundity measure responded positively to selection on wing length; long-winged, larger bugs also produced more early eggs. What was interesting was that there was no similar response in fecundity in the non-migratory population, also demonstrating that in this case the response in fecundity was not simply a consequence of altered overall size. The second correlated response in the migratory bugs, increased length of tethered flights, occurred only in the long-winged line; there was no difference in the average flight duration of the control and short-winged lines. Here again the response differed from the non-migratory population because in the latter we failed to change flight durations, which were already brief. These results indicate that: (i) there is a syndrome of traits involving flight, fecundity and wing length in the migratory bugs that is absent in the non-migratory bugs; (ii) because selection on wing length resulted in correlated responses in flight and fecundity, these traits share genes in common; and therefore (iii) natural selection has produced a distinct migratory syndrome in the migrant population based in genetic correlations.

Subsequent selection on flight itself confirmed the syndrome in the migratory bugs (Palmer and Dingle, 1989). Selection for long and short tethered flights was applied for two generations, enough to produce a strong divergent response (Fig. 8.1). This relatively brief period of selection was also sufficient to produce correlated responses in wing length and fecundity with longer wings and more eggs in the first 5 days of reproduction in the long flying lines. Direct selection on flight thus produced the same migratory syndrome as predicted from the results of selection on wing length, again indicating that the three traits shared pleiotropic genes.

In addition to demonstrating the genetic basis for migratory syndromes, it is also important to determine the proximate causes of the genetic correlations. The most likely candidates are hormones, because they lie between genes, behaviour and life history traits and thus can produce correlated

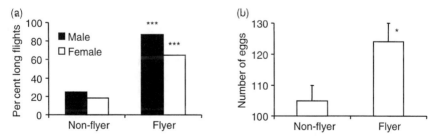

**Fig. 8.1.** (a) Selection for long flights (30 min) in the milkweed bug, *Oncopeltus fasciatus*. Flight is significantly longer in 'Flyer' bugs after two generations of selection. (b) Correlated response in fecundity, measured as eggs produced in the first 5 days of reproduction, after two generations of selection on flight. ***$P < 0.001$; *$P < 0.05$. Modified from Dingle (1996) after Palmer and Dingle (1989).

effects with potentially important fitness consequences (Ketterson and Nolan, 1992). Johnson (1969) and others had suggested that the oogenesis-flight syndrome was coordinated by juvenile hormone (JH) because of its role in the development of the reproductive system in adult female insects. This suggestion has been followed up for *Oncopeltus* by Rankin and her students and colleagues.

Using both topical applications of JH mimics and ablation and transplant experiments with corpora allata, Rankin and her associates demonstrated that JH stimulated migratory flight in *Oncopeltus* as well as in other insects (Rankin et al., 1986; Rankin, 1989, 1991). A typical pattern of effects is that JH application results in increased proportions of pre-reproductive individuals displaying long duration tethered flights characteristic of migratory insects. Once reproduction is initiated, JH has little or no influence on the proportion of long duration flights.

Because JH stimulates both migratory flight and reproduction, the question that arises is how the trade-off between these two activities is regulated. Rankin proposed that at intermediate titres migratory flight was stimulated, while at high titres it was reproduction that resulted; low levels of hormone would have no effect on either activity. The model was tested by using artificial selection (Rankin, 1978). Normally migration in *Oncopeltus* reared at around 23–24°C begins at about 12 days post-eclosion. Rankin selected for delayed flight so that migration did not start until adults were some 30 days old and then compared the delayed bugs with unselected controls (Fig. 8.2). There were two correlated responses to delayed flight. The first was delayed oviposition, and the second was a prolonged period of intermediate JH titres. Flight tests using the delayed bugs showed clearly that increased flight occurs at these intermediate JH titres, but that once the hormone reaches levels sufficient to induce oogenesis, most long duration flight ceases. More recently McNeil and colleagues (McNeil et al., 1995), working with the noctuid armyworm moth, *Pseudaletia unipunctata* (Haworth), have demonstrated that in addition to titres *per se*, changes in the ratio of different molecular forms of JH may also be involved in the regulation of migration. Ecdysteroids, which regulate the competence of cells to respond to JH (Gilbert et al., 1996), may also be necessary for the appropriate response. In any case, it is apparent that in migratory syndromes hormones, and especially juvenile hormone, mediate between the genes and behaviour, although the exact details of this mediation remain to be discovered.

## Wing Polymorphisms as Syndromes

Migration by flight is an important component of many insect life histories, but it imposes a cost. In the first place, energy and materials invested in wings and flight muscles are unavailable for other functions, in particular for reproduction. Second, there is often an increased rate of mortality during

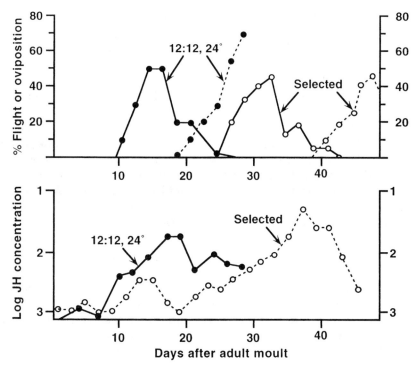

**Fig. 8.2.** Relations between migration, oviposition and juvenile hormone (JH) titre in milkweed bugs reared at LD 12:12 and 24°C. Top panel: tethered flight of more than 30 min (solid lines) or oviposition (dotted lines) in unselected bugs (labeled '12:12, 24°') versus bugs selected for delayed onset of flight. Bottom panel: JH titres in the selected versus unselected lines. Note that flight occurs at intermediate hormone titres and oviposition occurs at maximum levels. From Dingle (1996) after Rankin (1978).

migration. Because of these costs and risks it is not surprising that selection has favoured the evolution of alternative life styles when migration is not necessary. Frequently, these alternatives take the form of wing polymorphisms and polyphenisms, with a brachypterous or apterous morph occurring under conditions where migration is unnecessary or even selected against. Thus, short wings and high reproductive rates are favoured in stable or seasonally rich habitats, and long-winged migrants with lower reproductive rates occur or are produced in ephemeral habitats or under seasonal stress (Dingle, 1996).

Insect wing polymorphisms are attractive models for the investigation of syndromes involving flight and life history characteristics because costs are usually expressed as clear trade-offs between flying and flightless morphs (Dingle, 1996; Zera and Denno, 1997). Typically there is reduced reproductive effort in long-winged morphs. This is manifested as reduced fecundity and especially as delayed reproduction relative to apters or brachypters, although there are cases where selection has apparently acted to minimize the trade-off (Rankin and Burchsted, 1992). These influences on reproduction are so like the oogenesis-flight syndrome described for known migrants that it seems likely that the winged forms of many, if not most, wing polymorphic insects are in fact true migrants, although detailed studies such as those on *O. fasciatus* or *A. fabae* are lacking for those species where physiological and morphological trade-offs have been studied in detail. The energetic basis for the trade-off between flight and reproduction seems to lie in variation in the flight muscles even more than in the wings themselves because of the energy and materials required to maintain these muscles. The variation is expressed in histolysis before or after the occurrence of flight, in the properties of muscle fibres or metabolic enzymes, or as the absence of flight muscles even in individuals that are macropterous (Dingle, 1996).

The contribution of the wing muscles to the trade-off between morphs can be considerable. Not only is there much more muscle mass in long-winged flying forms, but the metabolism of muscle in these individuals is much greater. In the crickets, *Gryllus firmus* Scudder and *Gryllus assimilis* (F.), respiration rates for the active pink flight muscles of long-winged individuals can be double that of the reduced white muscle that is predominant in short-winged animals or in animals treated with JH mimics (Zera *et al.*, 1997, 1998). Activity differences in the mitochondrial enzyme citrate synthase were even more dramatic with 30-fold higher levels of whole-muscle activity in pink muscle in the polymorphic *G. firmus* (Zera *et al.*, 1997). Egg production was reduced in individuals with active pink muscle relative to those with white muscle, regardless of whether the latter had long or short wings. Muscle properties thus had more influence on the trade-off between flight and reproduction than did wing length. To complete the picture of trade-offs involved in wing polymorphic syndromes, it should be noted that a relation between wing length and reproduction can also occur in males. In *G. firmus*, short-winged males call for longer periods than their long-winged counterparts and attract more mates (Crnokrak and Roff, 1998).

The trade-off between reproduction and flight muscle maintenance is further suggested by flight-muscle histolysis which also seems to involve JH. In cotton stainer bugs, *Dysdercus* spp. (Davis, 1975; Nair and Prabhu, 1985), and bark beetles, *Ips* spp. (Borden and Slater, 1969; Unnithan and Nair, 1977), the protein from histolysed wing muscles is used for oocyte development with the process initiated following migratory flight. In *Dysdercus*, flight muscle histolysis is clearly a component of a migratory oogenesis-flight syndrome (Dingle and Arora, 1973). Applications of JH or its mimics can result in flight muscle degeneration. This relation is not always simple, however. In the Colorado potato beetle, *Leptinotarsa decemlineata* (Say), low JH titres in the autumn apparently result in flight-muscle degeneration prior to diapause, whereas in the spring, increasing JH levels are associated with regeneration of wing muscles, migration from diapause sites and eventually reproduction (de Kort, 1969; de Kort *et al.*, 1982). By histolysing flight muscles prior to diapause, these beetles avoid the high energetic cost of maintaining this active muscle tissue over the winter. These results suggest that JH may produce opposite responses in different insects, depending on the ecophysiology of the particular syndrome.

As with *Oncopeltus*, selection experiments and analyses of pedigrees have demonstrated that the wing polymorphic syndrome in crickets is based on evolved genetic correlations and thus on the sharing of genes among the contributing traits (Roff, 1994; Fairbairn and Yadlowski, 1997; Roff *et al.*, 1997, 1999). Selection for proportion of long-winged *G. firmus* resulted in selected lines with both higher and lower proportions of macroptery relative to controls, demonstrating the genetic basis for the trait. Furthermore, there was a correlated response in fecundity, although it was asymmetric. There was a statistically significant reduction in the egg output of macropterous females, but no corresponding increase in egg production in the micropters, demonstrating a genetic correlation between wing morph and fecundity. Roff and colleagues (1999) postulate that this asymmetry results because short-winged females always have small flight muscles, so that the impact of the trade-off between muscles and egg production cannot be dramatically changed in these morphs. Selection directly on fecundity confirmed the genetic correlation between this trait and wing morph. In lines selected for higher fecundity, there was a corresponding increase in the frequency of micropters, and conversely, when low fecundity was selected, there was an increase in the proportion of macropters. The cricket wing polymorphic syndrome is thus based on genetic correlations between wing form and fecundity, paralleling the flight fecundity relationship in migratory *Oncopeltus*.

The resemblance between the micropterous morph and juvenile stages led Kennedy and Stroyan (1959) and Southwood (1961) to posit that elevated juvenile hormone titre during development blocks growth and differentiation of wings and wing muscles in short-winged morphs. Concomitantly, the increased JH promotes earlier ovarian growth and higher fecundity. If this is the case, then JH serves as a mediator between genes and the manifestation of the polymorphic syndrome. There are ongoing efforts to test the model. For example, in *A. fabae* and the planthopper *Nilaparvata lugens* (Stål), topically applied JH or its analogues caused development to shift from the macropterous to the flightless morph (Hardie, 1980; Iwanaga and Tojo, 1986), although similar experiments have failed to produce an effect in other aphids (Zera and Denno, 1997). The interpretation of these results is complicated by the fact that the chemical structure of JH in Hemiptera has not been determined. In *Gryllus rubens* Scudder, individuals destined for macroptery can be redirected to the short-winged morph if JH-III is applied during the penultimate or last nymphal stadia (Zera and Tiebel, 1989). There is thus some support for the JH model.

Most attention regarding the model has, however, focused on juvenile hormone esterase (JHE), the enzyme that degrades and regulates JH in many insects (Hammock, 1985). The hypothesis here is that the production of JH itself may not be what directs development toward long or short wings, but rather its breakdown by JHE. The presumption is that levels of JHE differ between morphs, with higher JHE levels, especially in the last nymphal instar, resulting in greater degradation and hence reduced titres of JH. Thus macroptery should be associated with high JHE in the final nymphal stadium and microptery with low JHE. In *G. rubens*, JHE activity at the middle of the last instar was higher in individuals destined for macroptery than it was in those destined to be short winged (Zera and Tiebel, 1989, and Fig. 8.3). This critical modulation of the JH titre occurred after the major decline in JH that takes place early in the instar and permits metamorphosis into the adult. It is apparently the subtle elevation of JH in the middle of the instar produced when JHE is low that directs development to microptery (Zera and Tobe, 1990). There is also some evidence suggesting that JH interacts with a reduced titre of ecdysteroids (Zera et al., 1989).

The evidence thus strongly suggests that JH/JHE is part of the wing polymorphic syndrome in crickets. The question of whether the syndrome results from shared genes has been addressed directly with both selection experiments and pedigree analysis (Fairbairn and Yadlowski, 1997; Roff et al., 1997; Fairbairn and Roff, 1999). In the former, two replicate populations of *G. firmus* were selected for long and short wings, each with an appropriate control. In each case there was a correlated response in JHE in the final nymphal instar; selection for macroptery resulted in increased JHE activity as predicted from the data of Zera and Tiebel (1989) on *G. rubens*,

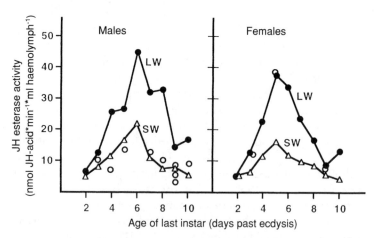

**Fig. 8.3.** Relation between JHE activity and occurrence of long- (LW) or short-winged (SW) individuals of *Gryllus rubens*. LW and SW are genetic stocks measured during last nymphal instar. Open circles are measurements on LW line of individuals that eclosed as SW adults. From Dingle (1996) after Zera and Tiebel (1989).

and selection for microptery resulted in decreased activity. Level of JHE activity is therefore genetically correlated with wing morph, and the two traits share pleiotropic genes in common. Pedigree analysis (Roff *et al.*, 1997) confirmed these genetic correlations and also demonstrated genetic correlations between JHE and fecundity (lower JHE, higher fecundity) and between fecundity and wing length (shorter wings, higher fecundity). The syndrome thus involves pleiotropically based lower JHE and higher fecundity for microptery and the opposite for macroptery. Levels of JHE are not the whole story for mediating between the genes and the polymorphism (Fairbairn and Roff, 1999; Zera, 1999), because if they were, genetic correlations between JHE and the other two traits would approximate unity, and they do not, but the genetic correlations do indicate that JHE contributes in significant ways to an evolved syndrome of wing polymorphism.

## The Wing Polymorphism of the Soapberry Bug

The soapberry bug, *Jadera haematoloma*, occurs across the southern tier of the USA from Arizona to Florida and as far north as Kansas, as well as on Caribbean islands and into Central America. Hosts are assorted bushes and trees belonging to the family Sapindaceae where the bugs feed on seeds occurring encapsulated in the fruits. Within the different parts of the range, the bugs feed almost exclusively on the seeds of a single species of host. The fruits of the different hosts vary considerably in size, but in all cases seeds occur at the centre of the fruit capsule and thus at different distances from the capsule wall, depending on species. One of the interesting characteristics of the host-specific *Jadera* populations is that the length of the mouthpart stylets (the 'beaks') is a positive function of the distance from fruit wall to seed in the host, a distance across which the beak must reach if the seed is to be penetrated (Carroll and Boyd, 1992). Beak length is relatively independent of body size.

A further interesting aspect of the biology of the soapberry bug is that, in parts of its range, exotic sapindaceous plants have been introduced, the bugs have transferred to these new hosts, and there has been a rapid evolution in both beak length and life-history characters on the introduced hosts (Carroll and Boyd, 1992; Carroll *et al.*, 1997, 1998). The best-studied example occurs in Florida. In the southern tip of Florida and in the Keys, the host plant is the native balloon vine, *Cardiospermum corindum*. This vine has a hollow fruit with a papery wall, and it is several millimetres between the wall and the seeds which are aligned centrally on septa. The resident population of soapberry bugs possesses relatively long beaks (*c.* 8+ mm). In contrast, populations of bugs in northern Florida have switched to feeding on the seeds of the 'flat-podded' goldenrain tree, *Koelreuteria elegans*, introduced from southeast Asia as a street tree shortly after World War II. The fruits of this tree are flattened, and seeds occur within a millimetre or two of the pod wall. Correspondingly beak lengths of these *Jadera* populations average only about 6.5 mm.

Detailed studies of the two Florida *Jadera* populations indicate they are more closely related to each other than to populations elsewhere in the range of the species (Carroll and Boyd, 1992). Furthermore, the introduction of the goldenrain tree can be dated with precision, because both the US and Florida departments of agriculture keep records of such introductions. Therefore the decrease in beak length in northern Florida, and an array of other differences in life history and host preferences, has evolved in the 50 or so years since goldenrain trees were introduced. The trees produce enormous quantities of fruit, in the order of thousands of pods, and it was probably of considerable advantage to switch to them, given the relative rarity of the balloon vine in this region. Balloon vine also produces few pods at any one time, seldom more than a hundred. Museum specimens of bugs also produce a sort of 'fossil record' and clearly show a transition in northern Florida from long beaks before 1950 to short beaks thereafter. There is thus little doubt that there has been a rapid evolution on the new host of a syndrome of traits involving beak length and life histories. Interestingly, in Texas, Oklahoma and Kansas there has been a similar evolution, but in the opposite direction from short to long beaks. There, the introduced goldenrain tree species, *Koelreuteria paniculata*, has larger more inflated fruits than the native host, the soapberry, *Sapindus saponaria*.

What is of particular interest here is that the soapberry bug is wing polymorphic, and differences in the biology of the polymorphism have evolved along with other life-history differences in the Florida populations feeding on different host plants. The polymorphism consists of four distinct morphs (Dingle and Winchell, 1997): (i) a fully winged form that retains the flight muscles and that flies throughout adult life (L+); (ii) a fully winged form that histolyses the wing muscles at about the time of mating (Lh); (iii) a fully winged form in which wing muscles are never present (L−); and (iv) a brachypterous form that also lacks wing muscles (S). Laboratory crosses within and between wing morphs demonstrated a genetic basis to the polymorphism, but it was not a simple one.

The frequencies of the different wing morphs differ between the 'ancestral' populations of bugs feeding on native balloon vine and 'derived' populations feeding on the goldenrain tree. This difference is most apparent in the absence of wing muscles among virgin macropters as shown in Table 8.1 which compares an ancestral population from Key Largo with a derived population from Leesburg in central Florida. A significantly higher proportion of the winged bugs in the derived population lacks wing muscles. What is interesting is that the frequency of the S form, also flightless, does not differ much; the evolution of a higher frequency of flightlessness has apparently evolved more conservatively by eliminating flight muscles but not necessarily by shortening wings, probably because little further is gained in energetic savings by the relatively small differences in wing morphology. Field studies suggest that flight is more likely to be favoured by selection in the ancestral populations because the host balloon vines are scattered and individual vines produce fruit only at low densities, in contrast to the synchronous mass fruiting of goldenrain trees. The ancestral habitat is thus patchy, and migration is likely to be a necessary part of a life-history syndrome (although ranging rather than true migration cannot be ruled out).

Life-history differences between ancestral and derived populations are also evident with respect to reproduction (Table 8.2). As with other insects displaying wing polymorphisms, the reproductive output of the brachypterous form is higher, but the contrast between morphs is much greater in the ancestral host race. In these ancestral populations, there is a large difference in age at first reproduction with brachypterous females initiating egg-laying much earlier; in derived populations the difference in age at first reproduction is much less and is not statistically significant. This difference between host races is probably the result of a higher proportion in the derived population of winged females that lack wing muscles (L–) and reproduce early. Note also that there are life-history differences between host races that occur irrespective of wing morph. In the ancestral population, egg mass is greater but egg production is lower than in the derived bugs. These

**Table 8.1.** Frequency of absence of wing muscles in virgin macropters of ancestral and derived populations of the soapberry bug *Jadera haematoloma*.

| | | Wing muscles | | |
|---|---|---|---|---|
| Sex | Population | Yes | No | $P$ value $\chi^2$ |
| Females | Ancestral | 44 | 14 | |
| | Derived | 40 | 35 | = 0.01 |
| Males | Ancestral | 47 | 9 | |
| | Derived | 19 | 14 | < 0.01 |

**Table 8.2.** Reproductive contrasts between long- and short-winged females of ancestral and derived populations of soapberry bugs.

| Population/trait | Long | Short |
|---|---|---|
| Ancestral | | |
| Age 1st reproduction | 23 ± 27 (23)*** | 4.8 ± 1.4 (19) |
| Egg weight (mg) | 7.2 ± 0.56 (20) | 6.8 ± 0.53 (15) |
| Total eggs | 142 ± 109 (24) | 127 ± 99 (18) |
| Derived | | |
| Age 1st reproduction | 7.3 ± 4.0 (18) | 5.1 ± 1.7 (11) |
| Egg weight (mg) | 5.6 ± 0.66 (24) | 5.4 ± 0.48 (15) |
| Total eggs | 264 ± 191 (18) | 288 ± 184 (11) |

Number in brackets = sample size.
***$P < 0.001$.

differences are consistent across various sampled populations of the two host races (Carroll *et al.*, 1997, 1998) and indicate that life-history differences between races have evolved in concert with differences in beak lengths and wing-morph frequencies. These differences also demonstrate physiological differences in the nature of trade-offs between wing form and reproduction, in particular because of the contrast in the reproductive performance of the morphologically identical S morphs. Note the much higher egg outputs of the derived bugs.

Physiological differences between ancestral and derived bugs are also evident in flight-muscle enzymes and in the response to topically applied analogues of JH (Dingle and Winchell, 1997; Winchell *et al.*, 2000). In the enzyme analysis, activities were measured for key enzymes in oxidative metabolism (citrate synthase), glycolysis (hexokinase, pyruvate kinase) and fatty acid oxidation (β-hydroxyl CoA dehydrogenase or HOAD) in long-winged bugs. I shall limit my discussion here to citrate synthase, which is representative of the results with all four enzymes. The bugs were first assessed for ability to fly by the simple criterion of tossing them into the air; flight-capable individuals opened the wings and flapped briefly before landing (a few flew for some distance). The results for citrate synthase are shown in Fig. 8.4. In the ancestral population there were significantly higher levels of activity in bugs that flew compared with those that did not ($P < 0.05$), whereas in the derived host race there was no difference between flyers and non-flyers. This difference between flyers and non-flyers in the ancestral but not the derived bugs was also true for the other enzymes. These measurements of enzyme activity were done on a per mg of bug basis; similar results were obtained when the measures were assessed per mg of tissue protein with the exception of HOAD where the difference disappeared.

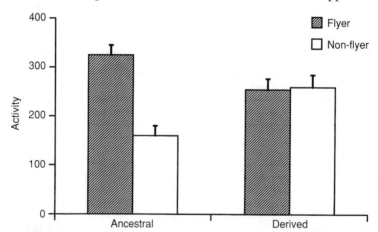

**Fig. 8.4.** Citrate synthase activity in wing muscles of long-winged *Jadera haematoloma* that either flew or failed to open the wings when tossed into the air. Ancestral and derived refer to the source of tested bugs (see text). Redrawn from Winchell (1998).

Responses to JH analogues also revealed host race differences. Methoprene was used in these experiments because of the uncertainty concerning the precise molecular structure of JH in hemipterans. In this experiment mating pairs were obtained from families (offspring of a single mated pair) that had displayed a high proportion of long-winged individuals in the expectation that their untreated offspring would likewise be mostly long winged. The progeny from each mated pair were divided between a control group, treated only with topical applications of acetone, and an experimental group, treated with methoprene dissolved in acetone (Dingle and Winchell, 1997). Families were not pooled; the identity of full-sibs was maintained throughout. All topical applications occurred during the middle of the fifth instar, and the resulting adults were scored for long or short wings, but not for muscle condition. The results indicated that in some, but not all, families methoprene treatment resulted in a higher frequency of long-winged individuals than in controls. Furthermore comparisons between host races in the number of families responding revealed that more derived families responded by producing S individuals than did families of the ancestral population ($G$ test, $P < 0.05$). Thus JH appears to have a role in determining adult wing morph, confirming results in other insects, but there is genetic variability both within and between populations in response to an exogenously applied JH analogue.

That morph phenotype was not solely due to genotype was demonstrated with a series of experiments in which samples from both host populations were subjected to varying densities, food levels and photoperiods (Dingle and Winchell, 1997; Winchell, 1998; Table 8.3). Here there were both treatment effects influencing the frequencies of the wing morphs and population × treatment interactions in response. An example of a treatment effect occurred with photoperiod in which there was an overall decline in frequency of macropters with shorter days (LD 14:10 to LD 10:14). Similarly there was an overall decline in macroptery when bugs were fed more seeds, but in this case there was also a population × seed supply interaction because the ancestral host race responded more strongly. Perhaps most interesting was the response to rearing density. Here there was no overall response to the experimental variation of density, but the interaction effect was significant

**Table 8.3.** Summary of significant sources of environmental variation ($P$ values from ANOVA) contributing to adult wing morph in the soapberry bug.

| Treatment | Source | | | P × T |
|---|---|---|---|---|
| | Population (P) | Family | Treatment (T) | |
| Density | 0.004 | 0.0002 | 0.33 | 0.046 |
| Photoperiod | 0.004 | 0.0001 | 0.0001 | 0.317 |
| Food level | 0.008 | 0.0001 | 0.0001 | 0.018 |

($P < 0.046$) with no change in wing-morph frequencies in ancestral bugs as rearing densities increased from 15 to 45 individuals per rearing container, but an increase in macroptery in the derived host race with this increase in density. In nature, density variation on host plants is much greater in the derived populations so the effect may reflect an adaptive response to leave overcrowded goldenrain trees.

A final manifestation of the wing polymorphism syndrome in soapberry bugs concerns the relation between beak length, body size and wing length. This was revealed by examining individuals from an ancestral population where it was apparent that wing morph tended to be a function of beak length, with long wings positively associated with longer beaks. A plot of both long- and short-winged phenotypes with body size (thorax width measured at the pronotum) as the independent variable and beak length as the dependent variable reveals this relationship and also indicates that it holds relatively independently of body size (Fig. 8.5). Examination of the graph reveals that long-winged bugs of the same size as short-winged still have substantially longer beaks. This means that selection for beak length,

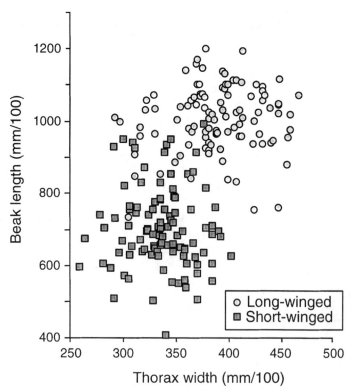

**Fig. 8.5.** Beak length versus size (measured by thorax width) in long- and short-winged *Jadera haematoloma* from the Florida Keys (ancestral). Note the relation of beak length to wing morph independent of size.

imposed by the size of the fruit capsule through which bugs must reach to pierce the seeds on which they feed, will also select for changes in the frequencies of the alary phenotypes and the syndrome associated with those phenotypes. Measurements on derived populations reveal the same relationship. The syndrome of migratory wing polymorphism of the soapberry bug thus includes the unexpected element of beak length. The evolution of the polymorphism cannot be understood without including the responses to selection engendered by fruit characteristics of the host plant on which a population feeds in addition to that engendered by the distribution of the host plants in space.

## Conclusions: the Nature of Migratory Syndromes

Once it is recognized that there is a distinction between migratory behaviour as a characteristic of individuals and the outcome of that behaviour as a characteristic of populations, it becomes possible to focus on individual traits that characterize migratory phenotypes. This in turn leads to questions concerning how selection acts on these phenotypes and what traits in particular might evolve to increase the fitness of migrants. Questions about the specific character of migratory behaviours lead into the realm of physiology, and questions concerning fitness require analysis of the genetics and place migration within the context of life histories. A full understanding of migration means assessing the range of traits that have evolved to function together to enhance fitness, or in other words, assessing what constitutes a migratory syndrome, the ecological context within which selection is acting on that syndrome, and the underlying genetic and developmental processes that determine the response to selection (Dingle, 1985).

One of the keys to understanding the physiology and development of migratory syndromes is surely JH. Even though details are not fully understood and hormones other than JH also play a role (Zera, 1999), studies from several insects indicate that varying titres of JH are associated either with migratory flight itself or with the development of wings, and especially flight muscle. Perhaps the most compelling evidence is that selection on migratory flight or on wing morph results in correlated responses in JH and its esterase, JHE, which breaks down JH and so regulates titre. This genetic correlation means that migratory traits and titres share genes in common and provides strong evidence of a relationship between hormonal titre and either migratory behaviour or wing form (Fairbairn and Roff, 1999). The physiology of flight, flight muscle and wing form extends to the level of the enzymes controlling the metabolic processes providing fuel for flight. The suite or syndrome of correlated phenotypic traits is thus mediated in important ways by hormones that act on several target tissues (Ketterson and Nolan, 1999).

At the level of morphology and life history, it is evident that there is consistency in some of the traits associated with migration. In the case of

*Oncopeltus*, migrants display higher fecundities than their non-migrant counterparts. In wing-polymorphic species there is a shift of reproductive functions to the brachypterous morph. In either case the relation between migration and reproduction is revealed by genetic correlations among the traits. The association of genetically correlated traits into syndromes implies selection acting on the correlations, either on their existence *per se*, on their sign (positive or negative) or on their strengths (Bradshaw, 1986). It also raises the question of whether multiple beneficial traits are all adaptations arising through the action of selection on the matrix of genetic correlations or whether some are exaptations or traits arising only as a result of their coexpression with traits responding to selection (Ketterson and Nolan, 1999). To answer questions about the action of selection and about which traits are likely to make up syndromes (and under what conditions) will require sophisticated analyses at multiple levels from genes and enzymes to ecology.

Answers to these questions are not a *priori* obvious. The association of beak length with the wing polymorphism of *Jadera* is both intriguing and humbling because it was certainly not an association we would have predicted. Nature is still full of surprises, and I suspect that we shall uncover many as we examine migratory syndromes in more detail. But surprises lead to insights, and that should be reason enough to explore the syndromes that insect migrants have evolved.

## Summary

C.G. Johnson and J.S. Kennedy defined insect migration in terms of an integrated set of physiological and behavioural traits, that is, a syndrome. This syndrome includes interactions between flight and settling and flight and reproduction and distinguishes migration from other types of movement. Studies of migratory milkweed bugs (*Oncopeltus*) indicate that migration syndromes consist of morphological, behavioural and life-history traits that share genes in common (i.e. are genetically correlated) and have evolved together. Selection on flight thus produces correlated responses in traits such as fecundity. Migration is stimulated by JH, which also influences other traits making up the evolved syndromes. Studies of wing polymorphisms, including flight-muscle variation, also support the notion of migratory syndromes and a trade-off between flight and reproduction. This trade-off is influenced by JH and its esterase and involves partitioning of energy and resources between flight-muscle maintenance and reproductive functions. Some syndromes consist of unexpected arrays of traits. In wing-polymorphic soapberry bug (*Jadera*) populations, rapidly evolving on introduced host plants, traits as varied as wing-muscle enzymes and trophic adaptations (stylet length) to host seed capsule size all evolve as a common syndrome. Understanding the evolution of insect migration thus requires placing behaviour in the context of ecology, genetics, physiology and life histories.

## Acknowledgements

This paper is dedicated to C.G. Johnson and J.S. Kennedy, two giants in the study of insect migration. I am grateful to the Royal Entomological Society for the opportunity to contribute to this Symposium. Roy Caldwell, Mary Ann Rankin, Jim Palmer and Ken Evans have contributed immeasurably over the years to understanding *Oncopeltus* migration. Scott Carroll introduced me to *Jadera*, and he and Ruth Winchell provided much of the data discussed here aided by an enthusiastic lab crew. I thank Marcel Holyoak for help with the figures and Maria Hope for her as always cheerful help in preparing the manuscript. The work in my laboratory has been supported by grants from the US National Science Foundation.

## References

Borden, J.H. and Slater, C.E. (1969) Flight muscle volume change in *Ips confusus* (Coleoptera: Scolytidae). *Canadian Journal of Zoology* 47, 29–32.

Bradshaw, W.E. (1986) Pervasive themes in insect life cycle strategies. In: Taylor, F. and Karban, R. (eds) *The Evolution of Insect Life Cycles*. Springer, New York, pp. 261–275.

Carroll, S.P. and Boyd, C. (1992) Host race radiation in the soapberry bug: natural history, with the history. *Evolution* 46, 1052–1069.

Carroll, S.P., Dingle, H. and Klassen, S. P. (1997) Genetic differentiation of fitness-associated traits among rapidly evolving populations of the soapberry bug. *Evolution* 51, 1182–1188.

Carroll, S.P., Klassen, S.P. and Dingle, H. (1998) Rapidly evolving adaptations to host ecology and nutrition in the soapberry bug. *Evolutionary Ecology* 12, 955–968.

Crnokrak, P. and Roff, D.A. (1998) The genetic basis of the trade-off between calling and wing morph in males of the cricket, *Gryllus firmus*. *Evolution* 52, 1111–1118.

David, C.T. and Hardie, J. (1988) The visual responses of free-flying summer and autumn forms of the black bean aphid *Aphis fabae*, in an automated flight chamber. *Physiological Entomology* 13, 277–284.

Davis, N.T. (1975) Hormonal control of flight muscle histolysis in *Dysdercus fulvoniger*. *Annals of the Entomological Society of America* 68, 710–714.

Dingle, H. (1965) The relation between age and flight activity in the milkweed bug, *Oncopeltus*. *Journal of Experimental Biology* 42, 269–283.

Dingle, H. (1981) Geographical variation and behavioral flexibility in milkweed bug life histories. In: Denno, R.F. and Dingle, H. (eds) *Insect Life History Patterns: Habitat and Geographic Variation*. Springer, New York, pp. 57–73.

Dingle, H. (1985) Migration and life histories. In: Rankin, M.A. (ed.) *Migration: Mechanisms and Adaptive Significance*. Contributions in Marine Science 27 (Suppl.), 27–42.

Dingle, H. (1994) Quantitative genetics of animal migration. In: Boake, C.R.B. (ed.) *Quantitative Genetic Studies of Behavioral Evolution.* University of Chicago Press, Chicago, pp. 143–164.

Dingle, H. (1996) *Migration: the Biology of Life on the Move.* Oxford University Press, New York.

Dingle, H. and Arora, G. (1973) Experimental studies of migration in bugs of the genus *Dysdercus. Oecologia* 12, 119–140.

Dingle, H. and Evans, K.E. (1987) Responses in flight to selection on wing length in non-migratory milkweed bugs, *Oncopeltus fasciatus. Entomologia Experimentalis et Applicata* 45, 289–296.

Dingle, H. and Winchell, R. (1997) Juvenile hormone as a mediator of plasticity in insect life histories. *Archives of Insect Biochemistry and Physiology* 35, 359–373.

Dingle, H., Evans, K.E. and Palmer, J.O. (1988) Responses to selection among life-history traits in a nonmigratory population of milkweed bugs (*Oncopeltus fasciatus*). *Evolution* 42, 79–92.

Fairbairn, D.J. and Roff, D.A. (1999) The endocrine genetics of wing polymorphism in *Gryllus*: a response to Zera. *Evolution* 53, 977–979.

Fairbairn, D.J. and Yadlowski, D.E. (1997) Coevolution of traits determining migratory tendency: correlated response of a critical enzyme, juvenile hormone esterase, to selection on wing morphology. *Journal of Evolutionary Biology* 10, 495–513.

Gatehouse, A.G. (1987) Migration: a behavioural process with ecological consequences? *Antenna* 11, 10–12.

Gilbert, L.I., Rubczynski, R. and Tobe, S. (1996) Endocrine cascade in insect metamorphosis. In: Gilbert, L.I., Tata, J.R. and Atkinson, B.G. (eds) *Metamorphosis.* Academic Press, San Diego, California, pp. 60–107.

Hammock, B.D. (1985) Regulation of juvenile hormone titre: degradation. In: Kerkut, G.C. and Gilbert, L.I. (eds) *Comprehensive Insect Physiology, Biochemistry and Pharmacology,* Vol. 7. Pergamon Press, Oxford, pp. 431–472.

Hardie, J. (1980) Juvenile hormone mimics the photoperiodic apterization of the alate gynopara of the aphid, *Aphis fabae. Nature* 286, 602–604.

Iwanaga, K. and Tojo, S. (1986) Effects of juvenile hormone and rearing density on wing dimorphism and oocyte development in the brown planthopper, *Nilaparvata lugens. Journal of Insect Physiology* 32, 585–590.

Jander, R. (1975) Ecological aspects of spatial orientation. *Annual Review of Ecology and Systematics* 6, 171–188.

Johnson, C.G. (1960) The basis for a general system of insect migration and dispersal by flight. *Nature* 186, 348–350.

Johnson, C.G. (1963) Physiological factors in insect migration by flight. *Nature* 198, 423–427.

Johnson, C.G. (1969) *Migration and Dispersal of Insects by Flight.* Methuen, London.

Jouventin, P. and Weimerskirch, H. (1990) Satellite tracking of wandering albatrosses. *Nature* 343, 746–748.

Kennedy, J.S. (1958) The experimental analysis of aphid behaviour and its bearing on current theories of instinct. *Proceedings 10th International Congress of Entomology, Montreal, 1956,* 2, 397–404.

Kennedy, J.S. (1961) A turning point in the study of insect migration. *Nature* 189, 785–791.

Kennedy, J.S. (1966) Nervous coordination of instincts. *Cambridge Research* 2, 29–32.
Kennedy, J.S. (1985) Migration, behavioral and ecological. In: Rankin, M.A. (ed.) *Migration: Mechanisms and Adaptive Significance. Contributions in Marine Science* 27 (suppl.), pp. 5–26.
Kennedy, J.S. and Booth, C.O. (1963a) Free flight of aphids in the laboratory. *Journal of Experimental Biology* 40, 67–85.
Kennedy, J.S. and Booth, C.O. (1963b) Co-ordination of successive activities in an aphid. The effect of flight on the settling responses. *Journal of Experimental Biology* 40, 351–369.
Kennedy, J.S. and Booth, C.O. (1964) Co-ordination of successive activities in an aphid. Depression of settling after flight. *Journal of Experimental Biology* 41, 805–824.
Kennedy, J.S. and Ludlow, A.R. (1974) Co-ordination of two kinds of flight activity in an aphid. *Journal of Experimental Biology* 61, 173–196.
Kennedy, J.S. and Stroyan, H.L.G. (1959) Biology of aphids. *Annual Review of Entomology* 4, 139–160.
Ketterson, E.D. and Nolan, V. Jr (1992) Hormones and life histories: an integrative approach. *American Naturalist* 140 (suppl.), s33–s62.
Ketterson, E.D. and Nolan, V. Jr (1999) Adaptation, exaptation, and constraint: a hormonal perspective. *American Naturalist* 154 (suppl.), s4–s25.
de Kort, C.A.D. (1969) Hormones and the structural and biochemical properties of the flight muscles in the Colorado beetle. *Medeleling Landbouwhogeschool Wageningen* 69, 1–63.
de Kort, C.A.D., Bergot, B.J. and Schooley, D.A. (1982) The nature and titre of juvenile hormone in the Colorado potato beetle, *Leptinotarsa decemlineata*. *Journal of Insect Physiology* 28, 471–475.
McNeil, J.N., Cusson, M., Delisle, J., Orchard, I. and Tobe, S.S. (1995) Physiological integration of migration in Lepidoptera. In: Drake, V.A. and Gatehouse, A.G. (eds) *Insect Migration: Physical Factors and Physiological Mechanisms*. Cambridge University Press, Cambridge, pp. 279–302.
Nair, C.R.M. and Prabhu, V.K.K. (1985) Entry of proteins from degenerating flight muscles into oocytes in *Dysdercus cingulatus* (Heteroptera: Pyrrhocoridae). *Journal of Insect Physiology* 31, 383–387.
Palmer, J.O. and Dingle, H. (1986) Direct and correlated responses to selection among life history traits in milkweed bugs (*Oncopeltus fasciatus*). *Evolution* 40, 767–777.
Palmer, J.O. and Dingle, H. (1989) Responses to selection on flight behavior in a migratory population of milkweed bugs (*Oncopeltus fasciatus*). *Evolution* 43, 1805–1808.
Rankin, M.A. (1978) Hormonal control of insect migratory behavior. In: Dingle, H. (ed.) *The Evolution of Insect Migration and Diapause*. Springer, New York, pp. 5–32.
Rankin, M.A. (1989) Hormonal control of flight. In: Goldsworthy, G.J. and Wheeler, C.H. (eds) *Insect Flight*. CRC Press, Boca Raton, Florida, pp. 139–163.
Rankin, M.A. (1991) Endocrine effects on migration. *American Zoologist* 31, 217–230.
Rankin, M.A. and Burchsted, J.C.A. (1992) The cost of migration in insects. *Annual Review of Entomology* 37, 533–559.

Rankin, M.A., McAnelly, M.L. and Bodenhamer, J.E. (1986) The oogenesis-flight syndrome revisited. In: Danthanarayana, W. (ed.) *Insect Flight: Dispersal and Migration.* Springer, Berlin, pp. 27–48.

Roff, D.A. (1994) Evolution of dimorphic traits: effect of directional selection on heritability. *Heredity* 72, 36–41.

Roff, D.A., Stirling, G. and Fairbairn, D.J. (1997) The evolution of threshold traits: a quantitative genetic analysis of the physiological and life-history correlates of wing dimorphism in the sand cricket. *Evolution* 51, 1910–1919.

Roff, D.A., Tucker, J., Stirling, G. and Fairbairn, D. J. (1999) The evolution of threshold traits: effects of selection on fecundity and correlated response in wing dimorphism in the sand cricket. *Journal of Evolutionary Biology* 12, 535–546.

Slansky, F. (1980) Food consumption and reproduction as affected by tethered flight in female milkweed bugs (*Oncopeltus fasciatus*). *Entomologia Experimentalis et Applicata* 28, 277–286.

Southwood, T.R.E. (1961) A hormonal theory of the mechanism of wing polymorphism in Heteroptera. *Proceedings of the Royal Entomological Society A* 36, 63–66.

Southwood, T.R.E. (1962) Migration of terrestrial arthropods in relation to habitat. *Biological Reviews* 37, 171–214.

Southwood, T.R.E. (1981) Ecological aspects of insect migration. In: Aidley, D.J. (ed.) *Animal Migration.* Cambridge University Press, Cambridge, pp. 196–208.

Taylor, L.R. (1961) Aggregation, variance, and the mean. *Nature* 189, 732–735.

Unnithan, G.C. and Nair, K.K. (1977) Ultrastructure of juvenile hormone induced degenerating flight muscles in a bark beetle, *Ips paraconfusus. Cell and Tissue Research* 185, 481–486.

Winchell, R. (1998) Wing morph determination and the associated life-history consequences in two populations of the wing polymorphic soapberry bug (*Jadera haematoloma*: Rhopalidae). PhD thesis, University of California, Davis.

Winchell, R., Dingle, H. and Moyes, C.D. (2000) Enzyme profiles in two wing-polymorphic soapberry bug populations (*Jadera haematoloma*: Rhopalidae). *Journal of Insect Physiology* 46, 1365–1373.

Zera, A.J. (1999) The endocrine genetics of wing polymorphism in *Gryllus*: critique of recent studies and state of the art. *Evolution* 53, 973–977.

Zera, A.J. and Denno, R.F. (1997) Physiology and ecology of dispersal polymorphism in insects. *Annual Review of Entomology* 42, 207–231.

Zera, A.J. and Tiebel, K.C. (1989) Differences in juvenile hormone esterase activity between presumptive macropterous and brachypterous *Gryllus rubens*: implications for the hormonal control of wing polymorphism. *Journal of Insect Physiology* 35, 7–17.

Zera, A.J. and Tobe, S.S. (1990) Juvenile hormone-III biosynthesis in presumptive long-winged and short-winged *Gryllus rubens*: implications for the endocrine regulation of wing dimorphism. *Journal of Insect Physiology* 36, 271–280.

Zera, A.J., Strambi, C., Tiebel, K.C., Strambi, A. and Rankin, M.A. (1989) Juvenile hormone and ecdysteriod titres during critical periods of wing morph determination in *Gryllus rubens. Journal of Insect Physiology* 35, 501–511.

Zera, A.J., Sall, J. and Grudzinski, K. (1997) Flight-muscle polymorphism in the cricket *Gryllus firmus*: muscle characteristics and their influence on the evolution of flightlessness. *Physiological Zoology* 70, 519–529.

Zera, A.J., Potts, J. and Kobus, K. (1998) The physiology of life-history trade-offs: experimental analysis of a hormonally-induced life-history trade-off in *Gryllus assimilis*. *American Naturalist* 152, 7–23.

# Orientation Mechanisms and Migration Strategies Within the Flight Boundary Layer

## R.B. Srygley[1,3] and E.G. Oliveira[2,3]

[1]*Department of Zoology, University of Oxford, Oxford OX1 3PS, UK;* [2]*Departamento de Biologia Geral, Instituto de Ciências Biológicas, Universidade Federal de Minas Gerais, 30161-970 Belo Horizonte, MG, Brazil;* [3]*Smithsonian Tropical Research Institute, PO Box 2072, Balboa, Republic of Panama*

## Introduction

Animal movements across long distances are one of the most spectacular behavioural phenomena (Brower and Malcolm, 1991). A striking example is the monarch butterfly's migration from southern Canada to a wintering site in central Mexico (see Brower, 1995 for a review). Indeed some insect species, whose flight is generally weak relative to prevalent winds, move in a directed fashion over vast distances. One means by which they accomplish self-directed, self-propulsed movements (Kennedy, 1961) is to fly near to the ground where wind speeds slacken. Here in the flight boundary layer (Fig. 9.1; Taylor, 1974), when and where wind speeds remain below the flight speed of the insect, insects are capable of self-direction. Hence, evaluating migrations of insects requires documentation of their flight speeds and directions with respect to ambient wind speeds and directions.

In the flight boundary layer, a migrating insect may be capable of orienting directionally either with local landmarks, celestial cues or a geomagnetic beacon. Goal orientation is a more sophisticated mechanism that refers to true navigation and goal finding. Goal orientation requires that the insect senses its current position relative to a destination site, and is able to fly to the destination regardless of displacement due to drift. True navigation results from use of either an innate vector programme or navigational maps. Although the ability of migrating birds to navigate has been studied in detail (for reviews, see Alerstam, 1990; Richardson, 1991; Berthold, 1993), the ability of insect migrants to navigate has received relatively little attention.

**Fig. 9.1.** Flight path of a migrating *U. fulgens* Walker moth flying in a neotropical rainforest. The track is in the boundary layer located just above ground level across the pasture where there is no vegetative structure and above tree level where there is canopy structure. Similar flight paths are observed in butterflies and dragonflies descending from the canopy to a lake or river (from Oliveira, 1998).

Migratory behaviour has evolved independently many times in terrestrial arthropods (Johnson, 1969; Dingle, 1980). Among insects, long-distance, self-propelled migrations occur more frequently in the Odonata (e.g. Dumont and Hinnekint, 1973; Dumont, 1977; Russell *et al.*, 1998), Orthoptera (Rainey, 1989) and Lepidoptera (Williams, 1930, 1957; Johnson, 1969; Baker, 1984; Dingle, 1996). These multiple origins of the migratory syndrome provide an advantage when investigating evolutionary forces that might lead to an adaptive behaviour (Harvey and Pagel, 1991).

Migration is an adaptation that allows insects to exploit ephemeral resources and reduce the environmental heterogeneity experienced throughout their lives. It is ultimately governed by the mosaic of resource availability through time and space, with individuals essentially tracking favourable conditions for survival and reproduction (Southwood, 1962, 1977; Dingle, 1984, 1996; Loxdale and Lushai, 1999). Because habitat quality is often associated with seasonal changes, insect migrations tend to be tightly linked to the seasonal cycles. In temperate habitats, extreme annual variation in solar radiation and temperature is the major environmental factor governing resource availability and consequently insect migratory movements. In the tropics, distribution of rainfall appears to be the principal factor underlying the phenology and direction of migrations (e.g. locusts in the Palaeotropics, Rainey, 1989; butterflies in the Neotropics, Oliveira, 1990).

Insect migratory syndromes comprise a suite of integrated, specialized behavioural and physiological processes (Johnson, 1969; Kennedy, 1985; Dingle, 1986, 1996). For example, migration may require insects to recognize favourable meteorological conditions to initiate flight or even be equipped with orientation and navigation mechanisms in order to assure that a particular geographical destination is reached. Migrants may also need to allocate energy and resources to sustain prolonged flight activity at the expense of growth and reproduction (Johnson, 1969; Zera *et al.*, 1998). The processing of such behavioural and physiological activities through time is coordinated by circadian and circannual rhythms to allow proper responses to occur in synchrony with the events happening in the environment.

Migration presents extreme energetic demands on flying organisms (Alerstam, 1990). Therefore it is interesting to consider how migrants might minimize the costs of getting from one location to another; in other words, to develop an optimal migration strategy. To date, these optimality models have been developed for migrating birds (see Alerstam and Hedenström, 1998, for a review). However, the more limited ability of insects to orient and navigate will ultimately place constraints on these models when adapted to insects. For example, their flight speeds are slower than birds and so ambient winds are likely to be an important factor determining when and where they migrate. Their compound eyes may constrain their ability to resolve landmarks from a distance (Wehner, 1981) and subsequently to use them for course correction. Relative to birds, their limited mental capabilities might impose constraints on their capacity to behave optimally. Finally, the limited life-spans of insects may constrain migration strategies to those that are genetically programmed and less modifiable by learning, and yet shorter generation times may permit them to track environmental changes more closely than birds (see Wilson, 1995, for other examples of proximate constraints). Hence we will first examine proximate mechanisms of orientation and control during insect migratory flights. Then we will provide examples from the literature and our own work that compare empirical data with theoretical models of behaviour that either maximize reproductive success or minimize energetic costs.

## Orientation Mechanisms

### Sun compass

In order to travel in a certain compass direction, many insects that move short distances have been shown to use a sun compass to orient (see Wehner, 1984, 1998, for reviews). In its simplest form, the insect maintains a constant heading relative to the sun's azimuth, which is the sun's position on the nearest horizon (Fig. 9.2). For long-distance migrants, whether directionally orienting or navigating, a sun compass should compensate for changes in the sun's position over the course of the day. Otherwise, the flight direction of the insects will change by, on average, $15° h^{-1}$ (which is the rotational velocity of the Earth).

Time compensation requires a temporal reference, and many organisms are known to use an endogenous chronometer to keep track of time. Hence, shifting the butterfly's internal circadian clock relative to the natural time should alter both the relative position of the sun and the butterfly's flight orientation. The predicted direction and magnitude of the shift in the butterfly's orientation are based on the time-shift applied and the local sun's azimuth versus time-of-day curve.

Two recent clock-shift experiments have demonstrated a time-compensated sun compass in migrating butterflies. In Panama, Oliveira *et al.* (1996, 1998) advanced the endogenous clocks of migrating sulphur butterflies (*Aphrissa statira* (Cramer) and *Phoebis argante* (Fabricius)) by 4 h and the butterflies' flight directions shifted in the predicted direction relative to control butterflies (Table 9.1).

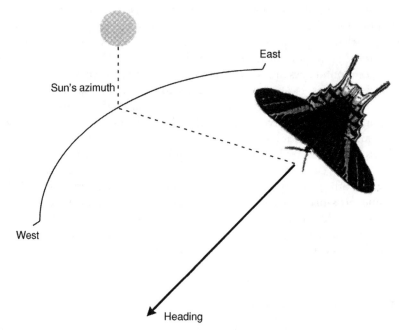

**Fig. 9.2.** Sun-compass orientation in a migrating insect. Without time compensation, the insect would maintain a constant angle relative to the sun's azimuth as the sun's position in the sky changed over the course of the day.

**Table 9.1.** Time compensation for the sun compass in long-distance migrating butterflies.

| Taxon (reference) | Clock-shift treatment | Observed (95% C.L.) | Mechanisms supported (with predicted degree-shift) |
|---|---|---|---|
| *Aphrissa statira* (Cramer) (Oliveira et al., 1998) | 4-h advance | 91° (56°, 126°) | Time-averaging (60°) or full (120°) compensation |
| *Phoebis argante* (Fabricius) (Oliveira et al., 1998) | 4-h advance | 83° (5°, 161°) | Time-averaging (60°) or full (120°) compensation |

Note 95% C.L. for the degree shift was calculated as the difference between the mean control direction and the mean experimental direction ± (0.5 (experimental C.I.) + 0.5 (control C.I.)), e.g. for *A. statira* (Cramer) 91° ± (23 + 12°)

We found that a recent clock-shift experiment on autumnally migrating monarch butterflies (*Danaus plexippus* L.) requires additional research to prove use of a sun compass. Perez *et al.* (1997) delayed the biological clocks by 6 h in Kansas, USA, and the authors concluded that the resulting heading of experimental butterflies was $287 \pm 46°$ ($n = 43$, mean $\pm 95\%$ confidence intervals, confidence intervals were calculated from published $r = 0.29$). By contrast, the mean heading of control butterflies was $211 \pm 22°$ ($r = 0.67$), a value similar to that for naturally flying migrants (200°). However, these statistics implicitly assume a unimodal distribution for flight orientations, a condition clearly not met by the data for the experimental group (Rayleigh, $r$(corrected) = 0.240, $n = 42$, $P > 0.05$; Hodges' and Ajne's test, $P > 0.50$). Apparently, clock-shifted butterflies were disoriented relative to controls, but displayed no systematic difference in flight orientation that would be consistent with a time-compensated sun compass (Srygley, unpublished results).

In fact, the sun's azimuth does not move at the same rate each hour. The azimuth is typically toward the east in the morning and toward the west in the afternoon. Near noon between these two positions, the azimuth rapidly sweeps across the south or north depending on the latitude and season (Fig. 9.3). Therefore, the most effective use of the sun as a compass would require full compensation for its daily motion. It would also require that the migrating insects possess an accurate ephemeris function for the sun at their current location and time of year. Because many migrating insects are newly emerged adults, the ephemeris function might be inherited or learned as immature insects. Alternatively, migrating insects might apply more robust rules to approximate the position of the sun. One such rule is to average the sun's positional changes over the course of the day, e.g. approximately $15°\ h^{-1}$. Another rule is a step-function that approximates the sun's position as lying in the east in the morning and then shifting abruptly to the west after noon (e.g. naive foraging honeybees; Dyer and Dickinson, 1994; Dyer, 1996).

Each time-compensation hypothesis had its own predicted result in the clock-shift experiments (Table 9.1). In Panama, *Aphrissa* and *Phoebis* would shift 120° under the full-compensation hypothesis, 60° under the time-averaging hypothesis and 180° under the step-function hypothesis. The resulting shift in heading was $91 \pm 35°$ relative to controls for *Aphrissa* (mean $\pm 95\%$ confidence intervals) and $83 \pm 78°$ for *Phoebis*. As a result, there was no evidence for the use of a step-function to compensate for the motion of the sun. However, the data do not eliminate either the time-averaging or the full-compensation hypotheses (Oliveira *et al.*, 1998). In Kansas, assuming an afternoon release, *Danaus* was predicted to shift approximately 120° with full-compensation, 90° with time-averaging compensation and 180° with step-function compensation. Because the resulting heading of the experimental group was not unimodal, the data did not eliminate any of the three time-compensation mechanisms. Hence, the experiments that definitively demonstrate that *Danaus* uses a sun compass and the mechanism of time compensation have yet to be done.

Another means of examining the time-compensation hypotheses is to associate the flight directions of migrants with the sun's apparent movement over the course of the day. Even within the Lepidoptera, there is variation in the ability to compensate for the sun's motion. For example, the flight directions of some migratory pierid butterflies (Baker, 1968a,b, 1969; Oliveira, 1990) and *Urania* moths (Oliveira, 1998) changed with time of day

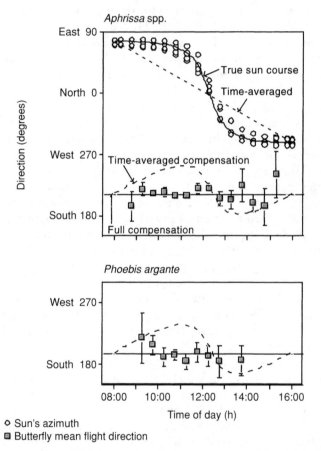

**Fig. 9.3.** The position of the sun's azimuth over the course of the day (solid line) and the butterfly's estimate of the sun's position under the time-averaged hypothesis (dashed line). The predicted flight directions of butterflies with full compensation is shown as a solid line equal to the mean flight direction over the course of the day. With time-averaged compensation, the predicted flight direction is shown as a dashed line. Mean flight directions of butterflies migrating across the Panama Canal (error bars show the 95% confidence limits) support the full compensation model, however a single orientation over the course of the day might also be maintained with a geomagnetic compass (adapted from Oliveira et al., 1998).

in accordance with the apparent movement of the sun. Hence no or only partial compensation was observed in these cases. However, other migrating butterflies maintain constant flight direction throughout the day (e.g. Williams, 1930; Nielsen, 1961; Johnson, 1969; Walker and Littell, 1994; Spieth et al., 1998; Fig. 9.3).

Each time-compensation hypothesis has a distinct prediction. Time-averaged compensation would result in variation in the orientation such that the maximum error would be at mid-morning and mid-afternoon (Fig. 9.3). Step-function compensation would result in maximum error at noon, and full compensation would not result in variation in the orientation. Although the track directions of naturally migrating butterflies support the full compensation model, directional orientation by other means such as use of a magnetic compass would result in identical observations. Hence, we now know that these migrating butterflies use a sun compass, but it is not yet clear what mechanism they adopt to compensate for the sun's movement. In addition, we do not know whether sun compasses are coordinated with other directional cues, which has been demonstrated in several migrating vertebrate (Able, 1980; Able and Able, 1999) and locally moving invertebrate species (e.g. *Cataglyphis* ants, Wehner et al., 1996; honeybees, Menzel et al., 1996).

## Orientation without the sun

Some locally moving insects orient by polarized skylight (e.g. bees, von Frisch, 1949; ants, Labhart, 1986; crickets, Labhart, 1999; flies, Hardie, 1984). The direction of polarisation is perpendicular to the plane defined by the positions of the observer, the sun and the observed point of the sky. Rossel and Wehner (1986) demonstrate in *Apis mellifera* L. that errors in the bees' orientation are consistent with the use of a mental template of the polarized light vectors. The honeybee matches this cognitive template to the orientation of light vectors in the sky which fits the template at dawn and dusk. At other times of day, the template and sky do not match perfectly, but discrepancies cancel one another unless only a small portion of the sky is visible. Hence, if half of the sky's hemisphere opposite to that of the sun is visible, polarized light may serve as a precise positional reference in the same manner as the sun. When no more than one-quarter of the sky is visible, then use of polarized light leads to a predictable error. Sensitivity to polarized light and its use as a directional reference has not been investigated in detail in migrating insects (but see Hyatt, 1993).

The abilities of migratory dragonflies (R.B. Srygley and E.G. Oliveira, personal observations) and butterflies (Schmidt-Koenig, 1985; Gibo, 1986; Walker and Littell, 1994) to maintain a preferred compass orientation in overcast skies without celestial cues indicate that the Earth's geomagnetic field may serve as an orientation cue. Jones and MacFadden (1982) demonstrated that monarch butterflies have magnetite particles, but their use of a magnetic compass had not been investigated further. Recently, Perez et al. (1999) provided some evidence that monarch butterflies are sensitive to strong magnetic fluxes. When approached with a magnet prior to release, the headings of magnetized butterflies were random whereas that of control butterflies were directed to the southwest. However, the evidence was weakened by the fact that the experimental butterflies' mean track direction (328°, including effects of the wind) was not different from that of the control group (326°). Hence headings may have differed between treatment groups as a result of compensation for different amounts of wind drift (Srygley et al., 1996).

Nocturnally active insects may also maintain a preferred orientation without the sun (Wehner, 1984; see also Riley and Reynolds, 1986). Baker and Mather (1982) reversed the orientation of caged *Noctua pronuba* L. moths with a magnetic field whose net effect was nearly a mirror image of the geomagnetic intensity and direction, and Baker (1987) reversed the orientation of caged *Agrotis exclamationis* (L.) moths. However, neither of these moths are known to be long-distance migrants. Baker (1987) concedes that the experiments lacked the proper controls, and outlines the appropriate evidence required for demonstrating geomagnetic orientation. Reversal of the geomagnetic field within a night should result in orientation reversal on each night, and reversal of the geomagnetic field after several nights should also result in orientation reversal. Hence, there is little evidence that long-distance migrating insects employ a magnetic sense for orientation to date, but the ability of insects that move short distances to sense the magnetic field (e.g. honeybees, Lindauer and Martin, 1968; sandhoppers, Ugolini et al., 1999) suggests that it is an exciting new area for research.

## Local landmarks and drift compensation

Flight near to the ground in the boundary layer provides an advantage in that nearby landmarks might be used for correcting for the tendency of local winds to blow a migrant off its preferred course. Over the ground, insects may use the relative motion of landmarks beneath their flight path as a means of correcting for wind drift (e.g. honeybees, Heran and Lindauer, 1963; von Frisch, 1967). This mechanism (hereafter called the ground reference) has been demonstrated experimentally in locusts (Preiss and Gewecke, 1991). Applying harmonic radar to track bumblebees, Riley et al. (1999; and Riley and Osborne, Chapter 7) recently confirmed its application in nature. Without drift, the ground and landmarks on it pass beneath the insect parallel to the body axis. With drift, the ground reference passes parallel to the track direction and at an angle oblique to the body axis (Fig. 9.4). Complete correction for wind drift off the intended track is accomplished if the insect, using the sun compass as a reference direction, turns into the wind until the component of its airspeed across the track counters the cross-track component of the wind (i.e. the crosswind drift; von Frisch, 1967).

Over a fluid surface, such as a lake or the sea, the problem of correcting for drift is more difficult. Indeed, in a previous volume from a Royal Entomological Society symposium, Lindauer (1976) described the inability of honeybees to compensate for crosswind drift when flying over a lake. With the experimental construction of a wooden bridge with planks spaced 3 m apart across the lake, the bees compensated by flying from one plank to the next in a scalloped track, as though each plank was a landmark (Heran and Lindauer, 1963).

Investigations of wind drift compensation when migrating insects were flying over bodies of water have provided additional information on the use of the ground reference and landmarks for course correction (Srygley *et al.*, 1996; Srygley, 2001a). Tracking individual insects and taking advantage of

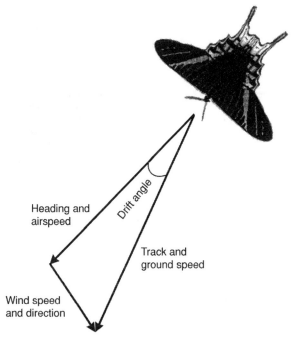

**Fig. 9.4.** An insect's airspeed is its flight speed generated by the insect's metabolic and aerodynamic power – independent of the wind. Addition of the wind vector to the insect's flight vector yields the ground vector with a length equal to ground speed and an orientation equal to track direction (or course over the ground). The difference between heading and track is called the drift angle. Without wind, heading and track direction would be the same and the ground would pass beneath the insect parallel to its longitudinal body axis. However, with wind, the ground passes beneath the insect at an angle relative to the body axis equal to the drift angle. One means of compensating for drift is for the insect to adjust its heading upwind until the component of its airspeed across the track counters the cross-track component of the wind.

leeward and windward regions of a lake, we measured the ability of butterflies, moths and dragonflies to compensate for changes in wind drift. We analysed the data with the objective of distinguishing between three hypothesized mechanisms for correcting for wind drift.

The first hypothesized mechanism is that the insects use ripples and waves on the water surface as a 'ground reference'. Whereas a solid ground reference passes parallel to the track direction when an insect is blown off-course (Fig. 9.4), a fluid ground reference is itself moving (usually downwind, but slower than the wind speed; Fig. 9.5). Hence, insects using this mechanism when flying over water can at best achieve only partial compensation for drift. Two lines of evidence would support this mechanism: the insect partially corrects for wind drift (for an example in birds, see Alerstam and Pettersson, 1976) and the track is a straight line across the lake in variable winds.

A second potential mechanism for course correction is that the insects maintain a heading toward a single landmark on the opposite shore. If an

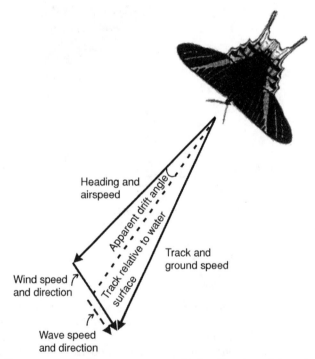

**Fig. 9.5.** Over water, ripples or waves move in the direction of the wind. Hence to the insect using the relative motion of the ground beneath it as a means to compensate for wind drift, the apparent drift angle is less than the true drift angle. The insect will compensate by adjusting its heading upwind by the apparent drift angle resulting in part compensation for wind drift (after Alerstam and Pettersson, 1976). If the waves were to move at the wind speed, the apparent drift angle would be zero, and no compensation could be achieved using a ground reference.

insect uses this method, we predict that it would drift downwind, but its heading would orient further upwind toward the landmark as it crossed the lake. The result is a curvilinear track across the lake with the concavity facing the wind (Nielsen, 1961).

A third putative mechanism is that the insects use two landmarks. The landmarks could be two in a line (called a 'range'), one at each side and held in parallax, or one at infinite distance such as the sun and another local landmark straight ahead. Use of this method would result in full correction for wind drift, and the track would be a straight line across the lake.

Using a boat on the Panama Canal, we paced butterflies and moths flying naturally over the water and measured the airspeed of the insect with an anemometer and the track direction with a hand-held compass (Srygley *et al.*, 1996). While crossing the lake, we took advantage of the changes in wind due to windward and leeward sides of land masses. From one side of the lake to the other, we measured the maximum change in wind drift and heading. If the insect is fully correcting for wind drift, the insect must change its heading so that the cross-track component of its airspeed counters the changing cross-track component of the wind speed. When correcting for wind drift, the insect's track remains straight.

We found that butterflies were capable of correcting for wind drift when crossing the lake with a mean compensation of 1.0. However, the variation was sufficiently great that the 95% confidence limits overlapped with part as well as full compensation. Hence, we could not distinguish between use of two landmarks or the relative motion of the lake surface to correct for drift. *Urania* moths, in contrast, did not correct for crosswind drift. However, the variance was sufficiently high that the confidence limits overlapped well into part compensation (Table 9.2).

A migration of *Pantala* dragonflies across the Panama Canal provided an opportunity to measure wind drift compensation in another insect Order. With commercially available sailboat navigation equipment, we measured the ground speed, track direction, wind speed and wind direction simultaneously

**Table 9.2.** Crosswind drift compensation in butterflies, diurnal moths and dragonflies.

| Taxon (reference) | Compensation (95% C.L.) | Predictions supported | Mechanisms supported |
|---|---|---|---|
| Butterflies (Nymphalidae and Pieridae, Srygley *et al.*, 1996) | 114% (63, 165) | Part compensation or full compensation | Ground reference or two landmarks |
| *Urania* moths (Srygley *et al.*, 1996) | 12% (−69, 93) | No compensation or part compensation | No reference or ground reference |
| *Pantala* dragonflies (R.B. Srygley, unpublished observations) | 54% (35, 72) | Part compensation | Ground reference |

as we followed individual dragonflies across the lake. Within individuals of two species, dragonflies compensated for changes in crosswind drift with a corresponding change in heading (R.B. Srygley, unpublished observations). However, their ability to compensate was significantly less than perfect (Table 9.2). Part compensation is strong evidence that dragonflies used a ground reference rather than landmarks for wind-drift compensation. Course correction also indicates that the migrating insects were attempting to fly toward a particular location. Hence, course correction is consistent with both directional orientation and true navigation in insects.

Wind speed has predictable effects on water surfaces as long as shorelines are sufficiently distant to have negligible influence. Ripples form at wind speeds greater than 0.5 m s$^{-1}$, small waves form at wind speeds greater than 3.5 m s$^{-1}$, and longer, higher waves form at wind speeds greater than 5.5 m s$^{-1}$ (Grocott, 1993). Migrating insects have flight heights that tend to be a characteristic of the species (over land, Oliveira 1990; over seas, Srygley, 2001a). At wind speeds below 5.5 m s$^{-1}$, the ripples or small waves that form may be useful as a ground reference for most insects that migrate in the flight boundary layer. However, at wind speeds greater than 5.5 m s$^{-1}$, the higher waves that form might be useless as a ground reference for those insects whose flight height is less than the wave height. For example, large waves may be useless as a ground reference for *Urania fulgens* Walker or *Ascia monuste* (L.), which migrate at heights between 0.25–1.5 m above the surface (R.B. Srygley and E.G. Oliveira, personal observations), whereas the same waves may be a ground reference for insects that fly at heights greater than the wave height.

Perhaps most importantly, investigations of migratory flyways and the ability of organisms to keep to the flyway will be relevant to their conservation. Long-distance, boundary-layer migrants require stopover sites for rest and nourishment. Our ability to predict the movement of insects and their energetic requirements en route will depend on their ability to compensate for changes in wind speed and direction. With continuing loss of habitat, those who make decisions on conservation priorities should take into account the ability of the insect to maintain a consistent flyway even when subjected to local variations in winds.

## Optimal Migration Strategies

### Optimal wind-drift compensation

In long-distance migration, energy is at a premium and natural selection for conserving energy is potentially strong. Animals may adopt behaviours while in transit to reduce the energetic cost of flight between a locale of origin and their destination. Insects migrating in the boundary layer provide an advantage for testing theoretical predictions: individual insects may be followed, and insect orientation, airspeed, and ambient wind direction and speed can be measured simultaneously.

Pennycuick (1978) based an optimization analysis of migrating birds on the U-shaped power curve. Without wind, the velocity (airspeed) that maximizes the range of the animal (i.e. the airspeed at which power per unit distance is at a minimum) is found by drawing a tangent from the origin to the power curve (Fig. 9.6). Because maximum-range velocity is based on ground speed, the addition of a tailwind to (or subtraction of a headwind from) the airspeed of the insect would result in a decrease (or an increase) in the optimal velocity that the insect should adopt. As a result, we would predict that if insects are adjusting their velocity for headwinds and tailwinds in order to maximize their flight range, we would see a reduction in the airspeed with a tailwind and an increase in the airspeed with a headwind. On the other hand, if the insects are not adjusting for winds, then ground speed should increase with tailwind speed in a one-to-one fashion.

Srygley *et al.* (1996) tested this prediction of optimal airspeed in a tailwind by measuring the airspeeds of *A. statira* butterflies on Lake Gatún,

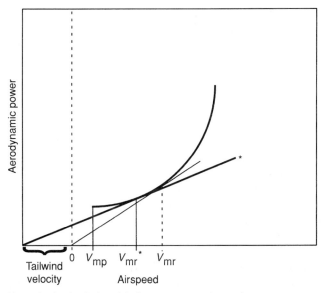

**Fig. 9.6.** Hypothetical relationship between aerodynamic power consumption and flight velocity (airspeed) for migrating insects (adopted from that for *Urania fulgens* Walker, a migrating moth (Dudley and DeVries, 1990) and *Sympetrum sanguineum* (Müller), a dragonfly (Wakeling and Ellington, 1997)). The airspeed giving a maximum range for a given energy ($V_{mr}$) is found by taking the tangent to the curve through the origin. If a tailwind is present the origin is displaced to the left and the airspeed which yields maximum range is lowered to $V_{mr}^*$ (after Pennycuick, 1978). The power relation below $V_{mp}$ (minimum power velocity) is not shown.

Panama. We intercepted the flight path of migrating butterflies flying over from one shore to another as they crossed the lake. We measured airspeed of the butterfly by pacing it with a boat and measuring the forward airspeed of the boat with an anemometer (for further details, see DeVries and Dudley, 1990; Dudley and DeVries, 1990). We then stopped the boat and measured wind speed and direction. Butterflies for which the wind was within 5° of head-on or from the tail were included (i.e. winds that incorporated a large crosswind drift component were excluded). The prediction is that if the butterflies are adjusting their airspeed for headwinds and tailwinds, then ground speed will regress on wind speed with a slope < 1. Because the resulting slope was not significantly different from 1, we concluded that *A. statira* butterflies were not adjusting airspeed for tailwinds. However, we did not examine differences between the sexes in this study.

In a second study, Srygley (2001b) followed male and female cloudless sulphur butterflies *Phoebis sennae* (L.) (Pieridae) migrating over the Caribbean Sea. *P. sennae* headed west in the morning and then turned to the south east to head downwind in the afternoon. As predicted from optimal migration theory, airspeeds of females decreased in a tailwind to minimize energy consumption. However, males did not show any compensation for tailwinds. Males may be minimizing the time to reach the destination site in order to maximize matings with newly arrived or newly emerged females. Orientation of females changed before that of males, presumably because their greater reproductive load imposed greater flight costs and limited flight fuels.

Alerstam (1979, 1990) looked at the optimal strategy of migrating birds to compensate for another aspect of wind: being blown off course by crosswind drift. Three features compose the vector analysis of crosswind drift (Fig. 9.4): the animal's heading and airspeed, the wind direction and wind speed, and the resultant vector of track direction and ground speed. Alerstam proposed that an animal that flew through a wind blowing at a constant direction would be most energetically efficient at arriving at its destination if it fully compensated for the wind when blown off course. This entails turning its heading partially upwind (i.e. flying into the wind). This results in an earlier arrival than the alternative of orienting continuously to the destination. The situation is more complicated if the wind is variable in direction. In this case, the animal would arrive earlier and with less energetic cost if it adapts to the variable wind such that at the end of each period it has minimized the distance to the goal. This results in an earlier arrival and less energetic cost than fully compensating continuously.

Tailwind drift compensation serves to maximize a migrant's flight distance on a given amount of energy, and crosswind drift compensation serves to hold a course true and minimize the distance flown. With full crosswind drift compensation, airspeed should increase with increasing drift angle. If the animal drifts, airspeed should be less than if the animal compensates fully; and if the animal compensates for crosswind drift in part, then the optimal airspeeed lies between that predicted with full compensation and that predicted with drift (Liechti et al., 1994; Liechti, 1995). Hence with full or part compensation, airspeeds are predicted to increase with greater crosswind drift.

The velocity and heading of *Pantala hymenaea* (Say) and *Pantala flavescens* (Fabricius) flying naturally over Lake Gatún, Panama, and the ambient winds were measured. The migrating dragonflies partly compensated for changes in crosswind drift as they crossed the lake (Table 9.2). As predicted from optimal migration theory, airspeed (ranging from 3 to 7 m s$^{-1}$) decreased with tailwind velocity both among individuals and within each individual. Following adjustment for tailwind drift, airspeeds of *P. hymenaea* and *P. flavescens* increased with crosswind drift (R.B. Srygley, unpublished observations). This result provides qualitative support for the theoretical predictions of Liechti (1995; Liechti et al., 1994); quantitative support will require evaluation of the power curve specific for each species.

## Optimal loading

Flight is fuelled by the metabolism of fats, carbohydrates and amino acids (Rankin and Burchsted, 1992), and yet additional mass increases the energetic cost of flight. As a result, there is an optimum fuel load whereby the distance travelled per mass of fuel is maximized. However, the optimum fuel load is also dependent on nutrient availability at refuelling sites, risk of predation during refuelling and the predictability of winds favourable for migration (Alerstam and Lindström, 1990).

The speed of migration incorporates the time spent at stopover sites for refuelling or taking refuge from unfavourable environmental conditions (Alerstam and Hedenström, 1998).

$$V_{migr} = V_{flight} \times P_{dep} / (P_{flight} + P_{dep})$$

where $V_{migr}$ is the average velocity during the migration, $V_{flight}$ is the flight velocity (airspeed), $P_{flight}$ is the power required to fly at $V_{flight}$, and $P_{dep}$ is the net power gained from feeding during a stopover. The velocity that minimizes migration time including stopovers is a function of the power required to fly at the flight velocity and the net energy gained during the stopovers (Fig. 9.7). Weber and Houston (1997; Weber et al., 1994) extended this model to estimate the optimal departure loading under stochastic nutrient and predation conditions at stopover sites. Models incorporating variability in winds during migration have also been developed (Liechti and Bruderer, 1998; Weber et al., 1998).

Loading the abdomen with flight fuel may also influence predation risk (Alerstam and Lindström, 1990). Loading repositions the centre of body mass further posterior, reducing manoeuvrability (Srygley and Dudley, 1993; Srygley and Kingsolver, 1998) and increasing the risk of predation (but see

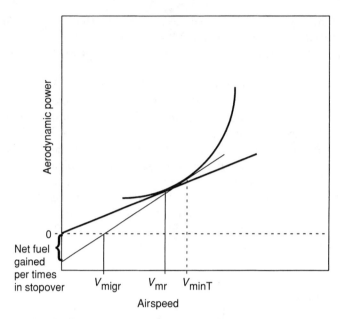

**Fig. 9.7.** With refuelling at stopover sites, the airspeed that minimizes time of migration $V_{minT}$ is estimated by a tangent to the power curve drawn through a point that is below the origin by a distance representing the rate of take-up of fuel energy at stopovers (after Alerstam and Lindström, 1990). $V_{minT}$ is greater than $V_{mr}$, which is estimated as in Fig. 9.6. The average speed of migration $V_{migr}$ is the x-intercept. Assuming that there are no differences in nutrient availability or predation risk among stopover sites, insects should depart from the stopover when the average migration speed declines to $V_{migr}$.

Kingsolver and Srygley, 2000; Srygley and Kingsolver, 2000). With a much lower risk of predation relative to palatable species, distasteful insects, such as monarch or queen butterflies *Danaus* (Ritland, 1998), may have the option of risk-prone behaviours for transport (Oliveira, 1998), such as heavier weight loading, soaring on thermal currents and gliding (e.g. Gibo and Pallett, 1979; Gibo, 1986). Palatable migrants, which have faster flight speeds than distasteful species (Srygley and Dudley, 1993), use powered flapping flight even when winds are favourable (E.G. Oliveira and R.B. Srygley, personal observations). In order to further reduce risks of predation during stopovers, palatable species are predicted to have an optimum loading that is lower than predicted by the time-minimization model of Alerstam and Lindström (1990).

Pennycuick (1978) also predicted that loading would influence flight velocity (airspeed). Assuming drag remains unchanged, water loss is negligible, the same fuel is consumed over the distance flown and fuel consumption is proportional to the total power consumption (mechanical power plus metabolic power; see Pennycuick, 1978), maximum range velocity ($V_{mr}$) is inversely proportional to the square root of body mass. Hence with distance flown, $V_{mr}$ is predicted to decrease. However, as Videler (1995) points out, this prediction is derived from the assumption that faster airspeed is the only option available for animals to increase lift with an increase in body mass. However, flight speeds of experimentally loaded animals are consistently lower than those of unweighted ones (e.g. kestrels, Videler et al., 1988; pigeons *Columba livia*, Gessaman and Nagy, 1988; Harris hawks *Parabuteo unicinctus*, Pennycuick et al., 1989; long-eared bats *Plecotus auritus*, Hughes and Rayner, 1991; honeybees *A. mellifera* L., Seeley, 1986; *Anartia fatima* Godart butterflies, Srygley and Kingsolver, 2000). Flying animals may compensate for heavier loads by increasing wing beat frequency or stroke amplitude.

## Conclusions, Future Questions and Directions

The role of sun-compass orientation in directional orientation over long distances is evident from clock-shift experiments performed on migratory butterflies. The sun may be an important positional reference for drift compensation, as well. Questions that require further investigation include: (i) do long-distance migrants learn the sun's daily motion or depend on hard-wired information?; (ii) is time-compensation achieved by averaging the movement of the sun over the course of the day, or do migrants compensate fully despite changes in the sun's rate of movement over the course of the day?; and (iii) how do long-distance migrants compensate for changes in the sun's ephemeris over the spatial distance that they travel and over seasons?

To date, there is some evidence for use of a magnetic compass in monarch butterflies, but additional work with controlled experiments is needed. Elaborate hypotheses have been proposed to explain how monarchs might encounter their overwintering site with a magnetic compass (Monasterio et al., 1984; Schmidt-Koenig, 1985), but the basic groundwork to demonstrate a magnetic compass is needed prior to addressing hypotheses that involve use of a magnetic compass for navigation over large spatial scales.

The observations: (i) that different butterfly species fly across the same area and at the same time with significantly different mean directions (e.g. Williams, 1930; Oliveira, 1990); (ii) that flight orientation changes depending on geographical location in a way that suggests these insects make 'turns' to follow a preferred migratory route towards their destination (e.g. Walker and Littell, 1994; Spieth and Kaschuba-Holtgrave, 1996; Srygley et al., 1996); and (iii) that flight direction changes according to season (e.g. Walker and Littell, 1994; Brower, 1995; Spieth and Kaschuba-Holtgrave, 1996) suggest that butterfly sun compasses are used in a flexible manner, and may interact with information received from other directional cues or features of the landscape. Spieth and Kaschuba-Holtgrave (1996) have recently developed an experimental set-up to study migratory orientation of European *Pieris brassicae* which for the first time enabled flight orientation of naturally migrating butterflies to be reproduced in circular cages. If this method is applicable to other species, it may prove useful for investigating environmental, genetic and developmental factors that influence orientation behaviour of migrating insects (see also Spieth and Kaschuba-Holtgrave, 1996; Spieth et al., 1998). Moreover, carefully controlled experiments could begin to tease apart the migrant's ability to evaluate and integrate directional information derived from the sun, polarized skylight, local landmarks, wind speed and direction, and geomagnetism.

Migrating insects compensate for drift to various degrees, and they appear to apply different mechanisms to correct their course when flying over water. Dragonflies use the sea surface as a reference, butterflies probably use local landmarks when available and use the sea surface when landmarks are unavailable. *Urania* moths may not compensate at all. This research has focused on individual responses to changes in wind drift, which provides information of the ability of insects to sense that they are being blown off-course and adjust their headings accordingly. At the population level, drift compensation may be even more complex. Our preliminary results indicate that individuals may respond to local weather that is predictable in the short term and climatic conditions that are predictable in the long term (R.B. Srygley, unpublished observations; Srygley, 2001b). Analysis of crosswind drift at both the population and individual levels will be required for a complete evaluation of optimal drift.

Mechanisms of orientation and the ability of insects to navigate to a destination site will place constraints on optimal migration models. Dragonflies and butterflies are capable of compensating for tailwind drift even when flying over water. Sexes may differ in behaviours that ultimately maximize reproductive success. At this point, kinematic and aerodynamic studies of migrating insects are sorely needed to quantify flight energetics and optimal drift compensation.

Insects that migrate in the boundary layer provide an additional advantage because they may be videotaped flying naturally for kinematic and aerodynamic analysis (Dudley and DeVries, 1990). Flight speeds recorded from insects in natural flight are consistently above those achieved in wind tunnels and insectaries (Srygley and Dudley, 1993; Dudley and Srygley, 1994). Because individual butterflies and dragonflies adjust their airspeed to compensate for changes in the tailwind, then the associated changes in wing kinematics and aerodynamic power may also be quantified in a single individual flying naturally for the first time.

Tracking of migrating insects over longer distances would enhance our understanding of orientation mechanisms. Technological advances that have recently been made in radar systems to track bee short-distance flights (Riley and Osborne, Chapter 7) give us hope that a similar method may improve the tracking of insects that perform long-distance migrations within the flight boundary layer. At present, use of motorboats to track insects migrating over lakes and seas is one alternative method. However, we are left with the question of whether orientation over lakes and seas is different from that over land.

## Acknowledgements

We thank T.R.E. Southwood and A.L.R. Thomas for commenting on the manuscript. The National Geographic Society provided financial support for the research. Additional support came from the University of Oxford (Varley-Gradwell Travel Fellowship to R.B.S.), the Royal Society (Developing World Travel Fellowship to E.G.O.), The Royal Entomological Society, Smithsonian Tropical Research Institute, CNPQ grant no. 20.0467/89-1 to E.G.O. and the Biotechnology and Biological Sciences Research Council grant no. 43/SO8664 to A.L.R. Thomas.

## References

Able, K.P. (1980) Mechanisms of orientation, navigation, and homing. In: Gauthreaux, S.A. Jr (ed.) *Animal Migration, Orientation, and Navigation.* Academic Press, New York, pp. 281–387.

Able, K.P. and Able, M.A. (1999) Evidence for calibration of magnetic migratory orientation in Savannah sparrows reared in the field. *Proceedings of the Royal Society of London B* 266, 1477–1481.

Alerstam, T. (1979) Wind as a selective agent in bird migration. *Ornis Scandinavica* 10, 76–93.

Alerstam, T. (1990) *Bird Migration.* Cambridge University Press, Cambridge.

Alerstam, T. and Hedenström, A. (1998) The development of bird theory. *Journal of Avian Biology* 29, 343–369.

Alerstam, T. and Lindström, A. (1990) Optimal bird migration: the relative importance of time, energy, and safety. In: Gwinner, E. (ed.) *Bird Migration: the Physiology and Ecophysiology.* Springer, Berlin, pp. 331–351.

Alerstam, T. and Pettersson, S.-G. (1976) Do birds use waves for orientation when migrating across the sea? *Nature* 259, 205–207.

Baker, R.R. (1968a) Sun orientation during migration in some British butterflies. *Proceedings of the Royal Entomological Society of London A* 143, 89–95.

Baker, R.R. (1968b) A possible method of evolution of the migratory habit in butterflies. *Philosophical Transactions of the Royal Society of London B* 253, 309–341.

Baker, R.R. (1969) Evolution of the migratory habit in butterflies. *Journal of Animal Ecology* 38, 703–746.

Baker, R.R. (1984) The dilemma: when and how to go or stay. In: Vane-Wright, R.I. and Ackery, P.R. (eds) *The Biology of Butterflies.* Academic Press, New York, pp. 279–296.

Baker, R.R. (1987) Integrated use of moon and magnetic compasses by the heart-and-dart moth, *Agrotis exclamationis. Animal Behaviour* 35, 94–101.

Baker, R.R. and Mather, J.G. (1982) Magnetic compass sense in the large yellow underwing moth, *Noctua pronuba* L. *Animal Behaviour* 30, 543–548.

Berthold, P. (1993) *Bird Migration: a General Survey*. Oxford University Press, Oxford.

Brower, L.P. (1995) Understanding and misunderstanding the migration of the monarch butterfly (Nymphalidae) in North America. *Journal of the Lepidopterists' Society* 49, 304–385.

Brower, L.P. and Malcolm, S.B. (1991) Animal migrations: endangered phenomena. *American Zoologist* 31, 265–276.

DeVries, P.J. and Dudley, R. (1990) Morphometrics, airspeed, thermoregulation, and lipid reserves of migrating *Urania fulgens* (Uraniidae) moths in natural free flight. *Physiological Zoology* 63, 235–251.

Dingle, H. (1980) Ecology and evolution of migration. In: Gauthreaux, S.A. Jr (ed.) *Animal Migration, Orientation, and Navigation*. Academic Press, New York, pp. 1–101.

Dingle, H. (1984) Behaviour, genes, and life histories: complex adaptations in uncertain environments. In: Price, P.W. Slobodchikoff, C.N. and Gaud, W.S. (eds) *A New Ecology: Novel Approaches to Interactive Systems*. John Wiley & Sons, New York, pp. 169–194.

Dingle, H. (1986) Evolution and genetics of insect migration. In: Danthanarayana, W. (ed.) *Insect Flight: Dispersal and Migration*. Springer-Verlag, New York, pp. 11–26.

Dingle, H. (1996) *Migration: the Biology of Life on the Move*. Oxford University Press, Oxford.

Dudley, R. and DeVries, P.J. (1990) Flight physiology of migrating *Urania fulgens* (Uraniidae) moths: kinematics and aerodynamics of natural free flight. *Journal of Comparative Physiology A* 167, 145–154.

Dudley, R. and Srygley, R.B. (1994) Flight physiology of Neotropical butterflies: allometry of airspeeds during natural free flight. *Journal of Experimental Biology* 191, 125–139.

Dumont, H.J. (1977) On migrations of *Hemianax ephippiger* (Burmeister) and *Tramea basilaris* (P. de Beauvois) in west and north-west Africa in the winter of 1975/1976 (Anisoptera: Aeshnidae, Libellulidae). *Odonatologica* 6, 13–17.

Dumont, H.J. and Hinnekint, B.O.N. (1973) Mass migration in dragonflies, especially in *Libellula quadrimaculata* L.: a review, a new ecological approach and a new hypothesis. *Odonatologica* 2, 1–20.

Dyer, F.C. (1996) Spatial memory and navigation by honeybees on the scale of the foraging range. *Journal of Experimental Biology* 199, 147–154.

Dyer, F.C. and Dickinson, J.A. (1994) Development of sun compensation by honey bees: how partially experienced bees estimate the sun's course. *Proceedings of the National Academy of Sciences USA* 91, 4471–4474.

von Frisch, K. (1949) Die Polarisation des Himmelslichtes als orientierender Faktor bei den Tänzen der Bienen. *Experientia* 5, 142–148.

von Frisch, K. (1967) *The Dance Language and Orientation of Bees*. Belknap Press of Harvard University Press, Cambridge, Massachusetts.

Gessaman, J.A. and Nagy, K.A. (1988) Transmitter loads affect the flight speed and metabolism of homing pigeons. *Condor* 90, 662–668.

Gibo, D.L. (1986) Flight strategies of migrating monarch butterflies (*Danaus plexippus* L.) in southern Ontario. In: Danthanarayana, W. (ed.) *Insect Flight, Dispersal and Migration*. Springer-Verlag, Berlin, pp. 172–184.

Gibo, D.L. and Pallett, M.J. (1979) Soaring flight of monarch butterflies, *Danaus plexippus* (Lepidoptera: Danaidae), during the late summer migration in southern Ontario. *Canadian Journal of Zoology* 57, 1393–1401.

Grocott, D.F.H. (1993) Navigation using visual features of the universe. In: *RIN 93: Orientation and Navigation: Birds, Humans, and other Animals: The 1993 International Conference of the Royal Institute of Navigation*. Royal Institute of Navigation, London, Paper no. 1.

Hardie, R.C. (1984) Properties of photoreceptor-R7 and photoreceptor-R8 in dorsal marginal ommatidia in the compound eyes of *Musca* and *Calliphora*. *Journal of Comparative Physiology* 154, 157–165.

Harvey, P.H. and Pagel, M. (1991) *The Comparative Method in Evolutionary Biology*. Oxford University Press, Oxford.

Heran, H. and Lindauer, M. (1963) Windkompensation und Seitenwindkorrektur der Bienen beim Flug über Wasser. *Zeitschrift für vergleichende Physiologie* 47, 39–55.

Hughes, P.M. and Rayner, J.M.V. (1991) Addition of artificial loads to long-eared bats *Plecotus auritus*: handicapping flight performance. *Journal of Experimental Biology* 161, 285–298.

Hyatt, M. (1993) The use of sky polarization for migratory orientation by monarch butterflies. PhD thesis, University of Pittsburgh, Pittsburgh, Pennsylvania.

Johnson, C.G. (1969) *Migration and Dispersal of Insects by Flight*. Methuen, London.

Jones, D.S. and MacFadden, B.J. (1982) Induced magnetization in the monarch butterfly *Danaus plexippus* (Insecta, Lepidoptera). *Journal of Experimental Biology* 96, 1–9.

Kennedy, J.S. (1961) A turning point in the study of insect migration. *Nature* 189, 785.

Kennedy, J.S. (1985) Migration, behavioral and ecological. In: Rankin, M.A. (ed.) *Migration: Mechanisms and Adaptive Significance. Contributions of Marine Sciences* 27 (suppl.), pp. 5–26.

Kingsolver, J.G. and Srygley, R.B. (2000) Experimental analyses of body size, flight and survival in pierid butterflies. *Evolutionary Ecology Research* 2, 593–612.

Labhart, T. (1986) The electrophysiology of photoreceptors in different eye regions of the desert ant, *Cataglyphis bicolor*. *Journal of Comparative Physiology A* 158, 1–7.

Labhart, T. (1999) How polarization-sensitive interneurones of crickets see the polarization pattern of the sky: a field study with an optoelectronic model neurone. *Journal of Experimental Biology* 202, 757–770.

Liechti, F. (1995) Modelling optimal heading and airspeed of migrating birds in relation to energy expenditure and wind influence. *Journal of Avian Biology* 26, 330–336.

Liechti, F. and Bruderer, B. (1998) The relevance of wind for optimal migration theory. *Journal of Avian Biology* 29, 561–568.

Liechti, F., Hedenström, A. and Alerstam, T. (1994) Effects of sidewinds on optimal flight speed of birds. *Journal of Theoretical Biology* 170, 219–225.

Lindauer, M. (1976) Foraging and homing flight of the honey-bee: some general problems of orientation. In: Rainey, R.C. (ed.) *Insect Flight*. John Wiley & Sons, New York, pp. 199–216.

Lindauer, M. and Martin, H. (1968) Die Schwereorientierung der Bienen unter dem Einfluss der Erdmagnetfelder. *Zeitschrift für vergleichende Physiologie* 60, 219–243.

Loxdale, H.D. and Lushai, G. (1999) Slaves of the environment: the movement of herbivorous insects in relation to their ecology and genotype. *Philosophical Transactions of the Royal Society of London B* 354, 1479–1495.

Menzel, R., Geiger, K., Chittka, L., Jasdan, J., Kunze, J. and Müller, U. (1996) The knowledge base of bee navigation. *Journal of Experimental Biology* 199, 141–146.

Monasterio, F.O., Sanchez, V., Liquidano, H.G. and Venegas, M. (1984) Magnetism as a complementary factor to explain orientation systems used by monarch butterflies to locate their overwintering areas. *Atala* 9, 14–16.

Nielsen, E.T. (1961) On the habits of the migratory butterfly *Ascia monuste* L. *Biologiske Meddelelser Det Kongelige Danske Videnskabernes Selskab* 23, 1–81.

Oliveira, E.G. (1990) Orientação de vôo de lepidópteros migratórios na região de Carajás, Pará. MSc. thesis, Instituto de Biologia, Universidade Estadual de Campinas, São Paulo, Brazil.

Oliveira, E.G. (1998) Migratory and foraging movements in diurnal neotropical Lepidoptera: experimental studies on orientation and learning. PhD thesis, University of Texas, Austin, Texas.

Oliveira, E.G., Dudley, R. and Srygley, R.B. (1996) Evidence for the use of a solar compass by neotropical migratory butterflies. *Bulletin of the Ecological Society of America* 775, 332.

Oliveira, E.G., Srygley, R.B. and Dudley, R. (1998) Do neotropical migrant butterflies navigate using a solar compass? *Journal of Experimental Biology* 201, 3317–3331.

Pennycuick, C.J. (1978) Fifteen testable predictions about bird flight. *Oikos* 30, 165–176.

Pennycuick, C.J., Fuller M.R. and McAllister, L. (1989) Climbing performance of Harris' hawks *(Parabuteo unicinctus)* with added load: implication for muscle mechanics and for radiotracking. *Journal of Experimental Biology* 142, 17–29.

Perez, S.M., Taylor, O.R. and Jander, R. (1997) A sun compass in monarch butterflies. *Nature* 387, 29.

Perez, S.M., Taylor, O.R. and Jander, R. (1999) The effect of a strong magnetic field on monarch butterfly *(Danaus plexippus)* migratory behavior. *Naturwissenschaften* 86, 140–143.

Preiss, R. and Gewecke, M. (1991) Compensation of visually simulated wind drift in the swarming flight of the desert locust *(Schistocerca gregaria)*. *Journal of Experimental Biology* 157, 461–481.

Rainey, R.C. (1989) *Migration and Meteorology: Flight Behaviour and the Atmospheric Environment of Locusts and other Migrant Pests*. Clarendon Press, Oxford.

Rankin, M.A. and Burchsted, J.C.A. (1992) The cost of migration in insects. *Annual Review of Entomology* 37, 533–559.

Richardson, W.J. (1991) Wind and orientation of migrating birds: a review. In: Berthold, P. (ed.) *Orientation in Birds*. Birkhäuser Verlag, Basel, pp. 226–249.

Riley, J.R. and Reynolds, D.R. (1986) Orientation at night by high-flying insects. In: Danthanarayana, W. (ed.) *Insect Flight: Dispersal and Migration*. Springer-Verlag, Berlin, pp. 71–87.

Riley, J.R., Reynolds, D.R., Smith, A.D., Edwards, A.S., Osborne, J.L., Williams, I.H. and McCartney, H.A. (1999) Compensation for wind drift by bumble-bees. *Nature* 400, 126.

Ritland, D.B. (1998) Mimicry-related predation on two viceroy butterfly (*Limenitis archippus*) phenotypes. *American Midland Naturalist* 140, 1–20.

Rossel, S. and Wehner, R. (1986) Polarization vision in bees. *Nature* 323, 128–131.

Russell, R.W., May, M.L., Soltesz, K.L. and Fitzpatrick, J.W. (1998) Massive swarm migrations of dragonflies (Odonata) in eastern North America. *American Midland Naturalist* 140, 325–342.

Schmidt-Koenig, K. (1985) Migration strategies of monarch butterflies. In: Rankin, M.A. (ed.) *Migration: Mechanisms and Adaptive Significance. Contributions of Marine Sciences* 27 (suppl.), 786–798.

Seeley, T.S. (1986) Social foraging by honeybees: how colonies allocate foragers among patches of flowers. *Behavioral Ecology and Sociobiology* 19, 343–354.

Southwood, T.R.E. (1962) Migration of terrestrial arthropods in relation to habitat. *Biological Review* 37, 171–214.

Southwood, T.R.E. (1977) Habitat, the templet for ecological strategies? *Journal of Animal Ecology* 46, 337–365.

Spieth, H.R. and Kaschuba-Holtgrave, A. (1996) A new experimental approach to investigate migration in *Pieris brassicae* L. *Ecological Entomology* 21, 289–294.

Spieth, H.R., Cordes, R.-G. and Dorka, M. (1998) Flight directions in the migratory butterfly *Pieris brassicae*: results from semi-natural experiments. *Ethology* 104, 339–352.

Srygley, R.B. (2001a) Compensation for fluctuations in crosswind drift without stationary landmarks in butterflies migrating over seas. *Animal Behaviour* (in press).

Srygley, R.B. (2001b) Sexual differences in tailwind drift compensation in *Phoebis sennae* butterflies (Lepidoptera: Pieridae) migrating over seas. *Behavioural Ecology* (in press).

Srygley, R.B. and Dudley, R. (1993) Correlations of the position of center of body mass with butterfly escape tactics. *Journal of Experimental Biology* 174, 155–166.

Srygley, R.B. and Kingsolver, J.G. (1998) Red-wing blackbird reproductive behaviour and the palatability, flight performance, and morphology of temperate pierid butterflies (*Colias*, *Pieris*, and *Pontia*). *Biological Journal of the Linnaean Society* 64, 41–55.

Srygley, R.B. and Kingsolver, J.G. (2000) Effects of weight loading on flight performance and survival of palatable Neotropical *Anartia fatima* butterflies. *Biological Journal of the Linnean Society* 70, 707–725.

Srygley, R.B., Oliveira, E.G. and Dudley, R. (1996) Wind drift compensation, flyways, and conservation of diurnal, migrant Neotropical Lepidoptera. *Proceedings of the Royal Society of London B* 263, 1351–1357.

Taylor, L.R. (1974) Insect migration, flight periodicity, and the boundary layer. *Journal of Animal Ecology* 43, 225–238.

Ugolini, A., Melis, C., Innocenti, R., Tiribilli, B. and Castellini, C. (1999) Moon and sun compasses in sandhoppers rely on two separate chronometric mechanisms. *Proceedings of the Royal Society of London B* 266, 749–752.

Videler, J.J. (1995) Consequences of weight decrease on flight performance during migration. *Israel Journal of Zoology* 41, 343–356.

Videler, J.J., Vossebelt, G., Gnodde, M. and Groenewegen, A. (1988) Indoor flight experiments with trained kestrels. I. Flight strategies in still air with and without added weight. *Journal of Experimental Biology* 134, 173–183.

Wakeling, J.M. and Ellington, C.P. (1997) Dragonfly flight. III. Lift and power requirements. *Journal of Experimental Biology* 200, 583–600.

Walker, T.J. and Littell, R.C. (1994) Orientation of fall-migrating butterflies in north peninsular Florida and in source areas. *Ethology* 98, 60–84.

Weber, T.P. and Houston, A.I. (1997) Flight costs, flight range and the stopover ecology of migrating birds. *Journal of Animal Ecology* 66, 297–306.

Weber, T.P., Houston, A.I. and Ens, B.J. (1994) Optimal departure fat loads and site use in avian migration: an analytical model. *Proceedings of the Royal Society of London B* 258, 29–34.

Weber, T.P., Alerstam, T. and Hedenström, A. (1998) Stopover decisions under wind influence. *Journal of Avian Biology* 29, 552–560.

Wehner, R. (1981) Spatial vision in arthropods. In: Autrum, H. (ed.) *Comparative Physiology and Evolution of Vision in Invertebrates. C: Invertebrate Visual Centers and Behavior II*. Springer-Verlag, Berlin, pp. 287–616.

Wehner, R. (1984) Astronavigation in insects. *Annual Review of Entomology* 29, 277–298.

Wehner, R. (1998) Navigation in context: grand theories and basic mechanisms. *Journal of Avian Biology* 29, 370–386.

Wehner, R., Michel, B. and Antonsen, P. (1996) Visual navigation in insects: coupling of egocentric and geocentric information. *Journal of Experimental Biology* 199, 129–140.

Williams, C.B. (1930) *The Migration of Butterflies*. Oliver & Boyd, Edinburgh.

Williams, C.B. (1957) Insect migration. *Annual Review of Entomology* 2, 163–180.

Wilson, K. (1995) Insect migration in heterogeneous environments. In: Drake, V.A. and Gatehouse, A.G. (eds) *Insect Migration: Tracking Resources Through Space and Time*. Cambridge University Press, Cambridge, pp. 243–264.

Zera, A.J., Potts, J. and Kobus, K. (1998) The physiology of life-history trade-offs: experimental analysis of a hormonally induced life-history trade-off in *Gryllus assimilis*. *American Naturalist* 152, 7–23.

# 10 Characterizing Insect Migration Systems in Inland Australia with Novel and Traditional Methodologies

V.A. Drake,[1] P.C. Gregg,[2] I.T. Harman,[1] H.-K. Wang,[1] E.D. Deveson,[3] D.M. Hunter[3] and W.A. Rochester[4]

[1]*School of Physics, University College, The University of New South Wales, Australian Defence Force Academy, Canberra, ACT 2600, Australia;* [2]*School of Rural Science and Natural Resources, University of New England, Armidale, NSW 2351, Australia;* [3]*Australian Plague Locust Commission, Agriculture Forestry and Fisheries Australia, GPO Box 858, Canberra, ACT 2601, Australia;* [4]*Department of Zoology and Entomology, University of Queensland, Brisbane, Queensland 4072, Australia*

## Introduction

Migration is now generally recognized to be an adaptation that allows an organism to exploit resources that vary in both space and time (Southwood, 1962, 1977). A contemporary perspective of migration in insects (Drake *et al.*, 1995) identifies four primary components of the broad migration phenomenon. These are: (i) the changing habitat *arena* in which migration occurs; (ii) the pattern of population movements through the arena that successive generations follow (the *pathway*); (iii) the *syndrome* of physiological and behavioural traits that allow the insects to make these migrations and that tend to steer them towards resources as these become available; and (iv) the *genetic complex* that underlies this syndrome. In this holistic view, the many interactions of these components include the process of contemporary natural selection, which acts especially through changes in the arena – weather and climate effects, variations in habitat quality, the incidence of natural enemies, etc. Selection acts continually as populations move along the pathway, adjusting the frequencies of alleles and (along with the normal processes of inheritance and sexual reproduction) maintaining sufficient variation within the population for it to survive and exploit the changing environment. It seems likely that the form of these processes, and of the genetic complex resulting from them, will depend significantly on the extent to which changes in the arena environment are predictable (Southwood, 1977; Wilson, 1995).

This perspective can help us to formulate questions about migration systems that may be empirically testable. One set of high-level questions examines the relationships between components of the conceptual model. Does the observed pathway match the arena characters? Will the observed syndrome characters steer populations along the pathway? Will the mix of alleles present produce the observed range of physiological and behavioural traits? Can observed changes in allele frequencies be accounted for by actual migrations and the resulting population partitioning (Drake *et al.*, 1995) and differential mortality? Empirical examination of questions such as these allows understanding of the functioning of a migration system to be developed. While many system components will be difficult to investigate, even a limited examination of the more accessible processes may be sufficient to reject some hypothesized mechanisms.

The semiarid inland of eastern Australia (Fig. 10.1), and neighbouring higher-rainfall areas nearer the coast, provide an arena that appears suitable for answering some of these questions. The inland is a region in which resources vary greatly in both space and time, and that is exploited by a number of species of migratory insects. It is also characterized by a high degree of unpredictability, with habitat quality determined primarily by rainfall – which exhibits high-to-extreme variability in this region (Bureau of Meteorology, 1986) – rather than the regular seasonal changes of temperature. Because two of the region's principal migrants, the Australian plague locust *Chortoicetes terminifera* (Walker) and the native budworm *Helicoverpa punctigera* (Wallengren), become serious economic pests when they move into adjoining cropping regions, there is a significant base of knowledge about the organisms' population processes (e.g. Farrow, 1979; Wright, 1987; Gregg *et al.*, 1995; Rochester, 1999), and survey infrastructures have already been developed. It has therefore been possible to make an initial investigation of the arena and pathway components of the migration systems of these two species by supplementing existing survey and monitoring programmes and by adding a programme of direct observations of 'migration-in-progress', using insect monitoring radars (IMRs; Drake, 1993; Smith *et al.*, 2000). This *Inland Insect Migration Project* (IIMP) commenced its operational phase in mid-1998, and a preliminary analysis of the first full year of observations is now available.

In this chapter we first discuss how migration outcomes arise from the interaction of each insect's migration syndrome and the particular environmental conditions it experiences. We then briefly describe the IIMP study region and the observing methods we are using, and follow this with a summary of the data available to date. The chapter concludes with a discussion of system characterization and the extent to which this has been achieved, and

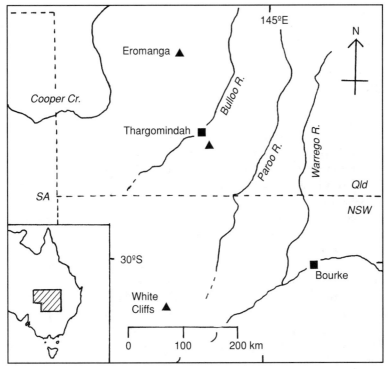

**Fig. 10.1.** The study region, showing the two insect monitoring radar (IMR) sites (■) and the locations of light-traps (▲). The inset shows the region's location within Australia.

of the use of characterization data for developing understanding of how a migration system functions.

## Migration Outcomes and the Factors Affecting Them

Migration results in relocation, and perhaps also aggregation or dispersal, of a population. Some migration – especially by vertebrates, but also by at least one insect, the Monarch butterfly *Danaus plexippus* (Linnaeus) (Malcolm and Zalucki, 1993) – ends at a particular, predetermined, destination, and therefore apparently incorporates goal-seeking behaviour. In most insect migrants, however, the process appears much less precise and only some general patterns in the distance and direction of movement can be identified. Migrations can be considered successful if suitable habitats are present at the destination; they are probably especially successful when the hosts there are just becoming available, and are thus largely free of competitors and natural enemies. When migrations are unsuccessful, the population will either perish immediately or fail to reproduce.

In many insect species, there appears to be a bias in the pattern of migratory relocations towards regions where suitable habitats will be present on arrival. For example, in the oriental armyworm *Mythimna separata* (Walker) movements in spring are towards higher latitudes and the population follows the growing or flowering season of its cereal hosts as this extends northwards (Drake and Gatehouse, 1996); a reverse southward movement occurs in autumn. Similarly, acridoids in tropical and subtropical North Africa move northwards and southwards with the belt of rains and the floodplain inundation associated with the seasonal movements of the intertropical convergence, or (in the case of the desert locust *Schistocerca gregaria* Forskål) move between temperate and tropical seasonal rainfalls north and south of the Sahara (Farrow, 1990). It is easy to envisage how natural selection might act to produce and maintain such biases when the set of traits required to effect such a 'to-and-fro' migration pattern is present in the population. Where host availability is less predictable, as in the study region described here, a seasonal migration pattern may provide little benefit. Migrants may instead develop adaptive responses to more ephemeral cues – e.g. 'disturbed weather' (Farrow, 1979, 1991) – or they may simply migrate 'at random'. In the latter case, many individuals will perish and a high level of variation in traits affecting migration will need to be maintained if some members of a population are to reach favourable habitats.

The vector displacement (i.e. direction as well as distance) resulting from a flight is determined by the airspeed and orientation of the insect, the duration of the flight, and the speed and direction of the wind. Wind varies significantly with height (Drake and Farrow, 1988), so the height at which an insect flies usually has a major effect on the migration outcome (i.e. the displacement achieved). Flight near the surface – within the 'flight boundary layer' (Taylor, 1974) – allows a migrant to maintain progress in a preferred direction even when there is a headwind. Flight at greater heights – typically 100–1000 m – allows migrants to move much more rapidly than their own airspeed, but only in directions close to that of the wind. Day-to-day variations in wind direction, however, may cause migratory displacements on successive days (or nights – much migration of this type is nocturnal) to be quite different. It is this latter type of migration – at altitude, nearly downwind and at night – that predominates in the IIMP study species and thus that we are concerned with here.

The factors affecting migration outcomes can be divided into those that that are determined in part by the insect's condition and its responses to cues, and those that arise solely from arena processes. Of principal importance among the former are: the date and time of commencement of each migratory flight; the number of nights on which migration occurs and the duration of each flight; the height of flight; and any responses to cues that result in flight being terminated. The insect's orientation and airspeed probably play a lesser role but may still be significant. External (i.e. arena) factors include wind speed and direction, and environmental conditions and cues such as photoperiod, temperature and weather that the insect responds to or that affect its capacity for flight. We will now examine some of these factors, and their effects on outcomes, in greater detail.

## Initiation of migration

Although *C. terminifera* is active by day and sometimes moves tens of kilometres downwind in low-altitude daytime flights, its principal migrations commence at dusk and are undertaken at heights of a few hundred metres (Clark, 1971; Drake and Farrow, 1983). The immediate factors influencing the initiation of these nocturnal migrations are illumination, temperature and wind. Take-off is synchronized by the low and falling light levels of dusk, but is suppressed by strong winds and by temperatures below ~20°C, with 'mass take-off' (involving a high proportion of the population and followed by sustained migratory flight) associated with temperatures above 25°C (Symmons, 1986) or a combination of disturbed weather and above-average temperatures (Farrow, 1979). Environmental conditions earlier in the day and during development may also influence a locust's capacity to migrate (e.g. Hunter *et al.*, 1981). *H. punctigera* is fully nocturnal and has been characterized as an obligate migrant (Gregg *et al.*, 1995). It also takes off at dusk, in temperatures down to about 5°C, and its peak migrations may be associated with disturbed weather (Gregg *et al.*, 1994; Rochester, 1999).

In both species, and as with most migrants, these nocturnal long-distance flights are made during the pre-reproductive period of the adult stage (Gatehouse and Zhang, 1995). Migratory flight probably commences on the first suitable night after the insect has ceased to be teneral. The 20°C temperature threshold for flight by *C. terminifera* is not usually limiting in summer (Symmons, 1986). However, at the beginning or end of the season, it may produce a 'rectification' effect (Kennedy, 1951) with migration occurring on warmer airflows – which in this situation will usually be from the north, carrying the migrants southwards – and suppressed when cooler southerlies are blowing.

## Ground speed and track

For migrants flying well above the surface, airspeeds – typically 3–6 m $s^{-1}$ for larger moths and acridoids (Schaefer, 1976; Westbrook *et al.*, 1994) – and heading directions (i.e. orientation angles) often play only a minor role in determining ground speeds and directions (Fig. 10.2). However, radar studies frequently show a migrating population to have some degree of common orientation (e.g. Riley, 1975; Schaefer, 1976; Drake, 1983), and the direction of orientation to have some degree of consistency from night to night (Riley

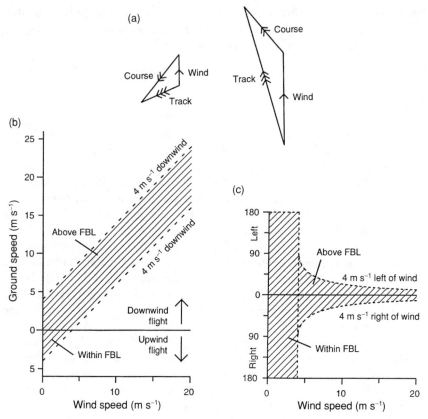

**Fig. 10.2.** Relationship between wind, a migrant's airspeed and course, and its track and ground speed. (a) Vector diagrams for a migrant flying within (left) and above (right) its flight boundary layer (FBL); airspeed of migrant 4 m s$^{-1}$, wind speed 2 m s$^{-1}$ (left), 6 m s$^{-1}$ (right). (b) Range of possible ground speeds (hatched area) for a migrant with an airspeed of 4 m s$^{-1}$ in a wind of up to 20 m s$^{-1}$. (c) Range of possible tracks (hatched area) for the migrant in (b).

and Reynolds, 1986; Wolf *et al.*, 1995), so it is possible that maintenance of a particular heading does have adaptive value.

## Migratory transport

The recognition that the migration outcomes of insects flying at altitude depend predominantly on the wind leads to the concept of windborne 'transport', i.e. that wind is a resource that migrants can exploit by using it as a conveyance. For a population to relocate to a new habitat, transport – i.e. a wind in the required direction combined with environmental conditions (temperature, precipitation, etc.) favourable for flight – must be available at some time during the period when the adults are in their pre-reproductive state. It could be hypothesized that if this combination of events occurs infrequently in a particular region or season, natural selection would produce migrants that fly low and maintain a fixed orientation, while if the combination is of frequent occurrence then ascent out of the migrant's flight boundary layer would be selected strongly but maintenance of orientation only weakly if at all.

The second aspect of transport that impacts directly on migration outcomes is the rate at which it produces displacement. However, we are concerned here not just with the wind speed but also with the straightness of the air's path. Both of these factors are properties of the arena and can be characterized (i.e. their statistics estimated) from ensembles of meteorological data.

## Duration of flight and number of nights of flight

For a specific flight, the migrant's trajectory can be estimated as the vector sum of a series of legs,

$$\Delta \mathbf{X} = \sum_k \mathbf{v}_k \Delta t_k, \tag{1}$$

where $\Delta t$ is the duration of each leg and $\mathbf{v}$ the migrant's velocity over the ground during that leg. The distance covered, $R$, is then simply

$$R = |\Delta \mathbf{X}|. \tag{1a}$$

This approach is valuable for understanding particular events (e.g. Gregg et al., 1994), but for more general insights into the functioning of a migration system, it is probably more useful to employ a simpler representation,

$$R \approx f \bar{v} T. \tag{2}$$

where $f$ ($0 \le f \le 1$) is a straightness factor, $\bar{v}$ the average wind speed, and $T$ the flight duration. This makes apparent that the distance covered depends directly on the duration of flight. If probability distributions of the two arena factors $f$ and $\bar{v}$ are available, the distances likely to result from flights of different durations can easily be estimated from Equation (2).

Flight durations have been measured in the laboratory in mill experiments for a number of species (e.g. Han and Gatehouse, 1993), and have been found to show significant variation (Gatehouse and Zhang, 1995). These laboratory durations, termed *flight capacities*, $\tau$, will tentatively be used to estimate the maximum durations $T$ of migratory flights, i.e.

$$T \le \tau,$$

with the inequality representing termination of flight by the migrant in response to some cue. (Flights longer than $\tau$ may occur in certain circumstances, e.g. during transoceanic flights (Rosenberg and Burt, 1999), but it is doubtful whether these are significant within the population's arena.)

Flight capacities vary between individuals, and probably also with temperature and other environmental conditions, and may be reduced on the second and subsequent nights of flight. They appear to represent a trait that both shows variation within a population and relates directly, through Equation (2), to migration outcomes.

A second trait that directly influences where a migrant will eventually relocate to is the *pre-reproductive period*, as this determines the number of nights on which migration can occur. A long pre-reproductive period may permit successive migrations on a series of nights, or allow greater scope for waiting until cues indicating transport in a favourable direction occur. The *migratory potential*, $M$, the overall capacity of the individual for migratory flight, could perhaps be represented as

$$M \equiv (\tau_1 ... \tau_N), \tag{3}$$

i.e. as the set of flight capacities for the sequence of pre-reproductive nights. Additional migratory flights will usually increase the total displacement distance, but differences in transport direction from night to night may be significant and will produce an overall trajectory that is far from straight. An appropriate extension of Equation (2) would appear to be

$$R_N \approx F_N \bar{v}_N \sum_N \tau_i \tag{4}$$

where $R_N$ is the potential displacement after $N$ nights and $F_N$ and $\bar{v}_N$ are the straightness factor and average speed for a sequence of $N$ successive nights. Estimation of $F_N$ and $\bar{v}_N$ requires calculation of trajectories starting from the previous night's endpoint, and has not yet been attempted. The total flight time $T_T$ provides a possible single-parameter measure of migratory potential, $M_1$,

$$T_T = \sum_N \tau_i = M_1 \tag{5}$$

and allows Equation (4) to be recast in the same form as Equation (2),

$$R_N \approx F_N \bar{v}_N T_T. \tag{4a}$$

Laboratory experiments on a number of migratory insects, including some noctuid moths, have shown a high level of variation in pre-reproductive period (Gatehouse and Zhang, 1995). For a few migrant species, a genetic basis for variation in both pre-reproductive period and flight capacity has been established (e.g. Han and Gatehouse, 1991, 1993). The limited evidence available suggests that these two components of migratory potential are inherited independently (Gatehouse and Zhang, 1995). Available relevant measurements on *C. terminifera* (Lambert, 1982) and *H. punctigera* (Coombs, 1992) suggest modest migratory potentials for both species, but the field evidence for their status as long-distance migrants appears irrefutable and little use can apparently be made of these results in interpreting the IIMP data.

## Study Area and Observing Methods

The aim of the IIMP observations program is to track the changing population patterns of the two study species, to monitor the varying availability of the resources on which they depend, to observe their migrations and to identify the wind-transport opportunities that make these migrations possible. As it is not practicable to do this over the species' full arenas – which extend over much of the continent – project activities have been focused on a more limited study area.

### Study area

Pre-existing knowledge of the migrations of *C. terminifera* and *H. punctigera* in inland Australia (e.g. Drake and Farrow, 1983; Wright, 1987; Gregg *et al.*, 1995) indicated that population movements occurred on a scale of at least several hundred kilometres, so the IIMP study area evidently needed to be at least this large. Other factors affecting the choice of the area included highly variable rainfall, the widespread presence of suitable hosts for both species – or, put alternatively, a historical record of significant populations developing there – and practicable access from the investigators' home bases. The area chosen (Fig. 10.1) meets these specifications well; it can be reached in 1–2 days of driving, and good locations for the IMRs were found that are accessible by all-weather roads. The two IMRs are located about 300 km apart and essentially define the core of the study region, which extends approximately 300 km outwards from this axis.

### Observing methods

The most practicable way of monitoring a region this large is by satellite remote sensing, supplemented by a limited programme of confirmatory surveys along roads. The Australian Plague Locust Commission (APLC) already monitors the area for locusts in these ways (Bryceson *et al.*, 1993; Deveson and Hunter, 2000), and is making its data available to IIMP. Because only habitats can be observed from space, APLC operates a small network of light-traps to detect adult *C. terminifera* populations, and IIMP has added two IMRs (Figs 10.1 and 10.6). The former provide an indication of locust population size, while the latter detect and quantify migration-in-progress. The IMRs (Drake, 1993) are fully automatic and observe almost throughout the hours of darkness, detecting insects passing overhead at altitudes between 200 and 1300 m. As well as measuring insect numbers, directions of movement, orientations and speeds, they can estimate the size, shape and wingbeat pattern of the targets and thus provide some information about their identities. Although the data obtained from the traps and radars strictly relates only to their particular location, extrapolation to a wider area usually appears reasonable. Atmospheric transport opportunities are inferred from the output of a limited area (i.e. higher resolution) numerical weather-forecasting model (Puri et al., 1998) run routinely by the Australian Bureau of Meteorology. Much existing knowledge of the study area, including the general favourability for the two species of the various habitats within it (inferred mainly from a soils database), is incorporated into a geographical information system (GIS) managed by APLC, into which survey and meteorological data is entered as it is received (Deveson and Hunter, 2000). Geostationary weather satellite images of cloud-top temperatures are also drawn upon to identify likely heavy-rainfall events (Bryceson et al., 1993).

Surveys are scheduled, and the routes selected, to coincide with regions where satellite or rainfall data, combined with climatic information, suggest that habitats are becoming suitable and where vegetation types are known to include the species' favoured hosts. (Generally, locusts occur in grasslands and on floodplains while native budworms, which breed on a variety of ephemeral forbs, especially *Asteraceae*, are found mainly in sparse shrublands.) Survey stops are made in favourable habitats at intervals of approximately 10 km (locust surveys) or at targeted locations where hosts are present (*Helicoverpa* surveys), and systematic ground searches – a 250-m flushing transect for locusts and 100 sweep-net sweeps for *Helicoverpa* larvae – conducted. Vegetation condition and the type of hosts present are also recorded, the former partly to 'ground truth' the satellite information. By these means, the full extent of the study region can be surveyed in a few days, though large gaps remain and there is very little coverage of areas thought unlikely to contain populations. Information on populations is also received at times from other sources, e.g. landholders.

## Initial Results

We summarize here the results of the first full year of IIMP observations, and some additional information on migrations in the study area over the last few years. The latter is drawn from APLC's records and an earlier study of *H. punctigera*. These preliminary findings will illustrate how characterization of migration by the two study species will proceed.

### Habitat availability

Habitat condition, as revealed by normalized difference vegetation index (NDVI) estimates from composites of NOAA (AVHRR) satellite images (Bryceson et al., 1993; Fig. 10.3), shows both patchiness on a broad range of scales – from about 10 km (along watercourses) to around 500 km (across the region) – and obvious changes over intervals of 2 months. Generally, areas

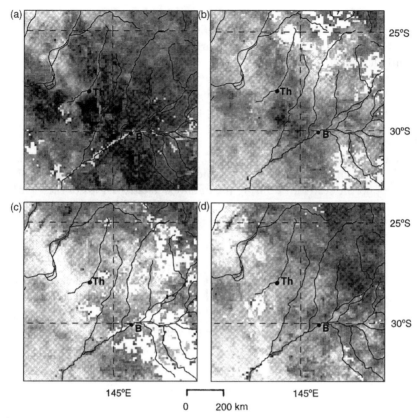

**Fig. 10.3.** Composite images of normalized difference vegetation index (NDVI) for 10-day periods during the months (a) October 1998, (b) December 1998, (c) January 1999 and (d) March 1999. Dark shading indicates fresh green vegetation, white areas are dry or, in (a) only, indicate standing water. The NDVI is relativized to the historically recorded range for each period.

with significant green vegetation are located where rainfall has been recorded 1–2 months earlier, or where floods have receded. Records of habitat conditions made during surveys generally confirm the NDVI information and inferences made from rainfall data. Ephemeral forbs, especially *Asteraceae*, are commonly found in early spring following winter rain and often host heavy populations of *Helicoverpa* larvae. Summer rains and subsequent inundation along watercourses are followed by growth of the tropical (C4) grasses that are the favoured hosts of *C. terminifera*.

### Population changes and migrations

The irregular nature of population changes in the semiarid inland of Australia is illustrated by the patterns of catches of locusts and of budworm moths in light-traps (Fig. 10.4). For locusts, the early and late-season catches in each year appear unrelated, and the correlation of catches between the two sites (Fig. 10.4c) is not strong. The more limited budworm data (Fig. 10.4b) is only slightly less erratic.

Locust migrations partly or wholly within the study region, identified by APLC from survey observations, light-trap records, landholder reports and meteorological (mainly wind-transport) information, occurred in all directions and over distances of 150–400 km (Fig. 10.5). The only documented *H. punctigera* migration within the region covered a similar distance, towards the southeast (Fig. 10.5; P.C. Gregg, unpublished data).

Population changes are also revealed by survey data. Surveys through and around the study area during the middle and late periods of the 1998/99

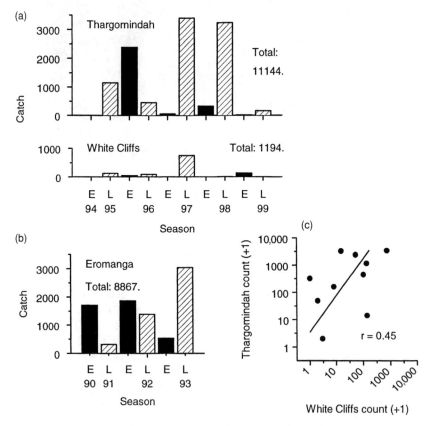

**Fig. 10.4.** Catches in light-traps at sites within or near the study region. (a) *C. terminifera* at Thargomindah and White Cliffs, 1994–1999; (b) *H. punctigera* at Eromanga, 1990–1993; (c) correlation of Thargomindah and White Cliffs locust catches. Catches have been combined into totals for the early (Sep–Dec, E, solid) and late (Jan–Apr, L, hatched) portions of each season. See Fig. 10.1 for trap locations. Data from APLC (locusts) and PCG (budworm), unpublished.

season (Fig. 10.6) showed that locusts were initially widespread, extending 1000 km north-to-south and about 500 km east-to-west. By autumn, densities were significantly higher and patchiness was more evident: most of the locusts were located within a band around 200 km wide, with a particularly dense infestation in the southeast of the region extending over an area of about 150 × 400 km. In the south, the main centre of population had shifted around 200 km eastward into a region becoming green after rain (Figs 10.3c,d). While phenological differences will also have contributed, it is likely that at least some of these changes arose through migration.

**Transport**

Transport has been estimated from the numerical weather-forecasting outputs (Puri *et al.*, 1998) via a trajectory model (Rochester, 1999) which uses the predictor-corrector algorithm of Scott and Achtemeier (1987) to calculate the successive locations of an air parcel over a period of up to 12 h, using a 0.5-h step size. The preliminary results presented here (Figs 10.7–10.9) are for trajectories starting at dusk from Bourke and continuing for 6 h, and are calculated for an altitude of 600 m. These height and duration values are comparable with those recorded for a moth migration in an earlier radar study at a location in the adjacent agricultural zone (Drake and Farrow, 1985). They are somewhat longer and higher than those for *C. terminifera* flights in a second such study (Drake and Farrow, 1983), but conditions in the inland of Australia would generally be more favourable for locust flight and a lower altitude was considered undesirable because of concerns about

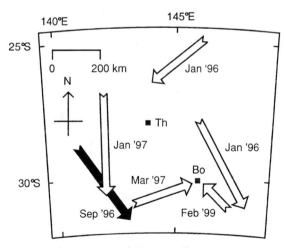

**Fig. 10.5.** Documented migrations of *C. terminifera* (open arrows) and *H. punctigera* (black arrow) in or near the study area, 1996–1999. Key: Bo – Bourke, Th – Thargomindah. Data from APLC, unpublished.

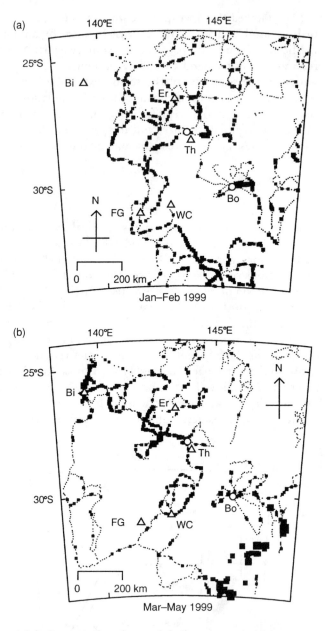

**Fig. 10.6.** Adult *C. terminifera* observed during surveys in (a) January and February and (b) March–mid-May 1999. Dots indicate point surveyed and none found; squares indicate presence at estimated densities ranging from 'isolated' (< 200 ha$^{-1}$, smallest squares) to 'medium-density swarm' (100,000–500,000 ha$^{-1}$, largest square). Locations of insect monitoring radars (IMRs) (○) and light-traps (△) are also shown. Key: Bi – Birdsville, Bo – Bourke, Er – Eromanga, FG – Fowlers Gap, Th – Thargomindah, WC – White Cliffs. Data from APLC, unpublished.

*Insect Migration Systems in Australia* 221

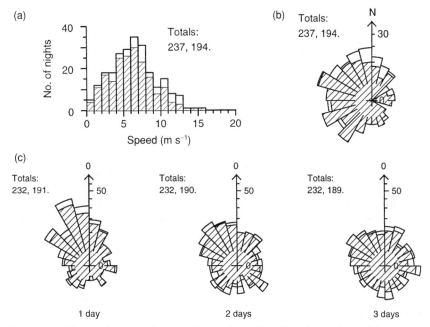

**Fig. 10.7.** Dusk wind speeds (a) and directions (b) at Bourke, estimated from outputs of a regional-scale numerical weather-forecasting model. Values are for an altitude of 600 m, and for the September–April 'insect flight season' of 1998/99. Values for nights when the dusk temperature exceeded 20°C are shown hatched. (c) Changes of dusk wind direction over 1, 2 and 3 days. Changes for which the dusk temperature of the final night exceeded 20°C are shown hatched. Source of data as for (a).

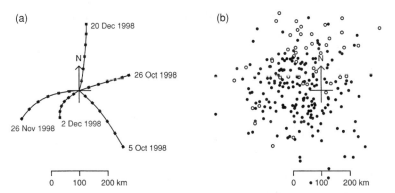

**Fig. 10.8.** (a) Some typical 6-h air-parcel trajectories from Bourke. Trajectories commence at dusk and are calculated for an altitude of 600 m; the temperature at dusk was > 20°C in all cases. The points along the trajectory indicate 1-h intervals. (b) Endpoints for all such 6-h trajectories for the 1998/99 September–April 'insect-flight season'; endpoints for nights when the temperature exceeded 20°C are shown solid.

the accuracy of extrapolating the numerical weather-forecasting outputs down to the surface. Graphs and statistics are for the single 'insect-flight season' (September–April inclusive, 242 nights – although for a small number of nights the data is missing) of 1998/99; elimination of the winter period in this way reduces the contribution of nights when few insects would be available for migration. Wind transport is of course only available to an insect if the temperature is above the insect's threshold for flight. The graphs here show distributions for all nights and for nights when the temperature exceeded 20°C, the latter corresponding approximately to the *C. terminifera* threshold.

The trajectory parameters that characterize transport are primarily the direction and distance of movement, and secondarily the trajectory's straightness. The former can be conveniently represented in terms of the initial or departure direction $\theta_d$ (calculated at about dusk) and the distance from the point of departure, $R_T$, after some appropriate standard time – in this case 6 h. The latter can be characterized by the trajectory straightness $f_T = R_T/L_T$, where $L_T$, the trajectory length, is estimated as the sum of the lengths of the 0.5-h steps.

The distributions of wind speeds and directions at dusk are shown in Fig. 10.7a, b. The median speed was 6.4 m s$^{-1}$, and directions were biased towards the west, i.e. winds were predominantly from the eastern sector. A study of potential windborne transport for *C. terminifera* that was based on routine meteorological upper-wind (i.e. weather-balloon) measurements over many summers (Symmons, 1986) also showed a directional bias towards the west or northwest in the vicinity of the study area.

The dusk wind directions also provide an indication of the likely effect on transport of deferring flight by 1 or more days. In Fig. 10.7c, the dusk direction is shown relative to its value 1, 2 and 3 days earlier. A bias towards backing (anticlockwise turning) of the wind is evident – a consequence of the location being predominantly to the north of the eastward-moving anticyclones of the subtropical ridge. A delay of 1 day will change the direction of transport by more than 45° on 34% of occasions, and after a delay of 3 days

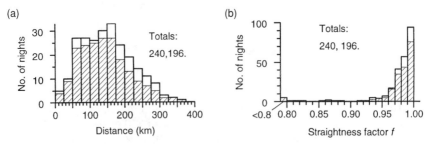

**Fig. 10.9.** Histograms of (a) displacement distance and (b) straightness factor *f* for the 6-h trajectories of Fig. 8(b). The hatched sections of the bars indicate nights when the temperature exceeded 20°C.

this proportion has risen to 61%. These results suggest that if a reliable cue is available, selective suppression of take-off would provide a highly effective means of 'steering' the migration towards a preferred direction – though only if airflows in that general direction occur with reasonable frequency.

A representative sample of calculated trajectories, starting from Bourke at dusk and continuing for 6 h, is shown in Fig. 10.8a. They indicate movement in a variety of directions, over distances of 100–200 km, along paths that are either simple curves or almost straight. The endpoints of all such trajectories for the 1998/99 flight season are presented in Fig. 10.8b. As is to be expected from Fig. 10.7b, movements are most frequently towards the west. Distributions of the direct distance covered in 6 h, and the straightness factors for the trajectories, are shown in Fig. 10.9. The median distance is 150 km, and straightness factors exceeded 0.95 on 90% of nights. These results show that single-night transport over distances of 100–200 km, and in a variety of directions, was frequently available. The 150–400-km displacements achieved in actual migrations (Fig. 10.5) fall in the upper part of the distribution of Fig. 10.9a, suggesting greater flight durations, higher wind speeds or a significant airspeed contribution on these occasions.

## Migration-in-progress

### Migration intensity

The number of larger insects overflying the Bourke IMR on 3 nights during a period when *C. terminifera* adults were plentiful in the surrounding area is

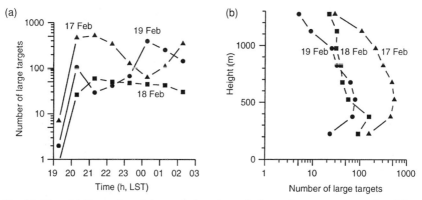

**Fig. 10.10.** (a) Examples of the variation through the night (19:00–02:30 h) of the number of targets of large-insect type ($0.5 < \sigma_{max} < 4$ cm$^2$) detected by the Bourke insect monitoring radar (IMR) at altitudes of 500–900 m. The same observing effort was used during each hour. (b) Variation with altitude of the number of these targets during the 3-h period following dusk (20:00–22:30 h).

shown in Fig. 10.10a. Numbers increase rapidly at dusk and then vary somewhat during the night, showing both rises and falls. On all 3 nights, numbers after midnight are broadly comparable with numbers immediately after dusk. The variation in the number of detected large targets with altitude is shown in Fig. 10.10b. While these do not provide an accurate profile (because the detection efficiency of the IMR, which varies significantly with altitude, has not been corrected for here), they do indicate that migration occurs throughout the bottom 1 km of the atmosphere. These results suggest rather longer and perhaps higher flights than those observed for *C. terminifera* in the agricultural belt (Drake and Farrow, 1983) and provide some support for the values of 6 h and 600 m chosen for the characterization of windborne transport (Figs 10.8 and 10.9).

The seasonal variation of insect numbers is shown in Fig. 10.11a. Few large targets were detected in winter (June–August), but numbers rose rapidly in early spring (September) and remained significant through to mid-autumn (April). Major peaks of activity occurred during the first half of spring (September–October) and the hottest months of January–March. A histogram of the log-transformed values of these numbers (Fig. 10.11b) has an almost symmetrical bell-shaped form, suggesting migration intensity may approximately follow the log-normal distribution.

## Direction and speed

IMR observations show that insects migrating at a particular height and time move in only a small range of directions and have speeds that fall into a quite narrow range (Fig. 10.12a,b). The variation during the night of the general direction of movement and the median speed is also frequently only slight (Fig. 10.12c,d). In contrast, night-to-night variations in direction are often large (Fig. 10.12a,c).

Almost all migration directions occur at some time during the flight season (Fig. 10.13a), but movement towards the north and west occurs on more nights than movement towards the south and east. The westward bias parallels that in the transport departure directions (Fig. 10.7b), but the latter lacks a significant northward component. An analysis that gives weight to the intensity of migration in each direction will perhaps reveal a rather different pattern. Speeds (Fig. 10.13b) are typically in the range 5–15 m s$^{-1}$, which corresponds to displacements of 100–300 km during a 6-h flight. This is broadly consistent with the 150–400 km of documented migrations (Fig. 10.5), as the latter may well have continued all night. The median speed (11.5 m s$^{-1}$, which corresponds to a displacement of 250 km over 6 h) is significantly higher than that for transport (6.4 m s$^{-1}$, Fig. 10.7a). Many technical factors could contribute to this difference, but while a firm conclusion cannot be drawn at this stage it does suggest that the contribution of the migrant's airspeed may be important.

## Discussion

The IIMP data-gathering programme described here has two primary aims: to characterize the migration systems of *C. terminifera* and *H. punctigera* (or at least some important components of them in a representative area), and to develop an understanding of how these systems function. By characterization we mean the acquisition of empirical statistical distributions of the key parameters of the system. For the arena and pathway components of the system that IIMP is focused on, this means developing statistics for the spatial and temporal variations of habitat quality and populations, for airborne transport, and for the migrations themselves. The significant advance this represents over the short-term observations typical of much previous

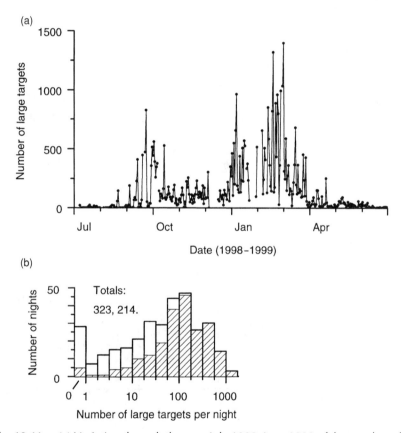

**Fig. 10.11.** (a) Variation through the year July 1998–June 1999 of the number of targets of large-insect type ($0.5 < \sigma_{max} < 4$ cm$^2$) detected at altitudes of 500–900 m by the Bourke insect monitoring radar (IMR) during the 3-h period (20:00–22:30 h) following dusk each night. Data available for 323 of the 365 nights (with scaling from 1 or 2 h of observations required for 14 nights). (b) Distribution of these numbers, with nights during the September–April 'insect-flight season' shown hatched.

radar-based migration research (e.g. Drake and Farrow, 1983, 1985) is that it provides information on the *range* of values of these parameters, i.e. of both their typical values (if any) and their variability, and of how these change through the season. Full characterization will require several seasons of observations, because the erratic rainfall of the study area produces habitats

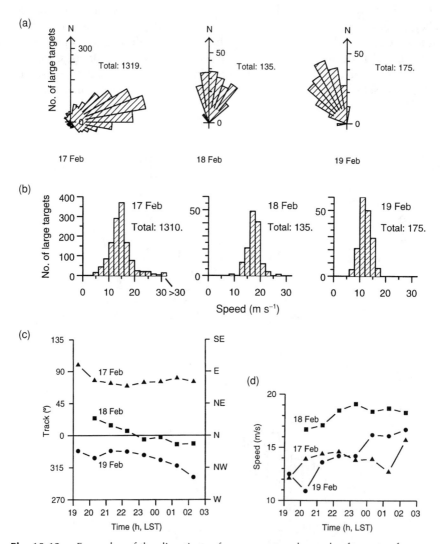

**Fig. 10.12.** Examples of the directions of movement and speeds of targets of large-insect type ($0.5 < \sigma_{max} < 4$ cm$^2$) detected at altitudes of 500–900 m by the Bourke insect monitoring radar (IMR) during the nights of 17–19 February 1999. (a, b) Distributions of directions and speeds for targets detected during the 3-h period (20:00–22:30 h) following dusk. (c, d) Variations of the circular-averaged direction and the median speed during the course of the night.

and populations that vary greatly from year to year. The IIMP observation programme is perhaps examining a more complex migration system than those that were the subjects of previous year-long (Beerwinkle *et al.*, 1995) and multiple-season (Chen *et al.*, 1995; Wolf *et al.*, 1995) studies, as these took place in regions where habitat changes are more predictable (being driven primarily by seasonal temperature changes) and migration is more one-dimensional (to higher latitudes in spring and back again in autumn).

The results available so far suggest that characterization of the arena and pathway of our study area is already starting to be achieved. In particular, quantitative empirical distributions for the scale, directionality, straightness and seasonality of transport opportunities (from the numerical weather-forecasting model, Figs 10.7–10.9) and for actual migration (from the IMRs, Figs 10.11 and 10.13) are already available, and are being steadily extended as the observation programme continues. Spatial data on habitat quality (from satellite imagery) and population distributions (from surveys) are also available – though in the case of populations the coverage is somewhat patchy. These are less straightforward to characterize quantitatively, but can certainly be used in qualitative analyses. Spatial variations in transport can be estimated from the numerical weather-forecasting outputs – distributions have already been calculated for Thargomindah as well as Bourke, though only those for the latter have been presented here. The second IMR will provide an indication of the extent of spatial variations in actual migration. Thargomindah is at a lower latitude than Bourke, and experiences greater aridity and rainfall variability. The preliminary IIMP results indicate that the separation of the two IMRs (~300 km) is comparable with typical displacement distances for a single-night migration (~150–400 km, Figs 10.5 and 10.13b) and with typical spatial scales of populations (~150–400 km,

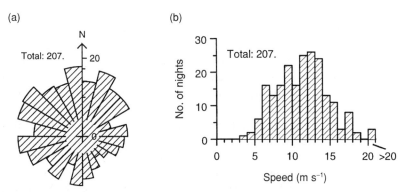

**Fig. 10.13.** The distribution for the 'insect flight season' September 1998–April 1999 of (a) circular-averaged directions and (b) median speeds of targets of large-insect type ($0.5 < \sigma_{max} < 4$ cm$^2$) detected by the Bourke insect monitoring radar (IMR). The statistics are calculated from all such targets detected at altitudes of 500–900 m during the 3-h period (20:00–22:30 h) following dusk, for nights when at least three such targets were detected.

Fig. 10.6). This suggests that population processes at the two sites will not be strongly synchronized, but that some linkage will exist and may become evident in a correlation analysis of the two sets of migration observations.

The second aim of IIMP, development of understanding of how the study area's migration systems function, represents a more ambitious and demanding agenda: it will require us to look at how the various components of the migration systems – including the physiological and genetic dimensions not investigated in IIMP – relate to each other. A system can be said to *function* if the set of processes that allow the population to exploit the spatio-temporally changing environment of its arena, and thus to persist, is maintained. This seems principally to involve interactions between the system's components: how a migrant's physiological characters affect its migration trajectories; how the various syndrome characters that may be required in different parts of the arena or at different seasons are maintained through the underlying genetic complex; how the frequencies of alleles for migration-related genes vary as the population moves along its pathway, etc. Perhaps the most amenable of these to investigation would be the relation of syndrome characters to migration outcomes. An individual's syndrome would be represented by its migratory potential $M$ (Equations (3) and (5)) and by a series of thresholds for initiating and terminating flight; similar characters for entering, maintaining and emerging from diapause – an alternative adaptation to a changing environment (Tauber *et al.*, 1986) – might also be needed. These could be determined in the laboratory (or perhaps some could be hypothesized), and then combined with characterization data for the arena – temperatures at dusk, availability of wind transport in different directions (Fig. 10.7b,c), etc. – and Equations (2) or (4a) to estimate whether migration would have occurred, and the range of locations where it would have terminated. Characterization data for the pathway (Figs 10.11 and 10.13) would then be employed for validation (or empirical hypothesis-testing). This approach allows the detail of the representation of each process to be tested, and the significance of variation between individuals to be assessed.

A more comprehensive but less precise approach, that seeks to investigate all system components together rather than one at a time, is to construct a population model that includes representations of a potentially complete set of processes, i.e. a set that allows the modelled population to cycle. The model would be of the individual-based type, using the 'i-state configuration' approach, and spatially explicit in the 'reaction–diffusion–advection' rather than the 'metapopulation' form (Murdoch and Nisbet, 1996). All available characterization data can be incorporated into the model, which would then be run to predict the system's future state. The model is validated by comparing the resulting forecasts with the populations and migrations that actually develop, and also by determining whether the model population persists and behaves in a biologically plausible way. Alternatively, a subset of the characterization data – e.g. the laboratory syndrome estimates and climatic and transport data – is incorporated into the model, which is then validated against the remainder of the characterization, e.g. the IMR observations of migration-in-progress or data from population surveys.

Simulation models have been constructed for both *C. terminifera* (Wright, 1987) and *H. punctigera* (Rochester, 1999). The former is quite comprehensive, but population-based. It has proved useful for analysing the development of outbreaks and as a forecasting aid, and in modified form has been incorporated into APLC's current decision-support system (Deveson and Hunter, 2000). The *H. punctigera* model is individual-based, but treats in detail only the migration process itself: calculating endpoints for given flight capacities and windfields, as in Fig. 10.8 (but with an airspeed and heading contribution from the simulated migrant). Model outputs are the migration's destination area and the relative likelihood of an endpoint falling in each part of it. Using this model, Rochester (1999) has identified the seasonal transport availability (in terms of the distribution of endpoints) for the entire continent, and tentatively identified some pathways that allow moths to cycle between the inland and the less arid continental periphery. He has also, as proposed here, validated the model both with data on migration-in-progress from an entomological radar (using a historical dataset) and by comparing the model's predictions of population relocations with observed population distributions (using data from a pheromone-trap network). The IIMP system characterizations would allow a more comprehensive validation of this model, and perhaps justify extending it to incorporate other processes.

The combination of characterization data and a comprehensive simulation model can also provide a framework for formulating testable hypotheses. The identification of the data required to test the hypotheses will guide the development of the characterization data-acquisition programme, and the results of the tests can be fed back into the model to improve its accuracy. For example, two alternative hypotheses could be formulated to account for the almost regular appearance of *H. punctigera* populations in the inland in early spring: one involving persistence through a summer quiescence and another involving an autumn return migration from the continental periphery. The former would lead to specific predictions for thresholds for entering and breaking diapause, and these could be tested in the laboratory (using appropriate field-derived stock). The latter, when combined with the transport characterization, might indicate a likely autumn–winter pathway into the inland, and guide the siting of an additional IMR capable of determining whether moths use it.

The more comprehensive, long-term approach to migration research that IIMP represents has been made possible largely through the appropriation of a series of rapidly developing non-biological technologies: satellite remote sensing, automated monitoring radars, boundary-layer numerical wind-trajectory forecasts, geographic information systems, and low-cost microcomputing and telecommunications. These have been combined with existing physiological, behavioural and ecological knowledge of the target organisms and established methodologies and infrastructures for surveying their populations and hosts. Although IIMP is not endeavouring to study all components of its subject migration systems, it has nevertheless adopted the multidisciplinary approach recommended in several reviews of insect migration research (Drake, 1991). The usually high costs of such comprehensive approaches to the study of migration have been contained, and the project thus made possible, by incorporating into it APLC's existing information-garnering and management infrastructure (Deveson and Hunter, 2000) and its locust surveys, i.e. by 'piggy-backing' the new research on to these already funded operational programmes.

## Acknowledgements

Our thanks to APLC survey staff, to A. del Socorro and M. Yee of UNE for assisting with *H. punctigera* surveys, to the electronics and mechanical workshop staff at ADFA School of Physics for their contribution to the construction and installation of the IMRs, and to Bourke and Bulloo Shire Councils for allowing the IMRs to be installed at their airports. The project is supported by a grant from the Australian Research Council.

## References

Beerwinkle, K.R., Lopez, J.D. Jr, Schleider, P.G. and Lingren, P.D. (1995) Annual patterns of aerial insect densities at altitudes from 500 to 2400 metres in east-central Texas indicated by continuously-operating vertically-oriented radar. *Southwestern Entomologist* Supplement No. 18, 63–79.

Bryceson, K.P., Hunter, D.M. and Hamilton, J.G. (1993) Use of remotely sensed data in the Australian Plague Locust Commission. In: Corey, S.A., Dall, D.J. and Milne, W.M. (eds) *Pest Control and Sustainable Agriculture*. CSIRO, Melbourne, pp. 435–439.

Bureau of Meteorology (1986) *Climate of Australia*, 1986 edn. Australian Government Publishing Service, Canberra.

Chen, R.L., Sun, Y.J., Wang, S.Y., Zhai, B.P. and Bao, X.Y. (1995) Migration of oriental armyworm *Mythimna separata* in East Asia in relation to weather and climate. I. Northeastern China. In: Drake, V.A. and Gatehouse, A.G. (eds) *Insect Migration: Tracking Resources through Space and Time*. Cambridge University Press, Cambridge, pp. 93–104.

Clark, D.P. (1971) Flights after sunset by the Australian Plague locust, *Chortoicetes terminifera* (Walk.), and their significance in dispersal and migration. *Australian Journal of Zoology* 19, 159–176.

Coombs, M. (1992) Environmental influences on the flight and migratory potential of *Helicoverpa punctigera* and *H. armigera* (Lepidoptera: Noctuidae). PhD thesis, The University of New England, Armidale, Australia.

Deveson, E.D. and Hunter, D.M. (2000) Decision support for Australian plague locust management using wireless transfer of field survey data and automatic internet weather data collection. In: Laurini, R. and Tanzi, T. (eds) *TeleGeo*

2000 – *Proceedings of the 2nd International Symposium on Telegeoprocessing*, 10–12 May 2000, Nice–Sophia Antipulis, France. École des Mines de Paris, Sophia Antipolis, Nice, France, pp. 103–110.

Drake, V.A. (1983) Collective orientation by nocturnally migrating Australian plague locusts, *Chortoicetes terminifera* (Walker) (Orthoptera: Acrididae): a radar study. *Bulletin of Entomological Research* 73, 679–692.

Drake, V.A. (1991) Methods for studying adult movement in *Heliothis*. In: Zalucki, M.P. (ed.) *Heliothis: Research Methods and Prospects*. Springer Verlag, New York, pp. 109–121.

Drake, V.A. (1993) Insect-monitoring radar: a new source of information for migration research and operational pest forecasting. In: Corey, S.A., Dall, D.J. and Milne, W.M. (eds) *Pest Control and Sustainable Agriculture*. CSIRO, Melbourne, pp. 452–455.

Drake, V.A. and Farrow, R.A. (1983) The nocturnal migration of the Australian plague locust, *Chortoicetes terminifera* (Walker) (Orthoptera: Acrididae): quantitative radar observations of a series of northward flights. *Bulletin of Entomological Research* 73, 567–585.

Drake, V.A. and Farrow, R.A. (1985) A radar and aerial-trapping study of an early spring migration of moths (Lepidoptera) in inland New South Wales. *Australian Journal of Ecology* 10, 223–235.

Drake, V.A. and Farrow, R.A. (1988) The influence of atmospheric structure and motions on insect migration. *Annual Review of Entomology* 33, 183–210.

Drake, V.A. and Gatehouse, A.G. (1996) Population trajectories through space and time: a holistic approach to insect migration. In: Floyd, R.B., Sheppard, A.W. and De Barro, P.J. (eds) *Frontiers of Population Ecology*. CSIRO Publishing, Collingwood, pp. 399–408.

Drake, V.A., Gatehouse, A.G. and Farrow, R.A. (1995) Insect migration: a holistic conceptual model. In: Drake, V.A. and Gatehouse, A.G. (eds) *Insect Migration: Tracking Resources through Space and Time*. Cambridge University Press, Cambridge, pp. 427–457.

Farrow, R.A. (1979) Population dynamics of the Australian plague locust, *Chortoicetes terminifera* (Walker), in central western New South Wales. I. Reproduction and migration in relation to weather. *Australian Journal of Zoology* 27, 717–745.

Farrow, R.A. (1990) Flight and migration in acridoids. In: Chapman, R.F. and Joern, A. (eds) *Biology of Grasshoppers*. John Wiley & Sons, New York, pp. 227–314.

Farrow, R.A. (1991) Migration strategies of insect pests of agriculture – a review of mechanisms. In: *Migration and Dispersal of Agricultural Insects. Proceedings of the International Seminar on Migration and Dispersal of Agricultural Insects*. National Institute of Agro-Environmental Sciences, Tsukuba, pp. 1–19.

Gatehouse, A.G. and Zhang, X.X. (1995) Migratory potential in insects: variation in an uncertain environment. In: Drake, V.A. and Gatehouse, A.G. (eds) *Insect Migration: Tracking Resources through Space and Time*. Cambridge University Press, Cambridge, pp. 193–242.

Gregg, P.C., Fitt, G.P., Coombs, M. and Henderson, G.S. (1994) Migrating moths collected in tower-mounted light traps in northern New South Wales, Australia: influence of local and synoptic weather. *Bulletin of Entomological Research* 84, 17–30.

Gregg, P.C., Fitt, G.P., Zalucki, M.P. and Murray, D.A.H. (1995) Insect migration in an arid continent. II. *Helicoverpa* spp. in eastern Australia. In: Drake, V.A. and Gatehouse, A.G. (eds) *Insect Migration: Tracking Resources through Space and Time*. Cambridge University Press, Cambridge, pp. 151–172.

Han, E.-R. and Gatehouse, A.G. (1991) Genetics of precalling period in the oriental armyworm, *Mythimna separata* (Walker) (Lepidoptera: Noctuidae), and implications for migration. *Evolution* 45, 1502–1510.

Han, E.-R. and Gatehouse, A.G. (1993) Flight capacity: genetic determination and physiological constraints in a migratory moth *Mythimna separata*. *Physiological Entomology* 18, 183–288.

Hunter, D.M., McCulloch, L. and Wright, D.E. (1981) Lipid accumulation and migratory flight in the Australian plague locust, *Chortoicetes terminifera* (Walker) (Orthoptera: Acrididae). *Bulletin of Entomological Research* 71, 543–546.

Kennedy, J.S. (1951) The migration of the desert locust (*Schistocerca gregaria* Forsk.). I. The behaviour of swarms. II. A theory of long range migrations. *Philosophical Transactions of the Royal Society of London* B 235, 163–290.

Lambert, M.R.K. (1982) Laboratory observations on the flight activity of the plague locust, *Chortoicetes terminifera* (Walker) (Orthoptera: Acrididae). *Bulletin of Entomological Research* 72, 377–389.

Malcolm, S.B. and Zalucki, M.P. (eds) (1993) *Biology and Conservation of the Monarch Butterfly*. Natural History Museum of Los Angeles County Science Series no. 38.

Murdoch, W.W. and Nisbet, R.M. (1996) Frontiers of population ecology. In: Floyd, R.B., Sheppard, A.W. and De Barro, P.J. (eds) *Frontiers of Population Ecology*. CSIRO Publishing, Collingwood, pp. 31–43.

Puri, K., Dietachmayer, G., Mills, G.A., Davidson, N.E., Bowen, R.A. and Logan, L.W. (1998) The new BMRC limited area prediction system, LAPS. *Australian Meteorological Magazine* 47, 203–223.

Riley, J.R. (1975) Collective orientation in night-flying insects. *Nature* 253, 113–114.

Riley, J.R. and Reynolds, D.R. (1986) Orientation at night by high-flying insects. In: Danthanarayana, W. (ed.) *Insect Flight: Dispersal and Migration*. Springer-Verlag, Berlin, pp. 71–87.

Rochester, W.A. (1999) The migration systems of *Helicoverpa punctigera* (Wallengren) and *Helicoverpa armigera* (Hübner) (Lepidoptera: Noctuidae) in Australia. PhD thesis, The University of Queensland, Brisbane.

Rosenberg, J. and Burt, P.J.A. (1999) Windborne displacements of desert locusts from Africa to the Caribbean and South America. *Aerobiologia* 15, 167–175.

Schaefer, G.W. (1976) Radar observations of insect flight. In: Rainey, R.C. (ed) *Insect Flight*. Blackwell Scientific, Oxford, pp. 157–197.

Scott, R.W. and Achtemeier, G.L. (1987) Estimating pathways of migrating insects carried in atmospheric winds. *Environmental Entomology* 16, 1244–1254.

Smith, A.D., Reynolds, D.R. and Riley, J.R. (2000) The use of vertical-looking radar to continuously monitor the insect fauna flying at altitude over southern England. *Bulletin of Entomological Research* 90, 265–277.

Southwood, T.R.E. (1962) Migration of terrestrial arthropods in relation to habitat. *Biological Reviews* 37, 171–214.

Southwood, T.R.E. (1977) Habitat: the templet for ecological strategies. *Journal of Animal Ecology* 46, 337–365.

Symmons, P.M. (1986) Locust displacing winds in eastern Australia. *International Journal of Biometeorology* 30, 53–64.

Taylor, L.R. (1974) Insect migration, flight periodicity and the boundary layer. *Journal of Animal Ecology* 43, 225–238.

Tauber, M.J., Tauber, C.A. and Masaki, S. (1986) *Seasonal Adaptations of Insects*. Oxford University Press, New York.

Westbrook, J.K., Wolf, W.W. and Lingren, P.D. (1994) Flight speed and heading of migrating corn earworm moths relative to drifting tetroons. In: *Preprints, 21st Conference on Agricultural and Forest Meteorology and 11th Conference on Biometeorology and Aerobiology*, 7–10 March 1994, San Diego, California. American Meteorological Society, Boston, Massachusetts, pp. 423–426.

Wilson, K. (1995) Insect migration in heterogeneous environments. In: Drake, V.A. and Gatehouse, A.G. (eds) *Insect Migration: Tracking Resources through Space and Time*. Cambridge University Press, Cambridge, pp. 243–264.

Wolf, W.W., Westbrook, J.K., Raulston, J.R., Pair, S.D. and Lingren, P.D. (1995) Radar observations of orientation of noctuids migrating from corn fields in the Lower Rio Grande Valley. *Southwestern Entomologist* Supplement No. 18, 45–61.

Wright, D.E. (1987) Analysis of the development of major plagues of the Australian plague locust *Chortoicetes terminifera* (Walker) using a simulation model. *Australian Journal of Ecology* 12, 423–438.

# Significance of Habitat Persistence and Dimensionality in the Evolution of Insect Migration Strategies

**Robert F. Denno, Claudio Gratton and Gail A. Langellotto**

*Department of Entomology, University of Maryland, College Park, MD 20742, USA*

## Introduction

In addition to the central position migration commands in the evolution of insect life-history strategies (Solbreck, 1978; Denno *et al.*, 1981, 1991, 1996; Denno, 1983, 1994a,b; Dingle, 1985, 1996; Roff, 1986, 1990; Roff and Fairbairn, 1991; Zera and Denno, 1997), migration also acts as a stabilizing force in metapopulation dynamics (Reddingius and den Boer, 1970; den Boer, 1981; Hastings, 1982; Hanski, 1991; Hanski and Kuussaari, 1995), influences species interactions (Denno and Roderick, 1992; Denno *et al.*, 2000a; Denno and Peterson, 2000), and directly affects gene flow and the genetic structure of populations (Taylor *et al.*, 1984; Bull *et al.*, 1987; Peterson and Denno, 1997, 1998; Mun *et al.*, 1999). Moreover, a high level of migration is characteristic of many of our severe agricultural and forest insect pests (Kenmore *et al.*, 1984; Berryman, 1988; Pedgley, 1993; Kisimoto and Rosenberg, 1994; Denno and Peterson, 2000). Thus, understanding those factors which foster or constrain the evolution of migration is critical to both population biology and pest management (Rabb and Kennedy, 1979; Denno *et al.*, 1991; Cappuccino and Price, 1995).

Traditionally, habitat persistence has been key in the development of life history theory and in predicting the evolution of insect migration strategies (Southwood, 1962, 1977; Solbreck, 1978; Harrison, 1980; Dingle, 1985; Roff, 1986, 1990, 1994; Denno *et al.*, 1991, 1996; Novotny, 1994). For example, migration is thought to be essential for the success of species which exploit ephemeral habitats (Southwood, 1962, 1977; Roff, 1986, 1990; Denno *et al.*, 1991, 1996). Nevertheless, there have been few rigorous assessments of the relationship between habitat persistence and the migratory ability of the inhabitants. Two factors have hindered studies investigating the relationship between habitat persistence and migration. First, accurately ageing habitats has proved difficult, and second, assessing the migratory capability of the inhabitants has been challenging as well (Denno et al., 1991, 1996). Moreover, other dimensional characteristics of the habitat, such as host-plant architecture and isolation, are also thought to influence the evolution of migration characters, but these too have been poorly investigated (Roff, 1990; Denno, 1994a,b). For instance, wings may be essential for negotiating highly structured, three-dimensional habitats (Reuter, 1875; Denno, 1978, 1994b; Waloff, 1983), and the constant loss of emigrants from very isolated habitats may select against flight capability (Roff, 1990).

For this contribution we have gathered information on the persistence, dimensionality and isolation of the habitats and host plants of delphacid planthoppers (Hemiptera: Delphacidae) in order to assess how these habitat-related factors influence the migratory capability and associated reproductive traits of this group of herbivorous insects. We also review the consequences of how selection for migratory capability may constrain other life-history traits such as siring ability in males, and fecundity and age to first reproduction in females. Specifically, we provide evidence for a phenotypic trade-off between migration and reproduction in both sexes of delphacid planthoppers. Lastly, we present a graphical model which integrates elements of habitat persistence and habitat dimensionality to predict levels of migration in phytophagous insects such as planthoppers.

## Benefits of Migratory Capability and Evidence for a Trade-off with Reproductive Success

Wing-dimorphic insects such as planthoppers are ideal for investigating the benefit of wings and the evolution of migration because volant and flightless forms are easily recognized (Denno et al., 1991). Brachypterous adults have reduced wings and cannot fly, whereas macropterous adults possess fully developed wings and can migrate long distances – distances of over 1000 km in some cases (Denno, 1994b; Kisimoto and Rosenberg, 1994; Kisimoto and Sogawa, 1995). Populations of most delphacid planthoppers contain both wing forms, but the proportion of each can vary tremendously among different species (Denno et al., 1991; Denno, 1994b), and geographically among populations of the same species (Iwanaga et al., 1987; Iwanaga and Tojo, 1988; Denno et al., 1996). Importantly, the fraction of macropters in a population can be used as a reliable index of the level of potential migration in the population (Denno et al., 1991).

Migration, although advantageous for exploiting temporary habitats, does not occur without cost in delphacid planthoppers (Denno et al., 1989; Denno, 1994b). The disadvantage of flight capability in female planthoppers can be seen by comparing the reproductive schedules of the two wing forms of planthoppers. For many species, macropterous females are both less fecund and reproduce later in life compared to brachypterous females (Denno et al., 1989; Denno, 1994b). As a consequence, the potential net replacement rate of macropters is only half as high as that for brachypters (Denno et al., 1989). The reproductive delay and reduced fecundity observed in the macropterous form of female planthoppers, and many other wing-dimorphic insects as well, support the view that flight capability is costly and that there are phenotypic trade-offs between flight and reproduction (Roff, 1986; Denno et al., 1989; Roff and Fairbairn, 1991; Zera and Denno, 1997).

Recent data suggests that there is a reproductive penalty associated with migratory ability in male planthoppers as well. For instance, macropterous males of the salt marsh-inhabiting *Prokelisia dolus* Wilson sire only half as many offspring as their brachypterous counterparts (Langellotto et al., 2000). For other species of planthoppers, macropterous males are less aggressive in male–male interactions (Ichikawa, 1982), develop slower (Novotny, 1995) or acquire matings less successfully (Novotny, 1995) than brachypterous males. Similar reproductive costs are associated with macroptery in the males of other insect species as well (Crespi, 1988; Fujisaki, 1992; Kaitala and Dingle, 1993; Crnokrak and Roff, 1995; Fairbairn and Preziosi, 1996). Thus, due to enhanced reproductive success, the brachypterous form appears to be advantageous in both sexes as long as conditions remain favourable for development and mates are locally available.

Wing form in planthoppers is determined by a developmental switch which responds to environmental cues (Denno, 1994b). The sensitivity of the switch, however, is heritable and under polygenic control, and the developmental switch is undoubtedly controlled by the level of a juvenile-like hormone (Iwanaga and Tojo, 1986; Denno, 1994b; Zera and Denno, 1997). Of all the proximate cues known to affect wing form in planthoppers, population density is by far the most influential for most species (Denno and Roderick, 1990). For many species, the production of macropterous wing forms is density dependent, is associated with crowded conditions, and is often escalated on nutritionally deficient host plants (Denno et al., 1985; Denno, 1994b). The density that triggers the production of macropterous forms, however, differs among species, among populations of the same species and even between the sexes of the same species (Denno et al., 1991). Variation in the wing-form response to crowding suggests that the benefit of possessing wings varies with population density and differs between the sexes for some species.

Wings allow for escape from deteriorating local conditions and the colonization of new habitats (Southwood, 1962, 1977). Wings can also function in mate location (Ichikawa, 1977; Hunt and Nault, 1991; Langellotto, 1997), and the density/wing-form response probably reflects the selective pressures associated with both mating success and migration to new habitats (Denno et al., 1991; Denno, 1994a). In planthoppers, only males actively search for stationary females by flying or walking (Ichikawa, 1977; Claridge and de

Vrijer, 1994; Langellotto et al., 2000). Both sexes communicate acoustically and locate each other using substrate-borne vibrations (Ichikawa et al., 1975; Claridge, 1985; Heady and Denno, 1991). Males move through vegetation, and after sensing a female's call, both sexes engage in duet-like bouts of communication during which time the male proceeds toward the female (Langellotto, 1997; Langellotto et al., 2000). After arrival at the female, courtship behaviour ensues, and mating ultimately occurs unless the female rejects the courting male (Claridge, 1985; Heady and Denno, 1991; Langellotto et al., 2000).

Planthopper species exploiting temporary habitats frequently experience extremely low densities following colonization and also after one successive generation (Kuno, 1979). Because females of most planthopper species mature sexually and mate after migration to the new habitat (Kisimoto, 1976; Kuno, 1979; Denno and Roderick, 1990), macroptery in males should facilitate locating virgin females, particularly when females are rare (Hunt and Nault, 1991; Langellotto, 1997). In fact, under low-density conditions or in sparse vegetation, males locate females by employing a 'fly-and-call' strategy, whereby males flit among plants searching for calling females (Hunt and Nault, 1991; Langellotto, 1997). Macroptery in females, however, should be a detriment at low densities due to the associated penalties of reduced fecundity and delayed reproduction (Denno et al., 1989). Thus, factors such as habitat persistence may influence the density/wing-form response of the sexes differently because the male response is shaped by additional selective pressures associated with mate location (Denno et al., 1991; Denno, 1994a).

## Habitat Persistence and the Incidence of Migration: an Interspecific Assessment

In general, selection in persistent habitats should favour reduced levels of migration due to the inherent reproductive costs associated with flight capability (Roff, 1986, 1990; Denno et al., 1989, 1991). In temporary habitats, however, wings should facilitate the tracking of changing resources (Southwood, 1962, 1977). Consequently, as habitats become more short-lived, the expectation was to observe higher levels of migration. This prediction was tested by analysing the wing-form composition and habitat duration of 35 species of planthoppers occupying a range of habitats varying in persistence (Denno et al., 1991). Data on the migratory capability and wing-form composition of planthopper populations were gathered from the ecological literature (Denno et al., 1991). Habitat persistence was expressed as the 'maximum number of generations a planthopper could attain during the existence of the habitat' (Denno et al., 1991). Thus, habitat persistence was estimated for each planthopper species by multiplying the number of generations each species completes annually by the age of its habitat in years. Data on habitat age (minimum estimate) were extracted from the agricultural and geological literature (Denno *et al.*, 1991). This analysis was confined to species exploiting low-profile vegetation (e.g. grasses and forbes < 1 m in height), because habitat dimensionality (host-plant architecture) can also affect the evolution of flight capability in insects (Denno 1978, 1994b; Waloff, 1983; Roff, 1990).

There was a significant negative relationship between habitat persistence and macroptery (%) for both female ($r_s = -0.78$, $P < 0.001$) and male planthoppers ($r_s = -0.79$, $P < 0.001$) (Fig. 11.1). The highest levels of macroptery (50% or greater in both sexes) were recorded for species inhabiting ephemeral agricultural crops (e.g. *Nilaparvata lugens* Stål) or natural disturbed habitats (e.g. *Prokelisia marginata* Van Duzee) that persist for < 1 year. The lowest incidence of macroptery (≤1% in both sexes) occurred in species associated with persistent habitats such as arctic bogs (e.g. *Javesella simillima* Linnavouri), freshwater marshes (e.g. *Megamelus paleatus* Van Duzee) and salt marshes (e.g. *Tumidagena minuta* McDermott) that have

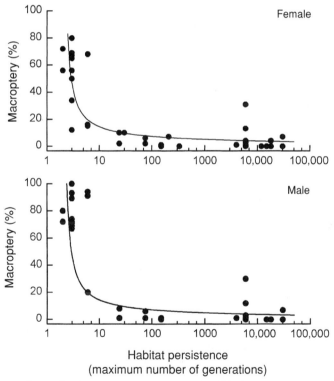

**Fig. 11.1.** Relationship between macroptery (%) and habitat persistence (the maximum number of generations attainable) for the females and males of 35 species (41 field populations) of planthoppers. Habitat persistence was estimated by multiplying habitat age (years) by the number of generations attainable per habitat per year for each species. Adapted from Denno *et al.* (1991).

existed in North America for 2000–12,000 years (Denno et al., 1991). With an increase in habitat persistence there was an abrupt decrease in macroptery (%), and habitats that provided the opportunity for at least 30 successive generations of planthoppers showed no higher levels of macroptery than habitats 1000 times more persistent. Using phylogenetically independent contrasts with congeneric pairs, the same result was obtained, suggesting that habitat persistence influences levels of migration independent of common ancestry (Denno et al., 1991).

In those species for which data were available, there was also a difference in the density/wing-form response between species in temporary and persistent habitats (Denno et al., 1991). For example, macropterous forms were triggered at a much lower density for species exploiting temporary habitats compared to species in persistent habitats. There was also a consistent difference in the density/wing-form response between the sexes for species in temporary habitats (Denno et al., 1991). Males of species inhabiting temporary habitats were macropterous at both low and high rearing densities, with brachypterous males appearing at intermediate densities (e.g. *N. lugens* and *P. marginata*). For the females of such species, macroptery was positively density dependent. By contrast, macroptery was positively density dependent in both sexes of species occupying more persistent habitats.

Sexual differences in wing-form response to density between species in temporary and permanent habitats probably contribute to the sexual differences in wing-form composition observed in the field (Denno et al., 1991). For instance, there was a significant negative relationship between the ratio of macroptery (%+1) in males to macroptery (%+1) in females and habitat persistence (maximum number of generations; $r_s = -0.60$, $P < 0.001$; Fig. 11.2).

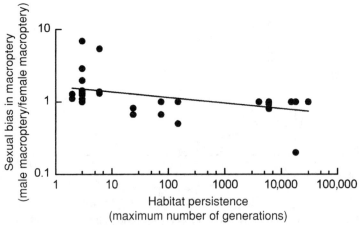

**Fig. 11.2.** Relationship between sexual bias in macroptery (macroptery (%+1) in males/macroptery (%+1) in females) and habitat persistence (maximum number of possible generations) for 35 species (41 field populations) of planthoppers ($r_s = -0.60$, $P < 0.001$). Adapted from Denno et al. (1991).

On average, males were 2.5 times as macropterous as females for species inhabiting the most temporary habitats, whereas macroptery was similar between the sexes (ratio ≈1.0) for species occupying the most persistent habitats. These data suggest that males are more macropterous than females in field populations of species which occur in temporary habitats, but not in species which reside in persistent habitats.

Thus, habitat persistence apparently influences the migratory capability of planthoppers in two ways: (i) by selecting for or against habitat escape, and (ii) by dictating the availability of mates (Denno et al., 1991). In persistent habitats, wings are less necessary for habitat escape and they are less frequently required for mate location. As a consequence, and because wings impose a reproductive penalty, flightlessness is favoured. In temporary habitats, wings are apparently favoured in males to locate females at low densities and are favoured in both sexes at high densities for reasons of habitat escape. Recent experimental evidence confirms that macropterous males of *P. dolus* are far more successful than are brachypterous males at locating females under low-density conditions (Langellotto, 1997).

## Habitat Persistence and the Incidence of Migration: an Intraspecific Assessment

Most insect studies have relied on interspecific comparisons to assess the relationship between habitat persistence and migration (Denno, 1978; Roff, 1990; Novotny, 1994), and are thus possibly confounded by phylogenetic non-independence (but see Denno et al., 1991). A less confounded approach is to study intraspecific variation in the incidence of migration among populations, and assess the underlying cause of this variation (Denno et al., 1996). *Prokelisia* planthoppers provide a unique opportunity to explore the effect of habitat persistence on migration because they exhibit intraspecific variation in migratory capability (Denno et al., 1996).

Both *P. marginata* and *P. dolus* co-occur broadly along the Atlantic and Gulf coasts of North America where they are restricted to intertidal marshes (Denno et al., 1996). Throughout their range, both planthoppers feed exclusively on the perennial cordgrass, *Spartina alterniflora*. Like most planthoppers, adults of both *Prokelisia* species are wing-dimorphic, making for an easy assessment of migratory potential (Denno, 1994b; Denno et al., 1996). Also, both planthoppers are multivoltine, and they both pass the winter as nymphs nestled in leaf litter, a resource which they require for successful overwintering (Denno et al., 1996).

*S. alterniflora* is restricted to the intertidal zone ranging from mean high water level to elevations as much as 2 m below (Redfield, 1972). Along this elevational gradient, *S. alterniflora* characteristically occurs in two growth forms, tall and short (Ornes and Kaplan, 1989). Many of the intertidal marshes in North America are characterized by expanses of short-form

*S. alterniflora* on the high marsh which abruptly intergrade into a fringe of tall-form plants bordering tidal creeks on the low marsh (Adams, 1963; Denno and Grissell, 1979; Mendelsshon *et al.*, 1981). However, regional differences in winter severity influence the level of persistence of *S. alterniflora* habitats (Denno *et al.*, 1996).

Along much of the Atlantic coast, short-form *S. alterniflora* on the high marsh reaches maximum live biomass in summer, plants senesce in fall and the dead rosettes persist through winter (Blum, 1968; Denno and Grissell, 1979; Denno *et al.*, 1996). In dramatic contrast, tall-form plants in low-marsh habitats, although robust and very nutritious during summer, are destroyed during winter by the action of winter tidewaters, winds and shifting ice leaving exposed creek banks often free of litter (Denno and Grissell, 1979). The selective winter destruction of tall-form *S. alterniflora* has been documented along most of the Atlantic coast (Denno *et al.*, 1996).

The growth dynamics and winter disturbance of *S. alterniflora* on the Gulf coast contrasts with that along the Atlantic coast (Denno *et al.*, 1996). The mild climate allows living shoots of both tall- and short-form plants to persist throughout winter and promotes year-round growth. Maximum live biomass occurs in early fall, after which above-ground portions of plants begin to senesce, but the understory of new shoots continues to grow. The winter destruction of low-marsh habitats along the Gulf coast is minimal due to reduced tidal range, an absence of ice scouring and reduced frequency of tidal submergence.

Denno *et al.* (1996) hypothesized that there should be an inverse relationship between the migratory capability of planthoppers and habitat persistence. In regions where planthoppers fail to remain through winter in their primary habitat for development, high levels of migration (% macroptery) should be evident in order to facilitate the recolonization of the primary habitat for development from overwintering sites elsewhere. Conversely, low levels of macroptery should occur in populations that persist through winter in the habitat where most summer development occurs.

These predictions concerning habitat persistence were tested by first examining regional patterns of low-marsh and high-marsh occupancy by the *Prokelisia* species and assessing their ability to remain in their primary habitat for development year-round (Denno *et al.*, 1996). The proportion of a species' population occupying its primary habitat for development during winter was used as an index of habitat persistence for each species. Secondly, intraspecific variation in migratory ability (% macroptery) for both species was determined within and between coastal regions. Lastly, the effect of habitat persistence on the incidence of migration was assessed using regression.

For both coastal regions, the proportion of the *P. marginata* population inhabiting the low marsh during the warm growing season was much higher than that for *P. dolus* (Fig. 11.3). These data establish that during the warm season, *P. marginata* develops primarily in the low marsh, whereas *P. dolus* remains mostly on the high marsh to reproduce. For *P. marginata*,

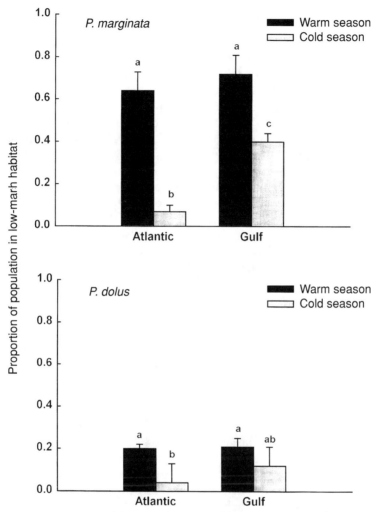

**Fig. 11.3.** Warm and cold seasonal patterns of low-marsh occupancy (proportion of population in low marsh compared to high marsh) by *Prokelisia marginata* and *P. dolus* at sites along the Atlantic where low-marsh habitats are destroyed during winter, and at locations along the Gulf coast where low-marsh habitats are less disturbed during winter. Note that for *P. marginata*, the proportion of the population remaining through the cold season in the low marsh is much lower for Atlantic than Gulf-coast locations, yet during the warm season most individuals occupy the low marsh on both coasts. In contrast, the proportion of the *P. dolus* population in the low marsh remains low during both seasons. Means with different letters are significantly different ($P < 0.05$, ANOVA followed by Sidak's adjustment for multiple comparisons). Adapted from Denno *et al.* (1996).

the proportion of the population in the low marsh during the warm season was high (65–75%) in both Atlantic and Gulf coast regions. During the cold season, however, the proportion of the population remaining in the low marsh plummeted to 7% on Atlantic marshes, but dropped to only 40% on Gulf-coast marshes. These data demonstrate that *P. marginata* is far better able to endure winter conditions on the low marsh along the Gulf coast than on the Atlantic coast. For *P. dolus* along both coasts, only a small proportion of the population (< 20%) occurred on the low marsh during either season.

To link coastal patterns of population persistence with levels of migration, geographic variation in the migratory capability of both *P. marginata* and *P. dolus* were determined (Denno et al., 1996). This was accomplished by extensively sampling planthoppers along the Atlantic and Gulf coasts. For each location, an index of migration for *P. marginata* and *P. dolus* populations was determined from the proportion of macropters in the local population. For *P. marginata*, populations with the greatest capacity for migration occurred along the Atlantic coast where the regional mean was 92% macroptery. In contrast, Gulf-coast populations averaged only 17% macroptery. The incidence of migration was much lower in *P. dolus* with levels of macroptery averaging only 8% in Atlantic populations and 6% along the Gulf.

The incidence of macroptery (%) was inversely related to a population's ability to persist through winter in its primary habitat for development ($Y = 89.8 - 90.8X$, $R^2 = 0.75$, $P = 0.001$; Fig. 11.4). For *P. marginata* along the Atlantic coast, large-scale migration is necessary to recolonize the low marsh, a habitat which is unsuitable for winter survival in this region. Such aerial migrations are commonly observed on Atlantic marshes in late spring (Denno, 1988; Denno et al., 1996). In contrast, the low levels of macroptery in most Gulf-coast populations of *P. marginata* reflect the ability of this species to remain on the low marsh year-round in this equitable region. Low levels of migration occur in all populations of *P. dolus*, a species which remains throughout the year in its primary habitat for development on the less-disturbed high marsh (Fig. 11.4). These data provide strong evidence that intraspecific variation in the migratory capability of planthoppers is inversely related to the persistence of their habitats.

## Habitat Dimensionality and the Incidence of Migratory Capability

Although wings facilitate escape from deteriorating local conditions and the colonization of new habitats, wings can also function in habitat negotiation, particularly in complex, three-dimensional vegetation (Reuter, 1875; Denno, 1978, 1994b; Waloff, 1983). For example, mate finding and relocation of feeding site following escape from a predator may prove difficult for flightless brachypters in arboreal habitats. On the other hand, the consequences of releasing hold of the host plant in low-profile vegetation (e.g. grasses) are probably minimal because resources can be easily relocated by walking. Consequently, selection may favour the retention of flight capability in arboreal species even though their habitats are persistent (Denno 1994a,b). This so-called habitat dimensionality/flight capability hypothesis was initially proposed in rough form by Reuter (1875), and was first tested rigorously by Waloff (1983) using the British cicadellid fauna.

The habitat dimensionality/flight capability hypothesis can be tested by comparing the wing-form composition of delphacid species with the growth form of their host plants (Denno, 1994b). Because most temperate delphacid species are monophagous on grasses and sedges and occupy low-profile vegetation (Wilson et al., 1994), few comparisons with arboreal species are possible. By contrast, most Hawaiian planthoppers, although monophagous, feed as a family on a diverse variety of host plants which vary in structure from herbs to tall trees (Giffard, 1922; Zimmerman, 1948). Importantly, most of the host plants of native Hawaiian delphacids occur in habitats that are relatively persistent such as wet forests (Zimmerman, 1948). Thus, our analysis was not confounded by major differences in habitat persistence.

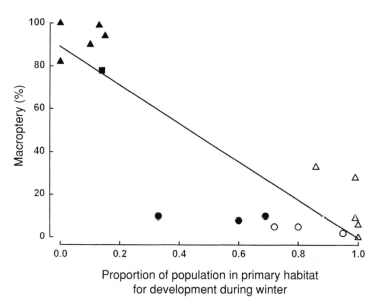

**Fig. 11.4.** Relationship between the level of macroptery (%) observed in field populations of *Prokelisia* planthoppers and the proportion of each species' population enduring through winter in the primary habitat for development ($Y = 89.8 - 90.8X$, $R^2 = 0.75$, $P < 0.001$). Populations which endure poorly through winter in their primary habitat exhibit high levels of migratory capability. *P. marginata* populations: Atlantic coast (▲), Gulf coast (●) and Pacific coast (■). *P. dolus* populations: Atlantic coast (△) and Gulf coast (○). Adapted from Denno et al. (1996).

Data on the wing-form composition of 150 species of delphacids in 12 genera, and the identity and growth form of their host plants, were obtained from the literature (Giffard, 1922; Zimmerman, 1948; Wagner et al., 1990). For simplicity, the height of the mature host plant was used as an index of host-plant architecture and habitat dimensionality (Denno, 1994b).

Wings were expected to be retained in species that exploited tall trees, because flight might better allow for the effective negotiation of these three-dimensional habitats. In contrast, we envisioned flightlessness to predominate only in species inhabiting short, low-profile host plants. Thus, with an increase in habitat dimensionality (plant height), we anticipated a commensurate increase in the frequency of macropterous species. Indeed this was the case. There was a positive relationship between host-plant height and the proportion of macropterous planthopper species in the assemblage associated with a particular plant-height category ($Y = 102.72(1-e^{-0.258X})$; $R^2 = 0.70$, $P = 0.0002$; Fig. 11.5). Macropterous species fed mostly on trees and large shrubs that averaged 7 m in height (Denno, 1994b). Brachypterous species occurred on much shorter vegetation, primarily small shrubs, herbs and grasses many of which contacted the ground and averaged < 1 m in height.

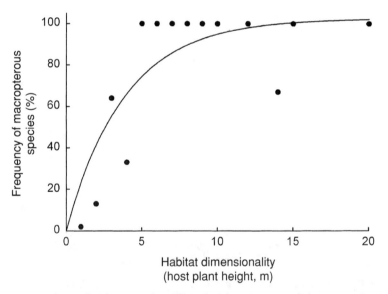

**Fig. 11.5.** Relationship between host plant height (metres) and the proportion of macropterous planthopper species in the assemblage associated with a particular plant-height category ($Y = 102.72(1-e^{-0.258X})$; $R^2 = 0.70$, $P = 0.0002$). Macropterous species ($n = 52$) fed mostly on trees and large shrubs that averaged 7 m in height, and brachypterous species ($n = 98$) occurred on shorter vegetation, primarily small shrubs, herbs and grasses, most of which averaged < 1 m in height. Data for analysis extracted from Giffard (1922), Zimmerman (1948) and Wagner et al. (1990).

A contrast of congeners within *Nesosydne* allowed for a rough assessment of the effects of plant dimensionality on wing form without the possible confounding effects of phylogeny (Felsenstein, 1985; Denno *et al.*, 1991). *Nesosydne* is a large genus in the Hawaiian archipelago with over 80 species feeding on a diverse assemblage of plant taxa ranging from ferns to monocots to dicots (Zimmerman, 1948). Our congeneric contrast was limited to the 60 species which feed on either low-profile plants (grasses, forbes or shrubs) or trees, all in persistent, wet-forest habitats (Zimmerman, 1948; Denno, 1994b). Planthopper species were sorted into one of three wing-form categories (brachypterous, wing-dimorphic or macropterous) and the frequency of species across the three categories was compared between low-profile (< 2 m in height) and high-profile vegetation (> 2 m in height).

On low-profile host plants there were 51 brachypterous species of *Nesosydne*, four wing-dimorphic species and no macropterous species. This distribution contrasted significantly with the distribution of species on high-profile vegetation: one brachypterous species, one wing-dimorphic species and three macropterous species ($\chi^2 = 19.53$, $P < 0.0001$, G test). Altogether, the Hawaiian delphacid fauna provide convincing support for the habitat dimensionality/flight capability hypothesis in that wings are retained in most arboreal species and flightlessness is restricted to species occupying persistent, low-profile vegetation.

## Habitat Isolation and the Evolution of Flightlessness

Darwin (1876) was among the first to reason that if suitable habitats were very isolated from one another, the constant loss of emigrants should select for reduced flight capability in the residents. He used as evidence for this notion the frequency of flightlessness in a variety of taxa inhabiting oceanic islands (see also Simon *et al.*, 1984). However, other habitat-associated factors, such as habitat persistence, also influence the evolution of flightlessness, and thus historical perspectives on the role of habitat isolation may be confounded (Roff, 1990; Denno, 1994b).

To minimize the effects of these confounding factors, we examined the impact of habitat isolation on flightlessness by comparing levels of migration between island (Hawaiian Islands) and mainland populations of planthoppers residing in persistent habitats on low-profile host plants (< 2 m in height). If habitat isolation were a contributing factor to migration strategy, then island populations should be more brachypterous than mainland populations, all else being equal. Data on the wing-form composition and host-plant architecture of planthoppers inhabiting both the mainland (North America) and oceanic islands (Hawaiian archipelago) were extracted from the literature (Giffard, 1922; Zimmerman, 1948; Denno *et al.*, 1991; Denno, 1994b). Only species inhabiting the most persistent habitats were compared.

Habitat age for the mainland species (> 3000 years) was roughly estimated from the geological literature on marshes, lakes and bogs (Denno et al., 1991). Similarly for the island-inhabiting planthoppers, only species occupying persistent habitats (forests) were selected for comparison (Giffard, 1922; Zimmerman, 1948). The major islands in the Hawaiian archipelago range in age from 700,000 (Hawaii) to 5.6 million years (Kauai) (Howarth, 1990). Needless to say, the forest habitats of the delphacids are much younger in age than the islands themselves. None the less, many of the forest habitats on the islands appear to be roughly comparable in age to the persistent mainland habitats we assessed (Howarth, 1990; Denno et al., 1991).

Our assessment of the migratory capability of planthoppers found no difference between mainland and island taxa. Macroptery (%) was equally low for species inhabiting both mainland ($1.50 \pm 1.15$, $n = 6$) and island habitats ($1.26 \pm 0.25$, $n = 91$; $t = 0.229$, $P = 0.82$). A similar result occurred when generic averages of macroptery (%) were compared between mainland ($1.25 \pm 0.75$, $n = 4$) and island taxa ($1.06 \pm 0.70$, $n = 5$; $t = -0.182$, $P = 0.86$). Both analyses suggest that habitat isolation does not result in a reduced level of migration beyond that explained by habitat persistence and dimensionality.

Spatial scale may be an issue here. It may be that isolation contributes to reduced migration only in the very smallest of habitat patches (Roff, 1990), such as the highly fragmented salt marshes in southern California. In these habitats, migration is virtually absent in planthopper (*P. dolus*) populations ($0.16 \pm 0.36\%$ macroptery, $n = 9$ populations; Denno et al., 1996), a level below that (5–8%) predicted by habitat persistence alone (Denno et al., 1991). We argue, however, that at the scale of expansive habitats on oceanic islands, isolation plays little role in the migration strategies of most insects. Moreover, for another wing-dimorphic planthopper (*Toya venilia* Fennah) inhabiting low-profile habitats in the British Virgin Islands, there was no evidence that the low incidence of macroptery (< 5%) was attributable to the effects of isolation on oceanic islands (Denno et al., 2000b). Rather, habitat persistence and structure best explained the low commitment to flight capability in this species. Similarly, in his extensive analysis of the effects of habitat isolation on the migration strategies of a wide diversity of insect taxa, Roff (1990) shows that oceanic islands do not have higher-than-expected incidences of flightlessness than mainland faunas.

# Interactive Effects of Habitat Persistence and Dimensionality on Migratory Capability

Of the three habitat-related factors we examined, persistence and dimensionality had far more impact on the migratory capability of planthoppers than did isolation. For this reason, we developed a graphical model which predicts only the interactive effects of habitat persistence (Fig. 11.1) and dimensionality (Fig. 11.5) on the incidence of migration (% macroptery) in both female and male planthoppers (Fig. 11.6).

As habitats become more persistent, the incidence of macroptery in populations plummets rapidly for both sexes (Fig. 11.6), but only for species occupying low-profile habitats. In very temporary habitats, males tend to be more macropterous than females on average, apparently for reasons of mate location under low-density circumstances. In more structurally complex vegetation, flight capability is completely retained in both sexes even if habitats are persistent. Thus, in highly structured habitats, the effects of habitat dimensionality override the impact of habitat persistence.

It is important to realize, however, that the majority of mainland delphacid species reside in low-profile vegetation, mostly on herbaceous monocots (Wilson *et al.*, 1994), where the effect of habitat persistence prevails. Thus, habitat persistence alone can be used to predict the migration strategy of most mainland species. On oceanic islands, however, delphacids depart from their conservative use of low-profile monocots (mostly grasses and sedges) where they exploit an immense taxonomic array of woody dicots with variable architecture (Zimmerman, 1948; Wilson *et al.*, 1994). In such habitats, the interactive effects of habitat dimensionality and persistence on the migration strategies of planthoppers can be seen (Fig. 11.6). The few

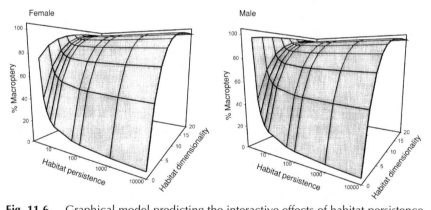

**Fig. 11.6.** Graphical model predicting the interactive effects of habitat persistence (temporary to permanent expressed as the maximum number of generations attainable) and dimensionality (low-profile to high-profile vegetation expressed as plant height in metres) on the incidence of migration (% macroptery) in female and male planthoppers. As habitats become more persistent, the incidence of macroptery in populations plummets rapidly for both sexes, but only for species occupying low-profile habitats. In very temporary and low-profile habitats, males tend to be more macropterous than females on average, apparently for reasons of mate location under low-density circumstances. In more structurally complex vegetation, flight capability is completely retained in both sexes even if habitats are persistent. In highly structured habitats, the effects of habitat dimensionality override the impact of habitat persistence. Data for model extracted from Giffard (1922), Zimmerman (1948), Wagner *et al.* (1990) and Denno *et al.* (1991).

mainland groups of delphacids which exploit both low-profile and high-profile host plants in persistent habitats (e.g. *Stobaera*) do have migration strategies consistent with the predictions of our model: arboreal species are more macropterous than those which feed on sprawling shrubs (Denno, 1994b).

## Conclusions

Flight facilitates numerous activities in insects such as the tracking of suitable habitats for development and overwintering, the negotiation of highly structured habitats and mate location (Roff, 1986, 1990; Denno *et al.*, 1991, 1996; Roff and Fairbairn, 1991; Langellotto *et al.*, 2000). Nevertheless, the loss of flight capability has evolved independently many times and is widespread throughout a diversity of insect orders (Roff, 1990; Wagner and Liebherr, 1992; Denno, 1994b). Discovering the factors which enhance or diminish migratory capability has proved difficult for many insects (Denno, 1994a). On the other hand, wing-dimorphic taxa such as delphacid planthoppers offer an ideal opportunity to explore the selective pressures underlying the retention and loss of flight capability (Roff, 1986, 1990; Denno *et al.*, 1991). Using planthoppers as a model group, our objective here was to examine the interactive effects of three habitat-related factors, persistence, dimensionality and isolation on the migration strategy of this diverse group of herbivorous insects.

Habitat persistence has a very strong influence on the incidence of migration in planthopper populations, an effect that is evidenced both among and within species (Figs 11.1 and 11.4). In habitats of short duration, planthoppers exhibit a high incidence of migration (> 80% macroptery). With an increase in habitat persistence, migratory capability (% macroptery) declines sharply, suggesting that when the demand for flight is relaxed, flight capability diminishes rapidly. Planthoppers existing in habitats where only 30 generations can be achieved are nearly as brachypterous (> 85%) as species occupying far more persistent habitats (Fig. 11.1). Coincidentally, macropterous and brachypterous lines of the planthopper *N. lugens* can be established after only 30 generations of selection in the laboratory on an initial population consisting of an equal mix of both wing forms (Marooka and Tojo, 1992). Thus, both field data on habitat persistence (Fig. 11.1) and selection experiments in the laboratory suggest that flight-related characters respond rapidly to selection (Marooka *et al.*, 1988; Marooka and Tojo, 1992). The reduction in migratory capability of planthoppers in persistent habitats is most certainly attributable to the reproductive advantage owned by flightless adults (Denno *et al.*, 1989; Zera and Denno, 1997; Langellotto *et al.*, 2000).

Despite the reproductive penalties imposed by flight, selection favours the retention of wings in species exploiting temporary habitats (Figs 11.1 and

11.4). Males in particular retain their flight capability in ephemeral habitats, apparently as a response to selection for mate location under low-density conditions (Denno et al., 1991; Langellotto et al., 2000; Fig. 11.2). Moreover, flight capability is preserved in three-dimensional habitats (Fig. 11.5), even if the habitat is very persistent. Thus, habitat dimensionality and persistence interact to influence flight capability, such that the predictions of habitat persistence are realized only in low-profile vegetation (Fig. 11.6).

One may ask if the effects of habitat persistence on the incidence of migration in delphacid planthoppers are representative of other insect taxa. In general, the answer is affirmative (reviewed in Roff, 1990). For instance, other taxa demonstrating an association between flightlessness and habitat persistence include water striders (Vepsäläinen, 1978; Spence, 1989), carabid beetles (den Boer et al., 1980), heteropterans (Brown, 1982) and orthopterans (Roff, 1990). Note that most of these taxa forage in two-dimensional space, and as a consequence, selection for flightlessness has not been compromised by the constraints associated with negotiating a highly structured habitat.

The planthopper model appears to be less predictive when considering the effect of habitat dimensionality on the incidence of migration in other insect groups. As with planthoppers, wings are retained in arboreal species of cicadellids, callaphidids, heteropterans and psocopterans, and flightlessness or wing dimorphism occur primarily in species occupying low-profile host plants (Southwood and Leston, 1959; New, 1974; Waloff, 1983). Arboreal tingids also show a greater commitment to flight than do forb-feeding species (Tallamy and Denno, 1981). Like planthoppers, most of these insects are small, actively search for food, and often exhibit escape responses such as jumping or dropping behaviour (Denno, 1994a). Thus, wings are essential in three-dimensional vegetation for foraging and the relocation of resources following escape from enemies.

In stark contrast to planthoppers, flightless females are the rule in several arboreal species of Lepidoptera, Phasmatodea and Coccoidea (Barbosa et al., 1989; Roff, 1990). Unlike planthoppers, the females of these flightless taxa either do not feed as adults (Lepidoptera), do not actively forage for food (Coccoidea) or exhibit cryptic defense strategies (Lepidoptera, Phasmatodea). The most critical difference between these taxa and planthoppers, however, is that migration and reproduction are partitioned between different life-history stages in the flightless Lepidoptera and Coccoidea with ballooning immatures as the primary means of migration (Barbosa et al., 1989; Roff, 1990; Denno, 1994a). Thus, enhancing reproduction by the loss of adult flight capability does not obviously compromise migration as it does in planthoppers and most other insects in which the adult is the dispersal stage (Denno et al., 1989). Notably, wings are retained in the males of Coccoidea and in arboreal species of lepidopterans with flightless females for reasons which undoubtedly concern mate location in highly structured habitats (Roff, 1990). Consequently, predictions concerning the evolution of migration in insects must take into account foraging behaviour, the nature of the escape response from predators and, in particular, the dispersal stage.

Habitat isolation has been also championed as a factor influencing the evolution of migration in insects (Darwin, 1876; Roff, 1990; Wagner and Liebherr, 1992). At the spatial scale we examined the effects of isolation on the incidence of migration in planthoppers (oceanic islands versus mainland), we found no significant impact. Similarly, Denno et al. (2000b) found no effect of habitat isolation on the migration strategy of another island-inhabiting planthopper, *T. venilia*. Roff (1990) reached the same conclusion for other insect groups and asserted that spatial scale was critical in detecting the effects of habitat isolation on migration. Isolation may explain the evolution of flightlessness, but only at a very small habitat scale where migration results in certain mortality (Roff, 1990). The evolution of flightlessness in insects residing in caves, sand dunes, mountain tops and subantarctic islands may be promoted to some extent by habitat isolation (Wagner and Liebherr, 1992), but many of these isolated habitats are also very persistent (Roff, 1990). Moreover, additional factors, such as extreme cold or windiness, may be operating which impose energetic costs on flight (Wagner and Liebherr, 1992). None the less, the collective effects of habitat persistence and dimensionality appear to be far more important than isolation in explaining the migration strategies of planthoppers.

Despite the few outliers, we feel that our conclusions concerning the influence of habitat persistence and dimensionality on the incidence of flight capability in planthoppers (Fig. 11.6) are broadly indicative of the pivotal role these two factors play in shaping the migration strategies of most insects. In our assessment, host-plant structure was controlled when testing for habitat persistence effects on migration and vice versa, and when possible, phylogenetic non-independence was taken into account. In most comparative studies of insect migration, habitat variables are often confounded and potential phylogenetic constraint is rarely considered (Roff, 1990; Denno et al., 1991, 1996). In this vein, delphacid planthoppers provide perhaps the strongest support to date for hypotheses that predict the retention of migratory capability in temporary and arboreal habitats.

## Summary

Delphacid planthoppers were used to test hypotheses concerning the effects of habitat persistence, dimensionality (host-plant structure) and isolation (oceanic islands) on the incidence of migration. Wing-dimorphic insects such as planthoppers are ideal for investigating the effects of habitat factors on migration because migratory adults (macropters with fully developed wings) and flightless adults (brachypters with reduced wings) are so easily recognized. Moreover, the proportion of macropters in a population can be used as a convenient index of migratory capability.

For planthoppers, there is an inverse relationship between habitat persistence and migratory capability (% macroptery), with volant species predominating in temporary habitats and flightless taxa occurring primarily in long-lived habitats. Habitat dimensionality and macroptery (%) are positively related, with flight reduction evident in species exploiting low-profile vegetation, and wing retention characteristic of arboreal species. Habitat persistence and dimensionality interact such that flight is retained in species exploiting three-dimensional habitats, even though habitats are persistent. Thus, the effect of habitat persistence on the incidence of flight capability is realized only for species occupying low-profile habitats. Habitat isolation, at the spatial scale of oceanic islands, does not appear to enhance flightlessness: flightlessness is no more prevalent in island species than mainland species when habitat persistence and dimensionality are controlled. The demonstrated importance of habitat persistence and dimensionality in explaining the incidence of flight capability in planthoppers is probably indicative of the pivotal role these two factors play in shaping the migration strategies of most insects.

## Acknowledgements

This chapter stems from a presentation by RFD at the Royal Entomological Society's 20th Symposium held at Imperial College on 13–14 September 1999. The authors gratefully acknowledge the organizers of this symposium (Ian Woiwod, Chris Thomas and Don Reynolds) for all their time and effort in assembling speakers for this stimulating conference and for editing this contemporary contribution. This research was supported by National Science Foundation Grants DEB-9209693 and DEB-9527846 to RFD.

## References

Adams, D.A. (1963) Factors influencing vascular plant zonation in North Carolina salt marshes. *Ecology* 44, 445–456.

Barbosa, P., Krischik, V. and Lance, D. (1989) Life-history traits of forest-inhabiting flightless Lepidoptera. *American Midland Naturalist* 122, 262–274.

Berryman, A.A. (1988) *Dynamics of Forest Insect Populations: Patterns, Causes, and Implications*. Plenum Press, New York.

Blum, J.L. (1968) Salt marsh spartinas and associated algae. *Ecological Monographs* 38, 199–221.

den Boer, P.J. (1981) On the survival of populations in a heterogeneous and variable environment. *Oecologia* 50, 39–53.

den Boer, P.J., Van Huizen, H.P.T., Den-Daanje, W., Aukema, B. and Den Bieman, C.F.M. (1980) Wing polymorphism and dimorphism in ground beetles as stages in an evolutionary process (Coleoptera: Carabidae). *Entomologica Generalis* 6, 107–134.

Brown, V.K. (1982) Size and shape as ecological discriminants in successional communities of Heteroptera. *Biological Journal of the Linnaean Society* 18, 279–290.

Bull, J.J., Thompson, C., Ng, D. and Moore, R. (1987) A model for natural selection of genetic migration. *The American Naturalist* 129, 143–157.

Cappuccino, N. and Price, P.W. (1995) *Population Dynamics: New Approaches and Synthesis*. Academic Press, London.

Claridge, M. (1985) Acoustic signals in the Homoptera: behavior, taxonomy, and evolution. *Annual Review of Entomology* 30, 297–317.

Claridge, M. and de Vrijer, P.W. (1994) Reproductive behavior: the role of acoustic signals in species recognition and speciation. In: Denno, R.F. and Perfect, T.J. (eds) *Planthoppers: Their Ecology and Management*. Chapman & Hall, New York, pp. 216–233.

Crespi, B.J. (1988) Adaptation, compromise, and constraint: the development, morphometrics, and behavioral basis on a fighter-flier polymorphism in male *Hoplothrips karnyi* (Insecta: Thysanoptera). *Behavioral Ecology and Sociobiology* 23, 93–104.

Crnokrak, P. and Roff, D.A. (1995) Fitness differences associated with calling behavior in the two wing morphs of male sand crickets, *Gryllus firmus*. *Animal Behavior* 50, 1475–1481.

Darwin, C. (1876) On the origin of species by means of natural selection, or the preservation of favoured races in the struggle for life. In: Barrett, P.H. and Freeman, R.B. (eds) *The Works of Charles Darwin*, Vol. 16. Pickering and Chatto, London.

Denno, R.F. (1978) The optimum population strategy for planthoppers (Homoptera: Delphacidae) in stable marsh habitats. *The Canadian Entomologist* 110, 135–142.

Denno, R.F. (1983) Tracking variable host plants in space and time. In: Denno, R.F. and McClure, M.S. (eds) *Variable Plants and Herbivores in Natural and Managed Systems*. Academic Press, New York, pp. 291–341.

Denno, R.F. (1988) Planthoppers on the move. *Natural History Magazine* 97, 40–47.

Denno, R.F. (1994a) The evolution of dispersal polymorphism in insects: the influence of habitats, host plants and mates. *Researches on Population Ecology* 36, 127–135.

Denno, R.F. (1994b) Life history variation in planthoppers. In: Denno, R.F. and Perfect, T.J. (eds) *Planthoppers: Their Ecology and Management*. Chapman & Hall, New York, pp. 163–215.

Denno, R.F. and Grissell, E.E. (1979) The adaptiveness of wing-dimorphism in the salt marsh-inhabiting planthopper, *Prokelisia marginata* (Homoptera: Delphacidae). *Ecology* 60, 221–236.

Denno, R.F. and Peterson, M.A. (2000) Caught between the devil and the deep blue sea, mobile planthoppers escape natural enemies and locate favorable host plants. *American Entomologist* 46, 95–109.

Denno, R.F. and Roderick, G.K. (1990) Population biology of planthoppers. *Annual Review of Entomology* 35, 489–520.

Denno, R.F. and Roderick, G.K. (1992) Density-related dispersal in planthoppers: effects of interspecific crowding. *Ecology* 73, 1323–1334.

Denno, R.F., Raupp, M.J. and Tallamy, D.W. (1981) Organization of a guild of sap-feeding insects: equilibrium vs. nonequilibrium coexistence. In: Denno, R.F. and

Dingle, H. (eds) *Insect Life History Patterns: Habitat and Geographic Variation.* Springer-Verlag, New York, pp. 151–181.

Denno, R.F., Douglass, L.W. and Jacobs, D. (1985) Crowding and host plant nutrition: environmental determinants of wing-form in *Prokelisia marginata. Ecology* 66, 1588–1596.

Denno, R.F., Olmstead, K.L. and McCloud, E.S. (1989) Reproductive cost of flight capability: a comparison of life history traits in wing dimorphic planthoppers. *Ecological Entomology* 14, 31–44.

Denno, R.F., Roderick, G.K., Olmstead, K.L. and Döbel, H.G. (1991) Density-related migration in planthoppers (Homoptera: Delphacidae): the role of habitat persistence. *The American Naturalist* 138, 1513–1541.

Denno, R.F., Roderick, G.K., Peterson, M.A., Huberty, A.F., Döbel, H.G., Eubanks, M.D., Losey, J.E. and Langellotto, G.A. (1996) Habitat persistence underlies the intraspecific dispersal strategies of planthoppers. *Ecological Monographs* 66, 389–408.

Denno, R.F., Peterson, M.A., Gratton, C., Cheng, J., Langellotto, G.A., Huberty, A.F. and Finke, D.L. (2000a) Feeding-induced changes in plant quality mediate interspecific competition between sap-feeding herbivores. *Ecology* 81, 1814–1827.

Denno, R.F., Hawthorne, D.J., Thorne, B.L. and Gratton, C. (2000b) Reduced flight capability in British Virgin Island Populations of a wing-dimorphic insect: role of habitat isolation, persistence, and structure. *Ecological Entomology* 26, 1–12.

Dingle, H. (1985) Migration. In: Kerkut, G.A. and Gilbert, L.I. (eds) *Comprehensive Insect Physiology, Biochemistry and Pharmacology*, Vol. 9, *Behavior*. Pergamon Press, New York, pp. 375–415.

Dingle, H. (1996) *Migration: the Biology of Life on the Move.* Oxford University Press, New York.

Fairbairn, D.J. and Preziosi, R.F. (1996) Sexual selection and the evolution of sexual size dimorphism in the water strider, *Aquarius remigis. Evolution* 50, 1549–1559.

Felsenstein, J. (1985) Phylogenies and the comparative method. *The American Naturalist* 125, 1–15.

Fujisaki, K. (1992) A male fitness advantage to wing reduction in the oriental chinch bug, *Cavelerius saccharivorus* Okajima (Heteroptera: Lygaeidae). *Researches on Population Ecology* 34, 173–183.

Giffard, W.M. (1922) The distribution and island endemism of Hawaiian Delphacidae (Homoptera) with additional lists of their food plants. *Proceedings of the Hawaiian Entomological Society* 1, 103–118.

Hanski, I. (1991) Single-species metapopulation dynamis: concepts, models and observations. In: Gilpin, M.E. and Hanski, I. (eds) *Metapopulation Dynamics: Empirical and Theoretical Investigations.* Academic Press, London, pp. 17–38.

Hanski, I. and Kuussaari, M. (1995) Butterfly metapopulation dynamics. In: Cappuccino, N. and Price, P.W. (eds) *Population Dynamics: New Approaches and Synthesis.* Academic Press, New York, pp. 149–171.

Harrison, R.G. (1980) Dispersal polymorphisms in insects. *Annual Review of Ecology and Systematics* 11, 95–118.

Hastings, A. (1982) Dynamics of a single species in a spatially varying environment: The stabilizing role of high dispersal rates. *Journal of Mathematical Biology* 16, 49–56.

Heady, S.E. and Denno, R.F. (1991) Reproductive isolation in *Prokelisia* planthoppers: acoustical differentiation and hybridization failure. *Journal of Insect Behavior* 4, 367–390.

Howarth, F.G. (1990) Hawaiian arthropods: an overview. *Bishop Museum Occasional Papers* 30, 4–26.

Hunt, R.E. and Nault, L.R. (1991) Roles of interplant movement, acoustic communication, and phototaxis in mate-location behavior of the leafhopper *Graminella nigrifrons*. *Behavioral Ecology and Sociobiology* 28, 315–320.

Ichikawa, T. (1977) Sexual communications in planthoppers. In: *The Rice Brown Planthopper*. Food and Fertilizer Technology Center for the Asian and Pacific Region, Taipei, Taiwan, pp. 84–94.

Ichikawa, T. (1982) Density-related changes in male-male competitive behavior in the rice brown planthopper, *Nilaparvata lugens* (Stål)(Homoptera: Delphacidae). *Applied Entomology and Zoology* 17, 439–452.

Ichikawa, T., Sakuna, M. and Ishii, S. (1975) Substrate vibrations: mating signal of three species of planthoppers which attack the rice plant (Homoptera: Delphacidae). *Applied Entomology and Zoology* 10, 162–171.

Iwanaga, K. and Tojo, S. (1986) Effects of juvenile hormone and rearing density on wing dimorphism and oöcyte development in the brown planthopper, *Nilaparvata lugens*. *Journal of Insect Physiology* 32, 585–590.

Iwanaga, K. and Tojo, S. (1988) Comparative studies on the sensitivities to nymphal density, photoperiod and rice plant stage in two strains of the brown planthopper, *Nilaparvata lugens* (Stål)(Homoptera: Delphacidae). *Japanese Journal of Applied Entomology and Zoology* 32, 68–74.

Iwanaga, K., Nakasuji, F. and Tojo, S. (1987) Wing dimorphism in Japanese and foreign strains of the brown planthopper, *Nilaparvata lugens*. *Entomologica Experimentalis et Applicata* 43, 3–10.

Kaitala, A. and Dingle, H. (1993) Wing dimorphism, territoriality and mating frequency of the water strider *Aquarius remigis* (Say). *Annales Zoologica Fennici* 30, 163–168.

Kenmore, P. E., Cariño, F.O., Perez, C.A., Dyck, V.A. and Gutierrez, A.P. (1984) Population regulation of the rice brown planthopper (*Nilaparvata lugens* Stål) within rice fields in the Philippines. *Journal of Plant Protection in the Tropics* 1, 19–37.

Kisimoto, R. (1976) Synoptic weather conditions inducing long-distance immigration of planthoppers, *Sogatella furcifera* Horvath and *Nilaparvata lugens* Stål. *Ecological Entomology* 1, 95–109.

Kisimoto, R. and Rosenberg, L.J. (1994) Long-distance migration in delphacid planthoppers. In: Denno, R.F. and Perfect, T.J. (eds) *Planthoppers: Their Ecology and Management*. Chapman & Hall, New York, pp. 302–322.

Kisimoto, R. and Sogawa, K. (1995) Migration of the brown planthopper *Nilaparvata lugens* and the white-backed planthopper *Sogatella furcifera* in East Asia: the role of weather and climate. In: Drake, V.A. and Gatehouse, A.G. (eds) *Insect Migration: Tracking Resources through Space and Time*. Cambridge University Press, Cambridge, pp. 67–91.

Kuno, E. (1979) Ecology of the brown planthopper in temperate regions. In: *Brown Planthopper: Threat to Rice Production in Asia*. International Rice Research Institute, Los Baños, Philippines, pp. 45–60.

Langellotto, G.A. (1997) Reproductive costs and mating consequences of dispersal capability in males of the wing-dimorphic planthopper *Prokelisia dolus* (Hemiptera: Delphacidae). MS thesis, University of Maryland, College Park, Maryland.

Langellotto, G.A., Denno, R.F. and Ott, J.R. (2000) A trade-off between flight capability and reproduction in males of a wing-dimorphic insect. *Ecology* 81, 865–875.

Marooka, S. and Tojo, S. (1992) Maintenance and selection of strains exhibiting specific wing form and body colour under high density conditions in the brown planthopper, *Nilaparvata lugens* (Homoptera: Delphacidae). *Applied Entomology and Zoology* 27, 445–454.

Marooka, S., Ishibashi, N. and Tojo, S. (1988) Relationships between wing-form response to nymphal density and black colouration of adult body in the brown planthopper, *Nilaparvata lugens* (Homoptera: Delphacidae). *Applied Entomology and Zoology* 23, 449–458.

Mendelsshon, I.A., McKey, K.L. and Patrick, W.H. Jr (1981) Oxygen deficiency in *Spartina alterniflora* roots: metabolic adaptation to anoxia. *Science* 214, 439–441.

Mun, J.H., Song, Y.H., Heong, K.L. and Roderick, G.K. (1999) Genetic variation among Asian populations of rice planthoppers, *Nilaparvata lugens* and *Sogatella furcifera* (Hemiptera: Delphacidae): mitochondrial DNA sequences. *Bulletin of Entomological Research* 6, 245–253.

New, T.R. (1974) Psocoptera. *Royal Entomological Society Handbooks for the Identification of British Insects*, Part 7. Blackwell, London.

Novotný, V. (1994) Relation between temporal persistence of host plants and wing length in leafhoppers (Hemiptera: Auchenorrhyncha). *Ecological Entomology* 19, 168–176.

Novotný, V. (1995) Adaptive significance of wing dimorphism in males of *Nilaparvata lugens*. *Entomologia Experimentalis et Applicata* 76, 233–239.

Ornes, W.H. and Kaplan, D.I. (1989) Macronutrient status of tall and short forms of *Spartina alterniflora* in a South Carolina salt marsh. *Marine Ecology Progress Series* 55, 63–72.

Pedgley, D.E. (1993) Managing migratory insect pests – a review. *International Journal of Pest Management* 1, 3–12.

Peterson, M.A. and Denno, R.F. (1997) The influence of intraspecific variation in dispersal strategies on the genetic structure of planthopper populations. *Evolution* 5, 1189–1206.

Peterson, M.A. and Denno, R.F. (1998) The influence of dispersal and diet breadth on patterns of genetic isolation by distance in phytophagous insects. *The American Naturalist* 152, 428–446.

Rabb, R.L. and Kennedy, G.G. (1979) *Movement of Highly Mobile Insects: Concepts and Methodology in Research*. North Carolina University Press, Raleigh, North Carolina.

Reddingius, J. and den Boer, P.J. (1970) Simulation experiments illustrating stabilization of animal numbers by spreading the risk. *Oecologia* 5, 240–284.

Redfield, A.C. (1972) The development of a New England salt marsh. *Ecological Monographs* 42, 201–237.

Reuter, O.M. (1875) Remarques sur le polymorphisme des hemipteres. *Annales of the Entomological Society of France* 5, 225–236.

Roff, D.A. (1986) The evolution of wing dimorphism in insects. *Evolution* 40, 1009–1020.

Roff, D.A. (1990) The evolution of flightlessness in insects. *Ecological Monographs* 60, 389–421.

Roff, D.A. (1994) Habitat persistence and the evolution of wing dimorphism in insects. *The American Naturalist* 11, 772–798.

Roff, D.A. and Fairbairn, D.J. (1991) Wing dimorphisms and the evolution of migratory polymorphisms among the Insecta. *American Zoologist* 31, 243–251.

Simon, C.M., Gagne, F.G., Howarth, F.G. and Radovsky, F.J. (1984) Hawaii: a natural entomological laboratory. *Bulletin of the Entomological Society of America* 30, 8–17.

Solbreck, C. (1978) Migration, diapause, and direct development as alternative life histories in a seed bug, *Neacoryphus bicrucis*. In: Dingle, H. (ed.) *The Evolution of Insect Migration and Diapause*. Springer-Verlag, New York, pp. 195–217.

Southwood, T.R.E. (1962) Migration of terrestrial arthropods in relation to habitat. *Biological Review* 37, 171–214.

Southwood, T.R.E. (1977) Habitat, the templet for ecological strategies. *Journal of Animal Ecology* 46, 337–365.

Southwood, T.R.E. and Leston, D. (1959) *Land and Water Bugs of the British Isles*. F. Warne, London.

Spence, J.R. (1989) The habitat templet and life history strategies of pond skaters (Heteroptera: Gerridae): reproductive potential, phenology, and wing dimorphism. *Canadian Journal of Zoology* 10, 2432–2447.

Tallamy, D.W. and Denno, R.F. (1981) Alternative life history patterns in risky environments: an example from lacebugs. In: Denno, R.F. and Dingle, H. (eds) *Insect Life History Patterns: Habitat and Geographic Variation*. Springer-Verlag, New York, pp. 129–147.

Taylor, C.E., Powell, J.R., Kekic, V., Andjelkovic, M. and Burla, H. (1984) Dispersal rates of species of the *Drosophila obscura* group: implications for population structure. *Evolution* 38, 1397–1401.

Vepsäläinen, K. (1978) Wing dimorphism and diapause in *Gerris*: determination and adaptive significance. In: Dingle, H. (ed.) *Evolution of Insect Migration and Diapause*. Springer-Verlag, Berlin, pp. 218–253.

Wagner, D.L. and Liebherr, J.K. (1992) Flightlessness in insects. *Trends in Ecology and Evolution* 7, 216–220.

Wagner, W.L., Herbst, D.R. and Sohmer, S.H. (1990) *Manual of the Flowering Plants of Hawaii*, Vols 1 and 2. Bishop Museum Special Publication 83. University of Hawaii Press and Bishop Museum Press, Honolulu, Hawaii.

Waloff, N. (1983) Absence of wing polymorphism in the arboreal, phytophagous species of some taxa of temperate Hemiptera: an hypothesis. *Ecological Entomology* 8, 229–232.

Wilson, S.W., Mitter, C., Denno, R.F. and Wilson, M.R. (1994) Evolutionary patterns of host plant use by delphacid planthoppers and their relatives. In: Denno, R.F. and Perfect, T.J. (eds.) *Planthoppers: Their Ecology and Management*. Chapman & Hall, New York, pp. 7–113.

Zera, A.J. and Denno, R.F. (1997) Physiology and ecology of dispersal polymorphisms in insects. *Annual Review of Entomology* 42, 207–231.

Zimmerman, E.C. (1948) *Insects of Hawaii*, Vol. 4, *Homoptera: Auchenorrhyncha*. University of Hawaii Press, Honolulu, Hawaii.

# Predation and the Evolution of Dispersal

## 12

Wolfgang W. Weisser*

*Zoology Institute, University of Basel, Rheinsprung 9, 4051 Basel, Switzerland*

## Introduction

Most organisms are preyed upon by a natural enemy and for most, dispersal is an important part of their life cycle. This chapter discusses the various effects that predation can have on the evolution of dispersal. I assume that in the environment it is possible to distinguish between discrete habitat patches, suitable for the focal species, and the rest of the environment (often referred to as the 'matrix', cf. Hanski and Simberloff, 1997). The habitat patches or habitats (*sensu* Southwood, 1981) are either empty or occupied by populations of the focal species. There are many, often contradictory, definitions of dispersal in the literature (for reviews, see e.g. Johnson, 1969; Stenseth and Lidicker, 1992; Dingle, 1996; Ims and Yoccoz, 1997). In this chapter, dispersal is understood broadly as the transfer of individuals from one habitat to another (cf. Howard, 1960; Ims and Yoccoz, 1997; Stenseth and Lidicker, 1992). An individual that moves from one habitat to another goes through the successive stages of leaving a population (emigration), travelling to the other habitat (transit) and arriving at this habitat (immigration). As will be shown below, predation and the risk thereof influence these three stages in different ways. The emphasis in this chapter is on top-down effects, i.e. the role of mortality caused by natural enemies on the dispersal of their prey. Bottom-up effects, i.e. the effects that prey have on the evolution of predator dispersal, are not discussed (see Rosenzweig (1991) for a review of

---

* Present address: Institute of Ecology, Friedrich-Schiller-University, Dornburger Str. 159, 07743 Jena, Germany.

©CAB *International* 2001. *Insect Movement: Mechanisms and Consequences*
(eds I.P. Woiwod, D.R. Reynolds and C.D. Thomas)

resource-dependent habitat selection of animals, and Bernstein et al. (1991) for an example involving insect host–parasitoid systems).

The chapter first reviews the effects of predation on the different stages of the dispersal process and the consequences of these effects for the evolution of dispersal. Each effect is illustrated with examples. The second half of the chapter deals with the evolution of predator-induced dispersal (PID), i.e. dispersal conditional on the presence of a predator or an increase in the risk of predation. Necessary conditions for the evolution of PID are derived by comparing it to other predator-avoidance strategies.

## The Effects of Predation on the Different Stages of Dispersal

### Emigration

Emigration is completed when an individual has departed from the population. In insects, both genetic and environmental factors have been shown to influence characters or behaviours related to dispersal and the departure of individuals from populations (Roderick and Caldwell, 1992). Predation, or the risk of it, may therefore be involved in both unconditional (genetic) and conditional (environment-dependent) dispersal strategies.

Most mathematical models on the evolution of dispersal consider unconditional, i.e. purely genetically determined, dispersal (Johnson and Gaines, 1990; Dieckmann et al., 1999; Ferriere et al., 2000; but see Ozaki, 1995; Ronce et al., 1998). Predation, or the risk of predation, is not normally discussed explicitly in these models, but predictions can still be derived from the existing theory. In the extreme case, predation leads to the extinction of the prey population. If predation increases the *risk* of population extinction, models would predict an increase in the fraction of individuals leaving the population (e.g. Johnson and Gaines, 1990; Dieckmann et al., 1999). This is analogous to the notion of 'temporary habitats' (Southwood, 1962, 1977; Roff, 1986, 1994) where succession or abiotic factors lead to population extinction, but even if predation does not cause extinction, it may influence the evolution of prey dispersal rates. In a variety of models it has been shown that temporal variability in population growth rates tends to select for higher dispersal rates, whereas spatial variability tends to select against dispersal (Johnson and Gaines, 1990). Thus, when there is spatial or temporal variability in the impact of predators on their prey resulting in spatial or temporal variability in prey population growth rates, models would predict an effect of predation on prey dispersal rates. Unfortunately, there is no empirical evidence that shows that any of these effects are at work in insect populations. There are a number of examples showing that dispersal is affected by habitat longevity (Southwood, 1962; Denno, 1994; Roff, 1994; Denno et al., 1996 and Chapter 11), but in these cases population extinctions are not due to the action of predators. Comparative studies of predator–prey metapopulations are needed that link the impact of predators on prey populations to the dispersal rates of prey.

While the empirical evidence for predation-related, unconditional dispersal strategies is scarce, emigration conditional on the presence of predators or an increase in the risk of predation in the habitat has been shown repeatedly. For example, spider mites, *Tetranychus urticae* Koch, have been shown to leave plants in the presence of predatory mites (Bernstein, 1984). Similarly, pea aphids, *Acyrthosiphon pisum* (Harris), are more likely to walk away from a host plant when a predatory ladybird is in the vicinity (Roitberg *et al.*, 1979). In these cases, individuals leave a population to reduce their immediate mortality risks. Examples can also be found in aquatic environments; for example, predatory stoneflies (Plecoptera) in streams have been shown to induce a variety of insect prey to drift and thus emigrate from habitat patches (Sih and Wooster, 1994). Behavioural responses to the immediate risk of being eaten are likely to be the most common way in which prey emigration is influenced by predation.

A special case of conditional emigration involves wing-dimorphic insects where only part of a population develops functional wings. Flight polymorphisms are widespread in insects, and in some groups such as planthoppers and aphids, wing dimorphism can be extreme; within a species both fully winged and unwinged forms can exist (Zera and Denno, 1997). Because flight polymorphisms represent easily quantified variation that directly affects dispersal, they are often used as model systems to study the ecology and evolution of insect dispersal (Roff, 1986, 1994; Denno *et al.*, 1996 and Chapter 11; Dingle, 1996; Zera and Denno, 1997). Recently, it has been shown that pea aphids responded to the presence of predatory ladybirds by increasing the proportion of their offspring that developed into winged morphs (Weisser *et al.*, 1999; Fig. 12.1). In this case, predation affected prey emigration by influencing the development of dispersal ability of an individual. In contrast to the behavioural responses discussed above, this effect involves a lag between the time when an increase in predation risk is perceived by the prey, and the time the individual leaves the population. The question why selection might favour such induction in the face of time-lags is discussed below. It is not known if predation has similar effects in other species with a dispersal polymorphism.

## Transit

In many organisms, dispersal is associated with a high risk of mortality. During the journeys from one habitat to another, the animals often travel long distances. However, even travel over short distances can entail significant mortality risks for an animal when predators are abundant. Quantitative estimates of predator-related mortality are difficult to obtain because of the often small size of insects, but the available evidence indicates that

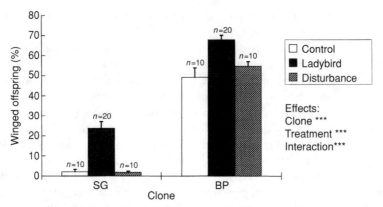

**Fig. 12.1.** Proportion of winged offspring produced by two clones of pea aphids, *Acyrthosiphon pisum*, when exposed to the presence of adults of the predatory two-spot ladybird, *Adalia bipunctata*, and to a disturbance treatment. The disturbance treatment (dropping the plant three times a day from a height of 16.5 cm) was intended to simulate the disturbance caused by foraging predators. Two-way ANOVA: clone $F_{1,74} = 4257.8$, $P = 0.0001$, treatment $F_{2,74} = 381.9$, $P = 0.0001$, interaction $F_{2,74} = 66.2$, $P = 0.0004$. Data from Weisser *et al.* (1999).

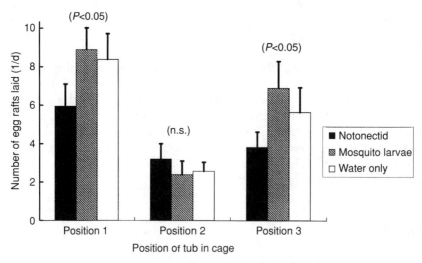

**Fig. 12.2.** Ovipositions of female mosquito *Culex pipiens* into tubs containing a predatory notonectid plus mosquito larvae, mosquito larvae or water (control). Because the position of a tub in the cage influenced mosquito oviposition behaviour, data are presented separately for each position. $N = 16$ for each position/treatment combination. Position 1: $F_{2,90} = 3.28$, $P < 0.05$; position 2: $F_{2,90} = 1.26$, $P > 0.05$; position 3: $F_{2,90} = 4.28$, $P < 0.05$. From Chesson (1984).

predator-related mortality during transit is high. For example, in the parasitic wasp *Pauesia unilachni* (Gahan) that attacks aphids on needles of pine, females leaving an aphid colony by flight have an 11% chance of dying in one of the spider webs that are very abundant between pine needles (Völkl and Kraus, 1996).

Theory makes very clear predictions about the effects of travel mortality on dispersal. Models of the evolution of dispersal predict that dispersal rates decrease if the cost of dispersal increases (Hamilton and May, 1977; Comins *et al.*, 1980). Similarly, optimal foraging models predict that travel mortality leads to longer patch residence times (e.g. Houston and McNamara, 1986). Unfortunately, there is no empirical evidence for these effects, partly because of the problems of quantifying mortality risks in insects. Thus, despite clear theoretical predictions, the link between dispersal rates and predation-related mortality during transit remains untested. What has been found, however, are behavioural adaptations of prey which reduce the risk of predation while travelling. An example is the flight behaviour of the aphid parasitoid *Aphidius rosae* Haliday that parasitizes aphids in rose bushes (Völkl, 1994). Rose bushes harbour many spider webs and *A. rosae* reduces the risk of getting caught in a spider web by departing only from the tops of shoots. Parasitoid females then fly about 30 cm out of the rose bush before they turn to land on the tip of the next shoot. This U-shaped flight behaviour reduces the mortality risks due to spider predation (Völkl, 1994).

## Immigration

Immigration involves the selection of suitable habitats and predation risk may be an important component of habitat suitability. Predation risks are often highest for juveniles. For an immigrant, reproductive success may therefore depend critically on the survival of the offspring in the new habitat. One would therefore predict that in species where this is the case individuals would select habitats based on the risk of predation for their offspring. Such behaviour has been observed in a number of species. A good example is habitat selection in mosquitoes. Chesson (1984) found that females of *Culex pipiens* L. avoided laying eggs in waters where predatory notonectids were present (Fig. 12.2). Predation risk-related habitat-selection behaviour has also been observed in another mosquito, *Culiseta longiareolata* Macquart, and in a variety of other taxa (Blaustein, 1998, 1999; Stav *et al.*, 1999).

## Conclusions

Three main effects of predation on the dispersal process can be identified: (i) predation influences whether individuals emigrate from a population; (ii) predation influences the habitat choice of immigrants; and (iii) predation influences the survival of individuals during the journey from one habitat to another. Predation is therefore an evolutionary force that may:

1. Select for dispersal, e.g. to escape predation risk;
2. Select against dispersal, e.g. when predators increase the cost of dispersal;
3. Select for conditional strategies, e.g. emigration conditional on the presence of predators in the habitat.

Thus it can be seen that predation may have conflicting consequences for the evolution of dispersal. While theoretical studies are needed that explore in more detail how dispersal might evolve under the influence of predation, there already is a body of theory that makes a number of predictions concerning specific effects of predation on the evolution of dispersal rate. More empirical work is needed to test these predictions. In particular, there is a need for comparative studies that link predation to prey population growth rates and rates of dispersal.

In the next sections, the example of predator-induced production of dispersal morphs in aphids is used to investigate more generally the conditions under which predator-induced dispersal may be expected to evolve.

## The Evolution of Predator-induced Dispersal (PID)

Conditional dispersal strategies are common among animals. External factors that have been shown to influence dispersal are population density, weather conditions or the social status of an animal. Among insects, population density and temperature are important environmental factors that are involved in a variety of dispersal-related phenomena, from the development of winged morphs in aphids and orthopterans to the swarming of ants and bees (Johnson, 1969; Kawada, 1987; Weisser *et al.*, 1997; Zera and Denno, 1997). PID, i.e. dispersal triggered by the presence of predators or by an increase in the risk of predation is a conditional strategy to escape from or reduce the risk of predation. Conceptually, PID is similar to cases of predator-induced defence where morphological, behavioural or life-history changes are employed to increase the survival chance of an organism (Tollrian and Harvell, 1998a). In the following, the theory developed to understand the evolution of predator-induced defences is used to investigate the conditions under which PID might be a successful strategy.

### Predator-induced defence

Induced structural or chemical defences against natural enemies have been shown for a wide range of plant–herbivore and predator–prey systems (Harvell, 1984, 1990; Schultz, 1988; Tollrian and Harvell, 1998a). Inducible defences are produced in response to stimuli from natural enemies and either deter further predator attack or decrease its impact (Karban and Baldwin, 1997; Tollrian and Harvell, 1998a). Examples are inducible neckteeth or helmets in waterfleas (*Daphnia* spp.), that are produced in response to kairomones released by predators such as fish or phantom midge larvae (Tollrian and Harvell, 1998a). However, predator-induced defences also include prey responses that represent escape strategies rather than defences in the strict sense, for example when zooplankton sink to deeper depths to avoid visually hunting predators (de Meester *et al.*, 1998). Current theory states that the following four conditions are a prerequisite for the evolution of an inducible defence (Tollrian and Harvell, 1998b):

1. Predator attacks have to be spatially or temporally intermittent. If the attacker is constantly present, the defence should not be inducible but constitutive, i.e. permanently present.
2. A reliable cue has to be available that signals the proximity of the threat. An attack itself can constitute this cue provided, of course, that it is non-fatal.
3. The defence has to incur a cost. Defence against predation is generally assumed to impose costs (Maynard Smith, 1972), and theory predicts that inducible defences evolve only if these costs can be saved in times when no protection is necessary (Lively, 1986).
4. The defence has to be efficient. This is a rather obvious condition but, interestingly, it has often been difficult to demonstrate this benefit, in particular in plant–herbivore–systems (Karban and Baldwin, 1997; Agrawal, 1998; Tollrian and Harvell, 1998a).

In the following sections, these conditions are applied to the case of PID. I then ask whether these conditions are likely to be fulfilled for aphids and their natural enemies and more generally in insect predator–prey systems.

## Variability in the risk of predation

Inducible defences are favoured when the selective pressure is variable (Tollrian and Harvell, 1998b). The selective agent has to have a variable impact which is sometimes low enough such that no defence is necessary. The higher the baseline of attack oscillations, and the more uniform the selective impact, the more likely it is that constitutive defences evolve (Tollrian and Harvell, 1998b). Variation in predator impact can be temporal or spatial. For example, many herbivores or predators are predictably seasonal, but the onset and magnitude of attack are variable. These conditions are sufficient to select for defence responses that are induced in times of high predation risk. Similarly, spatial variability in the impact of predators can lead to a defence that is induced only when the animal finds itself under predator attack (Tollrian and Harvell, 1998b).

In contrast to defences in the strict sense, dispersal implies a change of location. Thus, temporal variation in predator impact without any spatial variation cannot select for PID, because an individual would not improve its situation by changing habitats. If attacks are only temporally variable, other risk-avoidance strategies such as hiding or a change of microhabitat are more likely to evolve (Sih, 1997). Thus, for the evolution of PID, spatial variation in the risk of predation is required. PID can only be a successful strategy when animals improve their reproductive success by emigrating, i.e. when the impact of predation on fitness in the new habitat is lower than in the old habitat. This can be illustrated with a simple model for exponentially growing aphid populations.

In aphids, a good measure of clonal reproductive fitness is the number of diapausing eggs produced at the end of the season (Weisser and Stadler, 1994). This number is likely to be an increasing function of the number of sexual individuals produced, which in turn is likely to be positively related to population size at the end of the season. Assume that there is no intra-specific competition so that populations can grow exponentially until the end of the season. Let $r$ be the growth rate of the clone, $N_0$ the current population size and $T$ be the time left until the end of the season. In the absence of predation, the population size would be:

$$N(T) = N_0 \, e^{rT}.$$

Suppose predators can locally eradicate the entire clone and let $p$ be the chance that the local population is not found by the predators. Then expected population size at the end of the season is:

$$N(T) = p \, N_0 \, e^{rT}.$$

Now suppose that the clone sends out $M$ emigrants that have a chance $s$ of reaching a new habitat. Ignoring the fact that emigration decreases the local growth rate, and that emigrants will only reach the new habitat at a time closer to final time, the expected population size at the end of the season with dispersal is:

$$N(T) = p \, (N_0 - M) \, e^{rT} + q \, sM \, e^{rT}$$

where $q$ is the chance that the populations in the new habitats are not found by the predators before the end of the season. If $q = p$ and $s < 1$, then

$$N(T) = p \, (N_0 - M + sM) \, e^{rT} < p \, N_0 \, e^{rT}$$

Because dispersal is always associated with a risk of mortality during transit ($s < 1$), PID will only be favoured when the predation risk in the new habitat is lower than the predation risk in the old habitat (i.e. when $q > p$).

What is the evidence for spatial variation in the impact of predators? Most evidence comes from observational studies that show that the numbers of predators differ between habitats. For example, it is well known that generalist predators have habitat preferences (e.g. Honek, 1985; Cardwell et al., 1994; Costello and Daane, 1998; French and Elliott, 1999). Edge effects also influence the number of foraging predators, for example in field margins

(Dennis and Fry, 1992). For many predators it is known that they aggregate in areas of high prey density (Bryan and Wratten, 1984; Kareiva and Odell, 1987; Kareiva, 1990; Chen and Hopper, 1997). Finally, predator number may vary independently of environmental factors (Shukla and Pathak, 1987; Evans and Youssef, 1992). Thus, there is evidence for both prey-density dependent as well as prey-density independent spatial variation in the number of foraging predators. An increase in the *number* of foraging predators, however, does not necessarily lead to an increase in the *risk of predation* of the prey. For example, when predators show aggregation in response to prey density, the *per capita* risk of prey may remain unchanged or may even decrease at higher prey densities (Turchin and Kareiva, 1989). Thus, it often remains unclear if spatial variability in predator number translates into spatial variability in predator impact. In contrast to insect predator–prey systems, there is ample evidence from host–parasitoid associations that spatial variability in the risk of being killed by a natural enemy can be common in the field (Pacala *et al.*, 1990). One of the reasons why host–parasitoid systems have been studied in more detail is that rates of parasitism can be directly quantified by dissecting hosts. Bodies of dead prey are mostly not available for quantification so that direct measurements of rates of predation are difficult.

In aphids, predation risks for colonies are high, as indicated by predator-exclusion experiments (Way and Banks, 1968; Campbell, 1978; Frazer *et al.*, 1981a,b; Dennis and Wratten, 1991; Jervis and Kidd, 1996). Field studies on aphid–predator interactions show that the number of predators attacking aphid colonies varies both spatially and temporally (Hughes, 1963; Kareiva and Odell, 1987; Cappuccino, 1988; Farrell and Stufkens, 1988, 1990; Kfir and Kirsten, 1991; Chen and Hopper, 1997; Stewart and Walde, 1997). Chen and Hopper (1997) studied the population dynamics and the impact of natural enemies in the Russian wheat aphid, *Diuraphis noxia* (Mordvilko), in the Montpellier region of southern France. The density of predators varied spatially and increased with increasing aphid density (Fig. 12.3). Chen and Hopper also calculated the impact of predators on aphid growth rates. Predators had a significant effect on aphid population growth such that the *per capita* rate of increase of *D. noxia* was a decreasing function of local predator density (Fig. 12.4). Thus, in this example, there was significant spatial variation in the impact of predators on prey population growth. Aphids moving from an area of high predator impact to an area of low predator impact could therefore have increased their reproductive success. An open question is how quickly predators would adapt their distribution to a changed aphid distribution. The results of Chen and Hopper (1997) suggest, however, that significant spatial variation in the risk of predation can be found in insect predator–prey systems.

While pure temporal variability in the risk of predation cannot select for PID, temporal variability will affect whether PID evolves in response to spatial variability. At one extreme, there is no temporal variation in the risk of predation. Pure spatial variation in rates of reproduction is known to select against dispersal, because more individuals live in good habitats so that on average an individual will worsen its situation by changing habitats (Johnson and Gaines, 1990). Thus, if the risk of predation varies spatially but never

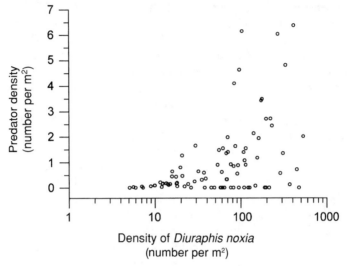

**Fig. 12.3.** Density of the predators of *Diuraphis noxia* versus *D. noxia* density during a 1992–1993 field study by Chen and Hopper (1997) in the Montpellier region, France. Each point represents the mean of 20 or 50 1-m² quadrats for a single field and date. Spearman $\rho = 0.43$; $n = 92$, $P = 0.0001$). From Chen and Hopper (1997).

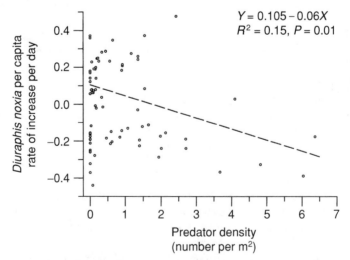

**Fig. 12.4.** Decline of *Diuraphis noxia* per capita rate of increase with increasing predator density in the field study by Chen and Hopper (1997). Each point represents a single field and sample interval. The dotted line and equation are linear regressions. From Chen and Hopper (1997).

changes, PID will not evolve. At the other extreme, predator impact in the habitats fluctuates at a time-scale shorter than the generation time of the prey. If fluctuations are very rapid such that during the life span of an individual predator impact is similar in all habitats, PID is selected against. Thus, for the evolution of PID, temporal variation in predation risk has to be autocorrelated at an intermediate time scale. A formal analysis is required to explore the combined effects of spatial and temporal variability in the risk of predation on the evolution of PID.

To summarize, PID is likely to evolve in only a subset of all possible forms of spatio-temporal variation in the risk of predation. While there is evidence for variability in the risk of predation in many insect predator–prey systems including aphids and their natural enemies, the scale of this variability needs to be quantified.

## Cues

In all inducible systems, cues are available that indicate the potential danger for the prey (Tollrian and Harvell, 1998a). The cues have to be timely and reliable, so that the defence is induced in time and only when it is really necessary. According to Tollrian and Harvell (1998b), cues may be the 'weak link' that determines whether an inducible defence can evolve. Defences can be induced by mechanical, tactile, visual or chemical cues. Chemical cues (kairomones), such as those released by *Chaoborus* larvae or fish, are the most spectacular type of cues. In many systems, little is known about their chemical identity or specificity (Tollrian and Harvell, 1998a). Because most predators leave many different types of traces, unravelling the exact nature of a cue is difficult. For example, it was shown in 1981 that kairomones from *Chaoborus* larvae induce neckteeth formation in *Daphnia pulex* (Krueger and Dodson, 1981). To date, more than 18 years later, the cue has still not been identified despite research effort in that direction.

In the case of aphid wing polymorphism, a number of cues could theoretically be used by the aphids to detect the presence of a predator (Table 12.1). One of the possibilities, that winged morphs are induced by the increased number of tactile stimuli an aphid receives from conspecifics when foraging ladybirds increase the level of disturbance in aphid colonies, was not supported by the experiments of Weisser *et al.* (1999; Fig. 12.1). Dixon and Agarwala (1999) tested whether aphids respond to chemical cues released by larvae of the two-spot ladybird, *Adalia bipunctata* L. The authors reported that tracks left by the searching larvae are used by pea aphids to detect predator presence. Weisser and Braendle (unpublished) performed a number of similar experiments using several aphid clones and two species of ladybird, and were unable to find evidence for the involvement of a chemical cue, even though in some experiments, the same aphid clones and the same predator species were used as in the experiments by Dixon and Agarwala (1999). As

**Table 12.1.** Possible cues that could be used by aphids to detect the presence of a predator.

---

1. *Chemical*
   Substance released by predator
   Substance released by attacked aphid
   Substance released by conspecific of attacked aphid
2. *Mechanical*
   Disturbance caused by predator
   Direct aphid–predator contact
   Vibrations caused by searching predator
3. *Visual*
   Aphid recognizes shape, colour of predator

---

Weisser *et al.* (1999) did not use larvae but adult ladybirds in their original experiments (Fig. 12.1), larval tracks cannot be the only cue in the aphid–ladybird system. Thus, at present, it remains unclear which cues are used by aphids in their response to the presence of a predator. Chemical cues cannot be ruled out, but the available evidence is still inconclusive.

## Costs

Costs play a central role in the theory of induced defences (Tollrian and Harvell, 1998a). The defended morph has to have a disadvantage compared to the undefended morph in the absence of the predator. If a defence incurs no costs, it is difficult to understand why the defence is not constitutive. Dispersal incurs a number of costs. Individuals that leave the population forego reproduction during their journey to the new habitat and they may show reduced fecundity as dispersal itself requires energy that cannot be used for reproduction. In addition, travelling is associated with a number of mortality risks. Finally, immigrants may be disadvantaged compared to residents, for example when there is competition for breeding sites.

In wing-dimorphic insects, it is well established that winged morphs have a number of disadvantages compared to unwinged morphs, because fat reserves and flight muscles divert energy from reproduction (Dingle, 1996; Zera and Denno, 1997). In aphids, winged morphs generally have a lower fecundity and a longer developmental time than wingless ones (Dixon and Wratten, 1971; Dixon, 1972; MacKay and Wellington, 1975; Dixon, 1998). The ancestral state in aphids is the winged phenotype and the costs associated with having a flight apparatus are thought to be the reason why the unwinged morph has been added to the life cycle (Heie, 1967, 1987; Dixon, 1998). Thus, the cost condition is likely to be fulfilled in most insect predator–prey systems.

## Benefits

This last condition seems rather obvious. If the defence does not reduce the risk of predation, why should it evolve in the first place? Benefits, however, may be difficult to demonstrate. For example, for many secondary compounds produced by plants in response to herbivory no clear-cut benefit has been shown (Karban and Baldwin, 1997). As discussed above, PID is only predicted to evolve if it results in a fitness increase. PID therefore requires that individuals that leave a population find a habitat where the risk of predation is lower. If a specialized dispersal morph is induced in the presence of a predator, a second condition to be satisfied arises from the delay between the time of induction of the dispersal morph and the time when this morph has matured and is ready to emigrate. For the induction of dispersal morphs to be a successful strategy, it must be demonstrated that the induced individuals have a chance to survive until they are ready to leave their current habitat.

In aphids, these two conditions have not yet been investigated in much detail. In the laboratory, single adults of the ladybird, *Coccinella septempunctata* L., have been shown to consume more than 100 aphids per day, and relative densities of more than two ladybirds per 100 pea aphids have been reported from the field (Hodek and Honek, 1996). Thus, there is a chance that predators will consume the entire aphid colony before the winged morphs have had time to develop. Other predators such as predatory gallmidge (cecidomyiid) or syrphid larvae are known to eradicate large aphid colonies within a few days (Markkula *et al.*, 1979; Chambers, 1988; Nijveldt, 1988). However, these are extreme scenarios. Although quantitative data are rare, field studies frequently report that aphid colonies are not completely eaten by the predators or that the time to extinction after the onset of predator attack is of the order of weeks rather than days (e.g. Chambers and Sunderland, 1982; Chambers *et al.*, 1983; Carroll and Hoyt, 1984). In Chen and Hopper's (1997) study, aphid intrinsic growth rates in areas with high predator pressure fell to $r = -0.2$. On average, a colony with an initial size of ten individuals would therefore go extinct after 11 days, and one with an initial size of 100 individuals after 28 days. Obviously, once colonies become smaller, population size may not decrease smoothly. It is likely, however, that in most cases at least some of the developing winged morphs have a good chance to escape from the host plant. If the impact of predators on prey growth rate was studied as a function of prey population size, the time to extinction and the probability of an individual surviving until maturity could be calculated.

Some other possibilities also remain to be explored. Weisser *et al.* (1999) showed a response of aphids to the presence of adult ladybirds. It is conceivable that aphids reacted not only to the predation risk emanating from adult beetles, but to the possibility that these adults, if they are female, lay a cluster of eggs. Eggs need some time to hatch, and small ladybird larvae only consume a few aphids (Hodek and Honek, 1996). The lag between oviposition and the time when larvae have grown to a size where they have a major impact on the colony may be sufficient for winged morphs to develop and emigrate.

Thus, the existence of spatial variability in predator impact on prey suggests that there is a chance for emigrants to settle in habitats with lower risks of predation. Clear-cut evidence that aphids leaving a habitat with a high risk of predation have a good chance of improving their situation is, however, lacking. For aphids, it should be possible to calculate the chances for winged morphs to develop successfully in the face of attack by different types of predators. A full understanding of aphid dispersal in the presence of predators requires, however, a clear demonstration of benefit for the individuals that leave the population.

## General Conclusions

This review has shown that our understanding of the interaction between the two fundamental ecological processes of predation and dispersal is still incomplete. Predation has been shown to have a number of conflicting consequences for the evolution of dispersal. More theoretical work is needed to understand when predation is expected to increase, and when it is expected to decrease, the dispersal rate of the prey. Also, more empirical work is needed to test the predictions made by the existing body of theory.

PID, i.e. dispersal conditional on the presence of a predator or an increase in the risk of predation, is present in many predator–prey systems. In most cases, it takes the form of a behavioural response to immediate mortality risks. The recent finding that aphids produce a dispersal morph in the presence of a predator emphasizes, however, that PID may also involve time-lags. It is likely that similar cases of PID exist in other insect predator–prey systems.

The evolution of PID can be investigated using theory developed to understand the evolution of predator-induced defences. Four necessary conditions for the evolution of PID can be identified. It is conceivable that these conditions are fulfilled in a number of insect predator–prey systems. However, as has been shown in the case of aphids and their predators, the empirical evidence is incomplete. For example, few studies have quantified the extent of spatial variability on the impact of predators on prey populations. As a consequence, it is difficult to calculate the benefits of PID because it is unclear whether individuals emigrating from habitats attacked by predators have a reasonable chance of immigrating into habitats with a lower risk of predation. For aphids, it is now known that PID exists, but more field work is needed to show that it is an adaptive predator-avoidance strategy.

The interdependence of dispersal and predation has so far received only limited attention. This review has shown that predation interacts with the process of dispersal in a number of ways, and that the dispersal rate is likely to be influenced by the various mortality risks imposed by predators. Studies investigating dispersal in relation to predation can be expected to reveal fascinating interactions in predator–prey systems.

## Summary

Predation influences all phases in the dispersal process: emigration, transit and immigration. In the first part of the chapter, the various effects of predation on the dispersal process are reviewed and illustrated with examples from insect predator–prey systems. Predation is shown to be an evolutionary force that may: (i) select for dispersal, (ii) select against dispersal and (iii) select for conditional dispersal strategies, i.e. dispersal conditional on predator attack or an increase in the risk of predation. More empirical and theoretical work is needed to understand these conflicting consequences of predation for the evolution of dispersal.

The second part of the chapter investigates dispersal conditional on the presence of a predator. This part is motivated by the recent finding that aphids respond to the presence of predatory ladybirds by increasing the proportion of winged dispersal morphs among their offspring. Four necessary conditions for the evolution of predator-induced dispersal (PID) are derived by comparing it to examples of predator-induced defences in animals and plants; the conditions are: (i) predator impact has to be spatially variable, (ii) a reliable cue has to be available that signals the proximity of predator attack, (iii) PID has to incur costs, and (iv) PID has to be beneficial for the prey. The available evidence from field studies indicates that of these four conditions, only the cost condition is likely to be fulfilled in all insect predator–prey systems. Future work should aim to quantify the variability in the impact of predators on the population growth rates of their prey.

## Acknowledgements

Comments by Christian Braendle, Tadeusz Kawecki, Don Reynolds, John Sloggett and two anonymous referees greatly improved this manuscript. This study was supported by grant no. 31-053852.98 of the Swiss Nationalfonds, by the Roche Research Foundation, and by the Novartis Stiftung.

## References

Agrawal, A.A. (1998) Induced responses to herbivory and increased plant performance. *Science* 279, 1201–1202.
Bernstein, C. (1984) Prey and predator emigration responses in the acarine system *Tetranychus urticae – Phytoseiulus persimilis*. *Oecologia* 61, 134–142.

Bernstein, C., Kacelnik, A. and Krebs, J.R. (1991) Individual decisions and the distribution of predators in a patchy environment. II. The influence of travel costs and structure of the environment. *Journal of Animal Ecology* 60, 205–225.

Blaustein, L. (1998) Influence of the predatory backswimmer, *Notonecta maculata*, on invertebrate community structure. *Ecological Entomology* 23, 246–252.

Blaustein, L. (1999) Oviposition site selection in response to risk of predation: evidence from aquatic habitats and consequences for population dynamics and community structure. In: Wasser, S.P. (ed.) *Evolutionary Theory and Processes: Modern Perspectives. Papers in Honour of Eviatar Nevo*. Kluwer Academic Publishers, Dordrecht, pp. 441–456.

Bryan, K.M. and Wratten, S.D. (1984) The responses of polyphagous predators to prey spatial heterogeneity: aggregation by carabid and staphylinid beetles to their cereal aphid prey. *Ecological Entomology* 9, 251–259.

Campbell, C.A.M. (1978) Regulation of the Damson-hop aphid (*Phorodon humili* Schrank) on hops (*Humulus lupulus* L.) by predators. *Journal of Horticultural Science* 53, 235–242.

Cappuccino, N. (1988) Spatial patterns of goldenrod aphids and the response of enemies to patch density. *Oecologia* 76, 607–610.

Cardwell, C., Hassall, M. and White, P. (1994) Effects of headland management on carabid beetle communities in Breckland cereal fields. *Pedobiologia* 38, 50–62.

Carroll, D.P. and Hoyt, S.C. (1984) Natural enemies and their effects on apple aphid, *Aphis pomi* DeGeer (Homoptera: Aphididae), colonies on young apple trees in central Washington. *Environmental Entomology* 13, 469–481.

Chambers, R.J. (1988) Syrphidae. In: Minks, A.K. and Harrewijn, P. (eds) *Aphids, Their Biology, Natural Enemies and Control*, Vol. B. Elsevier, Amsterdam, pp. 259–270.

Chambers, R.J. and Sunderland, K.D. (1982) The abundance and effectiveness of natural enemies of cereal aphids on two farms in southern England. In: Cavalloro, R. (ed.) *Aphid Antagonists*. A.A. Balkema, Rotterdam, pp. 83–87.

Chambers, R.J., Sunderland, K.D., Wyatt, I.J. and Vickerman, G.P. (1983) The effects of predator exclusion and caging on cereal aphids in winter wheat. *Journal of Applied Ecology* 20, 209–224.

Chen, K. and Hopper, K.R. (1997) *Diuraphis noxia* (Homoptera: Aphididae) population dynamics and impact of natural enemies in the Montpellier region of southern France. *Environmental Entomology* 26, 866–875.

Chesson, J. (1984) Effect of notonectids (Hemiptera: Notonectidae) on mosquitos (Diptera: Culicidae): predation or selective oviposition? *Environmental Entomology* 13, 531–538.

Comins, H.N., Hamilton, W.D. and May, R.M. (1980) Evolutionary stable dispersal strategies. *Journal of Theoretical Biology* 82, 205–230.

Costello, M.J. and Daane, K.M. (1998) Influence of ground cover on spider populations in a table grape vineyard. *Ecological Entomology* 23, 33–40.

Dennis, P. and Fry, G.L.A. (1992) Field margins: can they enhance natural enemy population densities and general arthropod diversity on farmland? *Agriculture, Ecosystems and Environment* 40, 95–115.

Dennis, P. and Wratten, S.D. (1991) Field manipulation of populations of individual staphylinid species in cereals and their impact on aphid populations. *Ecological Entomology* 16, 17–24.

Denno, R.F. (1994) The evolution of dispersal polymorphisms in insects: the influence of habitats, host plants and mates. *Researches on Population Ecology* 36, 127–135.

Denno, R.F., Roderick, G.K., Peterson, M.A., Huberty, A.F., Döbel, H.G., Eubanks, M.D., Losey, J.E. and Langellotto, G.A. (1996) Habitat persistence underlies intraspecific variation in the dispersal strategies of planthoppers. *Ecological Monographs* 66, 389–408.

Dieckmann, U., O'Hara, B. and Weisser, W.W. (1999) The evolutionary ecology of dispersal. *Trends in Ecology and Evolution* 14, 88–90.

Dingle, H. (1996) *Migration*. Oxford University Press, Oxford.

Dixon, A.F.G. (1972) Fecundity of brachypterous and macropterous alatae in *Drepanosiphum dixoni* (Callaphididae, Aphididae). *Entomologia Experimentalis et Applicata* 15, 335–340.

Dixon, A.F.G. (1998) *Aphid Ecology*. Chapman & Hall, London.

Dixon, A.F.G. and Agarwala, B.K. (1999) Ladybird-induced life-history changes in aphids. *Proceedings of the Royal Society London B* 266, 1549–1553.

Dixon, A.F.G. and Wratten, S.D. (1971) Laboratory studies on aggregation, size and fecundity in the black bean aphid, *Aphis fabae* Scop. *Bulletin of Entomological Research* 61, 97–111.

Evans, E.W. and Youssef, N.N. (1992) Numerical responses of aphid predators to varying prey density among Utha alfalfa fields. *Journal of the Kansas Entomological Society* 65, 30–38.

Farrell, J.A. and Stufkens, M.W. (1988) Abundance of the rose-grain aphid, *Metopolophium dirhodum*, on barley in Canterbury, New Zealand, 1984–87. *New Zealand Journal of Zoology* 15, 499–505.

Farrell, J.A. and Stufkens, M.W. (1990) The impact of *Aphidius rhopalosiphi* (Hymenoptera: Aphidiidae) on populations of the rose grain aphid (*Metopolophium dirhodum*) (Hemiptera: Aphididae) on cereals in Canterbury, New Zealand. *Bulletin of Entomological Research* 80, 377–383.

Ferriere, R., Belthoff, J.R., Olivieri, I. and Krachow, S. (2000) Evolving dispersal: where to go next? *Trends in Ecology and Evolution* 15, 5–7.

Frazer, B.D., Gilbert, N., Ivers, P.M. and Raworth, D.A. (1981a) Predator reproduction and the overall predator–prey relationship. *Canadian Entomologist* 113, 1015–1024.

Frazer, B.D., Gilbert, N., Ives, P.M. and Raworth, D.Λ. (1981b) Predation of aphids by coccinellid larvae. *Canadian Entomologist* 113, 1043–1046.

French, B.W. and Elliott, N.C. (1999) Temporal and spatial distribution of ground beetle (Coleoptera: Carabidae) assemblages in grasslands and adjacent wheat fields. *Pedobiologia* 43, 73–84.

Hamilton, W.D. and May, R.M. (1977) Dispersal in stable habitats. *Nature* 269, 578–581.

Hanski, I. and Simberloff, D. (1997) The metapopulation approach, its history, conceptual domain, and application to conservation. In: Hanski, I.A. and Gilpin, M.E. (eds) *Metapopulation Biology*. Academic Press, London, pp. 5–26.

Harvell, C.D. (1984) Predator-induced defense in a marine Bryozoan. *Science* 224, 1357–1359.

Harvell, C.D. (1990) The ecology and evolution of inducible defenses. *Quarterly Review of Biology* 65, 340.

Heie, O.E. (1967) Studies on fossil aphids (Homoptera: Aphidoidea). *Spolia Zoologica Musei Hauniensis* 26, 253–256.

Heie, O.E. (1987) Palaeontology and phylogeny. In: Minks, A.K. and Harrewijn, P. (eds) *Aphids, Their Biology, Natural Enemies and Control*, Vol. A. Elsevier, Amsterdam, pp. 367–391.

Hodek, I. and Honek, A. (1996) *Ecology of Coccinellidae*. Kluwer Academic Publishers, Dordrecht.

Honek, A. (1985) Habitat preferences of aphidophagous coccinellids (Coleoptera). *Entomophaga* 30, 253–264.

Houston, A. and McNamara, J. (1986) The influence of mortality on the behaviour that maximizes reproductive success in a patchy environment. *Oikos* 47, 267–274.

Howard, W.E. (1960) Innate and environmental dispersal of individual vertebrates. *American Midlands Naturalist* 63, 152–161.

Hughes, R.D. (1963) Population dynamics of the cabbage aphid, *Brevicoryne brassicae* (L.). *Journal of Animal Ecology* 32, 393–424.

Ims, R.A. and Yoccoz, N.G. (1997) Studying transfer processes in metapopulations: emigration, migration, and colonisation. In: Hanski, I.A. and Gilpin, M.E. (eds) *Metapopulation Biology*. Academic Press, London, pp. 247–292.

Jervis, M.A. and Kidd, N.A.C. (eds) (1996) *Insect Natural Enemies*. Chapman & Hall, London.

Johnson, C.G. (1969) *Migration and Dispersal of Insects by Flight*. Methuen, London.

Johnson, M.L. and Gaines, M.S. (1990) Evolution of dispersal: theoretical models and empirical tests using birds and mammals. *Annual Review of Ecology and Systematics* 21, 449–480.

Karban, R. and Baldwin, I.T. (1997) *Induced Responses to Herbivory*. The University of Chicago Press, Chicago, Illinois.

Kareiva, P. (1990) The spatial dimension in pest–enemy interactions. In: Mackauer, M., Ehler, L.E. and Roland, J. (eds) *Critical Issues in Biological Control*. VHC, New York, pp. 213–227.

Kareiva, P.M. and Odell, G. (1987) Swarms of predators exhibit 'prey taxis' if individual predators use area-restricted search. *American Naturalist* 130, 233–270.

Kawada, K. (1987) Polymorphism and morph determination. In: Minks, A.K. and Harrewijn, P. (eds) *Aphids, Their Biology, Natural Enemies and Control*, Vol. A. Elsevier, Amsterdam, pp. 299–314.

Kfir, R. and Kirsten, F. (1991) Seasonal abundance of *Cinara cronartii* (Homoptera: Aphididae) and the effect of an introduced parasite, *Pauesia* sp. (Hymenoptera: Aphidiidae). *Journal of Economic Entomology* 84, 76–82.

Krueger, D.A. and Dodson, S.I. (1981) Embryological induction and predation ecology in *Daphnia pulex*. *Limnology and Oceanography* 26, 219–223.

Lively, C.M. (1986) Canalization versus developmental conversion in a spatially variable environment. *American Naturalist* 128, 561–572.

MacKay, P.A. and Wellington, W.G. (1975) A comparison of the reproductive patterns of apterous and alate virginoparous *Acyrthosiphon pisum* (Homoptera: Aphididae). *Canadian Entomologist* 107, 1161–1166.

Markkula, M., Tiittanen, K., Hämäläinen, M. and Forsberg, A. (1979) The aphid midge *Aphidoletes aphidimyza* (Diptera, Cecidomyiidae) and its use in biological control of aphids. *Annales Entomologici Fennici* 45, 89–98.

Maynard Smith, J. (1972) *Evolution and the Theory of Games*. Cambridge University Press, New York.

de Meester, L., Dawidowicz, P., van Gool, E. and Loose, C. (1998) Ecology and evolution of predator-induced behavior of zooplankton: depth selection behavior and diel vertical migration. In: Tollrian, R. and Harvell, C.D. (eds) *The Ecology and Evolution of Inducible Defenses*, Princeton University Press, Princeton, New Jersey, pp. 160–176.

Nijveldt, W. (1988) Cecidomyiidae. In: Minks, A.K. and Harrewijn, P. (eds) *Aphids, Their Biology, Natural Enemies and Control*, Vol. B. Elsevier, Amsterdam, pp. 271–277.

Ozaki, K. (1995) Intergall migration in aphids – a model and a test of ESS dispersal rate. *Evolutionary Ecology* 9, 542–549.

Pacala, S., Hassell, M.P. and May, R.M. (1990) Host-parasitoid associations in patchy environments. *Nature* 344, 150–153.

Roderick, G.K. and Caldwell, R.L. (1992) An entomological perspective on animal dispersal. In: Stenseth, N.C. and Lidicker, W.Z. (eds) *Animal Dispersal. Small Mammals as a Model*. Chapman & Hall, London, pp. 274–290.

Roff, D.A. (1986) The evolution of wing dimorphism in insects. *Evolution* 40, 1009–1020.

Roff, D.A. (1994) Habitat persistence and the evolution of wing dimorphism in insects. *American Naturalist* 144, 772–798.

Roitberg, B.D., Myers, J.H. and Frazer, B.D. (1979) The influence of predators on the movement of apterous pea aphids between plants. *Journal of Animal Ecology* 48, 111–122.

Ronce, O., Clobert, J. and Massot, M. (1998) Natal dispersal and senescence. *Proceedings of the National Academy of Sciences USA* 95, 600–605.

Rosenzweig, M.L. (1991) Habitat selection and population interactions: the search for mechanisms. *American Naturalist* 137 (Supplement), S5–S28.

Schultz, J.C. (1988) Plant responses induced by herbivores. *Trends in Ecology and Evolution* 3, 45–49.

Shukla, R.P. and Pathak, K.A. (1987) Spatial distribution of corn-leaf aphid, *Rhopalosiphum maidis* (Fitch.) and its predator *Coccinella septumpunctata* Linn. *Indian Journal of Agricultural Sciences* 57, 487–489.

Sih, A. (1997) To hide or not to hide? Refuge use in a fluctuating environment. *Trends in Ecology and Evolution* 12, 375–376.

Sih, A. and Wooster, D.E. (1994) Prey behaviour, prey dispersal, and predator impact on stream prey. *Ecology* 75, 1199–1207.

Southwood, T.R.E. (1962) Migration of terrestrial arthropods in relation to habitat. *Biological Reviews* 27, 171–214.

Southwood, T.R.E. (1977) Habitat, the templet for ecological strategies? *Journal of Animal Ecology* 46, 337–365.

Southwood, T.R.E. (1981) Ecological aspects of insect migration. In: Aidley, D.J. (ed.) *Animal Migration*. Cambridge University Press, Cambridge, pp. 196–208.

Stav, G., Blaustein, L. and Margalith, J. (1999) Experimental evidence for predation risk sensitive oviposition by a mosquito, *Culiseta longiareolata*. *Ecological Entomology* 24, 202–207.

Stenseth, N.C. and Lidicker, W.Z. (1992) To disperse or not to disperse: who does it and why? In: Stenseth, N.C. and Lidicker, W.Z. (eds) *Animal Dispersal. Small Mammals as a Model*. Chapman & Hall, London, pp. 5–20.

Stewart, H.C. and Walde, S.J. (1997) The dynamics of *Aphis pomi* De Geer (Homoptera: Aphididae) and its predator, *Aphidoletes aphidimyza* (Rondani) (Diptera: Cecidomyiidae), on apple in Nova Scotia. *Canadian Entomologist* 129, 627–636.
Tollrian, R. and Harvell, C.D. (eds) (1998a) *The Ecology and Evolution of Inducible Defences*. Princeton University Press, Princeton, New Jersey.
Tollrian, R. and Harvell, C.D. (1998b) The evolution of inducible defenses: current ideas. In: Tollrian, R. and Harvell, C.D. (eds) *The Ecology and Evolution of Inducible Defences*. Princeton University Press, Princeton, New Jersey, pp. 306–321.
Turchin, P. and Kareiva, P. (1989) Aggregation in *Aphis varians*: an effective strategy for reducing predation risks. *Ecology* 70, 1008–1016.
Völkl, W. (1994) Searching at different spatial scales: the foraging behaviour of *Aphidius rosae* in rose bushes. *Oecologia* 100, 177–183.
Völkl, W. and Kraus, W. (1996) Foraging behaviour and resource distribution of the aphid parasitoid *Pauesia unilachni*: adaptation to host distribution and mortality risks. *Entomologia Experimentalis et Applicata* 79, 101–109.
Way, M.J. and Banks, C.J. (1968) Population studies on the active stages of the black bean aphid, *Aphis fabae* Scop., on its winter host *Evonymus europeus* L. *Annals of Applied Biology* 62, 177–197.
Weisser, W.W. and Stadler, B. (1994) Phenotypic plasticity and fitness in aphids. *European Journal of Entomology* 91, 71–78.
Weisser, W.W., Völkl, W. and Hassell, M.P. (1997) The importance of adverse weather conditions for behaviour and population ecology of an aphid parasitoid. *Journal of Animal Ecology* 66, 386–400.
Weisser, W.W., Braendle, C. and Minoretti, N. (1999) Predator-induced morphological shift in the pea aphid. *Proceedings of the Royal Society of London B* 266, 1175–1182.
Zera, A.J. and Denno, R.F. (1997) Physiology and ecology of dispersal polymorphism in insects. *Annual Review of Entomology* 42, 207–231.

# Evolution of Mass Transit Systems in Ants: a Tale of Two Societies

**Nigel R. Franks***

*Centre for Mathematical Biology, Department of Biology and Biochemistry, University of Bath, Bath BA2 7AY, UK*

## Introduction

This chapter has two goals. First, it aims to show, through a detailed comparison of two species, how social insects provide excellent material for studies of insect movement at many different levels. The levels of analysis are the movements of individuals, small groups (in the form of teams), swarms of hundreds of thousands and the mass movements of whole colonies. The second aspiration is to show in general that a multi-layered approach, combining mechanistic and strategic analyses, facilitates fuller understanding. This is because insights from each level and each approach are vital to the interpretation of adaptations at every other level.

The success of insects is often attributed to their small size and their great mobility – associated with flight (May, 1978, 1990; Southwood, 1978). Nevertheless, some of the most successful insects are those in which the vast majority of individuals walk – namely, the termites and the ants (Wilson 1990).

Here, I will discuss certain aspects of convergent evolution in two of the most remarkable groups of ants. They are the army ants of the Old World, the Dorylinae, and the army ants of the New World, the Ecitonninae. The swarm-raiding species of army ants may be the most polyphagous predators on earth. These army ants are also remarkable for their huge colony sizes: *Eciton burchelli* (Westwood) and *Dorylus* colonies may have 500,000 and 20

---

* Present address: School of Biological Sciences, University of Bristol, Woodland Road, Bristol BS8 1UG, UK.

million workers, respectively (Franks, 1985; Gotwald, 1995). They also have a relatively high degree of mobility; colonies emigrate frequently to new foraging sites where they substantially deplete populations of their insect prey. These army ants also have some of the most polymorphic workers of any ant species (Franks, 1985; Gotwald, 1995). This worker polymorphism is associated with a very sophisticated division of labour (Franks, 1985).

The division of labour within colonies of ants (and termites), and the associated efficiency with which they gather and utilize resources, is probably the single major reason for their ecological importance (Hölldobler and Wilson, 1990; Wilson, 1990) and seems to more than compensate for the constraints of a largely pedestrian existence. Of course, most ants and termites have winged males and females, so they do have a long-distance dispersal stage. They get, therefore, the best of both worlds: firstly, by having sexual forms that disperse well aerially; and secondly, because their workers can maintain communication with one another, and therefore cooperate beyond the nest, because they are obligate pedestrians.

Army ant colonies with huge populations of workers, swarm raids and high levels of worker polymorphism are believed to have evolved independently (and convergently) in the Old and New World tropics from a Gondwanaland common ancestor that already had the other army ant traits of wingless queens and frequent emigrations (Bolton, 1990; Gotwald, 1995; Brady, 1998; Franks et al., 1999). Army ants have wingless queens hence their populations have much lower dispersal than ant species in which both sexes fly. Army ant colonies cannot re-invade areas separated by water or inhospitable terrain. Hence, these army ants in Africa and in the Americas have had separate evolutionary histories for tens of millions of years.

In this chapter, I will discuss how self-organization can account, in part, for the extreme convergent evolution of the massive swarm raids of African and South American army ants. Such swarm raids enable these ants to capture tens of thousands or hundreds of thousands of insect prey per day. I will then discuss how army ants minimize the massive road-haulage problem associated with the retrieval of these myriad prey over considerable distances. I will review recent quantitative studies of super-efficient prey retrieval teams in New and Old World army ants. This will lead to a consideration of why, for example, *E. burchelli* has distinct physical castes of worker, whereas *Dorylus wilverthi* Emery has a continuous caste distribution and why *Eciton* colonies have a higher tempo and are much more mobile than *Dorylus* colonies.

There are remarkable similarities and also remarkable differences between Old World and New World army ants. As will become apparent, some of these differences can be attributed to differences in the population dynamics of their prey. These differences may have selected for differences in colony mobility and in turn for differences in caste ratios and worker tempo.

This chapter will address the closely coupled issues of how and why the movement patterns of social insects are organized in particular ways and it will provide insights into the evolution of two of the most remarkable of all social insect species.

## The Blind Leading the Blind and the Self-organization of Army Ant Swarm Raids

What is self-organization? The following definition will serve our purposes here. Self-organization is a process that creates a pattern at the global level (e.g. the colony level) through multiple interactions among the components (e.g. the workers). The components interact through local, often simple, rules that do not explicitly code for the global pattern (Camazine and Deneubourg, 1994; Bonabeau *et al.*, 1997; Sendova-Franks and Franks, 1999; Camazine *et al.*, 2001).

The 'global' patterns, whose generation we will examine, are the swarm raids of New World and Old World army ants (see Figs 13.1 and 13.2). Such swarm raids are massive compared to the size of the individuals that create them. An *E. burchelli* raid can be 20 m wide and 200 m long and employ 200,000 individuals (Franks, 1989). Given their much larger colony sizes,

**Fig. 13.1.** A swarm raid of *Dorylus*. A stylized drawing by Katherine Brown-Wing which first appeared in Wilson (1990). This illustration has been reproduced with the kind permission of E.O. Wilson.

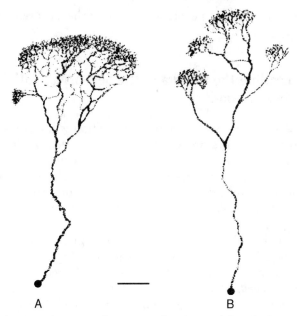

**Fig. 13.2.** The typical pattern of swarm raiding by (a) *E. burchelli* and (b) *E. rapax*. The scale bar represents 5 m. (a – redrawn from Rettenmeyer, 1963; b – redrawn from Burton and Franks, 1985.)

*Dorylus* raids are likely to be much larger still. Furthermore, an overview of the swarm raid pattern is not available to individual workers in either *Eciton* or *Dorylus*: the former have rudimentary eyesight and the latter are completely blind (Schneirla, 1971; Franks, 1989; Gotwald, 1995).

A computer simulation model of the self-organization of an army ant raid (see Fig. 13.3) has been based on the following small and simple set of rules (Deneubourg *et al.*, 1989).

**1.** Leading: each and every ant lays a followable pheromone trail wherever it goes (unless it is on a trail fully saturated with pheromone).
**2.** Randomness: if an ant is in virgin terrain it randomly goes left or right (at every bifurcation point in the computer simulation lattice).
**3.** Following: if an ant is in terrain already traversed by a nest mate it is most likely to follow the pheromone trail laid by that nest mate (e.g. it has a higher probability of turning left than right if the previous ant turned left rather than right, or vice versa, and marked that path). Because army ants follow one another's trails and reinforce them, trails can get stronger and stronger.
**4.** Speeding: up to a limit, the more trail pheromone present the faster each individual will run.
**5.** Crowding: an ant will not, however, enter an area that is already overcrowded by its nest mates.

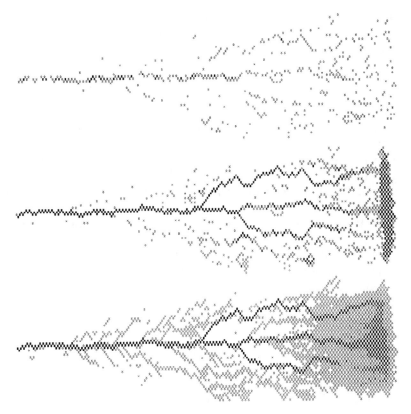

**Fig. 13.3.** A cellular automata model of the generation of a swarm raid of *Eciton burchelli*. The three illustrations are overlays of the same model raid. The top figure shows the density of returning ants; the middle figure shows the density of all ants, and the lower figure shows the density of pheromone trails. The darker the grey level the greater the density. See text and Deneubourg *et al.* (1989) and Franks *et al.* (1991).

**6.** Returning: ants only return home when they have encountered prey items. Returning ants obey the same rules for following the pheromone trail as outgoing ants, but they lay more trail pheromone than outgoing ants.
**7.** Flow: a constant number of ants leave the nest per unit time.

These seven golden rules are sufficient for the blind to lead the blind in the creation of a swarm raid of dazzling complexity and sophistication. It is imperative to note, however, that these seven qualitative rules must be executed in good quantitative agreement with the characteristics of the real army ants. The rate at which individual workers move matters: rates of deposition and evaporation of trail pheromones matter; relative rates of trail laying by outward bound and returning ants matter (Franks *et al.*, 1991). All these quantitative variables have been established by studying experimentally the movements of *E. burchelli* workers and they have been incorporated into the model (Fig. 13.3). That is, both the qualitative and quantitative assumptions of the model have been verified. Certain predictions of this modelling have also been tested (Franks et al., 1991).

The single most intriguing prediction of this modelling is that swarm raids have an active architecture. These are event-driven systems. The precise pattern of a raid depends on the distribution of prey encountered. Indeed, when the model is run with no prey and hence no returning ants (rule 6) the anastamosing series of columns behind the raid front is absent and a raid consists only of a broad swarm front and a principle trail. Furthermore, if prey are in large clumps that are few and far between, the model predicts that a swarm raid may break up into divergent sub-swarms. This occurs because strong return traffic flows from two (or more) directions and the outgoing ants part company. Such a raid pattern consisting of sub-swarms is seen in *Eciton rapax* Smith (Burton and Franks, 1985), a close relative of *E. burchelli*, which is a specialist predator of other social insect nests and thus encounters large clumps of prey that are few and far between (Fig. 13.2b). This prediction has been tested by presenting the normally cohesive swarm raids of *E. burchelli* with large clumps of prey that are few and far between. The result was that *E. burchelli* adopted a swarm raid pattern similar to that of *E. rapax* (Franks et al., 1991).

Some authors have claimed, or seem to wish to imply, that self-organization in biology represents order for free (see, e.g. Goodwin, 1994, 1998; Weber and Depew, 1996; Stuart, 1998). By this they seem to suggest that self-organizing biological patterns occur so automatically that natural selection has had little influence over the evolution of their structure. I believe that such a viewpoint is extremely misleading. In the case of self-organizing swarm raids of army ants, it is almost certain that natural selection has, for example, rather precisely tuned both the behavioural rules of thumb of the ants and the chemical properties (e.g. volatility) of their pheromones.

Furthermore, I am certain that self-organization theory does provide a major insight into mainstream evolutionary biology (see also Gerhart and Kirschner, 1997). The point is as subtle as it is important. Self-organization theory does not suggest that natural selection has had *no role* in the creation of certain patterns in biology – rather it suggests that natural selection has had *rather less to do* than one might expect given the complexity of the global structure. As illustrated by the swarm raids of Old World and New World army ants, natural selection may have had to select for a surprisingly small and simple set of rules to generate swarm-raiding patterns. Thus self-organization theory may help to explain why we observe such a high level of convergent evolution in certain biological structures.

Another fundamental misunderstanding of self-organization is tied up with the notion of *simple rules of behaviour*. These are equivalent to the rules of thumb typically elucidated in classical behavioural ecology (Stephens and

Krebs, 1986). Simple rules of thumb do not imply that the individuals that use them are either simple or stupid. Imagine the simple rule of thumb for a circus performer walking a tightrope – if you wobble one way, lean the other. The rule is simple, but the complexity and sophistication of the sensory apparatus, muscular coordination and the learning process needed to carry out such a simple rule of thumb is, for want of a more appropriate term, staggering.

The rules of thumb employed by swarm-raiding *E. burchelli* workers have been partly elucidated (Franks *et al.*, 1991). Almost nothing is known quantitatively about the rules of thumb employed by raiding *Dorylus* workers or the properties of their trail pheromones. This will be an important area for study in the future.

What is beyond doubt is that *E. burchelli* workers have rudimentary eyesight (their eyes have only a single facet) and that *Dorylus* workers are totally blind (Schneirla, 1971; Franks, 1989; Gotwald, 1995). Possibly their common ancestor was blind or nearly blind – perhaps because it lived almost entirely underground – or possibly both *Eciton* and *Dorylus* have newly evolved partial or total blindness. Such poor eyesight may be beneficial because it would cause workers to focus (sic) on local information (in the form of pheromone trails) and local interactions. In other words, it may actively facilitate the blind *leading* the blind.

## Super-efficient Teams

In both Old and New World army ants the outcome of swarm raiding is the capture and fragmentation of vast numbers of insect prey. Such prey fragments must then be carried back to the nest where they can be consumed. This represents a massive road-haulage problem. In the case of *E. burchelli* more than 30,000 prey fragments are retrieved per day over distances of up to 200 m (Franks, 1989). Given that *Dorylus* colonies may contain 20 million workers compared to the 500,000 workers in large *E. burchelli* colonies, prey retrieval by *Dorylus* will be an even larger undertaking.

Intriguingly, the size distribution of prey fragments in *D. wilverthi* and *E. burchelli* is almost identical (Fig. 13.4), and the overall size range of their workers is almost identical, yet 98% of *D. wilverthi* workers are within the size range of the smallest 25% of *E. burchelli* workers (Fig. 13.5). How do such small *D. wilverthi* workers cope with moving equally large prey items to those moved by the often much larger *E. burchelli* workers? The answer is that they form many more super-efficient teams (see later). Indeed, their raids teem with teams.

Franks (1986) demonstrated for the first time that *E. burchelli* workers formed super-efficient prey retrieval teams. That is, groups of workers cooperate to retrieve large items in such a way that their combined performance is more than the sum of its parts. For example, a team of three *E. burchelli*

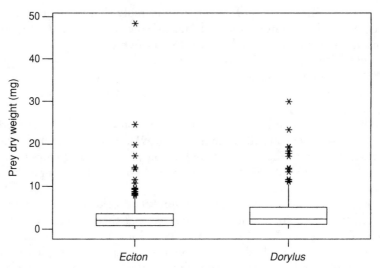

**Fig. 13.4.** Frequency/dry weight (mg) boxplots for a large sample of prey items of *E. burchelli* (n = 312) and *D. wilverthi* (n = 203) taken from principal raid columns. The box encompasses the interquartile range; the line across the box is the median; whiskers are drawn to the nearest value within 1.5 times the interquartile range; all remaining outlying points are marked with asterisks. Data from Franks *et al.* (1999).

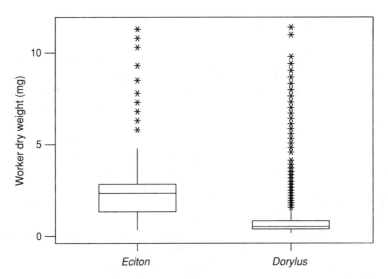

**Fig. 13.5.** Frequency/dry weight (mg) boxplots for a large sample of *D. wilverthi* workers taken from a raid column (n = 1754, *Dorylus* data originally published in Franks *et al.*, 1999) and for a sample of *E. burchelli* workers (n = 572, *Eciton* data originally published in Franks, 1985). The *Dorylus* data presented in this figure are from a different sample from those presented in Fig. 13.6. The meaning of the symbols in the boxplots is explained in the legend to Fig. 13.4.

workers can carry an item that is so large that however it was fragmented into three pieces the three ants would not be able individually to carry away all the pieces. Even more intriguingly, the allometric relationship between prey item dry weight and ant dry weight (for single porters) has a slope greater than 1 and plotted on the same graph the data for prey item dry weight versus the total ant (team) dry weight is a smooth seamless extrapolation of the same quantitative trend. Larger individuals can carry disproportionately heavy items and heavier teams can carry disproportionately heavy items and such teams have the performance expected of a single super-ant of a similar weight. The most likely reason that teams can be super-efficient is that when two or more individuals carry a bulky load between them rotational forces disappear (Franks, 1986; Franks et al., 1999).

Recently, Franks et al. (1999) demonstrated the existence of super-efficient teams in *D. wilverthi*. Here, as in *E. burchelli*, the performance curve for teams is a smooth, seamless extrapolation of the performance of individual porters of prey. Within either species, single porters and teams of porters run at the same speed, probably to minimize traffic congestion within a colony's raid system. This may give a clue as to how teams are organized. In either species, if a large prey item is taken away from a team and placed back in a foraging column, ants continue to join a team to move it until the item is being carried along at the standard retrieval speed. At such a speed a large item no longer represents a blockage to smooth traffic flow and hence the stimulus to join a team disappears. Such replacement teams weigh the same as the team they replaced.

*Dorylus* carry more per unit ant, or team, dry weight than does *E. burchelli* (Fig. 13.6). This can be explained because *Dorylus* move more slowly. Indeed, when weight and speed are taken into account *Eciton* are operating at a higher tempo (*sensu* Oster and Wilson, 1978: see Franks et al., 1999). *E. burchelli* workers are probably able to operate at a higher tempo than *D. wilverthi* workers because on average they are bigger. For example, *E. burchelli* has a specialist porter caste – the submajors – the largest ants in the colony capable of prey transport. They are only 3% of the colony's workforce but are 26% of its porters of prey (Franks, 1985). Employment of such a large ant in this way should help to minimize transport costs because transport costs, defined as energy consumed to move a unit weight a unit distance, decline with increasing body size (Jensen and Holm-Jensen, 1980; Schmidt-Nielsen, 1984; Franks, 1986). When submajors are not available to carry a large prey item in an *E. burchelli* raid the workers form super-efficient teams.

Workers of a similar size to *E. burchelli* submajors do occur in *D. wilverthi* but relative to other workers in their colony they are exceedingly rare (Fig. 13.5). *D. wilverthi* colonies substitute for the relative rarity of submajors in the raiding populations by forming many more teams than *E. burchelli*. In *E. burchelli*, 5% of all prey items are retrieved by teams; in *D. wilverthi* the corresponding figure is 39%. The percentages of the total

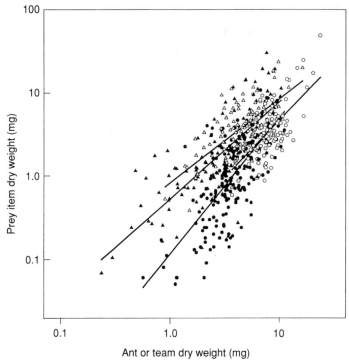

**Fig. 13.6.** The relationships between prey item dry weight and the dry weight of the ant or ants that were carrying that item for single porters (filled triangles) and teams of porters (open triangles) in *D. wilverthi* and for single porters (solid circles) and teams of porters (open circles) in *E. burchelli*. Redrawn from Franks *et al.* (1999). In the 'stack' of four regression lines in this figure the uppermost line is for *Dorylus* teams, then in descending order are lines for *Dorylus* single porters, *Eciton* teams and *Eciton* single porters. *The relationship for both sets of Dorylus* data combined can be described by $P = 0.57095\ W^{1.18}$ where $P$ is prey item dry weight in mg and $W$ is porter or total team dry weight in mg ($r^2 = 0.59$, $n = 223$, $P < 0.001$, SD for slope = 0.07). The relationship for both sets of *Eciton* data combined can be described by $P = 0.1212\ W^{1.61}$ where $P$ is prey item dry weight in mg and $W$ is porter or total team dry weight in mg ($r^2 = 0.62$, $n = 312$, $P < 0.001$, SD for slope = 0.07). (*Eciton* data originally published in Franks, 1986.) The gradients of the regression lines for single porters and teams of porters in *Eciton* are not significantly different from one another ($d = 1.20$, $P > 0.1$). The gradients of the regression lines for single porters and teams of porters in *Dorylus* are not significantly different from one another ($d = 1.09$, $P > 0.1$). The gradient of the regression line (not depicted) for all of the *Dorylus* data is significantly less than the gradient of the regression line (not depicted) for all of the *Eciton* data ($d = 4.25$, $P < 0.001$). Nevertheless, the gradient of the regression line (not depicted) for all of the *Dorylus* data is significantly greater than 1 ($t = 2.52$, $P < 0.02$; see Franks *et al.*, 1999).

weight of all prey items retrieved by teams in *E. burchelli* and *D. wilverthi* is 13 and 64%, respectively (Franks *et al.*, 1999).

## Mass Movements of Colonies: Emigration Frequency, Duration and Spatial Patterns

Why should *E. burchelli* have evolved to operate at a higher tempo than *D. wilverthi*? The answer is probably associated, at least in part, with the different prey dynamics of the two species. A substantial proportion of the prey of *E. burchelli* is very slow to recover from a raid (Franks, 1982; and see later). This in turn has probably selected for greater mobility on the part of *E. burchelli* colonies. The higher emigration frequency in *E. burchelli* means that they have less time in which to raid, because colonies do not emigrate and raid at the same time. Hence, raids need to proceed at a higher tempo.

*E. burchelli* colonies have 35-day activity cycles. For 20 days a colony nests in the same bivouac site from which it sends out swarm raids, on only about 13 days (Fig. 13.7). These statary raids are like the spokes of a wheel centred on the hub of the statary bivouac. However, the direction of raids are similar to the arrangement of leaves in spiral phyllotaxy. Statary raids are separated by an average angle of 123°, a close approximation to the optimal 126.4° to maximally separate 15 'spirally' arranged spokes (Franks and

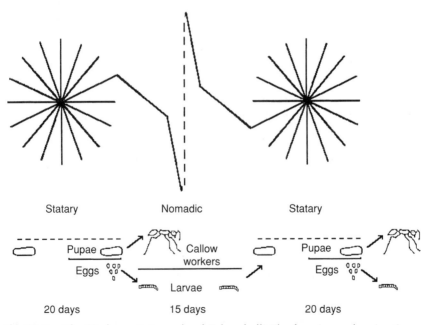

**Fig. 13.7.** The 35-day activity cycle of *E. burchelli*. The foraging and emigration pattern is diagrammatically represented above an outline of the brood cycle. Redrawn from Franks and Fletcher (1983). For further explanation see text.

Fletcher, 1983). The 15 raids (13 statary raids plus the nomadic raids leading into and out of the bivouac site) are thus separated in space and time. Raids neighbour one another only after several days in which the more mobile prey of *E. burchelli* can recover (Franks and Fletcher, 1983).

The reason that a colony can take holidays from raiding on certain statary days is that it does not have brood to feed. The 20-day statary phase coincides with the 20 days of pupal development. Half way through the statary phase, the single queen lays perhaps 100,000 eggs. Embryonic development takes 10 days. In synchrony, the pupae hatch into new workers and the eggs hatch into new larvae. This is associated with the end of the statary phase. The colony with its cohort of hungry larvae becomes nomadic, raiding every day for the next 15 days and emigrating at night down that day's raid path to a new bivouac site. Larval development takes 15 days. These larvae spin pupal cases in synchrony and the colony ends its nomadic phase to become statary once more. Successful colonies grow larger over perhaps 3 years and triggered by the onset of a dry season produce a brood of about 4000 males and a few queens. Such colonies reproduce by binary fission. Sometimes the old queen retains a daughter colony and sometimes both daughter colonies are headed by new queens that have been inseminated by males flying in from other reproductive colonies. Even when colonies are reproductive they still maintain 35-day activity cycles (Schneirla, 1971; Franks, 1989; Gotwald, 1995).

Britton *et al.* (1996) have shown that it is possible to analyse the population dynamics of *E. burchelli* by considering a patch occupancy model – where the patches have the dimensions of the areas occupied by, and raided over, in statary periods. Britton *et al.* (1996) have also shown that given that the predominantly social insect prey of *E. burchelli* are slow to recover from a raid, taking about 6 months to re-grow, it is important for colonies to navigate in their nomadic phase so that subsequent statary periods are pushed far apart and do not raid the same areas (Fig. 13.8). By migrating large distances and navigating in their nomadic phase (Franks and Fletcher, 1983), *E. burchelli* colonies are unlikely to land back on a patch they have personally denuded. Indeed, Britton *et al.* (1996) have shown that a hypothetical ancestral population of army ant colonies that have a short nomadic phase and a short nomadic path, so that they merely step into a neighbouring patch for their next statary phase, and have a good chance of stepping back into their own recently denuded patch, will be out-competed by a new mutant type of colony with longer migration. The corollary of this argument is that if prey populations were quick to recover, colonies that avoided all the attendant costs of long emigrations should be favoured. Such seems to be the case for *Dorylus* colonies. They prey upon the larvae of insects such as Lepidoptera and Heteroptera that are likely to recover much more quickly from raids.

*Dorylus* colonies do not have fixed activity cycles, emigrate at haphazard intervals to closely neighbouring areas and frequently return to the same areas (Fig. 13.9; Gotwald, 1995). Indeed, emigrations are an order of

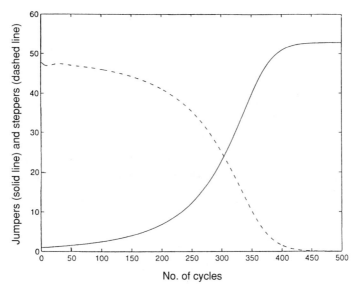

**Fig. 13.8.** The results of modelling to simulate the effect of introducing a single colony of mutant 'jumpers', i.e. army ant colonies that separate successive statary periods by a long distance as do colonies of *E. burchelli*, into a population of wild-type 'steppers', colonies that move one patch diameter, as do colonies of *Dorylus*. The y-axis is the number of colonies surviving in the model. The mutants drive the wild type to extinction within 500 modelling cycles (each cycle is 35 days hence this is less than 50 years) (redrawn from Britton *et al.*, 1996). In this simulation patches of prey are assumed to recover slowly, as they do for *E. burchelli*. If prey were to recover quickly steppers would not be at a disadvantage. See text for further explanation.

magnitude less frequent in certain *Dorylus* than they are in *E. burchelli* (Franks, 1989; Gotwald, 1995). Hence, in *Dorylus* emigrations do not impinge on raiding time as much as they do in *E. burchelli* and indeed *Dorylus* will raid day and night, not just diurnally as is the habit of *E. burchelli*. Hence, *Dorylus* raids can proceed at a lower tempo. Thus *Dorylus* can produce workers that are on average smaller than those of *E. burchelli* and can enjoy all the advantages of having more smaller workers rather than fewer larger ones. These advantages should include the higher productivity of modular growth when the modules (i.e. the workers) are smaller and can, all else being equal, be produced and become productive more quickly (Bourke and Franks, 1995; Franks *et al.*, 1999).

## Summary and Conclusions

The ancestors of New World and Old World army ants almost certainly parted company when Gondwanaland was slowly torn asunder, tens of

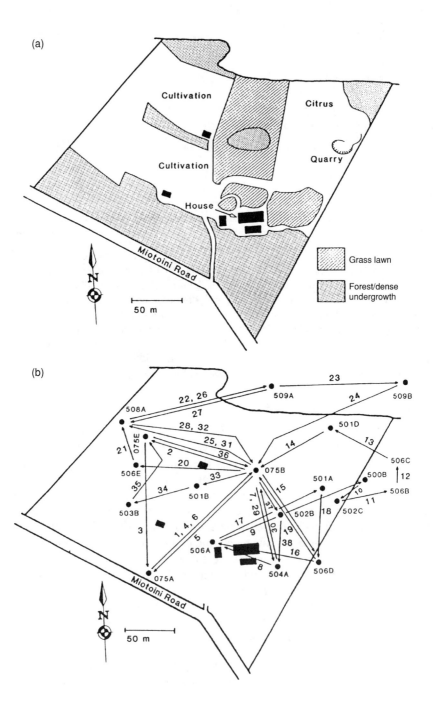

millions of years ago, during the earth-shattering scattering of its great tectonic plates. Army ant colonies, which stage huge swarm raids through rain forests and have vast populations of massively polymorphic workers, have evolved separately in the tropics of the Old and New World. *Eciton* in the New World, with 500,000 ants per colony, and *Dorylus* in the Old World, with 20 million ants per colony, are twin peaks in social evolution. Both types of society are maelstroms of ceaseless activity. They devastate prey populations locally and move on relentlessly. This chapter has discussed recent quantitative analyses, in Central America and sub-Saharan Africa, of the movement patterns of: (i) individual workers, (ii) swarm raids, (iii) super-efficient teams of workers and (iv) entire colonies. The convergent evolution of New World and Old World army ants, in part driven by the economic imperatives of mass transit, reveals the power of natural selection to shape striking similarities in social structure and behaviour.

The analysis of how the blind lead the blind in self-organizing swarm raids helps to explain why there has been such a high level of convergent evolution of huge army ant societies in the New World and Old World. In both cases, natural selection has homed in on a remarkably complex yet flexible pattern of raiding – which is nevertheless based on workers utilizing a surprisingly small and simple set of rules. Through the self-organization of these raids these army ants have become perhaps the most polyphagous of all predators on earth. Such raids also create an infrastructure of pheromone trails along which super-efficient teams of workers march home in triumph burdened with tens of thousands or hundreds of thousands of prey. The raids of *E. burchelli* in the Americas and *D. wilverthi* in Africa are populated with a similar size range of workers and retrieve a similar spectrum of sizes of prey items. Nevertheless, *Eciton* produces a sub-population of larger workers to act as porters of prey whereas *Dorylus* produces more small workers which form many more teams. Old World and New World army ants operate at a different tempo. *E. burchelli* workers are on average larger and quicker probably because of the selection pressures that have arisen because they have less time to raid since their colonies emigrate very frequently. This can in turn be attributed to the slow rate of recovery of the prey of *Eciton*. The prey of *Dorylus* recover more quickly and hence colonies can emigrate less frequently and raids can proceed more slowly.

**Fig. 13.9** (opposite).   The patterns of emigration of a colony of *Dorylus* (*Anomma*) *molestus* (Gerstaecker) observed for 432 consecutive days. Movements were limited to about 5 ha. The colony emigrated 38 times and revisited some nest sites many times. Dots represent nest sites, and emigrations are numbered sequentially. Arrows indicate emigration routes (redrawn from Gotwald, 1995). This colony behaved like a 'stepper', with short emigrations and frequent returns to earlier raiding areas (see Fig. 13.8). This illustration has been reprinted from William H. Gotwald Jr: *Army Ants: the Biology of Social Predation*. Copyright © 1995 by Cornell University. Used by permission of the publisher, Cornell University Press.

This chapter has discussed patterns of convergent and divergent evolution in Old World and New World army ants. What I hope is especially clear is the benefits of a muti-layered analysis, here from issues of individual size and biomechanics up to large-scale multi-population predator–prey dynamics. We have considered the ecology and evolution of the performance of individual workers and patterns of movement of colonies of up to 20 million workers. Social insects provide unique opportunities to investigate both mechanistic and strategic issues in the patterns of movement of both individuals and populations. In this way, they facilitate investigations into the relationship between the evolutionary play and the ecological theatre.

## Acknowledgements

I wish to thank the Smithsonian Tropical Research Institute, Panama, the Royal Society, and the Natural Environment Research Council for supporting my work on New World army ants and the Tropical Biological Association for supporting my work on Old World army ants. William Gotwald and Edward O. Wilson gave their kind permission to reproduce certain illustrations. Especial thanks go to Ana Sendova-Franks for her help with field-work and for commenting on an earlier draft of this chapter.

## References

Bolton, B. (1990) Army ants reassessed: the phylogeny and classification of the doryline section (Hymenoptera, Formicidae). *Journal of Natural History* 24, 1339–1364.
Bonabeau, E., Theraulaz, G., Deneubourg, J.L., Aron, S. and Camazine, S. (1997) Self-organization in social insects. *Trends in Ecology and Evolution* 12, 188–193.
Bourke, A.F.G. and Franks, N.R. (1995) *Social Evolution in Ants*. Monographs in Behavioral Ecology, Princeton University Press, Princeton, New Jersey.
Brady, S.G. (1998) The origin and evolution of army ants (Hymenoptera: Formicidae): a phylogenetic analysis using morphological and molecular data. In: Schwartz, M.P. and Hogendoorn, K. (eds) *Social Insects at the turn of the Millennium, Proceedings of the XIII International Congress of IUSSI*. Flinders University Press, Adelaide, p. 73.
Britton, N.F., Partridge, L.W. and Franks, N.R. (1996) A mathematical model for the population dynamics of army ants. *Bulletin of Mathematical Biology* 58, 471–492.
Burton, J.L. and Franks, N.R. (1985) The foraging ecology of the army ant *Eciton rapax*: an ergonomic enigma? *Ecological Entomology* 10, 131–141.
Camazine, S. and Deneubourg, J.L. (1994) What is self-organization? In: Lenoir, A., Arnold, G. and Lepage, M. (eds) *Les Insectes Sociaux*. Université Paris Nord, Paris, p. 228.

Camazine, S., Deneubourg, J.L., Franks, N.R., Sneyd, J., Theraulaz, G. and Bonabeau, E. (2001) *Self-Organization in Biology*. Princeton University Press, Princeton, New Jersey.

Deneubourg, J.L., Goss, S., Franks, N. and Pasteels, J.M. (1989) The blind leading the blind: modeling chemically mediated army ant raid patterns. *Journal of Insect Behaviour* 2, 719–725.

Franks, N.R. (1982) Ecology and population regulation in the army ants, *Eciton burchelli*. In: Leigh, E.G. Jr, Rand, A.S. and Windsor, D.W. (eds) *The Ecology of a Tropical Forest*. Smithsonian Institute Press, Washington, DC, pp. 389–395.

Franks, N.R. (1985) Reproduction, foraging efficiency and worker polymorphism in army ants. In: Lindauer, M. and Hölldobler, B. (eds) *Experimental Behavioural Ecology. Fortschritte der Zoologie* 31, 91–107. Fischer Verlag, Stuttgard, New York.

Franks, N.R. (1986) Teams in social insects: group retrieval of prey by army ants (*Eciton burchelli*, Hymenoptera, Formicidae). *Behavioral Ecology and Sociobiology* 18, 425–429.

Franks, N.R. (1989) Army ants: a collective intelligence. *American Scientist* 77, 138–145.

Franks, N.R. and Fletcher, C.R, (1983) Spatial patterns in army ant foraging and migration: *Eciton burchelli* on Barro Colorado Island, Panama. *Behavioral Ecology and Sociobiology* 12, 261–270.

Franks, N.R., Gomez, N., Goss, S. and Deneubourg, J.L. (1991) The blind leading the blind in army ant raid patterns: testing a model of self-organization. *Journal of Insect Behavior* 4, 583–607.

Franks, N.R., Sendova-Franks, A.B., Simmons, J. and Mogie, M. (1999) Convergent evolution, superefficient teams and tempo in Old and New World army ants. *Proceedings of the Royal Society of London B* 266, 1697–1701.

Gerhart, J. and Kirschner, M. (1997) *Cells, Embryos and Evolution*. Blackwell Science, Oxford.

Goodwin, B. (1994) *How the Leopard Changed its Spots*. Weidenfeld and Nicholson, London.

Goodwin. B. (1998) All for one – one for all. *New Scientist* 13th June, 32–35.

Gotwald, W.H. (1995) *Army Ants: the Biology of Social Predation*. Cornell University Press, Ithaca, New York.

Hölldobler, B. and Wilson, E.O. (1990) *The Ants*. Belknap Press of Harvard University Press, Cambridge, Massachusetts.

Jensen, T.F. and Holm-Jensen, I. (1980) Energetic cost of running in workers of three ant species, *Formica fusca* L., *Formica rufa* L., and *Camponotus herculeanus* L. (Hymenoptera, Formicidae). *Journal of Comparative Physiology* 37, 151–156.

May, R.M. (1978) The dynamics and diversity of insect faunas. In: Mound, L.A. and Waloff, N. (eds) *Diversity of Insect Faunas*. Blackwell, Oxford, pp. 188–204.

May, R.M. (1990) How many species? *Philosophical Transactions of the Royal Society of London B* 330, 293–304.

Oster, G.F. and Wilson, E.O. (1978) *Caste and Ecology in the Social Insects*. Monographs in Population Biology, Princeton University Press, Princeton, New Jersey.

Rettenmeyer, C.W. (1963) Behavioral studies of army ants. *University of Kansas Science Bulletin* 44, 281–465.

Schmidt-Nielsen, K. (1984) *Scaling: Why is Animal Size so Important?* Cambridge University Press, Cambridge.

Schneirla, T.C. (1971) *Army Ants: a Study in Social Organization*. Freeman, San Francisco.

Sendova-Franks, A.B. and Franks, N.R. (1999) Self-assembly, self-organization and division of labour. *Philosophical Transactions of the Royal Society of London B* 354, 1395–1405.

Southwood, T.R.E. (1978) The components of diversity. In: Mound L.A. and Waloff N. (eds) *Diversity of Insect Faunas*. Blackwell, Oxford, pp. 19–40.

Stephens, D.W. and Krebs, J.R. (1986) *Foraging Theory*. Monographs in Behavioral Ecology, Princeton University Press, Princeton, New Jersey.

Stuart, I. (1998) *Life's Other Secret: the New Mathematics of the Living World*. Penguin Books, London.

Weber, B.H. and Depew, D.J. (1996) Natural selection and self-organization. *Biology and Philosophy* 11, 33–65.

Wilson, E.O. (1990) *Success and Dominance in Ecosystems: the Case of Social Insects*. Ecology Institute Nordbünte 23, D-2124 Oldendorf/Luhe, Federal Republic of Germany.

# Dispersal and Conservation in Heterogeneous Landscapes

## 14

N.F. Britton,[1,2] G.P. Boswell[1,2] and N.R. Franks[1,3]*

[1]*Centre for Mathematical Biology,* [2]*Department of Mathematical Sciences and* [3]*Department of Biology and Biochemistry, University of Bath, Bath BA2 7AY, UK*

## Introduction

Dispersal parameters are crucial to the maintenance of diversity in ecological communities. We shall survey some models of dispersal and show why this is so, in general and in the context of partial habitat loss, which has dramatic consequences for dispersal. In particular, we shall look at the fugitive coexistence of competitors and how this might be mediated by a predator. We then consider a particular biological system, the army ant *Eciton burchelli* (Westwood) and its prey in the neotropical rainforest. We show that there is a danger of extinctions before total fragmentation of the habitat, and discuss what can be done to mitigate the effects of habitat loss.

Spatial effects have often been neglected in theoretical and empirical ecology because of their complexity (but see Tilman and Kareiva, 1997; Bascompte and Solé, 1998; Hanski, 1999). Yet they have been understood for many years to be crucial to species diversity. For example, the principle of competitive exclusion (Gause, 1935) states that when two species compete for a single resource, one will drive the other to extinction. Generalizing, $n$ species cannot exist on fewer than $n$ limiting factors. However, near the surface of the ocean there are few resources and many planktonic species. Hutchinson (1961) suggested that this 'paradox of the plankton' could be resolved if different species occupied different regions of space. This spatial heterogeneity could be and to some extent would have to be generated by the

---

* Present address: School of Biological Sciences, University of Bristol, Woodland Road, Bristol BS8 1UG, UK.

©CAB *International* 2001. *Insect Movement: Mechanisms and Consequences*
(eds I.P. Woiwod, D.R. Reynolds and C.D. Thomas)

species themselves through their interactions. As a second example, Huffaker (1958) studied herbivorous and predatory mites in a patchy environment consisting of oranges and rubber balls in a tray. He showed that the herbivorous mites survived in the absence of predators, but if predators were added they were liable to drive the prey to extinction and then to become extinct themselves, maybe after one or two predator–prey cycles. However, if the dispersal rates between the oranges were artificially reduced, especially for the predatory mites, then the populations could coexist through a mosaic of unoccupied patches, prey–predator patches heading for extinction, and thriving prey patches.

In both of these examples dispersal parameters are crucial in generating and maintaining the spatial heterogeneity that is necessary to maintain diversity, and dispersal is one of the themes of this article. Modelling approaches are discussed in the next section. Our other theme is the effect on biological systems of partial habitat destruction; approaches to modelling this are discussed in the third section. Partial habitat destruction and fragmentation is increasingly important in the light of current destruction of the earth's resources. It affects dispersal, and so disrupts the mechanisms promoting diversity, sometimes in very dramatic ways. In the fourth section, we shall draw our two themes together in a case study, and discuss the role of the army ant *E. burchelli* in maintaining diversity in the neotropical rainforest, one of the most diverse ecosystems on earth. We shall see how it may be threatened by loss of habitat, and discuss methods of reducing the threat.

# Models of Dispersing Populations

### Modelling approaches

Dispersal includes the scattering of seeds, of larvae and of adults. It may be in a prevailing direction, e.g. wind dispersal of seeds, but it will normally contain some random element, and may be completely random. In entomological systems, it is often the movement of adults that is the most important mechanism of dispersal. Models of dispersal may take space to be continuous or discrete. Continuous-space models include partial differential equations (PDEs) and integro-differential equations, depending on the dispersal process, but we shall not consider them here. Discrete-space models are often more appropriate for patchy habitats, with the spatial unit being the patch. In this case it may be useful to think of the population as a metapopulation, a term coined 30 years ago by Levins (1970) to describe a population consisting of many local populations on habitat patches, which persists through a balance between local extinction and re-colonization of these patches. In the original conception these patches were an intrinsic feature of the landscape, but the metapopulation approach can still be useful if the patchiness is generated by the species themselves, as long as local extinctions and re-colonizations occur.

The field of metapopulation ecology has recently been comprehensively reviewed by Hanski (1999). There are two main strands of metapopulation model development, considering space either explicitly or implicitly. Explicit models are often cellular automata (CA) or other lattice-based models (Caswell and Etter, 1993). More realistic spatial configurations may also be considered, e.g. by the incidence function method (Hanski, 1999), especially if the habitat is fully fragmented and the size, shape and isolation of the fragments is important. Population viability analysis packages are detailed computer simulation models of this type, and have recently been reviewed by Lindenmayer *et al.* (1995). However, we shall be concerned with the effects of partial habitat destruction on an initially fully connected landscape, and lattice-based models are sufficient for our purposes. Spatially explicit models typically consist of rules for the birth, death, interaction and dispersion of each individual (or group of individuals) in the population or metapopulation, which may be structured in various ways. They normally include stochastic effects, and often require detailed biological knowledge to set up. Such models are well adapted to stochastic events and spatial effects. Some results on the consequences of the rules may sometimes be found analytically, but the models are generally rather intractable, and computer simulations are essential.

If analytic progress can be made on such a model, it is often by making a mean-field approximation to it, extending it if necessary to take account of non-uniform spatial effects. This is analogous to a spatially implicit metapopulation model (Levins, 1969), which can either be derived directly or as an approximation to a spatially explicit model. It often leads to deep insights in a relatively simple way, stochastic effects including extinctions may be included (Nisbet and Gurney, 1982), and interpretation in terms of individual behaviour is straightforward. However, modification to deal with spatial distributions that are not statistically uniform or nearly uniform may be difficult, and its spatially implicit nature means that the effects of inaccessibility of parts of the domain cannot easily be incorporated. We shall start by describing these models.

## Spatially implicit metapopulation models

### Deterministic models

Consider a large number $K$ of potentially habitable sites, each of which is either occupied or unoccupied by an individual (or a population) of a particular species. Each of these sites is assumed to be identical and to be isolated from the others in an identical way. Let the fraction occupied at time $t$ be $p(t)$.

Let the chance of an occupied site becoming unoccupied in the next interval of time $\delta t$ be $e\delta t$, so that $e$ is a local extinction rate. Then the mean fraction of sites that become unoccupied in the next interval of time $\delta t$ is $ep(t)\delta t$. Let the chance of an unoccupied site becoming occupied in the next interval of time $\delta t$ be $cp(t)\delta t$, so that $c$ represents a colonization rate. Then the mean fraction of sites that are colonized in the next interval of time $\delta t$ is $cp(t)(1-p(t))\delta t$. We have, on average, taking the limit as $\delta t \to 0$,

$$\frac{dp}{dt} = cp(1-p) - ep \tag{1}$$

The true behaviour can be expected to track this mean behaviour closely if $K$ is large, and tracks it exactly in the idealized situation of an infinite number of patches. The critical parameter is the *basic reproductive ratio*, the number of sites an occupied site can expect to colonize before going extinct when the species is rare, given by $R_0 = c/e$. As intuition suggests, there is a threshold at $R_0 = 1$; if $R_0 > 1$ (colonization rate when rare greater than local extinction rate), then it is easy to show that $p(t) \to p^*$ as $t \to \infty$, where

$$p^* = 1 - \frac{e}{c}. \tag{2}$$

On the other hand, if $R_0 < 1$ (local extinction rate greater than colonization rate when rare), then $p(t) \to 0$ as $t \to \infty$. The dispersal, here represented by the colonization parameter $c$, must be sufficiently large ($c > e$) for the population to persist.

Those familiar with epidemic modelling will have noticed the parallel with the Kermack and McKendrick (1927) model for an *SIS* disease, i.e. a disease where all infected individuals eventually recover and become susceptible again. Here occupied patches are analogous to infected individuals, and unoccupied patches to susceptible individuals. The basic reproductive ratio $R_0$ is the number of infectious contacts an infected individual can expect to make when the disease is rare, and there is again a threshold at $R_0 = 1$, with $R_0 > 1$ necessary for the disease to remain endemic.

This model may be extended to two competing populations by considering that patches may be in one of four states, empty or occupied by one or other or both competitors (Levins and Culver, 1971). It is usual (Slatkin, 1974) to ignore local dynamics and to assume that the effects of competition are to lower colonization rates, or to increase extinction rates, or both. It can be shown that one species can exclude the other if it reduces its chances of colonization or increases its chances of extinction sufficiently. A limiting case was considered by Tilman (1994), who assumed that the superior competitor could immediately displace the inferior in a patch, so that there are no patches with both competitors present. The equations are

$$\frac{dp_1}{dt} = c_1 p_1 (1 - p_1) - e_1 p_1, \quad \frac{dp_2}{dt} = c_2 p_2 (1 - p_1 - p_2) - e_2 p_2 - c_1 p_1 p_2. \tag{3}$$

Species 1, whose occupancy fraction is $p_1$, is the superior competitor, and species 2 has no impact on it whatever. In contrast, species 2 can only colonize empty sites (the term $c_2 p_2 (1 - p_1 - p_2)$), and loses sites when species 1 colonizes them (the term $-c_1 p_1 p_2$). Species 1 goes extinct if $c_1$ is too small, but species 2 goes extinct if $c_1$ is too large (compared with $c_2$). The better competitor ousts the poorer if it is also a good disperser. On the other hand, if $c_2$ is sufficiently large compared with $c_1$, then species 2 survives, so the poorer competitor can survive by being a good disperser. Paradoxically, this mechanism for coexistence, known as fugitive coexistence, relies on local extinctions.

Nisbet and Gurney (1982, Chapter 10) present a spatially implicit metapopulation model for a predator–prey system, Huffaker's (1958) experiments of herbivorous and predatory mites on oranges. We used the subscript 2 to denote a competitor above and will continue to do so in the sequel; we use the subscript 3 here and later to denote a predator. The system consists of $K$ sites (oranges), and at time $t$ a fraction $p_1$ is in state 1, occupied by prey only, a fraction $p_3$ in state 3, occupied by predators only, a fraction $p_{13}$ in state 13, occupied by both prey and predators, and a fraction $p_0 = 1 - p_1 - p_3 - p_{13}$ in state 0, empty. Moreover, each orange moves from state to state in cyclic order, from 0 to 1 to 13 to 3 to 0. We must specify the probability that an orange moves from one state to the next in the next interval of time $\delta t$. For movement from state 0 to state 1 let this probability be $c_1 (p_1 + p_{13}) \delta t$, so that $c_1$ is the prey colonization rate, and from state 1 to state 13 let it be $c_3 (p_3 + p_{13}) \delta t$, so that $c_3$ is the predator colonization rate. From state 13 to state 3 let it be $e_{13} \delta t$, so that $e_{13}$ is the prey local extinction rate in the presence of predators, and from state 3 to state 0 let it be $e_3 \delta t$, so that $e_3$ is the predator local extinction rate in the absence of prey. The equations for the fraction of sites in each state are given by

$$\frac{dp_1}{dt} = c_1 p_0 (p_1 + p_{13}) - c_3 p_1 (p_3 + p_{13}),$$
$$\frac{dp_{13}}{dt} = c_3 p_1 (p_3 + p_{13}) - e_{13} p_{13}, \quad (4)$$
$$\frac{dp_3}{dt} = e_{13} p_{13} - e_3 p_3.$$

For $c_3$ too small, the predators die out and the prey persist. For $c_3$ larger, the model as it stands predicts stable coexistence between predators and prey. However, as $c_3$ increases further, the fraction of occupied patches becomes smaller. There are two ways in which this could lead to extinction in a real system. First, a delay in the system could lead to growing oscillations about the steady state, and population crashes to zero. Second, small population sizes could lead to stochastic effects becoming important.

We shall see how stochastic effects are dealt with in the next section. Before doing so, we shall combine models (3) and (4) to demonstrate how a predator can mediate coexistence between two competitors. The model becomes

$$\frac{dp_1}{dt} = c_1(p_0 + p_2)(p_1 + p_{13}) - c_3 p_1(p_3 + p_{13}) - e_1 p_1,$$

$$\frac{dp_2}{dt} = c_2 p_0(p_2 + p_{23}) - c_1 p_2(p_1 + p_{13}) - c_3 p_2(p_3 + p_{23}) - e_2 p_2,$$

$$\frac{dp_{13}}{dt} = c_3 p_1(p_{13} + p_{23}) + c_1 p_{23}(p_1 + p_{13}) - e_{13} p_{13}, \qquad (5)$$

$$\frac{dp_{23}}{dt} = c_3 p_2(p_{13} + p_{23}) - c_1 p_{23}(p_1 + p_{13}) - e_{23} p_{23},$$

$$\frac{dp_3}{dt} = e_{13} p_{13} + e_{23} p_{23} - e_3 p_3.$$

Again, species 1 is a better competitor than species 2, and can immediately displace it from a patch. With certain parameter values all three species can coexist. Essentially, the mechanism is fugitive coexistence, with the predator providing the necessary empty patches which the better disperser, species 2, can colonize and exploit. With no predators species 2, the poorer competitor, goes extinct in competition with species 1, the better (see Fig. 14.1).

## Stochastic effects

We have seen that extinction of a species may occur even if stochastic effects are neglected, so-called deterministic extinction. However, small populations are vulnerable to extinction through a run of bad luck, even if a deterministic model predicts a stable steady state. Stochastic analysis is required to decide how long a population might survive before such a run of bad luck overtakes it. We shall think of this as a gambler's ruin problem (e.g. Feller, 1968). Let us consider the simplest metapopulation model above. The population may be thought of as a gambler playing a coin-tossing game with the environment, with a fortune consisting of the number of sites it occupies. It wins a site whenever the coin comes down heads and a colonization occurs, and loses one whenever it comes down tails and a local extinction occurs. It can never win outright and become safe from extinction, but loses outright and becomes extinct by losing all its sites. Let $t_k$ be the expected time to extinction starting with $k = Kp$ sites, and let $k^* = Kp^*$ be the equilibrium number of sites occupied. Our ultimate aim is to calculate $t^* = t_{k^*}$, the expected time to extinction starting at equilibrium. Let $s_k$ be the expected time to the next coin toss when $k$ sites are occupied, and let $h_k$ be the probability that the next toss is heads, $1 - h_k$ the probability that it is tails. If the first toss is heads, then the game continues as if the initial number of colonies had been $k + 1$, and if it is tails, as if it had been $k - 1$. Hence

$$t_k = s_k + h_k t_{k+1} + (1-h_k) t_{k-1}, \tag{6}$$

for $k = 1, 2 \ldots K-1$. By definition $t_0 = 0$, and since no more than $K$ sites exist to be occupied, $h_K = 0$ and $t_K = s_K + t_{K-1}$. Coins come down heads at a rate $cp(1-p)$ and tails at a rate $ep$, so coin tosses are occurring at a rate $cp(1-p) + ep$, and the expected time to wait for the next toss is the reciprocal of this,

$$s_k = \frac{1}{cp(1-p)+ep} = \frac{K^2}{ck(K-k)+ekK}.$$

The probability of the coin coming down heads next time is

$$h_k = \frac{cp(1-p)}{cp(1-p)+ep} = \frac{c(K-k)}{c(K-k)+eK}.$$

The system of equations for $t_k$ may be solved explicitly or numerically, or approximate solutions may be found (Nisbet and Gurney, 1982). It can be shown that $t^*$ grows exponentially with $K$.

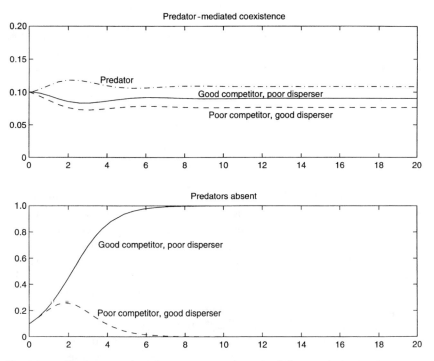

**Fig. 14.1.** Predator-mediated coexistence. The only difference between the two panels is in the initial conditions. In the first panel, initial conditions are chosen with the predator and each species of competitor occupying one-tenth of the patches. The system soon settles down to a stable coexistence state. In the second, the initial numbers of competitors are the same, but there are no predators. The inferior competitor is now displaced by the superior one. Parameter values are $c_1 = 1$, $c_2 = 1.3$, $c_3 = 6$, $e_1 = 0$, $e_2 = 0$, $e_3 = 2$, $e_{13} = 1$, $e_{23} = 1$.

Nisbet and Gurney (1982) have carried out a similar though algebraically more complicated analysis to determine the effects of stochasticity in the model of Huffaker's (1958) system. They show that population cycles may be driven by stochastic effects, and derive estimates of the period of the cycle and of times to extinction. They conclude that in Huffaker's small experimental system extinction is likely after a few cycles, in agreement with the (rather scarce) experimental results. Even in larger arenas, high predator dispersal rates would leave the system particularly vulnerable to extinctions.

## Spatially explicit models

An important assumption in spatially implicit metapopulation modelling is that the probability of a given unoccupied site becoming occupied in the next interval of time depends only on how many sites are occupied, but not on how close they are. This is true in the unlikely event that a propagule is equally likely to colonize any other unoccupied patch, and also for distributions which are statistically spatially uniform. However, many biological populations exhibit clumped distributions, so that such models are inadequate. One approach is to use moment methods to extract the essential information about the spatial distributions, but there are problems in obtaining a closed system to analyse (Levin and Pacala, 1997). Another is to use spatially explicit models (e.g. Caswell and Etter, 1993; Durrett and Levin, 1994; Dytham, 1994; Travis and Dytham, 1998). A lattice-based model for the extinction and colonization process described above makes space explicit by placing the sites on a lattice, which is usually regular rectangular or hexagonal. We shall refer to this as a cellular automata (CA) model, although some authors restrict this term to deterministic systems. At each time step there is a chance of any occupied site going extinct, or of colonizing an unoccupied site chosen at random from neighbouring sites. If the site chosen happens to be occupied already, colonization fails. The probability of colonization depends on the distance between the sites, and usually nearer sites have greater probability. The dispersal range may therefore be specified. CA models generally predict lower occupancy rates than the corresponding spatially implicit model, because the clumping implies that propagules are less likely to find an empty or a good site to colonize. If the range of dispersal in a CA is large, distributions are close to spatially uniform and the spatially implicit model is a good approximation to the CA model (Bramson et al., 1989; Durrett and Levin, 1994), but the approximation becomes worse as dispersal range decreases. However, the conclusions on dispersal rates drawn in the previous section still hold qualitatively in the CA context.

Summarizing, CA models are much more intractable than the corresponding spatially implicit models, but may pick up features that these cannot, especially on spatial distributions. Spatially implicit models give good insights, and give reliable quantitative predictions under certain circumstances, e.g. when dispersal is long range.

## Modelling Habitat Destruction

We have seen that if dispersal parameters are too high or too low, extinctions may occur, but what is the effect of habitat destruction on the ecological community? Of course it leads to smaller populations and hence greater chances of extinctions, but the effects are more far-reaching than that. In this section we shall look at the simple predictions made by spatially implicit models, and the phenomenon of percolation that cannot be captured by such models.

### Spatially implicit approach

Theoretically, we may investigate the effect of removal of some habitat in the spatially implict metapopulation context by modifying Equation (1). Removal of a fraction $D$ of habitat leads to the equation

$$\frac{dp}{dt} = cp(1-D-p) - ep, \tag{7}$$

where it has simply been assumed that any attempted colonization of a removed patch is unsuccessful and leads to immediate death of the propagules. The basic reproductive ratio $R_0 = c/e$ now has to be greater than before to maintain the population, the extinction threshold increasing from $R_0 = 1$ to $R_0 = 1/(1-D)$, so that extinction occurs when $D = D_c = 1 - (1/R_0)$. (The epidemiological analogy of this result is that only a fraction of susceptibles need to be vaccinated to eradicate a disease.) The critical fraction $D_c$ of sites to be deleted is the value of the steady state $p^*$ when $D = 0$. When the fraction of sites removed becomes equal to the original fraction occupied, the population becomes extinct (despite the fact that the sites are removed at random).

Tilman et al. (1994) looked at assemblies of competing populations using the spatially implicit metapopulation method, and assumed that poorer competitors are better dispersers. They showed that when habitat is destroyed, and under certain conditions on the parameters, it is the better competitors that go extinct first. As in the single species case, the critical fraction $D_c$ of sites to be deleted, when the best competitor goes extinct, is the fraction of sites occupied by the best competitor at steady state. Similar studies on prey–predator systems and food chains tend to suggest that the highest level predator will go extinct first in response to habitat destruction (May, 1994; Holt, 1997).

## Percolation

It is meaningless to talk about habitat fragmentation in a spatially implicit metapopulation model, since the quantities analysed are spatial averages. Percolation theory is the branch of mathematics that deals with the effect of clogging of channels in a porous medium, or alternatively with fragmentation of an environment. Consider a lattice of potentially habitable sites, and then choose sites at random one by one and remove them. At first the habitat will remain (at least for the most part) connected, but eventually it will split up into isolated fragments. Figure 14.2, obtained by simulating the process described above on a computer using a 25 × 20 rectangular lattice, shows how the size of the largest connected part of the environment decreases as sites are removed. At first the reduction is linear, because almost all remaining sites are still connected, but later habitat fragmentation leads to a deviation from linearity.

The surprise is how abrupt the transition is from a connected to a fragmented environment. It is a theorem of percolation theory that in the idealized case of an infinite lattice there is a critical fraction $D_c$ of sites

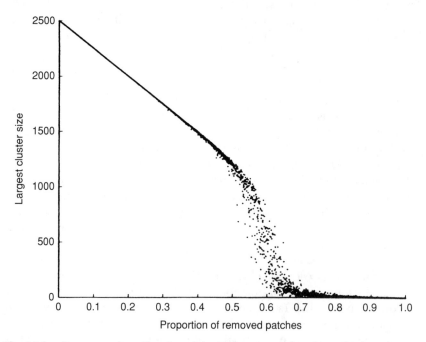

**Fig. 14.2.** Fragmentation. The size of the largest remaining cluster is plotted as a function of the fraction of patches removed (where patches are removed at random). Two patches are in the same cluster if there is a path between them via nearest and next-nearest neighbours. Even when half the patches are removed, the lattice still often consists of one large and a few very small clusters. The critical percolation value is at about 59% removal.

destroyed such that if $D < D_c$ the largest connected part is infinite, whereas if $D > D_c$ it is finite, with probability 1. This is termed a phase transition, by analogy with the transition of a substance from the solid to the liquid phase. The phase transition to fragmentation usually has a catastrophic effect on extinction rates. However, it is important to note that small habitat islands may become completely inaccessible, and larger areas difficult to access, before complete fragmentation, so that population levels may start to fall and extinction rates to rise before the phase transition occurs.

## Modelling *E. burchelli* in the Neotropical Rainforest

In this section we describe a model of a particular biological system. It is set up as a CA model (Britton et al., 1996), which has been applied to partially fragmented landscapes by Boswell et al. (1998, 2000). Some of the results we obtain will, however, be obtained by using a spatially implicit approximation of this (Partridge et al., 1996; Britton et al., 1999). The data on which the model is based were collected by Franks (1982a,b, 1989) and others (Rettenmeyer, 1963; Willis, 1967; Schneirla, 1971) on Barro Colorado Island (BCI) in Gatun Lake, which is part of the Panama Canal (Fig. 14.3).

The model is far more complex than the simple generic models above, but those models will inform the results that we obtain. The model will also be far simpler than the real biological system. It is a crucial part of the art of modelling to decide which features of the real system are essential to its behaviour, and which peripheral. Discarding the peripheral features, despite reducing the realism of the model, almost always repays the modeller with added insight into the essence of the system.

### Biology

More details of the biology are given by Franks (Chapter 13); we give a brief summary here. Army ants are among the most spectacular of social animals. *E. burchelli* live in organized colonies that may contain over half a million individuals. They are top predators, whose prey are mainly insects, some solitary but most social. Often the solitary insects are the better dispersers, and it is possible that the social insects are the better competitors. The army ants cause local extinctions or near-extinctions that may fuel the coexistence of these competing insects through a metapopulation-like process, as discussed above. Certainly they do promote diversity (Franks, 1982a; Franks and Bossert, 1983). Moreover about 50 species of ant-birds rely on them to flush prey from the undergrowth and cannot survive in their absence (Willis and Oniki, 1978). Many species of lizards and butterflies also depend on them.

*E. burchelli* colonies cycle between a statary and a nomadic phase. In the statary phase, which lasts about 21 days, they occupy a fixed nest site from

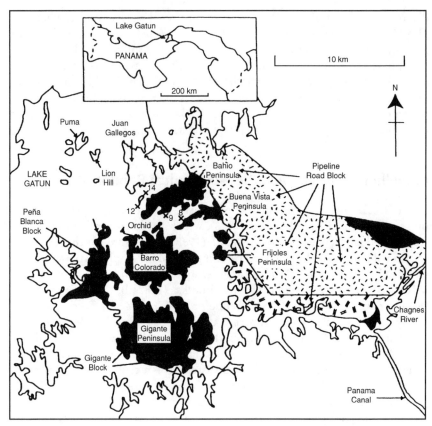

**Fig. 14.3.** Map of the study area. The key helps to identify various habitat islands mentioned in Table 14.1. The numbered crosses represent islands in Table 14.1 too small to be drawn to scale. The inset shows the location of the lake within Panama.

which on most days they raid areas of the surrounding forest. In the nomadic phase, which lasts about 14 days, they raid each day, and each night set up a new temporary nest site, or bivouac. Their raids are fan-shaped, and about 90 m long. At the end of a statary phase they have devastated an area of about 24,000 square metres around their nest site, which takes about 6 months to recover. In the nomadic phase they travel over 500 m in a randomly chosen direction, so that the new statary phase nest site is well removed from the old one in an undisturbed habitat. However, in the subsequent nomadic phase they may well return to spend the next statary phase very close to the old site.

Colonies are monogynous. If the queen dies, the colony may be able to replace her, or it may die. Colonies also die if they become too small. If they become too large they raise a new queen and reproduce by fission.

*E. burchelli* are good colonizers in a connected habitat, in the sense that they will always move to a new patch at the end of a cycle, but they are poor colonizers in a fragmented or partially fragmented habitat, being very reluctant to cross even very narrow areas of open ground. For this reason we expect them to be especially vulnerable to habitat loss.

## CA model

We first split the habitat up into a lattice of patches that are about the same size as the area devastated by an army ant colony during a statary phase. We shall assume that at each statary phase each colony in the system occupies one of these patches. We shall neglect the fact that in the real system these patches will not be arranged on a regular lattice. We shall split time up into 35-day periods, corresponding to the army ant cycles. We shall model the prey dynamics by assuming that a patch that has recently been raided throughout a statary phase by an army ant colony has a low prey population, whereas one that has not been raided for at least $n$ cycles has recovered a high prey population. We usually take $n$ to be 5 or 6, since patches take about 6 months to recover. We shall assume that the main determinant of a colony's success over a cycle is where it spends its statary phase. A nomadic phase ranges over a distance of 500 m, and generally passes through areas in different stages of ecological succession, so that any nomadic phase will consist of some good and some bad raiding days. However, in the statary phase the colony is forced to remain near its fixed nest site, and may suffer 21 consecutive days of poor raids. This will clearly affect its ability to replace workers who die, and we assume that such a colony decreases in size by a fixed amount. Alternatively, a good statary phase will allow the colony to grow, and we assume that such a colony increases in size by a fixed amount. If a colony becomes too small it dies, and if it becomes large enough it splits into two colonies of equal size. We also include the possibility that the queen dies and is not replaced, in which case the colony also dies. Finally, it remains to model army ant dispersal. The model only takes account of the patch where the colony spends its statary phase, and the next statary phase is chosen at random from those patches at the correct distance from the current one, as shown in Fig. 14.4.

The important parameters of the system are $K$, the size of the habitat in patches, $n$, the patch recovery time, about six cycles, $m$, the minimum time in cycles for an ant colony to split, about five cycles, and $Q$, the average queen lifetime, about 6 years. The CA model outlined above is simulated on a computer. For a simulated island the size of BCI, with $K = 500$, we obtain small amplitude oscillations of period about 50 cycles about a steady state which is very close to the value estimated from field observations. The biological data is not good enough to tell whether such oscillations actually occur on BCI. For an island the size of BCI we never get extinctions in the CA model, unless we also simulate the effects of catastrophes. However, simulations on smaller lattices do produce extinctions.

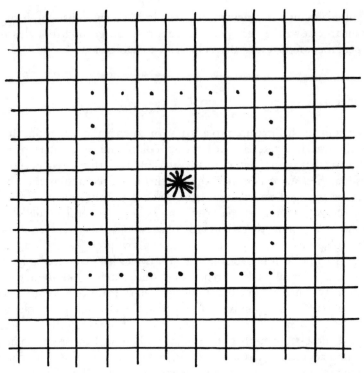

**Fig. 14.4.** Cellular automata (CA) model for dispersal of army ants. The central square is the current position of the colony in statary phase. It is assumed to move during the next nomadic phase to one of the dotted squares chosen at random, and to remain there for the next statary phase.

## Spatially implicit metapopulation model

In order to obtain some analytic results from the CA model, we approximated it by a spatially implicit metapopulation model. The approximation can be expected to be good because of the large distances (compared to patch sizes) that the colonies move in their nomadic phase (Bramson *et al.*, 1989; Durrett and Levin, 1994). It is far more complicated than the spatially implicit metapopulation models, mainly because of the number of states that a patch can be in. Rather than simply being occupied or not occupied, an occupied patch is in a different state depending on the size of the colony that occupies it, and an unoccupied patch is in a different state depending on how long it is since it was last occupied. The approach captures a good approximation to the steady state of the CA model but not to the (small amplitude) oscillations about it. To a good approximation, the steady state does not depend on queen lifetime $Q$ or minimum splitting time $m$ and is given by $k^* = Kp^* = K(1 - (1/2)^{1/n})$, about 55 for BCI, close to the value of about 50 obtained from field observations (Franks, 1982b).

## Times to extinction

Times to extinction are very long for large $K$ and it is not practicable to obtain them using the CA model, so that the spatially implicit metapopulation approach is essential. We use the method of the gambler's ruin described earlier, but there is an additional complication (Britton et al., 1999). We do not know the birth and death rates for single colonies because they are different for colonies of different sizes, so these have to be analysed first. This is done by considering the colony as a gambler playing against the environment with a fortune consisting of its workers. If the coin comes down heads, and the colony lands on a good patch, it wins some new workers, whereas if it comes down tails, and the colony lands on a bad patch, it loses some. It eventually wins outright, by gaining enough workers to split, or loses outright, by losing so many workers it dies. Queen death can also be included in this game.

Eventually we come up with a set of difference equations for times to extinction from different starting populations, analogous to but more complicated than Equation (6), which can be solved numerically. The results are summarized in Fig. 14.5. Predictions for survival on various habitat islands in Gatun Lake and its environs are given in Table 14.1. For further details see Partridge et al. (1996). It can be seen that they are very much in line with the data.

## Effects of habitat destruction

No data are available for this, although we plan to collect some in the near future. We make some predictions (Boswell et al., 1998) based on the CA model, which will be tested when the data are available. The results are summarized in Fig. 14.6.

It can be seen that quasi-steady state populations drop approximately linearly with habitat destruction, and we have deterministic extinction when a fraction $D$ of about 45% has been removed. The phase transition to fragmentation, discussed earlier, occurs when about 59% of habitat has been removed. This earlier extinction, preceded by falls in population levels, occurs as parts of the habitat become increasingly inaccessible. Essentially it is caused by a reduction in colonization rates as habitat is removed, until they can no longer overcome the local extinction rates. Pursuing the epidemiological analogy, the 'vaccination threshold' has been overcome.

## Corridors

It has been stated (Wilson and Willis, 1975) that 'extinctions will be lower when fragments can be connected by corridors of natural habitat, no matter how thin the corridors'. Figure 14.7 tests this contention and shows the effects of narrow corridors, one patch wide, and wider corridors, two patches wide, on expected survival times in a model population of *E. burchelli* in a dumbbell-shaped reserve (Boswell *et al.*, 1998).

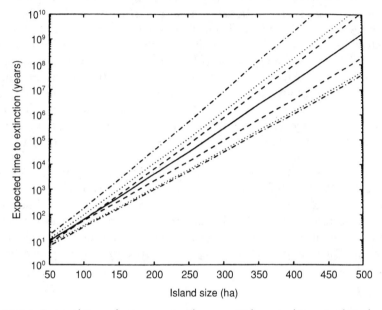

**Fig. 14.5.** Survival times for army ants. The expected survival time is plotted as a function of island size. The solid line represents our best estimate of the parameters, namely $m = 5$ cycles, $Q = 6$ years and $n = 6$ cycles. Dashed lines represent the effect of varying $m$, dotted lines $Q$, and dash-dotted lines $n$, each by 25%.

**Table 14.1.** Predicted survival times for some of the habitat islands in and around Gatun Lake.

| Habitat island | Size (ha) | Survival time (years) | Survival probability | Current status of *E. burchelli* |
|---|---|---|---|---|
| Island numbers 4, 8, 9, 12 and 14 | < 5 | < 0.5 | ≈ 0 | Absent |
| Orchid Island | 17 | 1.1 | ≈ 0 | Absent |
| Lion Hill Island | 51 | 13 | 0.002 | Absent |
| Puma Island | 77 | 45 | 0.16 | Absent (formerly partly in agriculture) |
| Juan Gallegos Island | 659 | $1.9 \times 10^{14}$ | ≈ 1 | Absent (formerly converted to pasture) |
| Peña Blanca block | 900 | $1.3 \times 10^{19}$ | ≈ 1 | Rumoured to be present |
| Barro Colorado Island | 1500 | $1.1 \times 10^{33}$ | ≈ 1 | Present |
| Gigante block | 2300 | $4.9 \times 10^{52}$ | ≈ 1 | Present |
| Pipeline Road block | 13100 | $8.0 \times 10^{277}$ | ≈ 1 | Present |

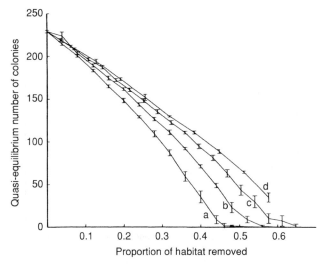

**Fig. 14.6.** Effect of destruction of habitat on population size. A proportion of habitat is randomly destroyed starting at zero and changing in small increments. One-tenth of the remaining habitat is then populated by colonies and the simulation is allowed to run for 500 cycles, with only the last 300 iterations being recorded; this allows a quasi-steady state to be reached. The habitat is removed in (a) 1 × 1 blocks, (b) 2 × 2 non-overlapping blocks, (c) 3 × 3 non-overlapping blocks and (d) 4 × 4 non-overlapping blocks. Each such simulation is repeated ten times and the mean and standard deviation are shown.

It can be seen that unless the patches are very close together narrow corridors may reduce metapopulation persistence times, confirming recent suggestions in the literature (Merriam, 1991; Soulé and Gilpin, 1991; Bonner, 1994; Boswell et al., 1998). In general, they may do so by allowing the transmission of disasters, but in our case they do so by acting as a demographic sink. Essentially, the army ant populations may fail to find the corridor exit, and subsequently starve in the resource-poor environment of the corridor itself. It may be important to take such effects into account when planning conservation strategies, especially for organisms that are poor dispersers in restricted habitats.

## Discussion

In this chapter we have surveyed some individual-based models of dispersal which can incorporate the effect of habitat loss. These are either spatially implicit or CA models. Spatially implicit models are more tractable than CAs but less realistic. They can be modified to deal with statistically non-uniform distributions, but it is difficult to gain any information on the effects of habitat inaccessibility or fragmentation from them. CAs predict lower

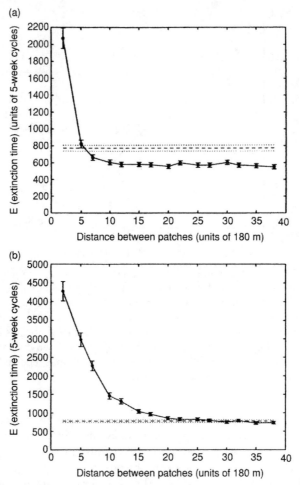

**Fig. 14.7.** Effect of narrow and wide corridors on survival times. The expected survival time is plotted as a function of the distance between the two patches, with (a) corridors 180 m wide and (b) corridors 360 m wide. The mean time of 1000 runs is shown for each corridor length (solid line) along with 95% confidence intervals. As a comparison, the mean survival time of 1000 runs in the absence of a corridor is also shown (dashed line), with 95% confidence intervals (dotted lines).

occupancy rates than spatially implicit models, but the differences are small if dispersal is long range.

The models show that low dispersal rates of good competitors are important in maintaining diversity. The mechanism involved is fugitive coexistence, whereby the poor competitors may keep one step ahead of the good competitors if they have good dispersal characteristics. This coexistence may be predator-mediated, with predators creating the empty patches essential for colonization. In such cases the predators themselves cannot disperse too widely, without running the risk of driving the whole ecosystem to extinction. The dispersal characteristics of army ants lead to the paradoxical conclusion that these predators may enhance diversity in the neotropical rainforests.

Many of the models presented here make common predictions on the effect of removing patches of habitat from a landscape, and some give more detail on different effects of different removal schemes. Common predictions are a reduction in diversity and in the fraction of the remaining habitat that is occupied. If habitat is to be removed, it is best to do it without creating difficulty of access to the remaining habitat, but even in the best scenario extinction rates increase exponentially as habitat is removed. If access to the remaining habitat does become difficult, dispersive is more important than competitive ability, and predators and better competitors will tend to go extinct first. As more habitat is removed, percolation theory shows that there is a phase transition where the habitat fragments, but CA models predict that poor accessibility to parts of a habitat can lead to low populations and extinctions even before such fragmentation takes place. Habitat corridors can mitigate the effect, but they have to be sufficiently wide that they do not themselves act as a demographic sink.

These predictions have yet to be tested experimentally. Such experiments on plant–insect associations could lead to results within a few years (Steinberg and Kareiva, 1997). It is also important to obtain field data, and we plan to gather some for the *E. burchelli* system. In the absence of such data it is hard to judge just how important spatial effects are in determining the dynamics of ecological communities. What evidence we have suggests that they are crucial to their conservation.

## Summary

Dispersal parameters are crucial to the maintenance of diversity in ecological communities. We surveyed some models of dispersal and showed why this is so, in general and in the context of partial habitat loss, which has dramatic consequences for dispersal. In particular, we looked at the fugitive coexistence of competitors and how this might be mediated by a predator. We then considered a particular biological system, the army ant, *E. burchelli*, and its prey in the neotropical rainforest. We showed that there is a danger of extinctions before total fragmentation of the habitat, and discussed what might be done to mitigate the effects of habitat loss.

## References

Bascompte, J. and Solé, R.V. (1998) *Modeling Spatiotemporal Dynamics in Ecology*. Springer, New York.

Bonner, J. (1994) Wildlife's roads to nowhere. *New Scientist* 143(1939), 30–34.

Boswell, G.P., Britton, N.F. and Franks, N.R. (1998) Habitat fragmentation, percolation theory and the conservation of a keystone species. *Proceedings of the Royal Society of London B* 265, 1921–1925.

Boswell, G.P., Franks, N.R. and Britton, N.F. (2000) Habitat fragmentation and swarm-raiding army ants. In: Gosling, L. and Sutherland, W. (eds) *Behaviour and Conservation*. Cambridge University Press, Cambridge, pp. 141–158.

Bramson, M., Durrett, R. and Swindle, G. (1989) Statistical mechanics of crabgrass. *Annals of Probability* 17, 444–481.

Britton, N.F., Partridge, L.W. and Franks, N.R. (1996) A mathematical model for the population dynamics of army ants. *Bulletin of Mathematical Biology* 58, 471–492.

Britton, N.F., Partridge, L.W. and Franks, N.R (1999) A model of survival times for predator populations: the case of army ants. *Bulletin of Mathematical Biology* 61, 469–482.

Caswell, H. and Etter, R.J. (1993) Ecological interactions in patchy environments: from patch-occupancy models to cellular automata. In: Levin, S., Powell, T. and Steele, J. (eds) *Patch dynamics*, number 96 in *Lecture Notes in Biomathematics*. Springer, Berlin, pp. 93–109.

Durrett, R. and Levin, S.A. (1994) Stochastic spatial models: a user's guide to ecological applications. *Philosophical Transactions of the Royal Society of London B* 343, 329–350.

Dytham, C. (1994) Habitat destruction and competitive coexistence: a cellular model. *Journal of Animal Ecology* 64, 145–146.

Feller, W. (1968) *An Introduction to Probability Theory and its Applications*, Vol. I, 3rd edn. John Wiley & Sons, New York.

Franks, N.R. (1982a) Ecology and population regulation in the army ant *Eciton burchelli*. In: Leigh, E., Rand, A. and Windsor, D. (eds) *The Ecology of a Tropical Forest: Seasonal Rhythms and Long-term Changes*. Smithsonian Institution Press, Washington, DC, pp. 389–395.

Franks, N.R. (1982b) A new method for censusing animal populations: the number of *Eciton burchelli* army ant colonies on Barro Colorado Island, Panama. *Oecologia, Berlin* 52, 266–268.

Franks, N.R. (1989) Army ants: a collective intelligence. *American Scientist* 77, 139–145.

Franks, N.R. and Bossert, W.H. (1983) The influence of swarm raiding army ants on the patchiness and diversity of a tropical leaf litter ant community. In: Sutton, E., Chadwick, A. and Whitmore, T. (eds) *The Tropical Rain Forest: Ecology and Management*. Blackwell, Oxford, pp. 151–163.

Gause, G.F. (1935) *The Struggle for Existence*. Williams and Wilkins, Baltimore.

Hanski, I. (1999) *Metapopulation Ecology*. Oxford University Press, Oxford.

Holt, R.D. (1997) From metapopulation dynamics to community structure: some consequences of spatial heterogeneity. In: Hanski, I. and Gilpin, M. (eds) *Metapopulation Biology*. Academic Press, San Diego, pp. 149–165.

Huffaker, C.B. (1958) Experimental studies on predation: dispersion factors and predator–prey oscillations. *Hilgardia* 27, 343–383.

Hutchinson, G.E. (1961) The paradox of the plankton. *American Naturalist* 95, 137–147.

Kermack, W.O. and McKendrick, A.G. (1927) Contributions to the mathematical theory of epidemics. *Proceedings of the Royal Society of Edinburgh A* 115, 700–721.

Levin, S.A. and Pacala, S.W. (1997) Theories of simplification and scaling of spatially distributed processes. In: Tilman, D. and Kareiva, P. (eds) *Spatial Ecology: the Role of Space in Population Dynamics and Interspecific Interactions*, number 30 in *Monographs in Population Biology*. Princeton University Press, Princeton, New Jersey.

Levins, R. (1969) Some demographic and genetic consequences of environmental heterogeneity for biological control. *Bulletin of the Entomological Society of America* 15, 237–240.

Levins, R. (1970) Extinction. *Lecture Notes in Mathematics* 2, 95–107.

Levins, R. and Culver, D. (1971) Regional coexistence of species and competition between rare species. *Proceedings of the National Academy of Sciences USA* 68, 1246–1248.

Lindenmayer, D.B., Burgmann, M.A., Akçakaya, H.R., Lacy, R.C. and Possingham, H.P. (1995) A review of the generic computer programs ALEX, RAMAS/Space and Vortex for modelling the viability of wildlife populations. *Ecological Modelling* 82, 161–174.

May, R.M. (1994) The effects of spatial scale on ecological questions and answers. In: Edwards, P., May, R. and Webb, N. (eds) *Large-scale Ecology and Conservation Biology*. Blackwell Scientific Press, Oxford, pp. 1–18.

Merriam, G. (1991) Corridors and connectivity: animal populations in heterogeneous environments. In: Saunders, D. and Hobbs, R. (eds) *Nature Conservation 2: the Role of Corridors*. Surrey Beatty and Sons, Chipping Norton, Australia, pp. 133–142.

Nisbet, R.M. and Gurney, W.S.C. (1982) *Modelling fluctuating populations*. John Wiley, Chichester.

Partridge, L.W., Britton, N.F. and Franks, N.R. (1996) Army ant population dynamics: the effects of habitat quality and reserve size on population size and time to extinction. *Proceedings of the Royal Society of London B* 263, 735–741.

Rettenmeyer, C.W. (1963) Behavioral studies of army ants. *Kansas University Science Bulletin* 44, 281–465.

Schneirla, T.C. (1971) Army ants. In: Topoff, H. (ed.) *A Study in Social Organization*. W.H. Freeman, San Francisco.

Slatkin, M. (1974) Competition and regional coexistence. *Ecology* 55, 128–134.

Soulé, M.E. and Gilpin, M.E. (1991) The theory of wildlife corridor capability. In: Saunders, D. and Hobbs, R. (eds) *Nature Conservation 2: the Role of Corridors*. Surrey Beatty and Sons, Chipping Norton, Australia, pp. 133–142.

Steinberg, E.K. and Kareiva, P. (1997) Challenges and opportunities for empirical evaluation of 'spatial theory'. In: Tilman, D. and Kareiva, P. (eds) *Spatial Ecology: the Role of Space in Population Dynamics and Interspecific Interactions*, number 30 in *Monographs in Population Biology*. Princeton University Press, Princeton, New Jersey.

Tilman, D. (1994) Competition and biodiversity in spatially structured habitats. *Ecology* 75, 2–16.

Tilman, D. and Kareiva, P. (eds) (1997) *Spatial Ecology: the Role of Space in Population Dynamics and Interspecific Interactions*, number 30 in *Monographs in Population Biology*. Princeton University Press, Princeton, New Jersey.

Tilman, D., May, R.M., Lehmann, C.L. and Nowak, M.A. (1994) Habitat destruction and the extinction debt. *Nature* 371, 65–66.

Travis, J.M.J. and Dytham, C. (1998) The evolution of dispersal in a metapopulation: a spatially explicit, individual-based model. *Proceedings of the Royal Society of London B* 265, 1117–1123.

Willis, E.O. (1967) The behavior of bicolored antbirds. *University of California Publications in Zoology* 79, 1–127.

Willis, E.O. and Oniki, Y. (1978) Birds and army ants. *Annual Review of Ecological Systems* 9, 243–263.

Wilson, E.O. and Willis, E.O. (1975) Applied biogeography. In: Cody, M. and Diamond, J. (eds) *Ecology and Evolution of Communities*. Harvard University Press, Cambridge, Massachusetts, pp. 522–534.

# Scale, Dispersal and Population Structure

## Chris D. Thomas

*Centre for Biodiversity and Conservation, School of Biology, University of Leeds, Leeds LS2 9JT, UK*

## Introduction

Movement determines the spatial structure of populations, and strongly influences the area over which we may need to apply management either to increase (conservation) or decrease (pest control) a species of interest. Interpretation of population structure depends on the rate of movement relative to the total area that a researcher is interested in, and relative to the size of the sample unit chosen to subdivide that total area. Since choices of total area (extent) and the level of subdivision (resolution) are necessarily somewhat subjective (e.g. Wiens, 1989), we must be concerned about the extent to which designations of population structure are also subjective. This scale-dependent diversity of ways in which one can look at the same set of individuals in one landscape may be partly responsible for the ever-increasing lexicon of terms used to describe the structure of animal populations (e.g. Pulliam, 1988; Harrison, 1991, 1994; Watkinson and Sutherland, 1995; Harrison and Taylor, 1997; Thomas and Kunin, 1999). This chapter aims to illustrate some of the diversity of types of population structure that exist, generated by different rates of movement relative to some scale of interest. I argue that the diversity of natural systems, and the scale-dependence of our conclusions, mean that we should seek a process-based framework for trying to understand population structure, rather than trying to force complex systems into descriptive categories they will rarely fit.

## Sources and Sinks

Movement results in a re-distribution of individuals. Individuals are normally born in one place and die in another, although these different places might be anywhere from millimetres to thousands of kilometres apart. Described at minute resolution, every single population exists as a source–sink system: some places generate individuals (sources) and others consume them (sinks) in a given generation. Described at a global resolution, no population acts as a source–sink system: place one vast quadrat over an entire distribution, and all births and deaths fall within the same sample unit. This simple argument illustrates why one should concentrate on the scale at which source–sink dynamics operate, rather than argue whether source–sink systems are any commoner or rarer than any other class of population dynamics.

Adult orange tip butterflies, *Anthocharis cardamines* (L.), range widely over kilometres of the lowland British landscape, with individual females laying eggs on scattered cruciferous host plants (Wiklund and Ahberg, 1978; Courtney, 1981; Courtney and Duggan, 1983; Dempster, 1997). The larvae feed on developing seed pods, with survival depending on the size of the plant and on the species of crucifer (surviving poorly on species with tough pods). Plant patches that contain large plant individuals of host species with relatively chewable pods are likely to act as local sources, whilst patches containing small individuals of hosts with tough pods are likely to act as relative sinks. In this case, variation in population-level productivity occurs at a spatial scale smaller than the average dispersal distances of individual adults.

An interesting question in these 'within-population' source–sink systems is why the adult butterflies do not evolve to avoid laying eggs on the bad patches. This could be achieved by the evolution of much reduced dispersal, such that adults stay in the patches where they emerge – in the case of *A. cardamines*, this is unlikely to be favoured because most good host-plant patches are small and ephemeral, and females must move around to exploit them (Dempster, 1997). Elimination of poor hosts could also be achieved by the refusal of females to lay on these plants when they encounter them – but because female fecundity is apparently limited by their ability to find hosts, there may be a selective advantage to laying some eggs on host plants with a low probability of larval survival when the alternative is for females to die leaving the eggs unlaid (Courtney, 1981). The butterflies continue to use the sinks apparently because individual selection favours movement and increased fecundity.

Variation in breeding success on resource patches at a much smaller scale than the movements of individuals is probably widespread among plant-feeding insects. It occurs among parasites too. The butterfly *Maculinea rebeli* (Hirschke) has larvae that become brood parasites of just one species of *Myrmica* ant, yet *M. rebeli* oddly lays its eggs in places with and without the right species of ant present (Thomas *et al.*, 1993; Clarke *et al.*, 1997;

Thomas, J.A. et al., 1998). Adult butterflies fly throughout the grassland habitat, with some places acting as local sources (places with the correct host ant), and others acting as sinks (places with other ants). Again the fascinating evolutionary question is why the females do not simply select those places with the correct ants – possible answers could be that they have been unable to evolve to identify different *Myrmica* species correctly or, more likely, that there is some trade-off, perhaps involving intraspecific competition (J.A. Thomas, personal communication) or adult fecundity. The adult butterfly might have to crawl down into the vegetation to identify the correct ant, which could be more time consuming and dangerous than the existing strategy of laying eggs on top of plants.

Source–sink dynamics may also take place at scales up to and sometimes exceeding the normal distances that individuals move, depending on how productive the source is, and how fast population densities decline within sink habitats (in the absence of immigration). Rodríguez et al. (1994) showed that larval survival varied in the mazarine blue butterfly, *Cyaniris semiargus* (Rottemburg), along a moisture gradient in a sand dune system in southern Spain. The phenology of butterfly and host plants were synchronized in the lower, moister areas, and larvae were able to complete the life cycle, whereas all larvae died in the slightly higher areas, where the host plants dried up before larval development was complete. Since the dryer host plants (sinks) were fatal to the larvae, sinks could only extend to a single dispersal distance from the source populations.

Thomas et al. (1996) described how a checkerspot butterfly, *Euphydryas editha* (Boisduval), bred highly successfully in one type of habitat, which acted as a population source, and poorly in another, which acted as a pseudosink. A pseudosink is like a sink (Table 15.1), but the butterflies could survive indefinitely at lower average density in this habitat, in the absence of immigration. Because the poorer habitat was not instantly fatal to the larvae, net movement away from the sources resulted in elevated densities in the pseudosink habitat at distances anywhere from one to several dispersal distances from the source populations: up to a few kilometres. In this case, the system was one that had been seriously perturbed by human activities, and the butterflies were gradually evolving increased use of resources in the source-type habitat, as might be expected (Singer et al., 1993; Singer and Thomas, 1996).

This landscape-level scale is probably what most people have in mind when they discuss source–sink population dynamics. Measured in terms of individual dispersal distances, however, the relevant 'landscape' could extend over entire continents for the most mobile species. However, the smaller the spatial scale, the more likely it is that movement will disrupt any local balance that would otherwise exist between birth and death, and there is a continuum from within-population to landscape-level source–sink dynamics. Virtually all systems are likely to contain elements of within-population source–sink dynamics because resources vary in quality. Therefore, we should abandon the question 'What proportion of systems have source–sink dynamics?' in favour of 'At what spatial scales do source–sink dynamics take place?' 'How do these scales relate to the movements of individuals?' 'How are sources and sinks maintained (ecology and evolution)?' 'What are the dynamic consequences of these processes at different spatial scales?' and 'What are the implications for practical management?'.

Conservation biologists often draw a useful distinction between effects of habitat quality (e.g. resource type/density within habitat remnants) and quantity (the number and size of habitat remnants in a fragmented landscape). Within-population source–sink dynamics affect habitat 'quality' whilst the landscape-level dynamics sometimes relate to habitat 'quantity'. Manipulating the *ratios* of good and bad places for breeding (and potentially the movements of individuals between good and bad places) may sometimes be as important as simply increasing the quantity of good breeding resources (Thomas, J.A. et al., 1998).

## The Compensation Axis

Patterns of birth, death, immigration and emigration can potentially be used to assign each sample area to one of a number of simple categories: a source, sink, pseudosink or classical population (Table 15.1). However, a category-based classification is rather limited. If we need quantitative data to assign each area to one of these categories, why should we throw away most of the information when describing how the system operates?

**Table 15.1.** Population and habitat units (from Thomas and Kunin, 1999, courtesy of the British Ecological Society).

|  | $B, D, I, E$ (actual or potential) | $I \Rightarrow$ zero[a] |
|---|---|---|
| Places with populations | | |
| Sources | $B > D$[b]   $I < E$ | Usually limited impact |
| Classical populations | $B = D$[b]   $I = E$ | Usually limited impact |
| Pseudosinks | $B < D$   $I > E$ | Decline |
|  | ($B > D$[b] at lower density) | |
| Sinks | $B < D$   $I > E$ | Go extinct |
| Places without populations | | |
| Potential sinks | $B < D$ | |
| Empty habitat[b,c] | $B > D$   $I \approx 0$ | |
| Sieves[c] | $B > D$   $I < E$ and $(E - I) > (B - D)$ | |

[a]Effect on local population size of reducing immigration rate to zero.
[b]Allee effects may result in $B < D$ at very low densities for populated areas (converting these areas into true sinks), and $B > D$ at higher densities, leading to alternative stable states. These could be regarded as four new categories.

Thomas and Kunin (1999) describe the 'Compensation Axis' to summarize one continuum for patterns of birth, death, emigration and immigration (Fig. 15.1). It is the equilibrium axis that describes all possible situations where birth plus immigration are equal to death plus emigration over the time scale considered (preferably one or more generations). A change in any one of the four population processes results in a compensatory change in one or all three of the other three processes until the factors increasing local population size (birth, immigration) are again in balance with the factors decreasing it (death, emigration). It describes a wide range of situations, where local population productivity varies from a net export of individuals (bottom right of Fig. 15.1), through the point at which immigration = emigration and birth = death (classical population), to pseudosinks and sinks that consume more individuals than they produce.

This is a simplified abstraction of reality since birth, death, immigration and emigration all vary stochastically, and systems may not be at equilibrium (Thomas and Kunin, 1999). However, it relates a number of categorical population types to a simple continuum. It also allows us to focus more on quantitative issues (what is the mean *and* variance of population productivity in a particular place?) than on qualitative ones (is it a source or a sink?). Rather than describe the proportions of the habitat/landscape that can be regarded as source or sink (a three-level world, if one includes non-habitat), we can start thinking in terms of a continuous topography of population productivity. In terms of conservation management it could also be useful. Can we increase the productivity of sites that are already sources? Can we reduce the rate at which sinks consume individuals? Are these changes best achieved by manipulating birth, death, immigration and/or emigration?

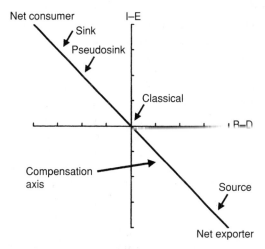

**Fig. 15.1.** The Compensation Axis. $B$ = birth, $D$ = death, $I$ = immigration, $E$ = emigration. From Thomas and Kunin, 1999, courtesy of the British Ecological Society.

## Patches, Metapopulations and Separate Populations

Movements of individuals link together the various locations where breeding is achieved or attempted. The amount of movement affects how we describe these habitat networks, the dynamics we might expect, and the extent to which local adaptations may evolve. It also affects the scale at which practical management may need to be applied – usually at increasingly large spatial scales for more mobile animals.

As with source–sink systems, it is pointless to ask 'What proportion of the population system falls into this or that category?'. The answer depends on the scale of analysis. Furthermore, individual habitat networks may contain a wide range of the different types of population and empty habitat (Table 15.1), so there is no good reason why one should expect a particular population system to fall into one or another category. Real population systems contain blends of different patterns and processes (Harrison, 1991; Sutcliffe *et al.*, 1997a), so it is not very helpful to apply strict definitions.

Harrison's (1991) population types represent a useful starting point. In patchy populations, most individuals move in and out of many small (relative to movement) patches during the course of their lives. In metapopulations, most individuals stay in their natal habitat patches, but some move between patches; in separate populations, virtually no individuals successfully move between populations. These are three regions on a continuous axis of mobility relative to patch size and spacing (below), so the categories inevitably blur. She and others have also drawn attention to the distinction between equilibrium (or quasi-stable – system-wide extinction is the only truly stable equilibrium) and non-equilibrium systems.

The ecology and dynamics of the silver-spotted skipper butterfly, *Hesperia comma* (L.), can be used to illustrate how one can describe a system in any way one wants, depending on the scale considered. Eggs are laid on small tufts of *Festuca ovina* grass, usually growing next to bare ground and/or in hollows (Thomas *et al.*, 1986). Grass tufts represent resource patches for egg-laying and larval development. At this resolution, a single area of grassland represents one patchy population, with egg-laying females flying freely across the habitat, visiting many such resource patches in their life. Suppose we now change the definition of patch to a continuous area of grassland that contains suitable tufts of grass at least every 10–20 m, and which is separated from other areas of breeding habitat by unsuitable breeding habitats (other types of grassland, scrub, woodland, arable fields). Mark–release–recapture of adult skippers in such a habitat network revealed levels of exchange consistent with a metapopulation interpretation across 1 km or more of fragmented grasslands (Hill *et al.*, 1996). This confirmed previous conclusions, based on patterns of local extinction and colonization (Thomas and Jones, 1993), but small patches within the network experienced a flow of individuals in and out of them, more characteristic of processes within a patchy population (Hill et al., 1996). The flow of individuals may be so great that when small patches of habitat are isolated, the drain of emigrants leaving the patch can be so high that they cannot be occupied, however suitable the breeding conditions (Thomas and Hanski, 1997; Thomas, C.D. et al., 1998). At the opposite extreme, and using the same definition of a habitat patch, *H. comma* has apparently survived as a single isolated population in Dorset since about 1970. Thus, depending on the spatial scale and region considered, we can describe this butterfly as occurring in all three of Harrison's major structural categories.

The issue of equilibrium is also important in this species, raising the question of the time scale over which dynamics are measured. *H. comma* has one generation per year. At a metapopulation level, the distribution of *H. comma* has remained relatively stable in Surrey since 1982, although there has naturally been much turnover of the small populations characteristic of this area (Thomas and Jones, 1993). The isolated population in Dorset similarly seems numerically quite stable (from transect count data), in as much as insect populations are ever stable. However, over the same time period, the East Sussex system has seen a massive expansion away from the single valley to which it was confined in the 1970s. Over a longer time period, the butterfly has certainly not been at equilibrium. In the first half the 20th century, it declined through fundamental habitat loss (ploughing) and reduced grazing of traditional pastures. Then, it declined dramatically between 1955 and the mid-1970s, after myxomatosis killed off the rabbits that had maintained the short, sparse grassland that this insect needs. For periods of a few decades, the populations in a few areas have remained quite stable, but this has not been the case in other areas, and instability has been the rule on a time scale of 100 years. Landscape instability has been the norm for Europe as a whole over the last 100 years and probably much longer, and this situation seems likely to continue for the foreseeable future. The spatial dynamics of *H. comma* and many other species are superimposed on a dynamically shifting distribution of suitable breeding areas (Thomas, 1994). For this reason, it is important for the applications of spatial modelling to concentrate on transient dynamics, rather than on imaginary equilibrium scenarios (Thomas, 1996).

The large body of theoretical work on different types of possible population system have provided insight into the sorts of spatial population processes that may be important. The broad conceptual messages are useful, and particular models have important practical implications at specific spatial scales (e.g. management of single fields, reserve acquisition, species re-introductions, etc.). Many of these insights are extremely helpful, but they often apply simultaneously in different places and in the same place at different spatial and temporal scales. It is a waste of time to attempt to force any or all *H. comma* systems into a single population category.

## The Mobility Axis

The example of *H. comma* highlights the difficulty in applying *any* strict definition of population structure. *H. comma* does not seem unusual in this regard. The logical answer is to treat mobility as a continuum relative to patch size and spacing (Harrison and Taylor, 1997). Either natural patches or arbitrary sample units can then be described in terms of the relative contributions of regional-level processes (immigration and emigration) versus local processes (birth and death). Each patch can be quantified along a 'mobility axis' defined as $(I+E) - (B+D)$ (Thomas and Kunin, 1999).

A patch that is visited briefly by large numbers of transient individuals would receive a high positive score, and a network entirely constructed of such patches (or sample units) might be considered to be a patchy population system. For populations that are isolated (some $E$ but no $I$), a negative score would be obtained – a traditional population. Metapopulation-like units fall somewhere in the middle.

The beauty of the approach is fourfold. Firstly, we can stop the pointless discussion of quite how much immigration and emigration is permissible before we decide that a particular system is or is not a metapopulation. We just quantify where a network of patches falls on the continuum, at the scale of analysis. Harrison's (1991, 1994; Harrison and Taylor, 1997) influential papers have argued that there are few truly metapopulation-like systems that exist in nature. The problem is that there are equally few empirical systems that fit *strict* definitions of patchy population systems, separate populations, mainland-island systems, or whatever. With strict definitions, no empirical system ever fits into any category. With broad categories, it is often just a matter of taste which category an author uses to describe a system.

Secondly, patches differ from one another. Each patch or sample unit can potentially be scored individually, and the network can be thought of in terms of the distribution of mobility scores, rather than as a single mean mobility for the system. This gets us around the issue we encountered with *H. comma* where some local breeding areas were metapopulation-like, but others were not.

Thirdly, a continuous axis of mobility increases our ability to think critically about the process of movement, and its consequences for system-wide dynamics. Models of individual movements and foraging theory/models usually deal with the patchy population end of the continuum – why and when do individuals leave patches? Alternatively, metapopulation models usually assume brainless movement, conveniently treating insects as randomly moving particles. For isolated populations, we ignore migration completely (or include an element of emigration). Arranging these systems along a continuum may allow us to think more explicitly about the interaction between movement behaviour and local population dynamics (see, e.g. Stamps *et al.*, 1987; Turchin, 1991, 1998; Kindvall, 1999).

Lastly, as described above, perception of population structure is entirely scale dependent. We can ask how position on the mobility axis varies with changes in resolution of analysis. Different processes assume different importance at different scales of analysis, but it is confusing to suggest that a single set of individuals moving in space simultaneously *is* two quite different categories of system.

## Patches are Not of Equal Importance to Persistence

The probability of persisting in a given patch of habitat depends on its size, isolation and quality (e.g. Hanski and Gilpin, 1997). A major distinction has often been made between Levins-type systems, in which all patches are equally important to persistence (Levins, 1969, 1970), and mainland-island-type systems, in which some patches are more equal than others (Boorman and Levitt, 1973). Movement is crucial to the former, through recolonization of empty patches, and the rescue of small populations teetering on the brink. Between-patch movement is not crucial to the persistence of mainland-island systems – all that matters is the persistence of the mainland.

Natural systems may contain various blends of small, intermediate and large patches, with a gradation of differences in probability of extinction. *H. comma* shows the typical pattern of an increased probability of extinction from small patches (Thomas and Jones, 1993). Because the distributions of patch sizes differ among regions, we can deduce that the dynamics will also vary. In some areas, most of the habitat patches are quite small (especially in Surrey), and most of the network is probably important to persistence – reminiscent of a Levins-type system. Other systems (e.g. East Sussex) are characterized by larger patches, which are more widely spaced, and the smaller peripheral patches in the same network are probably unimportant to persistence – more like a mainland-island system, though there are several large patches.

Patch size is not the only consideration. Kindvall (1996) found that local cricket populations were more resistant to extinction in heterogeneous patches of habitat (containing *both* short *and* long grass), than in equally sized patches that contain simpler vegetation (containing *either* short *or* long grass). The crickets thrived in the short grass when conditions were cool, but survived better in longer vegetation during drought conditions.

These patch-specific descriptions of probability of survival in relation to area or habitat heterogeneity do not answer the fundamental question 'How much longer do networks of patches survive than does the most persistent patch within the system?'. We lack proper empirical data on this question. In some systems, most notably the *Melitaea cinxia* (L.) metapopulation in Finland (e.g. Hanski *et al.*, 1994, 1995; Kuussaari *et al.*, 1998), it is clear that networks are crucial to persistence because even the largest populations in the system are quite small and prone to extinction.

Other empirical studies are far less conclusive, mainly because the time scale of each study is small relative to the extinction frequency of the largest patches. In most empirical studies, the largest populations are not observed to become extinct over periods of 2–10 years. Work on the silver-studded blue butterfly, *Plebejus argus* (L.), around the Great Orme in north Wales, showed that the largest populations all survived for a period of 7 years = generations (Thomas and Harrison, 1992; Thomas, 1996). Yet examination of turnover over a 30-year period has revealed that only the largest population, plus one other large population, have been populated throughout (C.D. Thomas, R.J. Wilson and O.T. Lewis, unpublished). We could either conclude that it is really a mainland-island system, dependent on the largest patch, or that persistence is dependent on the whole network because the largest population might itself go extinct every few hundred years, and would have to be recolonized from some of the slightly smaller, peripheral populations.

The theoretical argument that the persistence of population networks depends on the survival of the most persistent populations within the network (Gilpin, 1990; Harrison and Quinn, 1990) has gained support from the widespread empirical observation that population dynamics are often synchronized over large areas, especially in insects (Hanski and Woiwod, 1993; Pollard and Yates, 1993; Sutcliffe *et al.*, 1996). If extinction probability is also synchronized, and there is some evidence that it may be in small populations (Sutcliffe *et al.*, 1997b), all other local populations are likely to become extinct in the one year when conditions are so bad that the largest becomes extinct.

There are three important sources of synchrony: spatially correlated weather (including effects mediated through natural enemies and resources), spatially correlated changes in habitat and dispersal (including dispersal of hosts, prey and natural enemies). Correlated insect population dynamics over hundreds of kilometres, effects of land management (habitat change) on population dynamics, and effects of dispersal rate on the distance over which population dynamics are synchronized suggest that all of these factors may be important (Thomas, 1991; Hanski and Woiwod, 1993; Pollard and Yates, 1993; Sutcliffe *et al.*, 1996).

The problem is whether the factors that lead to the extinction of the largest populations are the same as those responsible for yearly population variation within populations, and the synchronized extinction of small populations. Weather may be largely responsible for within-population fluctuations and extinctions of small populations (say < 500 adult insects), but extinctions of much larger populations (> 5000) may be more likely to be caused by changes in habitat conditions (e.g. succession). These changes may be relatively local events, such that the entire system does not disappear when the largest population is lost. Long-term persistence is then promoted through dispersal – the ability of population systems to track changes in the spatial distributions of suitable habitats.

## Population Ensembles

If one can measure birth, death, immigration and emigration for each patch or sample unit within a population system, it is then possible to plot an entire population network against the two axes, compensation and mobility (Thomas and Kunin, 1999). This is difficult, but the same data are required to place each population unit in one of the coarse categories in Table 15.1. In practice, it is necessary to derive general relationships between easily measured patch characteristics and $B$, $D$, $I$ and $E$, from a subset of patches, and then extrapolate to all patches in the system. For example, immigration rate may be correlated with habitat isolation, area and relevant measures of patch quality in a subset of patches, based on analyses of mark–release–recapture data (Hanski et al., 2000). If the same habitat characteristics are measured for all other patches in the system, then it is possible to estimate approximate immigration rates for all other patches. The result is that we can describe any system in terms of the distribution of population processes, rather than having to make one semi-arbitrary choice of population category for the entire system.

To represent and describe population processes within networks, we do not necessarily want to treat each patch equally, given the unequal contribution of different patches to regional persistence. Schoener (1991) pointed out that, even in habitat networks with high rates of local extinction and recolonization, most individuals may live within the few very large patches that survive indefinitely: 90% turnover of patches might only affect 10% of the individuals in a system. When identifying the properties of a habitat network, it is very helpful, therefore, to provide both the traditional patch-level description *and* a description weighted by the numbers of individuals within each patch: 90% of patches being dominated by immigration might involve the movement of only 10% of individuals, if most individuals stay within a few very large patches.

Thomas and Kunin (1999) crudely described two *H. comma* networks in relation to the compensation and mobility axes, weighting the estimates for each patch by the relative population size within them (Fig. 15.2). We had to take various short-cuts and made dubious assumptions about equilibrium population dynamics because the data were not collected with this analysis in mind – it was intended solely for illustration. Previously, Thomas and Jones (1993) had described both networks as metapopulations. However, Thomas and Kunin's quantitative analysis revealed major differences between the systems. The Surrey system is characterized by many small patches (mean 0.33 ha) that are relatively close together (mean 75 m apart). Because *per capita* emigration rates are high from small patches (Hill et al., 1996), 52% of individuals are estimated to emigrate, across the system as a whole, and about 38% of these are likely to arrive successfully in another patch. Thus, about 20% of individuals experience more than one patch (c. 80% stay in the natal patch, or fail to find another patch after emigrating). In East Sussex, there

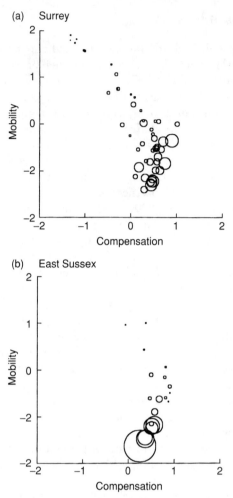

**Fig. 15.2.** *H. comma* population systems in Surrey and East Sussex (S. England), with each occupied patch plotted against the compensation and mobility axes (Thomas and Kunin, 1999, courtesy of the British Ecological Society). Symbols scale with approximate population size (Surrey 1–600 adults; E. Sussex 3–3720 adults).

are fewer, larger patches (mean 0.47 ha, but some are much bigger), that are further apart (mean 170 m apart). The largest patches have lower *per capita* emigration rates, but contain most of the population, so a much lower proportion of the entire population is predicted to emigrate (29%), and only about 4% of these are predicted to arrive successfully in other patches. In this system, only about 1% of individuals experience more than one patch. The greater flow of individuals in the Surrey system means that there is more spread along the compensation axis (i.e. a greater source–sink component to the dynamics) than in Sussex (Fig. 15.2).

Using a category-based system applied at the scale of grassland fragments, both systems can be described and modelled as metapopulations (Thomas and Jones, 1993; Hanski, 1994; Hanski and Thomas, 1994; Hill et al., 1996), but this common description hides differences in the relative importance of different population processes.

## Conclusions

The spatial structure and dynamics of animal populations have been described as belonging to a series of different categories. These categories are determined by different blends of birth, death, emigration and immigration. Different parts of the same population system may fall into different categories, and one system may be described as belonging to different categories, depending on the spatial scale at which it is studied. The answer is not to define more and more categories, but to move away from category-based descriptions. Existing population categories can be arranged along two axes that combine birth, death, emigration and immigration. This continuous process-based approach is theoretically more attractive and, in the applications of population ecology, it pinpoints the practical need to manipulate birth, death, emigration and immigration.

## Acknowledgements

Many thanks to Bill Kunin, Ilkka Hanski, Susan Harrison and Jeremy Thomas for their ideas.

## References

Boorman, S.A. and Levitt, P.R. (1973) Group selection on the boundary of a stable population. *Theoretical Population Biology* 4, 85–128.

Clarke, R.T., Thomas, J.A., Elmes, G.W. and Hochberg, M.E. (1997) The effect of spatial patterns in habitat quality on community dynamics within a site. *Proceedings of the Royal Society of London B* 264, 347–354.

Courtney, S.P. (1981) Coevolution of pierid butterflies and their cruciferous foodplants. III. *Anthocharis cardamines* survival, development and oviposition on different hostplants. *Oecologia* 51, 91–96.

Courtney, S.P. and Duggan, A.E. (1983) The population biology of the orange-tip butterfly *Anthocharis cardamines* in Britain. *Ecological Entomology* 8, 271–281.

Dempster, J.P. (1997) The role of larval food resources and adult movement in the population dynamics of the orange-tip butterfly (*Anthocharis cardamines*). *Oecologia* 111, 549–556.

Gilpin, M.E. (1990) Extinction of finite metapopulations in correlated environments. In: Shorrocks, B. and Swingland, I.R. (eds) *Living in a Patchy Environment*. Oxford Science Publications, Oxford, pp. 177–186.

Hanski, I. (1994) A practical model of metapopulation dynamics. *Journal of Animal Ecology* 63, 151–162.

Hanski, I.A. and Gilpin, M.E. (1997) *Metapopulation Biology: Ecology, Genetics, and Evolution*. Academic Press, San Diego.

Hanski, I. and Thomas, C.D. (1994) Metapopulation dynamics and conservation: a spatially explicit model applied to butterflies. *Biological Conservation* 68, 167–180.

Hanski, I. and Woiwod, I.P. (1993) Spatial synchrony in the dynamics of moth and aphid populations. *Journal of Animal Ecology* 62, 656–668.

Hanski, I., Kuussaari, M. and Nieminen, M. (1994) Metapopulation structure and migration in the butterfly *Melitaea cinxia*. *Ecology* 75, 747–762.

Hanski, I., Pakkala, T., Kuussaari, M. and Lei, G. (1995) Metapopulation persistence of an endangered butterfly in a fragmented landscape. *Oikos* 72, 21–28.

Hanski, I., Alho, J. and Moilanen, A. (2000) Estimating the parameters of survival and migration of individuals in metapopulations. *Ecology* 81, 239–251.

Harrison, S. (1991) Local extinction in a metapopulation context: an empirical evaluation. *Biological Journal of the Linnaean Society* 42, 73–88.

Harrison, S. (1994) Metapopulations and conservation. In: Edwards, P.J., May, R.M. and Webb, N. (eds) *Large Scale Ecology and Conservation Biology*. Blackwell Scientific Publications, Oxford, pp. 111–128.

Harrison, S. and Quinn, J.F. (1990) Correlated environments and the persistence of metapopulations. *Oikos* 56, 293–298.

Harrison, S. and Taylor, A.D. (1997) Empirical evidence for metapopulation dynamics. In: Hanski, I.A. and Gilpin, M.E. (eds) *Metapopulation Biology: Ecology, Genetics and Evolution*. Academic Press, San Diego, pp. 27–42.

Hill, J.K., Thomas, C.D. and Lewis, O.T. (1996) Effects of habitat patch size and isolation on dispersal by *Hesperia comma* butterflies: implications for metapopulation structure. *Journal of Animal Ecology* 65, 725–735.

Kindvall, O. (1996) Habitat heterogeneity and survival in a bush cricket metapopulation. *Ecology* 77, 207–214.

Kindvall, O. (1999) Dispersal in a metapopulation of the bush cricket, *Metrioptera bicolor* (Orthoptera: Tettigoniidae). *Journal of Animal Ecology* 68, 172–185.

Kuussaari, M., Saccheri, I., Camara, M. and Hanski, I. (1998) Allee effect and population dynamics in the Glanville fritillary butterfly. *Oikos* 82, 384–392.

Levins, R. (1969) Some demographic and genetic consequences of environmental heterogeneity for biological control. *Bulletin of the Mathematical Society of America* 15, 237–240.

Levins, R. (1970) Extinction. *American Maths Society* 2, 77–107.

Pollard, E. and Yates, T.J. (1993) *Monitoring Butterflies for Ecology and Conservation*. Chapman & Hall, London.

Pulliam, H.R. (1988) Sources, sinks, and population regulation. *American Naturalist* 132, 652–661.

Rodríguez, J., Jordano, D. and Fernández Haeger, J. (1994) Spatial heterogeneity in a butterfly–host plant interaction. *Journal of Animal Ecology* 63, 31–38.

Schoener, T.W. (1991) Extinction and the nature of the metapopulation. *Acta Oecologia* 12, 53–75.

Singer, M.C. and Thomas, C.D. (1996) Evolutionary responses of a butterfly metapopulation to human and climate-caused environmental variation. *American Naturalist* 148, S9–S39.
Singer, M.C., Thomas, C.D. and Parmesan, C. (1993) Rapid human-induced evolution of insect–host associations. *Nature* 366, 681–683.
Stamps, J.A., Buechner, M. and Krishnan, V.V. (1987) The effects of edge permeability and habitat geometry on emigration from patches of habitat. *American Naturalist* 129, 533–552.
Sutcliffe, O.L., Thomas, C.D. and Moss, D. (1996) Spatial synchrony and asynchrony in butterfly population dynamics. *Journal of Animal Ecology* 65, 85–95.
Sutcliffe, O.L., Thomas, C.D. and Peggie, D. (1997a) Area-dependent migration by ringlet butterflies generates a mixture of patchy-population and metapopulation attributes. *Oecologia* 109, 229–234.
Sutcliffe, O.L., Thomas, C.D., Yates, T.J. and Greatorex-Davies, J.N. (1997b) Correlated extinctions, colonizations and population fluctuations in a highly connected ringlet butterfly metapopulation. *Oecologia* 109, 235–241.
Thomas, C.D. (1991) Spatial and temporal variability in a butterfly population. *Oecologia* 87, 577–580.
Thomas, C.D. (1994) Extinction, colonization and metapopulations: environmental tracking by rare species. *Conservation Biology* 8, 373–378.
Thomas, C.D. (1996) Essential ingredients of real metapopulations, exemplified by the butterfly *Plebejus argus*. In: Hochberg, M.E., Clobert, J. and Barbault, R. (eds) *Aspects of the Genesis and Maintenance of Biological Diversity*. Oxford University Press, Oxford, pp. 292–307.
Thomas, C.D. and Hanski, I. (1997) Butterfly metapopulations. In: Hanski, I.A. and Gilpin, M.E. (eds) *Metapopulation Biology: Ecology, Genetics and Evolution*. Academic Press, London, pp. 359–386.
Thomas, C.D. and Harrison, S. (1992) Spatial dynamics of a patchily-distributed butterfly species. *Journal of Animal Ecology* 61, 437–446.
Thomas, C.D. and Jones, T.M. (1993) Partial recovery of a skipper butterfly (*Hesperia comma*) from population refuges: lessons for conservation in a fragmented landscape. *Journal of Animal Ecology* 62, 472–481.
Thomas, C.D. and Kunin, W.E. (1999) The spatial structure of populations. *Journal of Animal Ecology* 68, 647–657.
Thomas, C.D., Singer, M.C. and Boughton, D.A. (1996) Catastrophic extinction of population sources in a butterfly metapopulation. *American Naturalist* 148, 957–975.
Thomas, C.D., Jordano, D., Lewis, O.T., Hill, J.K., Sutcliffe, O.L. and Thomas, J.A. (1998) Butterfly distributional patterns, processes and conservation. In: Mace, G.M., Balmford, A. and Ginsberg, J.R. (eds) *Conservation in a Changing World*. Cambridge University Press, Cambridge, pp. 107–138.
Thomas, J.A., Thomas, C.D., Simcox, D.J. and Clarke, R.T. (1986) The ecology and declining status of the silver-spotted skipper butterfly (*Hesperia comma*) in Britain. *Journal of Applied Ecology* 23, 365–380.
Thomas, J.A., Elmes, G.W. and Wardlaw, J.C. (1993) Contest competition among *Maculinea rebeli* butterfly larvae in ants nests. *Ecological Entomology* 18, 73–76.
Thomas, J.A., Clarke, R.T., Elmes, G.W. and Hochberg, M.E. (1998) Population dynamics in the genus *Maculinea* (Lepidoptera: Lycaenidae). In: Dempster, J.P.

and McLean, I.F.G. (eds) *Insect Populations in Theory and in Practice*. Kluwer, Dordrecht, pp. 261–290.

Turchin, P. (1991) Translating foraging movements in heterogeneous environments into the spatial distribution of foragers. *Ecology* 72, 1253–1266.

Turchin, P. (1998) *Quantitative Analysis of Movement*. Sinauer Associates, Sunderland, Massachusetts.

Watkinson, A.R. and Sutherland, W.J. (1995) Sources, sinks and pseudo-sinks. *Journal of Animal Ecology* 64, 126–130.

Wiens, J.A. (1989) Spatial scaling in ecology. *Functional Ecology* 3, 385–397.

Wiklund, C. and Ahberg, C. (1978) Host plants, nectar source plants and habitat selection of males and females of *Anthocharis cardamines*. *Oikos* 40, 53–63.

# Gene Flow

**16**

## James Mallet

*Galton Laboratory, Department of Biology, University College London, 4 Stephenson Way, London NW1 2HE, UK*

## What is Gene Flow?

'Gene flow' means the movement of genes. In some cases, small fragments of DNA may pass from one individual directly into the germline of another, perhaps transduced by a pathogenic virus or other vector, or deliberately via a human transgenic manipulation. However, this kind of gene flow, known as horizontal gene transfer, is rare. Most of the time, gene flow is caused by the movement or dispersal of whole organisms or genomes from one population to another. After entering a new population, immigrant genomes may become incorporated due to sexual reproduction or hybridization, and will be gradually broken up by recombination. 'Genotype flow' would therefore be a more logical term to indicate that the whole genome is moving at one time. The term 'gene flow' is used probably because of an implicit belief in abundant recombination, and because most theory is still based on simple single locus models: it does not mean that genes are transferred one at a time. The fact that gene flow is usually caused by genotype flow has important consequences for its measurement, as we shall see.

## Two Meanings of 'Gene Flow'

We are often taught that 'dispersal does not necessarily lead to gene flow'. The term 'gene flow' is then being used in the sense of a final *state* of the population, i.e. successful establishment of moved genes. This disagrees somewhat with a more straightforward interpretation of gene flow as *actual movement* of genes. The tension between actual movement and successful establishment is at the heart of many misunderstandings of the term 'gene flow' in studies of genetic variability at marker loci (Slatkin and Barton, 1989). Under some stringent assumptions (see below) we can measure genetic variability (a description of the *state* of populations) and obtain an estimate of 'gene flow' in the form of $N_e m$ (the product of population size and fraction of migrants), but this tells us almost nothing about *actual* fraction of the population moving, $m$. Actual gene flow, $m$, could be very high or very low, depending on the value of $N_e$.

'Reproductive isolation', as employed in the biological species concept, can be viewed as a kind of inverse of gene flow, and has a similar ambiguity. 'Reproductive isolation' is a combination of some very different things: pre-mating isolation, or sexual behaviour which inhibits *actual* gene flow, and post-mating isolation, selection against hybrids which determines whether genes that do flow survive in their new genomic environment. Reproductive isolation is therefore most straightforwardly interpreted as a stable but strongly divergent balance between gene flow and selection, rather than as complete isolation from actual gene flow. As with 'gene flow', the term 'reproductive isolation' is used ambiguously in many publications, sometimes to mean pre-mating isolation, sometimes to mean post-mating isolation, and sometimes both. It is therefore advisable to distinguish hybrid inviability, sterility and mate choice rather than lumping all into a single term 'reproductive isolation' or 'isolating mechanisms' (Mallet, 2000).

Of course, there is nothing wrong with ambiguous terminology provided we know what we mean, but it is all too easy to become deluded into measuring 'gene flow' via $N_e m$ and then to attempt to use this knowledge in ways which require knowledge of actual gene flow, $m$, between populations, as in the retardation of adaptation. Similarly, to say that sexual species are defined by their 'reproductive isolation' is fine provided one realizes at the same time that species may not *actually* be reproductively isolated: horizontal gene transfer rates can be non-zero.

## Which 'Gene Flow' do we Want to Measure?

Both types of gene flow are of course worth measuring. However, the balance version of gene flow simply measures the state of the population, rather than the process of gene flow leading to that state. Thus, it is normally more useful to measure actual gene flow because it is true of all loci, whether under selection or not, and is of interest in all models of evolution. For instance, supposing we were able to measure actual gene flow distance, $\sigma_x$ (the standard deviation of parent–offspring distances), from studies of neutrally drifting genes (in practice, we usually cannot; see below), we could then apply this measure in ecological and population genetic problems unrelated to genetic drift. As an ecological example, the velocity of spread of an invading species is expected to be $v_x = \sqrt{2r\sigma_x^2}$, where $r$ is the intrinsic rate of increase (Andow et al., 1993). In contrast, 'gene flow' in the sense of the balance of gene flow and drift given by $N_e m$ is of use only in understanding population differentiation under drift.

The remainder of the chapter will examine the measurement of the balance state and of actual gene flow between local populations, geographic races and sympatric host races or species.

## The Balance Between Gene Flow and Genetic Drift

One of the simplest effects of gene flow is on genetic drift in interconnected populations. The probability of identity by descent increases in a population by a factor $F_{ST} = 1/(2N_e)$ in each generation. $N_e$ is the 'effective size' of an ideal diploid population that loses heterozygosity at the same rate as the actual population of size $N$. Because some individuals in a population can monopolize resources or reproduction, $N_e$ is often less than, although of the same order as, $N$ (Nunney, 1993). Frankham (1995) reviewed data which suggest that $N_e$, may typically be as low as 5–10% of the value of $N$. In the absence of gene flow, drift would eventually cause fixation at neutral alleles in all subpopulations: if the initial allele frequency in all the populations were $\bar{q}$, subpopulations would diverge so that a fraction $\bar{q}$ of the populations eventually become fixed for the allele, while the remaining $1 - \bar{q}$ will be fixed for some alternative allele. Dispersal, however, reduces this effect by carrying alleles between populations, and a balance between drift and gene flow can result. The probability of identity by descent can be measured by the variance in gene frequency $s_q^2$ between populations expressed as a fraction of its theoretical maximum $\bar{q}(1 - \bar{q})$ when all populations are fixed, i.e.

$$F_{ST} = \frac{s_q^2}{\bar{q}(1-\bar{q})}.$$

$F_{ST}$ can reach an equilibrium when drift, which causes divergence of neutral allele frequencies and local homozygosity, is balanced by the homogenizing influence of gene flow. A simple but spatially unrealistic model of metapopulation structure, the 'island model' (Fig. 16.1a), has an immigration fraction $m$ per generation into each subpopulation, or 'island', of size $N_e$; migrants can come from anywhere else in the metapopulation or from a 'mainland', and therefore have a constant gene frequency $\bar{q}$. Under these circumstances, it can be shown (Wright, 1969: 291) that

$$F_{ST} \approx \frac{1}{1+4N_e m}$$

will result at equilibrium between gene flow and drift. This relation can remain approximately valid even in more realistic population structures, such as the 'stepping stone' model, in which migrants are exchanged mainly with

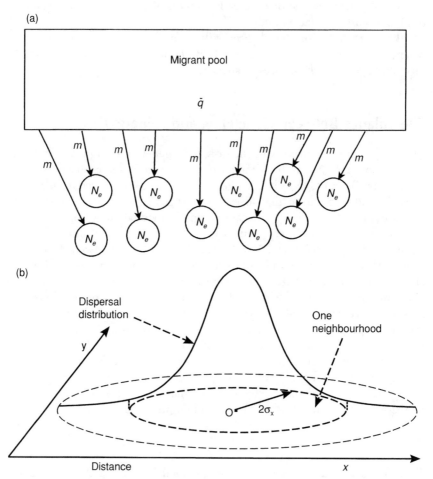

**Fig. 16.1.** Models of population structure. (a) Wright's island model. A large number of 'islands' each with population size $N_e$ are assumed. A fraction $m$ of each subpopulation is replaced per generation from a universal pool of migrants having neutral gene frequency $\bar{q}$ equivalent to the average over all subpopulations. Under this model, the potential for local drift then depends on the combined parameter $N_e m$. (b) Wright's isolation by distance model. In this model, population structure is more spatially realistic so that individuals are distributed continuously in a two-dimensional plane. If the dispersal is $\sigma_x$ (the standard deviation of parent-offspring distances), the potential for local drift can be shown to depend on the 'neighbourhood population size', numerically the number of individuals within a circular neighbourhood of radius $2\sigma_x$. The neighbourhood size, $N_b = 4\pi \rho_e \sigma_x^2$, is the product of the area of this circle and the effective population density $\rho_e$. $N_b$ is again a product of gene flow and population numbers, and it plays a role in isolation by distance equivalent to $N_e m$ in the island model.

local subpopulations (Slatkin and Barton, 1989). If, in addition to local dispersal, a small fraction of migrants are exchanged over very long distances, $F_{ST}$ may stabilize over the whole metapopulation structure as in the island model. If there is no long-distance dispersal, mutation or weak selection to restore polymorphisms, $F_{ST}$ will increase with distance as drift within subpopulations becomes more and more independent (Wright, 1969; Slatkin, 1993).

If populations are small and/or migrants are rare so that $N_e m \ll 1$, drift on the islands outweighs gene flow, and gene frequencies in subpopulations become nearly fixed at $q = 0$ or $q = 1$, so $F_{ST} \to 1$. Conversely, if populations are large and/or migrants are common so that $N_e m \gg 1$, migration outweighs drift and all gene frequencies converge on $\bar{q}$, so that $F_{ST} \to 0$. The value $N_e m \approx 1$ is therefore a kind of cusp between these two regimes, where populations can have almost any gene frequency with equal probability (Wright, 1969).

Wright (1969: 292) points out that $N_e m$, as a product of population size and fraction of the population migrating, is numerically equivalent to the number of immigrants that arrive in any subpopulation in each generation. $N_e m$ has therefore been interpreted as a kind of 'gene flow', $\hat{M}$, especially by recent authors (Slatkin, 1987, 1993). This 'gene flow' can be estimated by studying allele frequencies in natural populations, because

$$\hat{M} = N_e m \approx \frac{1}{4}\left(\frac{1}{F_{ST}} - 1\right).$$

However, it is worth remembering that this 'gene flow' in fact consists of a ratio between two dimensionless probabilities (actual gene flow, $m$, and genetic drift, $1/2N_e$); each consists of a fraction of INDIVIDUALS per total INDIVIDUALS). The contributions of drift and gene flow to gene frequency variation are both, of course, relative to a single generation, and therefore also have units TIME$^{-1}$. $N_e m$ itself is formed from a ratio of these probabilities, and its units (TIME$^{-1}$)$^{-1}$ × TIME$^{-1}$ cancel to give a quantity that is dimensionless (Barton and Rouhani, 1991: 501), as it must be if it is to be proportional to $1/F_{ST}$. In a sense, then, the interpretation of $N_e m$ as a number-based form of gene flow is illusory: although numerically correct, this parameter actually has no units at all. Quite apart from the obvious difficulties with assumptions under the unrealistic island model (Weir, 1996; Bossart and Pashley Prowell, 1998; Whitlock and McCauley, 1999), $N_e m$ is not a very useful quantity because the equation merely transforms our data, in the form of $F_{ST}$, rather than solving for the gene flow parameter of interest, $m$, valid for use where drift is not involved. We might of course employ $N_e m$ to 'predict' equilibrium levels of $F_{ST}$ under neutral drift, but this is of very limited use because we already know what the $F_{ST}$ is: it is the information from which we obtained our estimate of $N_e m$. In many cases, it will be more sensible, and perhaps more honest, simply to cite values of $F_{ST}$ from which this 'gene flow' is derived rather than making the transformation to $N_e m$ or $\hat{M}$, whether based on $F_{ST}$, rare alleles (Slatkin, 1987) or gene genealogies (Slatkin, 1989, 1993; Slatkin and Maddison, 1989).

Wright (1969) also developed a more lifelike model of continuous population structure known as 'isolation by distance'. This model is far from tractable (Felsenstein, 1975), explaining its unpopularity in genetic marker studies. However, its spatial realism makes isolation by distance potentially far more useful than the island model. The metapopulation is assumed more or less continuous, rather than discrete, and migration is modelled as a diffusion process with a bivariate Gaussian distribution: any individual born at the origin O has a dispersal distribution that is symmetrical around the birth place, and its own offspring will be born along the $x$ dimension with probability given by the variance of the dispersal distribution $\sigma_x^2$ (Fig. 16.1b). This variance is related to the 'diffusion parameter', $D$, much used in ecology (e.g. Kareiva, 1982; Andow et al., 1993): in fact, $D = \sigma_x^2 / 2$. Assuming that dispersal in $x$ and $y$ dimensions are both given by $\sigma_x^2$, 86.5% of offspring are born within a circular 'neighbourhood' of $\pm 2\sigma_x$ around the birthplace of the parent (Fig. 16.1b – note that the familiar 95.4% probability integral of the Gaussian distribution at $\pm 2\sigma_x$ applies only in one dimension). The equilibrium $F_{ST}$ of neutral alleles under drift is controlled by the 'neighbourhood population size', $N_b = 4\pi\sigma_x^2\rho_e$, which is the area of the neighbourhood multiplied by the effective population density $\rho_e$ (Wright, 1969). Drift enters as population density $1/\rho_e$, a probability term with units DISTANCE$^2$, while gene flow, $\sigma_x^2$, represents the probability of movement also with units DISTANCE$^2$; the two again cancel. Because both drift and gene flow effects are per generation, TIME$^{-1}$ also enters as a unit in each, and again cancels in the ratio. Thus 'neighbourhood population size', $N_b$, although numerically the effective number of individuals within a circle of radius $2\sigma_x$, is another dimensionless quantity, the continuous population equivalent of $N_em$ (Slatkin and Barton, 1989; Barton and Rouhani, 1991). Curiously, $N_b$ has usually been seen as a kind of 'effective population size' (e.g. Hartl and Clark, 1989), rather than 'gene flow' as for $N_em$, but both these incomplete interpretations are equally misleading. $N_em$ and $N_b$ are neither measures of gene flow nor of effective population size. The truth is somewhere in between: both are dimensionless ratios which determine the relative strength of gene flow and drift. As with the stepping stone model, the distances across which $F_{ST}$ is measured will determine its magnitude (Wright, 1969; Slatkin, 1993; Rousset, 1997) unless some additional 'restoring force' of mutation, long-distance migration or weak stabilizing selection controls the average gene frequency across the whole metapopulation of neighbourhoods (Slatkin and Barton, 1989).

In conclusion, while it is possible to estimate the dimensionless quantities $N_em$ and $N_b$, it is hard to infer anything very useful about actual gene flow from these estimates alone (see also McCauley and Eanes, 1987; Bossart and Pashley Prowell, 1998; Whitlock and McCauley, 1999). This may seem surprising, because this is exactly what a rather large number of studies have apparently attempted (see the useful reviews by Roderick, 1996; Peterson and Denno, 1998). Of course, comparative studies of $F_{ST}$ do give *some* idea of gene flow, but inferences are rather weak: a species with a high $F_{ST}$ is likely to have lower levels of gene flow ($m$ or $\sigma_x^2$) than a different species with low $F_{ST}$, provided their population densities are comparable. But in many cases, we cannot make this assumption. In fact, it is rare that we can assume $F_{ST}$ is even at equilibrium with respect to drift and gene flow. Somewhat embarrassingly, I myself have co-authored a paper that purports to estimate gene flow via $N_e m$ (Korman *et al.*, 1993). I defend this paper on the grounds that we generated some information about the spatial scale of gene flow, as well as $N_e m$ (see below).

## Multilocus Data and Interlocus Correlations

Because dispersal causes a flow of whole genotypes, or 'genotype flow', it will lead to a correlation between immigrant alleles at different loci, 'linkage disequilibria' (Fig. 16.2), as well as to eventual homogenization of gene frequencies. Thus, a balance between gene flow, drift and recombination will control linkage disequilibria within populations, just as a balance between drift and gene flow controls $F_{ST}$. In principle, therefore, combined data on

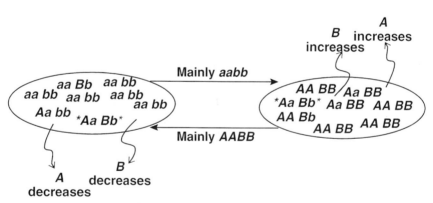

Fig. 16.2. Gene flow and gametic correlations. A pair of populations are assumed to exchange individuals. If the two populations come to differ in gene frequency at each of two loci (*A, B*), whether caused by selection or drift (bent arrows), exchange of individuals will ensure that there are gametic correlations, or linkage disequilibria, within each population. For example, the left-hand population consists mainly of *ab* gametes, but immigrants will bring in mainly *AB* gametes. If these immigrants mate, their offspring will be mainly AB/ab genotypes (starred in the diagram). Thus the presence of an immigrant allele *A* makes it more likely that another immigrant allele *B* will also be present in the same genome: the two loci will be correlated. This principle applies also to spatially separated areas within continuous populations, and is particularly important in clines and hybrid zones.

$F_{ST}$ and linkage disequilibria can be used separately to estimate the gene flow ($m$ or $\sigma_x^2$) required for the maintenance of disequilibria, as well as the neighbourhood size ($N_e m$ or $N_b$) at equilibrium: two independent effects are known, allowing us to solve for two unknowns (Vitalis and Couvet, 2000). In what is in effect the same method, multilocus information may be used to infer the source area from which migrants originated (Cornuet *et al.*, 1999). This genotypic identification technique may prove especially useful for determining sources of exotic pests such as the Mediterranean fruit fly, *Ceratitis capitata* (Weidemann) (Davies *et al.*, 1999a,b). However, it is as yet unclear whether linkage disequilibria expected under drift (which are, after all, weak second-order effects) provide sufficient information with reasonable sample sizes.

## Spatial Information in Gene Flow/Drift Balance

Spatial information present in the measurement of gene frequencies under neutral drift may in principle be used to give spatial information about gene flow. In a continuous population, the magnitude of local gene frequency fluctuations ($F_{ST}$) is controlled by $N_b$, while the spatial extent of these fluctuations is controlled by $\sigma_x^2$ (Barton and Clark, 1990; see also Fig. 16.3). The

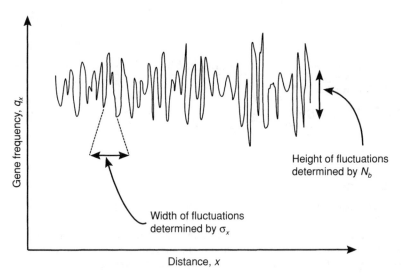

**Fig. 16.3.** Drift in continuous populations. In a continuous population, gene frequencies will fluctuate under genetic drift due to finite neighbourhood size $N_b$. Locally, these fluctuations will be correlated due to dispersal, $\sigma_x$. Thus the spatial scale of the decline in gene frequency correlation with distance (in other words, the width of the gene frequency fluctuations) will depend on $\sigma_x$, while the variability of gene frequencies will depend on $N_b$.

distance at which gene frequency fluctuations in local populations become more or less independent should therefore be useful for understanding the spatial extent of gene flow as measured by $\sigma_x$. However, because local fluctuations change between generations, one should carry out all local sampling as a 'snapshot' within a single generation to preserve this spatial information. Needless to say, it has rarely been recognized that spatial information is destroyed when sampling is done over multiple generations. There are many studies where hierarchical analyses of $F_{ST}$ have been used to investigate the relative importance of gene flow at different spatial scales (Peterson and Denno, 1998), but few explicitly state that samples were obtained as a snapshot. In the important cotton pest *Heliothis virescens* (F.) we collected all samples within a single generation, and showed that allozyme frequency variation was significant at the lowest spatial scale (8 km); this implies that dispersal is restricted at this scale, even though $F_{ST}$ was very low overall ($\approx 0.002$) across the entire southeastern USA (Korman *et al.*, 1993). On the basis of this evidence for restricted gene flow ($\sigma_x \leq 10$ km gen.$^{-1/2}$), we argued that insecticide resistance in *Heliothis* is manageable on a local or countywide scale of ~10 km. The relationship of $F_{ST}$ (or its inverse, $\hat{M} = N_e m$) with distance can similarly be used to give an idea of the degree of isolation by distance (Slatkin, 1993).

It is possible to use 'spatial autocorrelation' (Sokal *et al.*, 1989) to determine the distance at which gene frequency correlations between subpopulations (again sampled as a snapshot) drop to an insignificant level. Recently, simulations have shown how $\sigma_x$ affects patterns of spatial genetic autocorrelation (Epperson and Li, 1997; Epperson *et al.*, 1999), but direct analytical methods are not yet available. This distance should give a rough idea of $\sigma_x$ (see also Rousset, 1997). An interesting feature of drift in two-dimensional populations is that it will generally outweigh balancing selection locally. Drift around a mean gene frequency determined by selection could in principle be as informative as completely neutral drift (Barton and Clark, 1990). In fact, drift around any equilibrium gene frequency could be used: for example, variation around the gene frequency cline expected for selection against heterozygotes can also give estimates of $N_b$ and $\sigma_x$ (Barton and Clark, 1990; N.H. Barton, unpublished analyses).

## Direct Estimates of Dispersal and 'Slatkin's Paradox'

A puzzling phenomenon in empirical studies of population structure is known as 'Slatkin's paradox'. Direct estimates of effective population size and dispersal distance often suggest low values of $N_e m$ or $N_b$ (~0.1–10; e.g. see Frankham, 1995) which predict high expected equilibrium levels of $F_{ST}$. In contrast, actual measures of $F_{ST}$ at marker loci such as allozymes, especially in flying insects, are often very low (<< 0.1), implying large neighbourhood sizes ($N_b$, $N_e m$ >> 10; for reviews of insect cases see McCauley and Eanes,

1987; Roderick, 1996; Peterson and Denno, 1998). For example, the butterfly *Euphydryas editha* (Boisduval) differs little in allozyme frequencies between the Rocky mountains and the Sierra Nevada in the western USA, even though it is almost inconceivable that this weakly flying, colony-forming butterfly ever disperses across the vast Great Basin desert (Slatkin, 1987).

Possible explanations of Slatkin's paradox include three major candidates:

1. The lack of allelic fixation might be due to balancing selection on allozyme loci (e.g. Avise, 1994). However, in cases where $F_{ST}$ at different loci are similar, it is not really tenable to expect that the strengths of balancing selection are also so similar at every locus.
2. Slatkin (1987) himself argued that these low $F_{ST}$ values resulted from frequent colonization and extinction, and that founders from distant populations would cause greater spatial homogeneity than expected given local movement measurable in mark–recapture studies. In other words, gene frequencies are similar across large areas because drift and gene flow equilibrate on a time scale longer than that of population turnover. In northern latitudes, genetic homogeneity could be a consequence of events as long ago as $10^4$ years. Recolonization occurred after the last retreat of the ice, and there have been high effective population numbers since then. Under these circumstances, an equilibrium between slow drift and low existing levels of gene flow will not yet have been reached over the entire range. On a more local scale, analysis of populations of the fungus beetle *Phalacrus substriatus* (Gyllenhal) shows clearly that long-distance colonizations cause a much lower $F_{ST}$ than expected on the basis of the mainly local post-colonization gene flow and population sizes observed (Ingvarsson and Olsson, 1997). However, colonization/extinction cycles may also inflate $F_{ST}$ compared with the island model, instead of reducing it, depending on the number and genetic similarity of original colonists in each subpopulation (Whitlock and McCauley, 1990). If populations go through frequent bottlenecks when colonizing new habitat, and if members of each founder group are from similar population sources, $F_{ST}$ will be higher than that predicted from post-colonization gene exchange equilibrium, as found by Whitlock (1992) in another fungus-feeding beetle, *Bolitotherus cornutus* (Panzer). Thus, consideration of population turnover can explain Slatkin's paradox but does not necessarily do so.
3. Another highly probable reason for Slatkin's paradox is that long-range movement of marked individuals will usually be underestimated because study sites are finite. Small numbers of very long-range dispersers which have a large genetic effect on $\sigma_x^2$ and lead to strongly lowered $F_{ST}$ may escape the study site and so be missed in field studies. Thus Dobzhansky and Wright measured dispersal in *Drosophila pseudoobscura* Frolova as $\sigma_x = 59\text{--}81$ m day$^{-1/2}$ depending on the study site, leading to an expected 7-day lifetime dispersal of only $\sigma_x = 157\text{--}218$ m gen.$^{-1/2}$ (Wright, 1978). It therefore appeared

that local genetic drift could be quite extensive in this species. However, later studies showed that oases were recolonized every year by *D. pseudoobscura* females dispersing many tens of kilometres across desert habitat (Jones et al., 1981; Coyne et al., 1982). Of course, dispersal over desert may be enhanced compared with the habitats in which earlier studies were done. None the less, the existence of such long-range movements casts grave doubt on the original estimates.

Explanations (2) and (3) are in fact related, because long-distance colonists are also those that fall outside the normal dispersal range of the majority. The distribution of dispersal found in many mark–recapture studies is not Gaussian but markedly leptokurtic (i.e. most individuals disperse near the parent, but there is a long-distance 'tail' containing significantly more individuals than expected under the Gaussian model). It can easily be shown that the strength of leptokurtosis (measured by $\sigma_x^4$) has little effect on overall $F_{ST}$ at neutral loci; use of $\sigma_x^2$ alone in the Gaussian model usually suffices (Wright, 1969: 303–307). Instead, and more importantly, the practical effect of leptokurtosis is to cause the value of $\sigma_x^2$ itself to be greatly underestimated in mark–recapture studies, when a long tail escapes detection beyond the edge of the study site. Thus leptokurtosis will homogenize distant populations, leading to much stronger reduction of equilibrium $F_{ST}$ over large areas than expected from local observations of dispersal (Rousset, 1997).

All in all, therefore, it remains extremely difficult to obtain useful measures of gene flow either directly, via mark–recapture, or indirectly, using allele frequency variation due to genetic drift.

## Gene Flow – Selection Balance

One reason why inferring gene flow indirectly is so difficult is that, in reasonably abundant mobile species (such as flying insects), $N_e$ and $m$ are large, so equilibrium levels of drift measured by $F_{ST}$ or disequilibria are expected to be very small, leading to large errors in its measurement. Selection, sampling or even laboratory scoring errors can dominate the weak signal due to drift. We found clear examples of laboratory errors by previous workers, almost certainly due to scoring problems, when we showed $F_{ST}$ to be an order of magnitude less than previously recorded in *Heliothis* (Korman et al., 1993). In contrast to drift, selection in natural populations is often a much stronger force: inferences about gene flow from the balance between gene flow and selection will be more robust than those from the balance between gene flow and drift (Lenormand et al., 1998).

The frequency $\hat{q}$ of an allele at equilibrium between immigration $m$ and counterselection $s$ ($s$ is the fractional increase of mortality of genotypes bearing the allele) within a population is given by $\hat{q} \approx m/s$, providing that $m < s$ (Haldane, 1930). This has potential applications for the possibility of adaptation, for example in insecticide resistance: gene flow will prevent the evolution of resistance in a treated population until the selection for resistance reaches a critical level, whereupon the treated population flips to a new state of high resistance caused both by selection and also by the attainment of a higher population density more resistant to genetic swamping (Comins, 1977). The flip corresponds to a 'cusp catastrophe' which occurs at a critical value of $m/s$ (Comins, 1977).

## Minimum Size of an Area for Adaptation

The balance between selection and gene flow is also tractable when we consider the problem spatially. The extent to which an insect is a 'slave of the environment' depends to a large extent on its rate of dispersal (Loxdale and Lushai, 1999). Thus, insecticide resistance (or any other adaptation) will evolve locally if the area treated has a radius much greater than about $\sigma_x/\sqrt{s}$. The reason for this, as originally shown by Haldane and Fisher, is that the equilibrium width of a cline in allele frequency, where selection on an allele with fitnesses $(1+s)$ and $(1-s)$ in two adjacent areas is given by $w \approx \sqrt{(3\sigma_x^2/s)}$ (Fig. 16.4; Slatkin, 1973; Endler, 1977; Roughgarden, 1979). This analytical theory of clines depends on a diffusion approximation, and so requires weak selection, but the results remain approximately valid for $s \leq 0.2$. For insecticide resistance to evolve, the critical diameter $d_c$ of an area for adaptation requires at least a small multiple of cline width

$$d_c > \pi\sqrt{\frac{\sigma_x^2}{8s}}$$

(Nagylaki, 1975; Roughgarden, 1979; Barton and Clark, 1990), so that insecticide resistance can attain a high frequency inside the treated area (Fig. 16.4; see also Lenormand and Raymond, 1998).

## Indirect Measurement of Gene Flow in Clines

Although theory relates cline width to gene flow $\sigma_x$ and selection $s$ in a potentially constructive way, it is again hard to reverse the equation. We will often be able to measure cline widths, but would prefer to estimate $\sigma_x$ and $s$ separately rather than their ratio. An estimate of $\sigma_x/\sqrt{s}$, being merely a reformulation of cline width, is as useless, in its way, as the combined drift/gene flow quantity $N_e m$. If we had direct measures of either $\sigma_x$ or $s$, we could of course solve for the other. Endler (1977) used field estimates of dispersal $\sigma_x$ to estimate selection pressures acting in clines using this method. Many of the selection pressures Endler estimated were so low ($10^{-5}$–$10^{-9}$) as to be incredible, for instance in the case of chromosomal or butterfly warning colour hybrid zones. These low estimates of selection are probably due to Slatkin's paradox: a tenfold underestimate of $\sigma_x$ would generate a 100-fold underestimate of s, which enters into the cline equation as a square root. Even under ideal circumstances, there will often be situations where $\sigma_x$, and $s$ are both poorly known.

A possible exception occurs when a cline is moving as a wave of advance. Turelli and Hoffmann (1991) estimated dispersal in *Drosophila simulans* Sturtevant from data on a moving cline of maternally inherited *Wolbachia* parasites. The *Wolbachia* infection, which causes unidirectional sexual incompatibility, was spreading north in California at a rate of about 175 km year$^{-1}$. The gene flow of $\sigma_x \approx 45$–$60$ km gen.$^{-\frac{1}{2}}$ was estimated using cline theory, the measured speed of the wave of advance of *Wolbachia*, and laboratory estimates of selection pressures. Again in accordance with Slatkin's paradox, the observed measurements were nearly two orders of magnitude larger than those estimated in Dobzhansky and Wright's experiments on *D. pseudoobscura* (Wright, 1969).

One might be tempted to employ pairwise measures of $N_e m$ based on marker alleles across clines or hybrid zones as 'gene flow' estimates, as advocated by Porter (1990). This should be avoided. Estimating $N_e m$ from $F_{ST}$ assumes that drift and gene flow have reached equilibrium. However, the

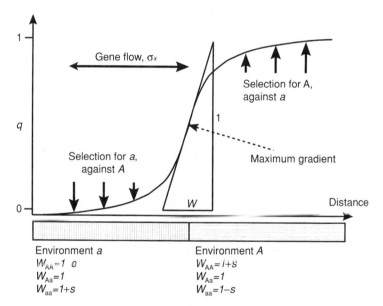

**Fig. 16.4.** Selection in continuous populations. Selection, $s$, in a continuous population may favour allele $a$ on the left of the diagram, and allele $A$ on the right. At equilibrium, the gene frequencies will form a sigmoid cline over the boundary between the environments, whose width, $w$, is given by a ratio between gene flow and selection ($\sigma_x / s$). The case of different environments is shown here; however, similar clines are formed in the case of intrinsic or frequency-dependent selection, for example in contact zones between races differing in an underdominant chromosomal rearrangement, or in warning colour pattern.

hiatus in gene frequency across a cline implies either a temporary situation where gene flow and drift have not equilibrated as required, or, more probably, that selection is maintaining allelic divergence (albeit possibly indirectly), so that a selection/gene flow balance rather than a drift/gene flow balance is the correct model.

## Gene Flow in Multiple Locus Clines

Often, multiple locus clines occur together in hybrid zones. For example, in the butterfly *Heliconius erato* (L.), three loci determine colour pattern differences across a hybrid zone near Tarapoto, Peru. *Heliconius* are warningly coloured, which causes purifying frequency-dependent selection against rare morphs. Analytical solutions for this type of selection suggest that cline widths at equilibrium should stabilize at $w \approx \sqrt{(8\sigma_x^2/s)}$ to $w \approx \sqrt{(12\sigma_x^2/s)}$, depending on dominance (Mallet and Barton, 1989a).

Dispersal across a multiple locus hybrid zone causes the immigration of whole genotypes, rather than flow of single genes. This genotype flow between two populations with distinct gene frequencies should cause a non-independence in the form of linkage disequilibria or correlations between genes within genotypes sampled in these populations (Fig. 16.2). As first shown by Barton (1982), genotype flow across such clines of allele frequency will lead to a stable maximum of correlations $R_{AB}$ between a pair of genes, $A$ and $B$ of approximately

$$R_{AB} \approx \frac{4\sigma_x^2}{c_{AB} w_A w_B}$$

near the centre of the hybrid zone where gene frequencies are 50%. Here, $R_{AB}$ is the interlocus correlation coefficient, $w_A$ and $w_B$ are the widths of clines at loci $A$ and $B$, and $c_{AB}$ is the recombination rate. The correlation equilibrates due to a balance between immigration, which increases disequilibrium, and recombination, which reduces it. By sampling genotypes, we can measure $w_A$ and $w_B$ as well as $R_{AB}$ directly. Gametic correlations can also, of course, be affected directly by selection for particular gene combinations, or epistasis. In *Heliconius*, fitnesses at the different loci were assumed multiplicative, i.e. non-epistatic. Additional epistasis is indeed likely in colour pattern combinations. However, provided that selection is not too strong, this second order effect on disequilibria will be weak in comparison to gene flow (Mallet and Barton, 1989a; Barton and Shpak, 2000). Given our equations for equilibrium width of a cline and disequilibria between loci, we now have two equations and two unknowns, $\sigma_x$ and $s$, which we can therefore solve. Analysis of three unlinked colour pattern genes in a hybrid zone in *H. erato* led to estimates of $\sigma_x \approx 2.6$ km gen.$^{-\frac{1}{2}}$, and $s \approx 0.15$–0.33 gen.$^{-1}$ per locus (Mallet *et al.*, 1990, 1998a). An estimate of selection based on mark–recapture agreed approximately with the indirect measure from cline theory (Mallet and Barton, 1989b), but again in accordance with Slatkin's paradox, dispersal estimated via mark–recapture (Mallet, 1986, $\sigma_x \approx 0.296$ km gen.$^{-½}$) was an order of magnitude less than that estimated in clines. The discrepancy was presumably due to the difficulty of tracing long-distance movers in the field dispersal study.

Recently, work with insecticide resistance clines in the mosquito *Culex pipiens* L. has used similar techniques to obtain robust estimates of selection and gene flow from genotypic data. This mosquito has been treated for many years with insecticides during the summer near Montpellier in southern France. The coastal strip where insecticides are applied is approximately 20 km wide. This has led to high summer coastal frequencies of resistance alleles at two major loci: target site insensitivity *Ace*-1, and amplified esterase *Ester*. These loci are linked with recombination rate $c \approx 14.5\%$. Resistance is selected against in untreated areas inland, leading to the formation of clines in the two resistance loci. Analysis of cline widths and disequilibria of the two loci gave estimates of resistance allele selective advantages of 30 and 16% in the presence of insecticides, and disadvantages of 12 and 5.5% in their absence, for *Ace*-1 and *Ester* respectively. Gene flow was estimated as $\sigma_x \approx 6.6$ km gen.$^{-½}$ (Lenormand *et al.*, 1998).

Further analyses of *Culex* were able to show variation in selection and dispersal through the season. In the autumn, migration was inland, towards overwintering sites. In the winter, no spraying is performed and the adults roost in caves, dispersing little, and there is strong selection against resistant alleles in the females. However, there is little or no effective selection against resistance alleles in the males during winter, since the females store sperm after mating in the autumn. Breeding in spring therefore causes a rise in resistance frequency as these unselected sperm unite with female gametes to form offspring (Lenormand *et al.*, 1999; Lenormand and Raymond, 2000). In accordance with Slatkin's paradox, gene flow was generally higher than expected on the basis of mark–recapture estimates, probably because long-distance dispersers were missed in field studies; selection estimates were consistent with laboratory data (Lenormand and Raymond, 1998).

The practical value of these kinds of estimates of gene flow and selection can be profound. Lenormand *et al.* (1999) recommend that if the coastal insecticide treatment area was halved to about 10 km wide, gene flow would swamp the selection for insecticide resistance, and there would be little problem controlling the mosquito. This year (2000), health authorities in the south of France are attempting to test this prediction for public benefit. The *Culex* story is perhaps the most complex and informative analysis of selection and gene flow ever achieved in the field, and the results are of potential economic and public health importance. This work shows the power that can be achieved through indirect measures of gene flow via cline theory and genetic data.

## Gene Flow Between Host Races or Species

As we move up the evolutionary continuum from populations through to species, hybridization rates tend to decrease. For pairs of taxa near the species boundary, we can use mass-action population genetic theory similar to that discussed above (Barton and Gale, 1987), rather than investigating the results of occasional historical events to study gene flow. For example, in the apple maggot *Rhagoletis pomonella* (Walsh), disequilibrium measurements suggest that gene flow between genetically distinct apple and haw host races was about $m \approx 20\%$ per generation (Barton et al., 1988; Feder et al., 1988, 1990). Direct measurements of habitat and mate choice between apple and haw host races suggest a slightly lower level of $m \approx 6\%$ (Feder et al., 1994). However, these results are not too dissimilar, given that gene flow must be highly contingent on the particular spatial and temporal arrangements of hawthorn and apple fruiting trees.

Hybridization and gene flow between species is something that, under the biological species concept, apparently ought not to happen. There is now increasing realization that hybridization, while rare, does occur between clearly marked taxonomic species in the wild. Among insects, hybridization seems especially prevalent in Lepidoptera, perhaps because strong colour pattern divergence makes hybrids easily identifiable, as in birds. Among European butterflies, 12% of all species are known to hybridize with at least one other species (Guillaumin and Descimon, 1976), but the rate is probably similar elsewhere, such as in North American swallowtails (6%, Sperling, 1990) and heliconiines (25%, Mallet et al., 1998a). Few *Drosophila* species are known to hybridize, but it is notable that those that do are within the *melanogaster* Meigen, *pseudoobscura* Frolova, Hawaiian picture-wing, and other groups that are genetically well studied (Gupta et al., 1980). It seems not unlikely that interspecific hybridization among related *Drosophila* is as frequent, though less evident, as in Lepidoptera. However, while species that hybridize are fairly common, the rates at which hybrids occur are very low, almost always, in insects, less than 1/1000 individuals per sympatric species pair.

We should avoid the fallacy that $N_e m$ measures 'gene flow' between species, host races or parapatric populations separated by hybrid zones (see also above), as is sometimes recommended (e.g. Porter, 1990). Supposing a newly arisen pair of species never hybridizes ($m = 0$), a very long time indeed is required before all marker loci drift to diagnostic fixation so that $F_{ST} = 1$ and $N_e m = 0$ is inferred.

To be identifiably different, host races or sibling species must have multiple genetic differences and, in sexual taxa, this implies little gene flow. Therefore, individual hybrids may be identifiable at multiple loci, and mass-action models become less practical for estimating rates of gene flow. For example, in the narrow contact zone between *Heliconius erato* and

*Heliconius himera* Hewitson, hybrids identified via colour pattern were clearly also identifiable using molecular data (Jiggins *et al.*, 1997). In a direct analysis of hybridization in nature, backed up by laboratory studies of mate choice, it was shown that the production of hybrids was similar to the approximately 5% frequency of $F_1$ hybrids in the centre of the contact zone (Jiggins *et al.*, 1997; McMillan *et al.*, 1997; Mallet *et al.*, 1998b). In another pair of species, *Heliconius charithonia* (L.) and *Heliconius peruvianus* Felder & Felder, hybrids were identified via allozyme genotypic signatures (Jiggins and Davies, 1998). As already mentioned, this genotypic identification technique also has potential for determining the geographic sources of conspecific immigrants (Cornuet *et al.*, 1999; Davies *et al.*, 1999b). We have recently used genotypic identification to estimate hybridization rates of ~2% between sympatric larch and pine host races of the larch budmoth *Zeiraphera diniana* Guenée (Emelianov *et al.*, 1995). These estimates are similar to those based on direct analyses of cross-attraction of host races in the field (Emelianov *et al.*, 2000).

If hybrids between species are rare, and when formed often sterile or inviable, can 'gene flow' between species ever occur? Hybridization itself is a form of actual gene flow in the sense I have advocated here, but a valid question is: can genes be transferred via backcrossing between populations normally considered as separate species? This is difficult to answer, because the rarity of hybridization itself makes large-sample direct observations almost impossible. Instead, accurate typing and identification of alleles using sensitive DNA-based techniques will usually be required. Phenotypic manifestations alone, as in colour pattern or allozyme banding patterns, will usually be insufficient evidence. In several cases, evidence strongly supports gene transfer. Molecular homologies suggest that transposable P-elements in some strains of *D. melanogaster* originated recently from elements common in the unrelated *Drosophila willistoni* group, and were transferred, perhaps by parasitic mite vectors (Houck *et al.*, 1991). Some individuals of *Drosophila mauritiana* Tsacas & Davis, a species in the *simulans* group endemic to the island of Mauritius, share mtDNA haplotypes with the worldwide species *D. simulans* Sturtevant, while other *D. mauritiana* have a more divergent mtDNA. This suggests recent introgression from *D. simulans* into *D. mauritiana*, presumably via hybridization (Solignac and Monnerot, 1986). Very different gene genealogies for *Adh* compared with other genes in the *D. pseudoobscura* group suggest transfer of *Adh* genes between closely related species (Wang *et al.*, 1997). If these interpretations are correct, gene transfer may have the utmost importance for systematics: estimated molecular phylogenies, or indeed any phylogenies of closely related species, may depend on which characters are used in their construction, not only because of errors in phylogeny estimation and ancestral polymorphisms, but also because different genes may have genuinely different histories within a single group of species.

## Prospects for Estimating Gene Flow

The use of selection/gene flow balance and disequilibrium across clines to estimate gene flow has been achieved in only a handful of cases. Apart from the insect examples cited here, only three other taxa, all vertebrates, have been studied in this way (Barton, 1982; Szymura and Barton, 1991; Sites et al., 1994). In every case, development of new population genetic theory and new multidimensional likelihood analyses has been required. There may be some general models that can be applied via standardized computer programs, for example in multilocus clines where the precise nature of selection on each locus is less important than aggregate polygenic effects on cline width and disequilibria (Szymura and Barton, 1991; Stuart Baird et al., unpublished). Every other example studied so far had a unique mode of selection or spatial situation, which caused difficulties for standardized methods. Similarly, analyses of hybridization and gene flow between species require a good understanding of the theory of gene flow/selection balance between a pair of demes (Barton and Gale, 1987) or of coalescence theory (Wang et al., 1997). Practitioners of these newer gene flow methods will have to develop theoretical skills and gain a good understanding of likelihood fitting or analysis of deviance. Gene flow/selection balance is complex. To study it, we must all become more numerate.

Various readers of this chapter have commented on the grim picture I paint for the estimation of gene flow from gene frequency data using the drift/gene flow equilibrium assumption. Problems range from the error-proneness of estimating very small values of $F_{ST}$, the virtual uselessness of the resultant $N_e m$ combined parameter, the slow attainment of equilibrium between gene flow and selection over large areas, the possibility of selection, and laboratory scoring errors. What if no selection/dispersal balance with multiple locus clines exists in the species? Is it hopeless to attempt to estimate gene flow indirectly? I fear that the answer is often 'yes'. At minimum, anyone seeking to calculate $N_e m$ from their data on $F_{ST}$ must avoid unrealistic expectations that the 'gene flow' they measure thereby will be very useful. Such expectations have existed since the 1930s, when Dobzhansky's hopes for estimating gene flow from the allelism of lethals in D. pseudoobscura were dashed by Sewall Wright (Provine, 1986: 367). It is especially unwise to estimate 'gene flow' using $N_e m$ when a cline or other situation clearly indicates that a drift/gene flow equilibrium is unlikely. Only in well justified comparative analyses between species or populations, and where spatial and temporal information about gene frequency samples are preserved, may some inferences about gene flow be made. Anyone attempting to estimate gene flow from marker data should consider carefully whether their inferences will be as useful as they would like. Many previous studies have used an $N_e m$ route to estimate 'gene flow', but this popularity does not ensure utility.

## Conclusions

Estimating gene flow, or its converse, reproductive isolation, is essential for understanding evolution and diversification. In this chapter, I have reviewed the somewhat confused current state of the field. In particular, an impression of limited dispersal is often based on field observations of local insect movement; this conflicts with genetic evidence for high rates of gene flow. I attempt to provide some hope that we may in future comprehend these areas of evolutionary biology more clearly. Molecular advances have enabled considerable improvements on 'gene flow' measured as combined parameters such as $N_e m$, or 'reproductive isolation' in the sense of pre- and post-mating isolation. Incorrect application of population genetic theory to these new data has hitherto held back our understanding. In the first few decades of this century, improvements in theory and data analysis will allow us to topple gene flow from its current almost mystical status as a dimensionless quantity to being an observable, measurable parameter in everyday use. Unless better methods for analysing drift/gene flow models are developed, the information may have to come from situations where selection/gene flow balance can be observed. Because of the complexity of selection/dispersal balance in continuous populations, estimating gene flow will rarely be as easy as grinding up a few bugs, running a few gels, and then analysing the data using an off-the-shelf computer program. However, these new methods will ultimately provide far more insight into evolution, particularly the important case of evolution under natural selection.

## Summary

Insect populations have provided many of the best studies of gene flow, a parameter which plays a central role in topics as diverse as population structure, adaptation (e.g. insecticide resistance), conservation genetics and speciation. In the literature, 'gene flow' often refers to $N_e m$ (effective population size × migration rate) estimated from frequencies of genetic markers, typically allozymes, by assuming an equilibrium between gene flow and neutral genetic drift. $N_e m$ and its continuous population analogue, 'neighbourhood population size', $N_b$, are actually both dimensionless parameters that measure the relative strength of gene flow and drift, rather than being simple measures of 'gene flow' or 'population size' alone. Inferring gene flow using this method therefore has many pitfalls. A more robust procedure is to exploit equilibria between gene flow and selection when suitable geographic variation in selection is present. New population genetic theory makes selection-based methods particularly useful, and selection is often a stronger force than genetic drift in many insect populations, leading to estimates with a lower margin of error. Finally, gene flow may also occur between sympatric host races or closely related species that hybridize, albeit rarely. I give examples from ongoing work with insects, especially fungus-eating beetles, *Heliconius* butterflies (Nymphalidae) and other Lepidoptera, *Culex* mosquitoes (Culicidae), and *Drosophila* (Drosophilidae). These newer methods provide the power to estimate actual gene flow ($m, \sigma_x^2$), rather than merely dimensionless gene flow measures such as $N_e m$ and $N_b$.

## Acknowledgements

This work was supported by NERC and BBSRC grants. I am very grateful for useful comments and discussions on this manuscript by Nick Barton, Roger Butlin, Chris Jiggins, Thomas Lenormand, Hugh Loxdale, Russ Naisbit, Graham Stone, Chris Thomas and Mike Whitlock. I am particularly grateful to Nick Barton for pointing out the dimensionlessness of $N_e m$ and $N_b$, and for the prototype of Fig. 16.3 which traces to an unpublished grant proposal, c. 1984.

## References

Andow, D.A., Kareiva, P.M., Levin, S.A. and Okubo, A. (1993) Spread of invading organisms: patterns of spread. In: Kim, K.C. and McPheron, B.A. (eds) *Evolution of Insect Pests: Patterns of Variation*. John Wiley & Sons, New York, pp. 219–242.

Avise, J.C. (1994) *Molecular Markers, Natural History and Evolution*. Chapman & Hall, London.

Barton, N.H. (1982) The structure of the hybrid zone in *Uroderma bilobatum* (Chiroptera: Pyllostomatidae). *Evolution* 36, 863–866.

Barton, N.H. and Clark, A.G. (1990) Population structure and processes in evolution. In: Wohrmann, K. and Jain, S. (eds) *Population Genetics and Evolution*. Springer-Verlag, Berlin, pp. 115–174.

Barton, N.H. and Gale, K.S. (1993) Genetic analysis of hybrid zones. In: Harrison, R.G. (ed.) *Hybrid Zones and the Evolutionary Process*. Oxford University Press, New York, pp. 13–45.

Barton, N.H. and Rouhani, S. (1991) The probability of fixation of a new karyotype in a continuous population. *Evolution* 45, 499–517.

Barton, N.H. and Shpak, M. (2000) The effect of epistasis on the structure of hybrid zones. *Genetical Research, Cambridge* 75, 179–198.

Barton, N.H., Jones, J.S. and Mallet, J. (1988) No barriers to speciation. *Nature* 336, 13–14.

Bossart, J.L. and Pashley Prowell, D. (1998) Genetic estimates of population structure and gene flow: limitations, lessons and new directions. *Trends in Ecology and Evolution* 13, 202–206.

Comins, H.N. (1977) The development of insecticide resistance in the presence of migration. *Journal of Theoretical Biology* 64, 177–197.

Cornuet, J.-M., Piry, S., Luikhart, G., Estoup, A. and Solignac, M. (1999) New methods employing multilocus genotypes to select or exclude populations as origins of individuals. *Genetics* 153, 1989–2000.
Coyne, J.A., Boussy, I.A., Prout, T., Bryant, S.H., Jones, J.S. and Moore, J.A. (1982) Long-distance migration of *Drosophila*. *American Naturalist* 119, 589–595.
Davies, N., Villablanca, F.X. and Roderick, G.K. (1999a) Bioinvasions of the medfly *Ceratitis capitata*: source estimation using DNA sequences at multiple intron loci. *Genetics* 153, 351–360.
Davies, N., Villablanca, F.X. and Roderick, G.K. (1999b) Determining the source of individuals: multilocus genotyping in nonequilibrium population genetics. *Trends in Ecology and Evolution* 14, 17–21.
Emelianov, I., Mallet, J. and Baltensweiler, W. (1995) Genetic differentiation in the larch budmoth *Zeiraphera diniana* (Lepidoptera: Tortricidae): polymorphism, host races or sibling species? *Heredity* 75, 416–424.
Emelianov, I., Drès, M., Baltensweiler, W. and Mallet, J. (2001) Host plant effects on assortative pheromone attraction by host races of the larch budmoth. *Evolution* (in press).
Endler, J.A. (1977) *Geographic Variation, Speciation, and Clines*. Princeton University Press, Princeton, New Jersey.
Epperson, B.K. and Li, T.Q. (1997) Gene dispersal and spatial genetic structure. *Evolution* 51, 672–681.
Epperson, B.K., Huang, Z. and Li, T.Q. (1999) Measures of spatial structure in samples of genotypes for multiallelic loci. *Genetical Research* 73, 251–261.
Feder, J.L., Chilcote, C.A. and Bush, G.L. (1988) Genetic differentiation between sympatric host races of the apple maggot fly *Rhagoletis pomonella*. *Nature* 336, 61–64.
Feder, J.L., Chilcote, C.A. and Bush, G.L. (1990) The geographic pattern of genetic differentiation between host associated populations of *Rhagoletis pomonella* (Diptera: Tephritidae) in the Eastern United States and Canada. *Evolution* 44, 570–594.
Feder, J.L., Opp, S.B., Wlazlo, B., Reynolds, K., Go, W. and Spisak, S. (1994) Host fidelity is an effective premating barrier between sympatric races of the apple maggot fly. *Proceedings of the National Academy of Sciences USA* 91, 7990–7994.
Felsenstein, J. (1975) A pain in the torus: some difficulties with the model of isolation by distance. *American Naturalist* 109, 359–368.
Frankham, R. (1995) Effective population size/adult population size ratios in wildlife: a review. *Genetical Research, Cambridge* 66, 95–107.
Guillaumin, M. and Descimon, H. (1976) La notion d'espèce chez les lépidoptères. In: Bocquet, C., Génermont, J. and Lamotte, M. (eds) *Les Problèmes de l'Espèce dans le Règne Animal*, Vol. I. Société zoologique de France, Paris, pp. 129–201.
Gupta, J.P., Dwivedi, Y.N. and Singh, B.K. (1980) Natural hybridization in *Drosophila*. *Experientia* 36, 290.
Haldane, J.B.S. (1930) A mathematical theory of natural and artificial selection. 6. Isolation. *Proceedings of the Cambridge Philosophical Society* 26, 220–230.
Hartl, D.L. and Clark, A.G. (1989) *Principles of Population Genetics*, 2nd edn. Sinauer, Sunderland, Massachusetts.
Houck, M.A., Clark, J.B., Peterson, K.R. and Kidwell, M.G. (1991) Possible horizontal transfer of *Drosophila* genes by the mite *Proctolaelaps regalis*. *Science* 253, 1125–1129.

Ingvarsson, P.K. and Olsson, K. (1997) Hierarchical genetic structure and effective population sizes in *Phalacrus substriatus*. *Heredity* 79, 153–161.
Jiggins, C.D. and Davies, N. (1998) Genetic evidence for a sibling species of *Heliconius charithonia* (Lepidoptera: Nymphalidae). *Biological Journal of the Linnaean Society* 64, 57–67.
Jiggins, C.D., McMillan, W.O., King, P. and Mallet, J. (1997) The maintenance of species differences across a *Heliconius* hybrid zone. *Heredity* 79, 495–505.
Jones, J.S., Bryant, S.H., Lewontin, R.C., Moore, J.A. and Prout, T. (1981) Gene flow and the geographical distribution of a molecular polymorphism in *Drosophila pseudoobscura*. *Genetics* 98, 157–178.
Kareiva, P.M. (1982) Experimental and mathematical analysis of herbivore movement: quantifying the influence of plant spacing and quality on foraging discrimination. *Ecological Monographs* 52, 261–282.
Korman, A., Mallet, J., Goodenough, J.L., Graves, J.B., Hayes, J.L., Hendricks, D.E., Luttrell, R.G., Pair, S.D. and Wall, M. (1993) Population structure in *Heliothis virescens* (F.) (Lepidoptera: Noctuidae): an estimate of gene flow. *Annals of the Entomological Society of America* 86, 182–188.
Lenormand, T. and Raymond, M. (1998) Resistance management: the stable zone strategy. *Proceedings of the Royal Society of London Series B* 265, 1985–1990.
Lenormand, T. and Raymond, M. (2000) Analysis of clines with variable selection and variable migration. *American Naturalist* 155, 70–82.
Lenormand, T., Guillemaud, T., Bourguet, D. and Raymond, M. (1998) Evaluating gene flow using selected markers: a case study. *Genetics* 149, 1383–1392.
Lenormand, T., Bourget, D., Guillemaud, T. and Raymond, M. (1999) Tracking the evolution of insecticide resistance in the mosquito *Culex pipiens*. *Nature* 400, 861–864.
Loxdale, H.D. and Lushai, G. (1999) Slaves of the environment: the movement of herbivorous insects in relation to their ecology and genotype. *Philosophical Transactions of the Royal Society of London B* 354, 1–18.
Mallet, J. (1986) Dispersal and gene flow in a butterfly with home range behaviour: *Heliconius erato* (Lepidoptera: Nymphalidae). *Oecologia* 68, 210–217.
Mallet, J. (2001) Species, concepts of. In: Levin, S (ed.) *Encyclopedia of Biodiversity*, Vol. 5. Academic Press, New York, pp. 427–440.
Mallet, J. and Barton, N.H. (1989a) Inference from clines stabilized by frequency-dependent selection. *Genetics* 122, 967–976.
Mallet, J. and Barton, N.H. (1989b) Strong natural selection in a warning color hybrid zone. *Evolution* 43, 421–431.
Mallet, J., Barton, N.H., Lamas, G., Santisteban, J., Muedas, M. and Eeley, H. (1990) Estimates of selection and gene flow from measures of cline width and linkage disequilibrium in *Heliconius* hybrid zones. *Genetics* 124, 921–936.
Mallet, J., McMillan, W.O. and Jiggins, C.D. (1998a) Mimicry and warning color at the boundary between races and species. In: Howard, D.J. and Berlocher, S.H. (eds) *Endless Forms: Species and Speciation*. Oxford University Press, New York, pp. 390–403.
Mallet, J., McMillan, W.O. and Jiggins, C.D. (1998b) Estimating the mating behavior of a pair of hybridizing *Heliconius* species in the wild. *Evolution* 52, 503–510.
McCauley, D.E. and Eanes, W.F. (1987) Hierarchical population structure analysis of the milkweed beetle *Tetraopes tetraophthalmus* (Forster). *Heredity* 58, 193–201.

McMillan, W.O., Jiggins, C.D. and Mallet, J. (1997) What initiates speciation in passion-vine butterflies? *Proceedings of the National Academy of Sciences USA* 94, 8628–8633.

Nagylaki, T. (1975) Conditions for the existence of clines. *Genetics* 80, 595–615.

Nunney, L. (1993) The influence of mating system and overlapping generations on effective population size. *Evolution* 47, 1329–1341.

Peterson, M.A. and Denno, R.F. (1998) Life-history strategies and the genetic structure of phytophagous insect populations. In: Mopper, S. and Strauss, S.Y. (eds) *Genetic Structure and Local Adaptation in Natural Insect Populations.* Chapman & Hall, New York, pp. 263–322.

Porter, A.H. (1990) Testing nominal species boundaries using gene flow statistics: the taxonomy of two hybridizing admiral butterflies (*Limenitis*: Nymphalidae). *Systematic Zoology* 39, 131–148.

Provine, W. (1986) *Sewall Wright and Evolutionary Biology.* University of Chicago Press, Chicago, Illinois.

Roderick, G.K. (1996) Geographic structure insect populations: gene flow, phylogeography, and their uses. *Annual Review of Entomology* 41, 325–352.

Roughgarden, J. (1979) *Theory of Population Genetics and Evolutionary Ecology: an Introduction.* Macmillan, New York.

Rousset, F. (1997) Genetic differentiation and estimation of gene flow from $F$-statistics under isolation by distance. *Genetics* 145, 1219–1228.

Sites, J.W., Barton, N.H. and Reed, K.M. (1994) The genetic structure of a hybrid zone between two chromosome races of the *Sceloporus grammicus* complex (Sauria, Phrynosomatidae) in central Mexico. *Evolution* 49, 9–36.

Slatkin, M. (1973) Gene flow and selection in a cline. *Genetics* 75, 733–756.

Slatkin, M. (1987) Gene flow and the geographic structure of natural populations. *Science* 236, 787–792.

Slatkin, M. (1989) Detecting small amounts of gene flow from phylogenies of alleles. *Genetics* 121, 609-612.

Slatkin, M. (1993) Isolation by distance in equilibrium and nonequilibrium populations. *Evolution* 47, 264–279.

Slatkin, M. and Barton, N.H. (1989) A comparison of three indirect methods for estimating average levels of gene flow. *Evolution* 43, 1349–1368.

Slatkin, M. and Maddison, W.P. (1989) A cladistic measure of gene flow inferred from the phylogenies of alleles. *Genetics* 123, 603–614.

Sokal, R.R., Jacquez, G.M. and Wooten, M.C. (1989) Spatial autocorrelation analysis of migration and selection. *Genetics* 121, 845–855.

Solignac, M. and Monnerot, M. (1986) Race formation, speciation and introgression within *Drosophila simulans, D. mauritiana,* and *D. sechellia* inferred from mtDNA analysis. *Evolution* 40, 531–539.

Sperling, F.A.H. (1990) Natural hybrids of *Papilio* (Insecta: Lepidoptera): poor taxonomy or interesting evolutionary problem? *Canadian Journal of Zoology* 68, 1790–1799.

Szymura, J.M. and Barton, N.H. (1991) The genetic structure of the hybrid zone between the fire-bellied toads *Bombina bombina* and *B. variegata*: comparisons between transects and between loci. *Evolution* 45, 237–261.

Turelli, M. and Hoffmann, A.A. (1991) Rapid spread of an inherited incompatibility factor in California *Drosophila. Nature* 353, 440–442.

Vitalis, R. and Couvet, D. (2001) Estimation of effective population size and migration rate from one- and two-locus identity measures. *Genetics* (in press).

Wang, R.L., Wakeley, J. and Hey, J. (1997) Gene flow and natural selection in the origin of *Drosophila pseudoobscura* and close relatives. *Genetics* 147, 1091–1106.

Weir, B.S. (1996) *Genetic Data Analysis II*. Sinauer Associates, Sunderland, Massachusetts.

Whitlock, M.C. (1992) Nonequilibrium population structure in forked fungus beetles: extinction, colonization, and the genetic variance among populations. *American Naturalist* 139, 952–970.

Whitlock, M.C. and McCauley, D.E. (1990) Some population genetic consequences of colony formation and extinction: genetic correlations within founding groups. *Evolution* 44, 1717–1724.

Whitlock, M.C. and McCauley, D.E. (1999) Indirect measures of gene flow and migration: $F_{ST} \neq 1/(4Nm+1)$. *Heredity* 82, 117–125.

Wright, S. (1969) *Evolution and the Genetics of Populations*, Vol. 2. *The Theory of Gene Frequencies*. University of Chicago Press, Chicago, Illinois.

Wright, S. (1978) *Evolution and the Genetics of Populations*, Vol. 4. *Variability Between and Among Natural Populations*. University of Chicago Press, Chicago, Illinois.

# Use of Genetic Diversity in Movement Studies of Flying Insects

## Hugh D. Loxdale[1] and Gugs Lushai[2]*

[1]*Entomology and Nematology Department, IACR-Rothamsted, Harpenden, Herts AL5 2JQ, UK;* [2]*Biodiversity and Ecology Division, School of Biological Sciences, Basset Crescent East, Southampton University, Southampton SO16 7PX, UK*

## General Introduction

The power of flight probably evolved in semi-aquatic insects in the Middle and Upper Devonian when watery habitats regularly dried out (Wigglesworth, 1976; Kurkalova-Peck, 1978). This necessitated dispersal in search of new habitats, resources and mates and may have selected for morphs using articulated gill-plates as wings (Wigglesworth, 1976; Wootton, Chapter 3). Flight dispersal to other localities also served to mix populations, thereby increasing the likelihood of favourable genotypes arising from these distinct, quite literal gene pools in a constantly changing environment.

Whatever the mode of displacement, movement from an established resource to another potential site is often a gamble that is necessitated by abiotic and biotic factors that select against continual habitation of the original location. The 'gamble' that dispersing individuals face is that they may fail to reach new habitats or resources, particularly if these are distant and separated by geographical barriers, due to insufficient fuel sources to maintain movement, or navigational errors, or if they are predated *en route*. Even if they do reach a suitable niche, sexually reproducing organisms still need to find a mate to begin to establish a new population, unless they have mated prior to dispersal. Over geological time, there has been a selection for optimal habitat-resource and mate location, although, interestingly, there are only a few insects which show 'true' migration (in the sense of Thomson,

---

\* Present address: Crop Protection Program, Eastern Cereal and Oilseed Research Centre, K.W. Neatby Building, C.E.F., Ottawa, Ontario, Canada K1A OC6.

©CAB *International* 2001. *Insect Movement: Mechanisms and Consequences* (eds I.P. Woiwod, D.R. Reynolds and C.D. Thomas)

1926), i.e. innate, directed, long-distance two-way movements. In contrast, the majority of insect species undertake changes of spatial distribution where individual control of direction and distance is rather weak. These movement events, which we class as 'dispersal', often occur over several generations (Loxdale and Lushai, 1999).

With winged insects, there is often a low chance of successfully recapturing marked individuals after population movements have occurred. This is especially so if: (i) the population size is small; (ii) the spatial scale over which migrants move is geographically extensive, or (iii) their movement involves 'passive' transport at wind speeds greater than their flight speed in still air. As a consequence, individuals are usually collected from the field with little knowledge of where they have come from, thereby reducing what can be learnt from these samples.

In the present article, we discuss how patterns of genetic variation derived from the use of molecular markers can be used to infer the extent and nature of population movements in flying insects. We highlight how in some cases this information provides insights into biological processes as well as how these patterns are influenced over time in an evolutionary sense. We confine ourselves to descriptions of aerial movement which allow the insect population concerned to find new habitats, rather than local movements such as foraging activities (e.g. as with bees).

## Markers Used to Study Genetic Variation and to Infer Population Movement

Amongst the methods used to explore variation across an insect population are those that can be considered non-molecular 'tags', i.e. that describe phenotypic polymorphisms. These include behavioural differences, e.g. insect song (Henry, 1994); wing pattern (Dowdeswell and Ford, 1953; Roskam and Brakefield, 1999); wing size (Zera and Denno, 1997); eye colour (Dobzhansky and Wright, 1943); and body pigments (Odonald and Majerus, 1992). Use has also been made of element markers (Sherlock et al., 1985), radio-labelling (Taimr and Kríz, 1978) and other innovative markers such as pollen from unique sites (Westbrook et al., 1998), fluorescent dust dyes (Dobzhansky and Powell, 1974; Byrne et al., 1996), as well as radar monitoring (Smith and Riley, 1996; see Reynolds et al., 1997 for a review of these methodologies). Pheromone lures (Suckling et al., 1999) are, physiologically speaking, molecular substances, but they differ from the next group of markers in that population data may be collected with little further processing. Molecular 'tags' require the collection of samples and their subsequent molecular processing before a population can be typed. These include proteins, especially allozymes, and DNA markers, both of which are separated electrophoretically; chromosomal markers (White, 1978; Bullini and Nascetti, 1989), e.g. chromosome number, nuclear organizing regions

(NORs) and C-banding (Berry *et al.*, 1992 and references therein), and fluorescent *in situ* hybridization (FISH) markers (Blackman *et al.*, 1995; Bizzaro *et al.*, 1996). Allozyme and DNA-based markers are the most commonly used molecular tags in entomological studies of population structure and movement and hence we concentrate on these, whilst paying tribute to the important findings made using the aforementioned other markers. Some of the most informative studies have utilized a two-pronged approach, for example using mark–release–recapture (MRR) protocols in conjunction with molecular analysis of the re-trapped samples to study movement (e.g. Lewis *et al.*, 1997).

Of the two main categories of molecular marker, allozymes (Richardson *et al.*, 1986) are the products of genes (alleles) coding for enzymes and for many insects provide very useful Mendelian markers. However, the number of allelic variants recorded per locus is often not large (< 10). In some insect groups, i.e. Hymenoptera and Aphididae, allozyme loci are found to be generally invariant (Loxdale, 1994). When allozyme variability is found to be low over a range of loci in the insect species under study, DNA markers are usually sought (Carvalho, 1998; Loxdale and Lushai, 1998). DNA analysis is based on six basic practices of molecular biology: (i) acquisition of DNA (DNA release or extraction); (ii) cutting (restricting) the DNA; (iii) joining (ligating) the DNA; (iv) probing the DNA with a labelled sequence to investigate sequence homology between samples; (v) amplifying the DNA in a polymerase chain reaction (PCR) and (vi) sequencing the DNA.

Involved with one or a number of these basic processes are the following molecular marker types: restriction fragment length polymorphisms (RFLPs); minisatellites (DNA 'fingerprinting' using single and multilocus probes); PCR involving mitochondrial DNA (mtDNA) and nuclear DNA (nDNA), and scrutinized using primer pairs which amplify hypervariable regions, e.g. rDNA, introns and microsatellites (single sequence repeats, SSRs); random amplified polymorphic DNA markers (RAPDs); amplified fragment length polymorphisms (AFLPs), and transposons. Each marker has its advantages and disadvantages for population studies (as discussed in detail elsewhere, i.e. Hoy, 1994; Loxdale *et al.*, 1996; Carvalho, 1998; Karp *et al.*, 1998; Loxdale and Lushai, 1998), in particular whether the marker produced is dominant or co-dominant, the latter thereby allowing heterozygotes to be screened for analysing population structure and inbreeding, etc. (Loxdale and Lushai, 1998).

Allozymes mutate at rates of around $10^{-6}$–$10^{-9}$ per gene per generation whereas with microsatellites, rates are two to three orders of magnitude greater (Jarne and Lagoda, 1996; Luikart and England, 1999). As a consequence of this relatively high mutation rate, microsatellites are less useful for assessing historical patterns of population structure and dynamics (Jarne and Lagoda, 1996). Within this group of rapidly evolving marker, the different repeat classes (e.g. di- and tri-nucleotides) have differential rates of evolution thereby complicating the picture derived when using a range of these markers in conjunction (Rubinsztein et al., 1999). Similarly, mtDNA has several regions which evolve at different rates (Zhang and Hewitt, 1997) and this has to be taken into account when making conclusions derived from such markers. Zhang and Hewitt (1996) have also highlighted another problem, that of multiple gene-copies for a given mtDNA region, with some of these integrating within the nuclear genome. Even so, mitochondrial markers are considered extremely versatile because of their predominantly maternal inheritance, low recombination rate (Wallis, 1999) and fast rate of evolution (Avise, 1994; Zhang and Hewitt, 1997). They have been employed in studies where genetic bottlenecks may have had a role in shaping population structure (Loxdale et al., 1996; Carvalho, 1998). However, whilst they are often found to show low variation in migration studies, this could be as a direct result of the populations tending to move together, i.e. the 'moving-deme' hypothesis (Loxdale and Lushai, 1999) rather than the influence of a bottleneck. Concerted evolution occurs where multiple genes exist as if they represent a single locus. Hence, homogeneity may occur among repeats within individuals and populations, but heterogeneity among the tandem arrays from divergent populations and species (Rand, 1994). Examples are found in nuclear rDNA genes (Black, 1993) and variable number of tandem repeats (VNTRs) in the control region of mitochondrial genes (Rand, 1994). However, in some cases they have intraspecific, even intraclonal variation at these sites, which can give as much variation within an individual as in a population (Shufran et al., 1992).

Bossart and Pashley-Prowell (1998) discuss the shortcomings of various molecular markers and their interpretations in population genetic studies including estimates of gene flow (views that have been challenged by Bohonak et al., 1998). Suffice to say that there are examples where different markers, i.e. DNA fingerprinting, RAPDs and microsatellites describe the same population genetic phenomena, but reveal increasingly greater ecological and evolutionary information. Different markers may provide a contrasting picture of the ecological and evolutionary forces shaping the genetic patterns observed, as demonstrated in the peach-potato aphid, *Myzus persicae* (Sulzer) by Brookes and Loxdale (1987) and Fenton et al. (1998) using allozymes and rDNA-IGS (intergenic spacer) markers, respectively. Interestingly, minisatellites, RAPDs and microsatellites appear to show host-plant-based adaptation in aphids, whereas the IGS markers do not. This may be because the former markers are integrated at random along the genome, closely linked by 'hitchhiking' (Maynard Smith and Haigh, 1974) to genes under selection (Lushai et al., 2001). In contrast, the rDNA-IGS regions are confined to the ends of the arms of the X chromosomes (Blackman and Spence, 1996) where it is possible that they are not under the same kind of molecular selection.

When markers are employed to elucidate patterns of movement, assessment always involves a survey of sample populations and specific gene and genotype frequencies. In non-molecular genetic methodologies, it is often assumed that a single gene or very closely associated groups of genes (under similar selection pressure) are associated with a particular trait. Although allozyme and DNA-based allelic markers are thought to be neutral (not selected for; Jarne and Lagoda, 1996), they are also often assumed to be genetically independent, i.e one-gene-one-trait. This can be statistically proved by testing for linkage disequilibrium, i.e. whether alleles at different loci screened are independently assorted or linked (Parkin, 1979; Richardson et al., 1986). Only allozymes, microsatellites and RFLPs of nDNA provide heterozygote banding patterns from which, in the case of the former two markers, deviations of Hardy–Weinberg (H-W) expectations (Richardson et al., 1986), and heterozygosity per locus and over all loci (average heterozygosity) can be deduced (Luikart and England, 1999). Daly (1989) and Slatkin (1985a, 1987, 1993, 1995) discuss how estimates of gene flow are derived from allele frequency data, e.g. contingency $\chi^2$ tests, $F$-statistics ($F_{ST}$) and $R_{ST}$ in microsatellites, $Nm$ (effective population size, $N \times$ migration rate, $m$), Slatkin's (1985b) rare allele method, genetic distances, etc. All these analyses can be related to the extent of movement (see also Mallet, Chapter 16). Measures of direct sequence similarity-relatedness or nucleotide divergence within and between groups can also give useful inference to movement. Genotypes may be assumed to be moving to or from an area if a known genotype of a sample population group is traced from a source by subsequent re-trapping away from the origin. These approaches depend on the ability to differentiate genotypes in a statistical manner. If genotypes are demographically distinct, for example, an invasion of an insecticide-resistant genotype into an insecticide-susceptible population, then a more definitive statement about movement can be made, assuming that such a genotype has not secondarily evolved by mutation in the residing population. Some caution is needed in analysing across developmental stages and morph types as these may show different genetic patterns within an individual lineage (cf. Loxdale, 1994; Loxdale et al., 1996; Lushai et al., 1997).

The above detail provides a framework of suitable markers along with analytical approaches to address topics of spatial scale and gene flow. In the next section we show how these have been applied to insect examples which reveal patterns of spatial distribution across various geographic scales.

## Genetic Variation: Patterns of Range Expansion and Factors Affecting Spatial Distribution

Theoretical aspects of spatial and temporal genetic patterns of animal populations in relation to gene flow have been recently detailed by Baur and Schmid (1996) and in insects, by Roderick (1996; see also Waser and Strobeck, 1998 and Dieckmann et al., 1999). We briefly describe contemporary methods of spatial sampling and concentrate on environmental and behavioural factors that influence patterns of aerial displacements in insects.

Collection of insects in the field has been performed by hand, by vacuum sampling, or by using nets or sticky, pheromone and fixed suction traps, or some suction trap networks (e.g. Rothamsted Insect Survey, which is national and operated daily). Many of these methods have been employed for molecular ecological studies in aphids by Shufran et al. (1991), De Barro et al. (1995a,b) and Llewellyn et al. (1997, 1999), and have involved collecting insects at local and regional scales over a growing season to assess spatial and temporal trends. Recently, an ingenious MRR method has been employed to examine the flight behaviour of individual white fly, *Bemisia tabaci* (Gennadius), at local scales in the field (Isaacs and Byrne, 1998). MRR studies carried out in conjunction with molecular markers by Lewis et al. (1997) have also provided effective ways of mapping genetic variation over different spatial and temporal scales. For more accurate spatial experiments, a concentric collection approach from a fixed point spreading out at geometric distances is optimal (e.g. Byrne et al., 1996; Byers, 1999). Even with all this innovative methodology, sampling small flying insects is always a problem. Whilst sometimes small insects like aphids are carried by fast directional jet streams above 900 m in some parts of the world (e.g. USA and India), more usually they are borne on lower altitude winds. Since these air masses may quickly change direction within a few hours, it is difficult to make typical sampling regimes comprehensive (Riley et al., 1995; see also Gatehouse, 1997), and indeed thereby be sure of direction and scale of displacement.

In general, the self-directed type of migratory behaviour is the exception rather than the rule (Loxdale and Lushai, 1999). The geographical displacements of most insects produce rather random changes in distribution, sometimes over several generations, and this pattern is thus the basic one for most species. Unfortunately, there is little direct evidence from molecular marker data demonstrating the genetic consequences of such invasions.

One study indicating displacement, and not migration, concerns the movement of the cynipid Knopper gall wasp, *Andricus quercuscalicis* (Burgsdorf), which host alternates between Turkey oak, *Quercus cerris* and the pedunculate oak, *Quercus robur*, in spring and summer, respectively (Stone and Sunnucks, 1993). The wasp has tracked its obligate host, Turkey oak, in a typical *radiating* pattern of colonization since this was planted throughout Western Europe outside its native range over the last 300–400 years. Since the planted Turkey oak trees are rarer and more widely scattered outside the native range than within it, this was seen as likely to influence the population structure of the insect. Allozyme analysis of 39 populations failed to detect any new alleles in the invaded range, although allelic diversity and mean heterozygosity decreased significantly with distance from the original refugia (Stone and Sunnucks, 1993). Spatial analysis revealed that genetic differences were greater in the invaded compared with the native area. This suggested that genetic changes across Europe were probably the result of strong directional movement followed by limited gene flow between populations, rather than due to selection. There was also evidence that genetic patterns of populations were founded sequentially from the east spreading westward by a process of random invasion, from which new populations invaded further, with a concomitant dilution of genetic diversity. Another recent study tracking invasion is described by Taberner et al. (1997). Here RAPD markers were used to deduce the spread of a pest sugarbeet weevil, *Aubeonymus mariaefranciscae* (Roudier), in southern Spain. A combination of phylogenetic and genetic distance versus geographical distance approaches enabled the spread of the beetle to be ascertained in relation to its ecology and evolution. This genetic information affirmed that the beetle populations had arisen from a unique colonization event, probably dating from 1979.

When geographic barriers are present such as mountains, expanses of water, deserts or other inhospitable regions, the movement of individuals between populations can be prevented and hence gene flow influenced to the extent that populations may not show homogenous gene frequencies over time. In the case of the Knopper gall wasp, the English Channel and Irish Sea have proven to be partial barriers to the rate of insect colonization, ~80 years in the former case (Stone and Sunnucks, 1993). With the grain aphid, *Sitobion avenae* (F.) populations from Britain and Spain separated by 800 km of sea and the Pyrenees mountains showed a degree of genetic differentiation in terms of allozyme-derived genetic distances (Loxdale et al., 1985).

Clouded Apollo butterflies, *Parnassius mnemosyne* (L.), in the Alps appear to consist of 'closed' populations as shown following behavioural observations (Napolitano et al., 1988). Survey of four such populations using 24 allozyme loci showed these populations to have differing levels of heterozygosity and significant allele frequency differences at some polymorphic loci, suggesting that migration was restricted among these colonies separated by mountainous terrain (Napolitano et al., 1988). In a later study, 24 populations of the butterfly collected over a wider area in southern France revealed that distinct populations from the Pyrenees and the Massif Central were similar to those collected in the Prealps. Genetic and clustering analysis indicated that these far spread populations may have been colonized from northeastern Europe via the central Prealps during warm interglacial periods and thereafter, become isolated during subsequent climate cooling (Napolitano and Descimon, 1994). Thus, the genetic structuring of these outlying populations represents the probable historical pattern of gene flow. However, the fact that the average $F_{ST}$ over all 24 loci was high (0.135), revealing significant population heterogeneity and hence substructuring comparable with a low dispersal rate, emphasized the role of geographical and ecological barriers in shaping population structure at small geographic scales (~30 km; Napolitano and Descimon, 1994). Meglecz et al. (1998, 1999) have also studied this butterfly in northeast Hungary using both MRR and molecular markers (allozymes and microsatellites) and utilizing field and historical collections. Their work showed that even large colonies appear to have small effective reproductive population sizes with a lack of heterozygotes compared with H-W expectations, indicative of uneven sex ratio, recent bottlenecks and/or founder effects. Therefore, the genetic approach differentiates between subpopulations from some regions having metapopulation structure (Hanski and Gilpin, 1997); some with homogeneous genetic structure within regions; and some geographically widespread populations that are genetically differentiated, indicative of restricted movement and genetic drift.

Population genetic structure and gene flow is also influenced by elevation. Leibherr (1988) studied beetles at different altitudes in North America, analysing allele frequency data using $F_{ST}$ and private allele approaches. He showed that in five carabid species of varying vagility, $F_{ST}$ values were positively correlated with elevation from 0 to 1500 m above sea level, indicating more population substructuring with increased elevation. At first instance this was thought to be a result of low vagility at greater elevation, but closer analysis revealed that one of the key factors influencing this trend was greater habitat fragmentation with increasing altitude. It was concluded that lower levels of gene flow associated with greater population subdivision occurred in upland regions due to habitat fragmentation, i.e. topographical diversity and habitat persistence leading to a reduced population extinction rate (Liebherr, 1988). Again, this study highlights the fact that although population genetic markers reveal a lot of ecological detail, correct interpretation of this depends on full assessment of flight behaviour correlated with the various ecosystem parameters.

For some species, the apparent high estimates of gene flow are artefactual rather than real and probably reflect historical patterns of movement. This leads to a very important point of gene flow versus vagility. For many insects, even truly migratory species, the further populations are away from each other, even without the problems of physical barriers to gene flow, the more likely they are to diverge genetically, i.e. show more heterogeneous genetic population patterns.

The relationship of isolation by distance (IBD) and dispersal ability of insects has recently been studied in detail by Peterson and Denno (1998a). They analysed genetic data versus IBD for 43 phytophagous species and races of insects and tested two opposing hypotheses. In the first, IBD slopes in log plots of gene flow versus distance were unchanged by distance yet the intercepts increased with mobility. In the second, IBD slopes varied with dispersal ability. They found that mobility trends tended to fit the second hypothesis. In sedentary and highly mobile species, IBD appeared weak over distances from 10 to 1000 km and both species categories had shallow slopes and markedly different intercepts. In moderately mobile species, the relationship was more pronounced, i.e. there was greater gene flow with distance. From the analysis of the genetic data they concluded that divergence amongst most populations of sedentary species over the distances tested was due to limited gene flow. The genetic homogeneity found in intermediate dispersers over small spatial scales was due to appreciable gene flow, whilst the heterogeneity at larger spatial scales was thought to be a result of its restriction. The homogeneous genetic structure of highly mobile species at all spatial scales was considered to be due to high gene flow. Hence, genetic structure was to some degree directly correlated with the ability of the insect to disperse providing valuable insights and demonstrating the usefulness of the genetic marker approach in arriving at these conclusions. However, there are numerous factors that influence population genetic structure, some of which have already been outlined (cf. also Peterson and Denno, 1998b). In some instances, genetic structuring is strongly influenced by biotic factors. In the next section we relate evolutionary and biological processes of various host–herbivore interactions, host ranges and fidelity to genetic variation.

## Genetic Patterns and Biotic Factors

There is a growing body of evidence for host association of adaptive-genetic patterns in insects at a variety of spatial scales (e.g. lepidopteran leaf miners; Mopper, 1998), both small (e.g. scale insects; Alstad and Corbin, 1990) and large (e.g. wood boring beetles; Stock *et al.*, 1979; but see Peterson and Denno, 1998b). Whether this apparent association is due to differential survival on hosts rather than to differential host preference of genotypes *per se* (Langor and Spence, 1991) needs to be verified. Whilst it is difficult to relate these patterns to the beginning of evolutionary divergence below the species level, i.e. biotypes, race formation, ecotypes, etc. (Cobb and Witham, 1998), host-dependent traits suggest coevolutionary trade-offs (Thompson, 1994) and may be the fundamental mechanisms of such divergence (Nuismer *et al.*, 1999).

Population genetic homogeneity in highly dispersive species may relate to the fact that the transient presence of the host plants selects for enhanced dispersive behaviour compared with related insects that colonise perennial hosts. Hence, with high dispersers, host adaptation seems less likely to evolve. Even so, in at least one case, the tiger swallowtail butterfly, *Papilio glaucus* L., there is evidence that even in such a vagile species, differential selection for host plant has contributed to significant genetic differentiation of populations. Bossart and Scriber (1995) have shown that in North American *P. glaucus* populations sampled in southern Ohio, north central Georgia and southern Florida from three hosts (*Liriodendra tulipifera*, *Magnolia virginiana* and *Prunus serotina*), differentiation existed among populations for both oviposition preference and larval performance. However, using a range of allozyme markers, no evidence was found for population differentiation matching these traits. This argues for significant gene flow between populations on the various hosts (Bossart and Scriber, 1995), but alternatively, other markers such as microsatellites could perhaps resolve the host-based populations.

In aphids with high vagility, plant-host-related differentiation is less likely to evolve. It also appears that in species which do not host alternate, i.e. utilize a single plant group, even spatial scale of the niche seems to have little effect on genetic structuring. An example is the sycamore aphid, *Drepanosiphum platanoidis* (Schrank), which lives all year around on sycamore, *Acer pseudoplatanus*, in both asexual and sexual phases. In this species, allozyme studies revealed a range of genotypes (eight at the phosphoglucomutase, PGM, locus) and genetic homogeneity at both local and national scales in southern England (Wynne et al., 1994). The fact that as much genotypic variation was found at the leaf scale as the national scale suggests that 'trivial' flight, i.e. local flight within or close to the canopy of the host tree itself (Dixon, 1969), prevents clonal formation on individual sycamore leaves, as well as homogenizing aphid genetic patterns between clumps of trees and trees in the same area. Whilst the aphid appears to be highly dispersive, since it is commonly found in the summer in Rothamsted 12.2-m-high suction trap samples (Woiwod et al., 1988), the homogeneous national genetic pattern observed may reflect a historical colonization rather than inter-population migration over seasons (Wynne et al.,1994). The green bug, *Schizaphis graminum* (Rondani), a serious predominantly asexual aphid pest of cereals in North America studied using rDNA-IGS markers, also showed population genetic homogeneity at various spatial scales (Shufran et al., 1991). Numerous genotypes were detected, yet the populations appeared as genetically diverse at the lowest spatial (sorghum leaf) scale as at the county scale (Kansas), possibly the result of long-distance dispersal (see also Fenton et al., 1994).

If insects have low vagility, they are perhaps more likely to be subject to local differentiating selective forces (Cobb and Whitham, 1998) and produce host-adapted demes with little gene flow between them (Mopper and Strauss, 1998). There is some genetic evidence that this is occurring in the holocyclic (with sexual phase) cereal aphid, *Sitobion fragariae* (Walker). Local British populations of this species (< 30 km apart) showed genetic heterogeneity and stable genotype frequencies at an allozyme locus (glutamate-oxaloacetate transaminase, GOT) over a 5-year sampling period during the summer asexual phase of reproduction (Loxdale and Brookes, 1990).

Sometimes genetic patterns are disrupted such that a discontinuity or transition occurs in gene frequencies between populations. Usually, such patterns have a spatial dimension although occasionally, they have a temporal one. At the extreme end of the spectrum, discontinuous allele frequency patterns may relate to the presence of two adjacent species populations. An example of this situation is found in two noctuid moth species, the native hop vine borer, *Hydraecia immanis* (Guenée), and the introduced potato stem borer, *Hydraecia micacea* Esper. These populations are separated by the North American Great Lakes plant community ecotone through Wisconsin and Michigan (Scriber et al., 1992). The former species is more specialized in its diet and generally occurs south of the plant community transition zone, whereas the latter species is polyphagous and occurs in corn north of this zone. Of 19 enzyme loci screened, six showed fixed or nearly fixed allelic differences. Hybridization of the two *Hydraecia* species 'may contribute to wider larval host plant use abilities, altered voltinism patterns, reduced developmental temperature thresholds of the larva and possibly changes in oviposition preferences' (Scriber *et al.*, 1992, and references therein). Here migration and subsequent colonization of either species into adjacent ecosystems is restricted, unless moths can adapt to the new ecological challenges presented. In other cases, such separation has been shown to be due to the presence of morphologically cryptic species, e.g. crickets of the genus *Allonemobius* Hebard (Howard, 1983) or indeed to the phenomenon of sympatric speciation. Somewhere in this spectrum of diversification is also the phenomenon of hybridization. The influence of hybrid zones on genetic patterning and gene flow is detailed by Mallet (Chapter 16). Boecklen and Howard (1997) discuss the choice of molecular marker, number required and their power of resolution when studying such zones.

Much more rarely, as supported by scientific evidence, population divergence occurs sympatrically instead of resulting from para- and allopatric separation (White, 1978). In the tortricid larch budworm moth, *Zeiraphera diniana* Guenée, 11 larch- and pine-feeding populations were studied in western Europe using 24 allozyme loci (Emelianov *et al.*, 1995). From these data, genetic analysis showed that the larch and pine forms of the moth were either host races or sympatric species that hybridized rarely.

The 'classic' example of allelic discontinuities in geographically coexisting populations is demonstrated in tephritid fruitflies of the genus *Rhagoletis* Loew. Sympatric populations on apple/hawthorn and blueberry appeared to be reproductively distinct, probably the result of host-plant recognition acting as pre-mating barriers to gene flow, as shown by allozyme separation at 11 out of 29 loci indicating host-specific alleles. The molecular separation supports the status of these two host-adapted forms as *Rhagoletis pomonella* (Walsh) and *Rhagoletis mendax* (Curran) respectively, despite their close morphological similarity, overlapping geographic distributions and interfertility in laboratory crosses (Feder *et al.*, 1989). In addition, populations of *R. pomonella* showed geographic differentiation. Populations from the eastern USA, the native range of the fly, were highly polymorphic and showed homogeneity within and among populations. In contrast, populations from the western USA where it has colonized recently (*c.* 1980) generally lacked genetic variation (McPheron, 1990), although at the remaining polymorphic loci showed spatial heterogeneity suggestive of founder events. Other data by Feder *et al.* (1990; six enzyme loci) also showed that flies (*R. pomonella*) not only differed at loci between hosts (apple or hawthorn) but that allele frequencies were influenced by latitude. This produced a north–south cline in the eastern USA and Canada superimposed on inter-host patterns. It appears that adult flies do not migrate between host species, but rather, tend to infest the same host that they colonized as larvae. Hence genetic markers were able to differentiate individuals with specific host preference and conditioning that caused host fidelity, whilst selection may be responsible for geographic patterns observed. Therefore, these fly populations are strong contenders for the sympatric speciation model because they are most likely to successfully colonize new hosts in the area of their birth, and this must restrict the dispersal ambit of these insects (Feder et al., 1998).

Patterns of genetic diversity and gene flow of insects are influenced by niche patchiness and behaviour-related patch fidelity even when obvious plant-host influences appear to be absent. Some butterflies with low vagility such as the Glanville fritillary, *Melitaea cinxia* L., show metapopulation structure. Genetic studies on island populations in southwestern Finland using both allozyme and microsatellite markers (Saccheri et al., 1998) have revealed that the risk of extinction of subpopulations increased significantly with decreasing heterozygosity, a sign of inbreeding. These deleterious effects were manifest in larval survival, adult longevity and egg hatch rate, which in turn appeared to be the fitness components responsible for the relationship between inbreeding and extinction (Sutcliffe et al., 1997; Saccheri et al., 1998). In a lycaenid blue butterfly, *Euphilotes enoptes* (Boisduval), subpopulations from Washington State and Oregon, USA, were studied using allozyme markers. The apparent low vagility of the butterfly was thought to be the cause of genetic heterogeneity found between subpopulations (Peterson, 1996). This was described by clustering analysis of genetic distances as well as a regression analysis of gene flow versus geographical distance. The two parameters were, as expected, negatively correlated. Even so, estimates of gene flow among population pairs separated by more than 100 km were greater than ten individuals per generation, a value higher than predicted from the butterfly's apparent aerial dispersion (rarely more than 1 km). Yet neighbourhood size was estimated at around 40 individuals, consistent with poor vagility (Peterson, 1996). Thus it appears that in sedentary insects, the magnitude of gene flow described by the 'stepping stone' hypothesis (Peterson, 1996) has a significant influence on genetic homogenization over large spatial scales. This is precisely the scenario in the Knopper gall wasp. These studies again emphasize the potential utility of molecular markers for deriving behavioural and ecological trends. Spatial heterogeneity within a population can also have a temporal component to it. As described below, these include life cycle synchrony, with annual (short-term) temporal aspects, and very much longer historical patterns of genetic variation.

## Temporal Changes in Genetic Variation (Not Independent of Spatial Variation)

In the satyrid butterfly, the inornate ringlet, *Coenonympha tullia* Müller ssp. *inornata* Edwards, populations in the northern USA and Canada showed a step-latitudinal cline in some allozyme frequencies monitored as well as wing characters (Wiernasz, 1989). In the old, more northern part of the range, the insects are predominantly univoltine, but in the southwestern and eastern United States they are bivoltine. The eastern region is an area of recent range expansion. The level of genetic variability is strongly correlated with life history so that bivoltine populations tended to have higher levels of polymorphism and heterozygosity with a higher average number of alleles than univoltine populations, and were differentiated using genetic distance and wing pattern data. Although interpopulation gene flow (movement) is high, selection is the probable cause maintaining the cline. In bivoltine populations, asynchrony between larval diapause and parents leads to alternating selection, thereby increasing the level of genetic variation present (cf. Wiernasz, 1989).

Aphids also show latitudinal distribution of certain life cycle strategies (obligate asexual and sexual) correlated with climate. In French populations of the bird cherry-oat aphid, *Rhopalosiphum padi* (L.), studied using allozyme markers, sexual populations were found to be polymorphic early in the season, whereas asexual populations were relatively less varied, and possibly monoclonal in origin (Simon et al., 1996a). On seasonal analysis within a year using mtDNA markers, sexual insects with haplotypes II and III were differentiated from asexuals, most of which were of haplotype I (Simon et al., 1996b; Martinez-Torres et al., 1996, 1997). This last type showed a north–south latitudinal cline in frequency in France, decreasing from the southwest of the country (c. >90%) to the north (c. 15–45%) in various samples. This trend coincides with a climatic gradient that may reflect selection in the north for aphids with the sexual phase. These forms are favoured because the egg stage is more resistant compared with living aphids to the harsher winter conditions prevailing there (cf. Martinez-Torrez et al., 1997). Recent microsatellites analysis at five polymorphic loci in *S. avenae* by Simon et al. (1999) has revealed widespread genotypes throughout France and demonstrated similar life cycle structuring. Presumably, the genotypic structuring of aphids of different life cycle strategy impacts on the dispersal ability of this species. If too great a displacement occurs, climatic parameters will select against forms with maladapted behavioural responses including life cycle synchrony in tune with particular photoperiodic responses (Smith and MacKay, 1990; Loxdale et al., 1993; Lushai et al., 1996). Both the aforementioned butterfly example and aphid work exemplify the degree of resolution obtainable using molecular markers, which enables intraspecific variation to be teased apart in relation to biological and ecological parameters. It also emphasizes the complexity that has to be unravelled if the biology of a given species is to be better understood.

Aphids also exemplify another aspect of short-term seasonal variation. In southern British populations of *S. avenae*, analysis of patterns of genetic variation indicated that population movements were very localized (De Barro et al., 1995a; Sunnucks et al., 1997). In these studies, the range sampled covered mainly asexual clones showing a strong genetic differentiation between various species of *Gramineae* early in the field season (De Barro *et al.*, 1995b,c; Sunnucks *et al.*, 1997). The host-based genotype heterogeneity dissipated as the growing season progressed, probably associated with increasing local movements, both within and between fields (De Barro *et al.*, 1995b). Similarly, genetic variation was also seen to fall within a field as the result of adaptive performance, measured by increases in fitness, which selected for particular genotypes which came to dominate greater areas as they spread from founder sites (De Barro *et al.*, 1995a). In the damson-hop aphid, *Phorodon humuli* (Schrank), sampled within the commercial-hop growing regions of England, highly elevated esterase variants (associated with resistance to insecticides) were common in spring populations on the overwintering host, *Prunus* spp. compared with summer populations on wild, uncultivated hops, *Humulus lupulus* (Loxdale *et al.*, 1998). These data suggested that autumn fliers leave cultivated hop gardens and move only a short distance to the primary host whereas summer fliers moving in the opposite direction come from relatively further afield. This result, and the local heterogeneity at esterase loci and at another enzyme locus (6-phosphogluconate dehydrogenase, 6-PGD), support both suction trap data and wind tunnel experiments that this aphid is a short-range migrant (probably usually < 20 km; Loxdale *et al.*, 1998). Comparisons of the life cycle history and host selection data confirm the importance of long-term studies in order to move away from 'time-slice' analysis of ecosystems which may fail to resolve informative population detail.

An example of such detail is given by the diurnal behavioural responses of migratory butterflies. Changes in enzyme phenotype or activity at particular loci as a function of flight periodicity have been recorded in the Monarch butterfly, *Danaus plexippus* L. A higher proportion of heterozygotes at the PGI (phosphoglucose isomerase) and PGM loci were found in butterflies captured earlier than later in the day (Carter *et al.*, 1989). Further, using PGI alone, Hughes and Zalucki (1993) showed that animals with a particular allele (M) were more likely to fly at low temperatures, and so were active earlier in the day. Differences were also observed between males and females in the effect of PGI alleles on flight activity. Work on clouded yellow butterflies of the genus *Colias* F. demonstrated that young adults in the early part of their non-overlapping generations showed genotype frequencies in H-W equilibrium, although a heterozygote excess developed as the insects aged (Watt, 1977; Watt *et al.*, 1983). Of four frequent alleles carried by butterflies in culture, major differences were observed in heat stability and various kinetic parameters among the ten possible genotypes, with heterosis for kinetic parameters in some cases (Watt, 1977). Later, it was shown that certain heterozygotes began flights earlier in the day compared with others. Among the most common genotypes, these could be arranged in order of heterotic advantage, time of flight initiation, duration and overall flight density throughout the day. Under heat-stress conditions, butterflies with the most thermally stable enzyme phenotypes survived the best. These experiments suggest that selection operates on these glycolytic enzymes, which clearly influences the time and temperature at which butterflies first become active, and in migratory species, when they are liable to disperse. They also appear to affect populations in relation to genotype and sex and hence, these enzymes play a pivotal functional role in determining population structuring and dynamics. So clearly without molecular markers, and even more so, the right molecular marker (in this case a relevant functional one), such biological intricacy for flight behaviour would allude us.

In the context of historical trends, allele frequency patterns are sometimes seen to be rather homogeneous, even at large spatial scales. Such patterns may or may not relate to concomitant large-scale aerial movements of the insects concerned. For example, the fruit fly, *Drosophila willistoni* Sturtevant, displays similar allele frequencies at a range of allozyme loci over much of Mexico, Florida, Central America, the West Indies and tropical South America (Ayala et al., 1972). Likewise, two noctuid moth species examined also display homogenous among-population allele frequencies, i.e. the velvetbean caterpillar, *Anticarsia gemmatalis* Hübner, in Louisiana and middle America (Pashley and Johnson, 1986), and the African armyworm, *Spodoptera exempta* (Walker), in east Africa (Den Boer, 1978), as does the cabbage butterfly, *Pieris rapae* (L.), along the eastern seaboard of the USA (Vawter and Brussard, 1983). In the case of *D. willistoni*, this homogeneity of pattern is thought to be the result of locally differentiating forces, i.e. stabilizing selection rather than the result of dispersal counteracting the effects of drift and mutation (e.g. Nevo et al., 1998). Although *Drosophila* are known to be capable of dispersing up to 15 km across inhospitable desert in California (Coyne et al., 1982), it is unlikely that the homogeneity observed is due to long-range movements. The effect here is probably a historical pattern of expansion and colonization, reinforced to some degree by selection and occasional migrants between populations (Slatkin, 1987). In the velvetbean caterpillar, etc., the homogeneity observed is considered to be predominantly the result of large-scale inter-population movement over long distances, and high levels of gene flow.

In the geometrid November moth, *Epirrita dilutata* Denis & Schiffermüller, a weakly flying woodland species, different inferences can be drawn from genetic pattern depending on spatial scale and the sample size examined (Wynne, 1997). At the national scale in Britain, a range of enzyme markers tested showed subpopulations to have homogeneous allele frequency distributions. However, when subpopulations from woods on the Rothamsted estate (330 ha) were tested using large sample sizes ($n > 200$), significant differentiation was observed, even at the commonest alleles. This suggested a restriction of inter-population gene flow between populations less than 2 km apart, supported by MRR techniques. It was concluded that only in years of large population size was there enough gene flow to lead to homogenization of populations locally. At the national scale, the similarity of population gene frequencies probably reflects a historical pattern of colonization. These data exemplify the necessity of examining spatial patterns using *representative* sample sizes for the scale involved, so that the molecular markers utilized are able to provide a credible measure. Historical events have also been further elucidated using mitochondrial molecular markers. The invasion of the medfly, *Ceratitis capitata* (Weidemann), into California in the late 1980s was thought to have originated in Hawaii; however, this premise was not supported by mtDNA evidence (Sheppard *et al.*, 1992). Similarly, in the brown rice planthopper, *Nilaparvata lugens* (Stål), it was shown, again using mtDNA markers, that populations which annually invade Korea most probably arise from populations in China which share haplotypes in common, rather than those from the Indochina Peninsula, which do not (Mun *et al.*, 1999).

The population genetics of the Monarch, *D. plexippus*, and how this is influenced by historical patterns of movement is also of considerable interest. The butterfly annually undergoes innate, long-distance, two-way directed movement, i.e. true migration from overwintering sites in southern California (western population) and Mexico (eastern population) northward to southern Canada and back (Urquhart and Urquhart, 1977). Here the insect is tracking its milkweed host, *Asclepias* spp. Interestingly, allozyme and mtDNA analysis of populations have shown a generally low level of genetic diversity (Eanes and Koehn, 1978; Brower and Boyce, 1991; Brower, 1996). This invariance probably reflects the fact that the two populations, separated geographically by the Rocky Mountains, move *en masse* as large, discrete panmictic populations (Eanes and Koehn, 1978), and with gene flow apparently restricted between them (Brower and Boyce, 1991). A low rate of mtDNA mutation, natural selection, a high effective fecundity per female and possible population bottleneck effects may all contribute to the genetic pattern observed (Brower and Boyce, 1991). Another possible explanation for the low genetic variation is that the two populations are effectively 'moving demes' which breed together in close groups. As a consequence such lack of variance, tantamount to specialization, is indicative of a potential adaptive dead-end and the species is thus 'trapped' in an evolutionary sense because of its very specific behaviour and life history (Loxdale and Lushai, 1999).

## Concluding Remarks

In this synthesis, we have endeavoured to present some of the clearest examples of the use of molecular markers in the interpretation of movement of flying insects. From these, we have attempted to emphasize the complexity of biological systems which molecular markers can assist in elucidating. It is only now in hindsight, looking back at studies of insect migration and dispersal over the last century, that new challenges and objectives can be advanced in a field of study which is, after all, still relatively young.

Study of the aerial movements of insects in relation to their behaviour, ecology and population biology perhaps rank amongst the most difficult in zoology. This is especially so for small insects that cannot be readily tagged (except using radio-labelling) making it difficult to estimate flight duration/distance, direction and speed. With larger insects where direct physical tagging is possible, dilution effects and displacement by air currents still hinder recapture so that such studies can also be inconclusive over large spatial scales. Consequently molecular markers are our best solution to attempt to resolve population structure and dynamics. Nevertheless, the analysis of molecular genetic patterns still has to be performed in the light of knowledge of the natural history of the insect concerned, otherwise, as outlined in this synthesis, there is tremendous room for misinterpretation.

Prior to the 1970s, even though a wealth of information was collected and analysed for various moving insects (e.g. Johnson, 1969), molecular tools were generally not available. Today, however, a plethora of such markers exist. Despite this utility and availability, whilst there is copious and random gathering of information, yesterday's questions remain largely unanswered, even for the most well studied migratory species, including various international pests. Major questions that are largely unresolved are: (i) what is the direction and defining range of a moving population of a given species; (ii) are there, for given species, migratory genotypes and populations, separate from genotypes of established perennial populations of the same species; (iii) what are the rates of movement of species, individuals and populations; (iv) to what extent do flying insects travel individually or collectively, initiating gene flow between isolated populations; and (v) does mating take place in some groups producing moving demes? It is not even well known in most instances whether homogeneous allelic patterns reflect the influence of recent gene flow or are the result of stabilizing selection on historical demographic patterning. Whilst it may be argued that these are the fundamental and applied questions that we strive and seek to answer to illuminate adaptation and evolution, it may also be timely to reflect upon the words of Sir Isaac Newton (1642–1727), '... I seem to have been only like a boy playing on the sea-shore and diverting myself in now and then finding a smoother pebble or a prettier shell than ordinary, whilst the great ocean of truth lay all undiscovered before me' (Partington, 1992). With the advent of the Millennium, our endeavours should perhaps be focused on the above questions more clearly in a case-specific way to move the science forward, e.g. Lenormand et al. (1999). Our knowledge of the movement of flying insects is far from complete in spite of enormous efforts, particularly those over the last 50 years. Further advances will undoubtedly result from the application of the right molecular markers and a careful consideration of the questions asked.

## Acknowledgements

This paper was supported by the Ministry of Agriculture, Fisheries and Food, MAFF (H.D.L) and The Leverhulme Trust (G.L.; F/180/AP). IACR-Rothamsted receives grant-aided support from the BBSRC. We are most grateful to Ian Woiwod, Ian Denholm and Jim Mallet for their helpful comments on the manuscript.

## References

Alstad, D.N. and Corbin, K.W. (1990) Scale insect allozyme differentiation within and between host trees. *Evolutionary Ecology* 4, 43–56.

Avise, J.C. (1994) *Molecular Markers, Natural History and Evolution*. New York, Chapman & Hall.

Ayala, F.J., Powell, J.R., Tracey, M.L., Mourão, C.A. and Pérez-Salas, S. (1972) Enzyme variability in the *Drosophila willistoni* Group. IV. Genic variation in natural populations of *Drosophila willistoni. Genetics* 70, 113–139.

Baur, B. and Schmid, B. (1996) Spatial and temporal patterns of genetic diversity within species. In: Gaston, K.J. (ed.) *Biodiversity*. Blackwell Science, Oxford.

Berry, R.J., Crawford, T.J. and Hewitt, G.M. (eds) (1992) *Genes in Ecology*. Blackwells Scientific, Oxford, pp. 487–490.

Bizzaro, D., Manicardi, G.C. and Bianchi, U. (1996) Chromosomal localization of a highly-repeated EcoRI DNA fragment in *Megoura viciae* (Homoptera, Aphididae) by nick-translation and fluorescence *in-situ* hybridization. *Chromosome Research* 4, 392–396.

Black, W.C., IV (1993) Variation in the ribosomal RNA cistron among host adapted races of an aphid (*Schizaphis graminum*). *Insect Molecular Biology* 2, 59–69.

Blackman, R.L. and Spence, J.M. (1996) Ribosomal DNA is frequently concentrated on only one X chromosome in permanently apomictic aphids, but this does not inhibit male determination. *Chromosome Research* 4, 314–320.

Blackman, R.L., Spence, J.M., Field, L.M. and Devonshire, A.L. (1995) Chromosomal location of the amplified esterase genes conferring resistance to insecticides in *Myzus persicae* (Homoptera: Aphididae). *Heredity* 75, 297–302.

Boecklen, W.J. and Howard, D.J. (1997) Genetic analysis of hybrid zones: numbers of markers and power of resolution. *Ecology* 78, 2611–2616.

Bohonak, A.J., Davies, N., Roderick, G.K. and Villablanca, F.X. (1998) Is population genetics mired in the past? *Trends in Ecology and Evolution* 13, 360.

Bossart, J.L. and Pashley-Prowell, D. (1998) Genetic estimates of population structure and gene flow: limitation, lessons and new directions. *Trends in Ecology and Evolution* 13, 202–206.

Bossart, J.L. and Scriber, J.M. (1995) Maintenance of ecologically significant genetic variation in the tiger swallowtail through differential selection and gene flow. *Evolution* 49, 1163–1171.

Brookes, C.P. and Loxdale, H.D. (1987) Survey of enzyme variation in British populations of *Myzus persicae* (Sulzer) (Hemiptera: Aphididae) on crops and weed hosts. *Bulletin of Entomological Research* 77, 83–89.

Brower, A.V.Z. (1996) Parallel race formation and the evolution of mimicry in *Heliconius* butterflies: a phylogenetic hypothesis from mitochondrial DNA sequences. *Evolution* 50, 195–221.
Brower, A.V.Z. and Boyce, T.M. (1991) Mitochondrial DNA variation in Monarch butterflies. *Evolution* 45, 1281–1286.
Bullini, L. and Nascetti, G. (1989) Speciation by hybridization in insects. In: Loxdale, H.D. and den Hollander, J. (eds) *Electrophoretic Studies on Agricultural Pests*. The Systematics Association, Special Volume No. 39, Clarendon Press, Oxford, pp. 317–339.
Byers, J.A. (1999) Effects of attraction radius and flight path on catch of scolytid beetles dispersing outward through rings of pheromone traps. *Journal of Chemical Ecology* 25, 985–1005.
Byrne, D.N., Rathman, R.J., Orum, T.V. and Palumbo, J.C. (1996) Localized migration and dispersal by the sweet potato whitefly, *Bemisia tabaci*. *Oecologia* 105, 320–328.
Carter, P.A., Hughes, J.M. and Zalucki, M.P. (1989) Genetic variation in a continuously breeding population of *Danaus plexippus* L.: an examination of heterozygosity at 4 loci in relation to activity times. *Heredity* 63, 191–194.
Carvalho, G.R. (1998) Molecular ecology: origins and approach. *Advances in Molecular Ecology* 306, 1–23.
Cobb, N.S. and Whitham, T.G. (1998) Prevention of deme formation by the Pinyon needle scale: problems of specializing in a dynamic system. In: Mopper, S. and Strauss, S.Y. (eds) *Genetic Structure and Local Adaptation in Natural Insect Populations*. Chapman & Hall, New York, pp. 37–63.
Coyne, J.A., Boussy, I.A., Prout, T., Bryant, S.H., Jones, J.S. and Moore, J.A. (1982) Long-distance migration of *Drosophila*. *American Naturalist* 119, 589–595.
Daly, J.C. (1989) The use of electrophoretic data in a study of gene flow in the pest species *Heliothis armigera* (Hübner) and *H. punctigera* Wallengren (Lepidoptera: Noctuidae). In: Loxdale, H.D. and den Hollander, J. (eds) *Electrophoretic Studies on Agricultural Pests*. Clarendon Press, Oxford, pp. 115–141.
De Barro, P.J., Sherratt, T.N., Carvalho, G.R., Nicol, D., Iyengar, A. and Maclean, N. (1995a) An analysis of secondary spread by putative clones of *Sitobion avenae* within a Hampshire wheat field using the multilocus $(GATA)_4$ probe. *Insect Molecular Biology* 3, 253–260.
De Barro, P.J., Sherratt, T.N., Brookes, C.P., David, O. and Maclean, N. (1995b) Spatial and temporal variation in British field populations of the grain aphid *Sitobion avenae* (F.) (Hemiptera: Aphididae) studied using RAPD-PCR. *Proceedings of the Royal Society B* 262, 321–327.
De Barro, P.J., Sherratt, T.N., David, O. and Maclean, N. (1995c) An investigation of the differential performance of clones of the aphid *Sitobion avenae* on two hosts. *Oecologia* 104, 379–385.
Den Boer, M.H. (1978) Isoenzymes and migration in the African armyworm *Spodoptera exempta* (Lepidoptera, Noctuidae). *Journal of the Zoological Society of London* 185, 539–553.
Dieckmann, U., O'Hara, R. and Weisser, W. (1999) The evolutionary ecology of dispersal. *Trends in Ecology and Evolution* 14, 88–91.
Dixon, A.F.G. (1969) Population dynamics of the sycamore aphid, *Drepanosiphum platanoides* (Schr.) (Hemiptera: Aphididae): migratory and trivial flight activity. *Journal of Animal Ecology* 38, 585–606.

Dobzhansky, T. and Powell, J.R. (1974) Rates of dispersal of *Drosophila pseudoobscura* and its relatives. *Proceedings of the Royal Society of London B* 187, 281–298.
Dobzhansky, T. and Wright, S. (1943) Genetics of natural populations. X. Dispersion rates in *Drosophila pseudoobscura*. *Genetics* 28, 304–340.
Dowdeswell, W.H. and Ford, E.B. (1953) The influence of isolation on variability in the butterfly *Maniola jurtina* L. In: *Evolution, 7th Symposium of the Society of Experimental Biology*, Cambridge University Press, Cambridge, pp. 254–273.
Eanes, W.F. and Koehn, R.K. (1978) An analysis of genetic structure in the Monarch butterfly, *Danaus plexippus* L. *Evolution* 32, 784–797.
Emelianov, I., Mallet, J. and Baltensweiler, W. (1995) Genetic differentiation in *Zeiraphera diniana* (Lepidoptera, Tortricidae), the larch budworm polymorphism, host races or sibling species. *Heredity* 75, 416–424.
Feder, J.L., Chilcote, C.A. and Bush, G.L. (1989) Are the apple maggot, *Rhagoletis pomonella*, and blueberry maggot, *R. mendax*, distinct species? Implications for sympatric speciation. *Entomologia Experimentalis et Applicata* 51, 113–123.
Feder, J.L., Chilcote, C.A. and Bush, G.L. (1990) The geographic pattern of genetic differentiation between host associated populations of *Rhagoletis pomonella* (Diptera: Tephritidae) in the eastern United States and Canada. *Evolution* 44, 570–594.
Feder, J.L., Berlocher, S.H. and Opp, S.B. (1998) Sympatric host-race formation and speciation in *Rhagoletis* (Diptera: Tephritidae): a tale of two species for Charles D. In: Mopper, S. and Strauss, S.Y. (eds) *Genetic Structure and Local Adaptation in Natural Insect Populations*. Chapman & Hall, New York, pp. 408–441.
Fenton, B., Birch, A.N.E., Malloch, G., Woodford, J.A.T. and Gonzalez, C. (1994) Molecular analysis of ribosomal DNA from the aphid *Amphorophora idaei* on an associated fungal organism. *Insect Molecular Biology* 3, 183–190.
Fenton, B., Woodford, J.A.T. and Malloch, G. (1998). Analysis of clonal diversity of the peach-potato aphid, *Myzus persicae* (Sulzer), in Scotland, UK and evidence for the existence of a predominant clone. *Molecular Ecology* 7, 1475–1487.
Gatehouse, A.G. (1997) Behaviour and ecological genetics of wind-borne migration by insects. *Annual Review of Entomology* 42, 475–502.
Hanski, I.A. and Gilpin, M.E. (eds) (1997) *Metapopulation Biology: Ecology, Genetics and Evolution*. Academic Press, San Diego.
Henry, C.S. (1994) Singing and cryptic speciation in insects. *Trends in Ecology and Evolution* 9, 388–392.
Hoy, M.A. (1994) *Insect Molecular Genetics*. Academic Press, London.
Howard, D.J. (1983) Electrophoretic survey of eastern North American *Allonemobius* (Orthoptera: Gryllidae): evolutionary relationships and the discovery of three new species. *Annals of the Entomological Society of America* 76, 1014–1021.
Hughes, J.M. and Zaluki, M.P. (1993) The relationship between the Pgi locus and the ability to fly at low temperatures in the Monarch butterfly *Danaus plexippus*. *Biochemical Genetics* 31, 521–532.
Isaacs, R. and Byrne, D.N. (1998) Aerial distribution, flight behaviour and egg load: their inter-relationship during dispersal by the sweet potato whitefly. *Journal of Animal Ecology* 67, 741–750.
Jarne, P. and Lagoda, P.J.L. (1996) Microsatellites, from molecules to populations and back. *Trends in Ecology and Evolution* 11, 424–429.

Johnson, C.G. (1969) *Migration and Dispersal of Insects by Flight*. Methuen, London.
Karp, A., Isaac, P.G. and Ingram, D.S. (eds) (1998) *Molecular Tools for Screening Biodiversity*. Chapman & Hall, London.
Kukalova-Peck, J. (1978) Origin and evolution of insect wings and their relation to metamorphosis, as documented by the fossil record. *Journal of Morphology* 156, 53–125.
Langor, D.W. and Spence, J.R. (1991) Host effects on allozyme and morphological variation of the mountain pine beetle, *Dendroctonus ponderosae* Hopkins (Coleoptera: Scolytidae). *Canadian Entomologist* 123, 395–410.
Lenormand, T., Bourguet, D., Guillemaud, T. and Raymond, M. (1999) Tracking the evolution of insecticide resistance in the mosquito *Culex pipiens*. *Nature* 400, 861–864.
Lewis, O.T., Thomas, C.D., Hill, J.K., Brookes, M.I., Crane, T.P.R., Graneau, Y.A., Mallet, J.L.B. and Rose, O.C. (1997) Three ways of assessing metapopulation structure in the butterfly *Plebejus argus*. *Ecological Entomology* 22, 283–293.
Liebherr, J.K. (1988) Gene flow in ground beetles (Coleoptera: Carabidae) of differing habitat preference and flight-wing development. *Evolution* 42, 129–137.
Llewellyn, K.S., Brookes, C.P., Harrington, R., Loxdale, H.D. and Sunnucks, P. (1997) Using microsatellite DNA to study aphid genetic heterogeneity in the U.K. *V. International Symposium on Aphids*, León, Spain, 15–19 September, p. 18.
Llewellyn, K.S., Loxdale, H.D., Harrington, R. and Brookes, C.P. (1999) Migration and microsatellite variation in the grain aphid. *Entomology '99, Third Meeting of the Royal Entomological Society, Imperial College, London, 15 September 1999.* (Abstract).
Loxdale, H.D. (1994) Isozyme and protein profiles of insects of agricultural and horticultural importance. In: Hawksworth, D.L. (ed.) *The Identification and Characterization of Pest Organisms*. CAB International, Wallingford, UK, pp. 337–375.
Loxdale, H.D. and Brookes, C.P. (1990) Genetic stability within and restricted migration (gene flow) between local populations of the blackberry-grain aphid *Sitobion fragariae* in south-east England. *Journal of Animal Ecology* 59, 495–512.
Loxdale, H.D. and Lushai, G. (1998) Molecular markers in entomology (review). *Bulletin of Entomological Research* 88, 577–600.
Loxdale, H.D. and Lushai, G. (1999) Slaves of the environment: the movement of herbivorous insects in relation to their ecology and genotype. *Philosophical Transactions of the Royal Society B* 354, 1479–1495.
Loxdale, H.D., Tarr, I.J., Weber, C.P., Brookes, C.P., Digby, P.G.N. and Castañera, P. (1985) Electrophoretic study of enzymes from cereal aphid populations. III. Spatial and temporal genetic variation of populations of *Sitobion avenae* (F.) (Hemiptera: Aphididae). *Bulletin of Entomological Research* 75, 121–141.
Loxdale, H.D., Hardie, J., Halbert, S., Foottit, R., Kidd, N.A.C. and Carter, C.I. (1993) The relative importance of short- and long-range movement of flying aphids. *Biological Reviews* 68, 291–311.
Loxdale, H.D., Brookes, C.P. and De Barro, P.J. (1996) Application of novel molecular markers (DNA) in agricultural entomology. In: Symondson, W.O.C. and Liddell, J.E. (eds) *The Ecology of Agricultural Pests: Biochemical Approaches*. Systematics Association, Special Volume No. 53. Chapman & Hall, London, pp. 149–198.

Loxdale, H.D., Brookes, C.P., Wynne, I.R. and Clark, S.J. (1998) Genetic variability within and between English populations of the damson-hop aphid, *Phorodon humuli* (Hemiptera: Aphididae), with special reference to esterases associated with insecticide resistance. *Bulletin of Entomological Research* 88, 513–526.

Luikart, G. and England, P.R. (1999) Statistical analysis of microsatellite DNA data. *Trends in Ecology and Evolution* 14, 253–256.

Lushai, G., Hardie, J. and Harrington, R. (1996) Inheritance of photoperiodic response in the bird cherry aphid, *Rhopalosiphum padi*. *Physiological Entomology* 21, 297–303.

Lushai, G., Loxdale, H.D., Brookes, C.P., von Mende, N., Harrington, R. and Hardie, J. (1997) Genotypic variation among different phenotypes within aphid clones. *Proceedings of the Royal Society B* 264, 725–730.

Lushai, G., Loxdale, H.D., Markovitch, O. and Brookes, C.P. (2001) Host-based genotype variation in insects revisited. *Bulletin of Entomological Research*, in press

Martinez-Torres, D., Simon, J.C., Fereres, A. and Moya, A. (1996) Genetic variation in natural populations of the aphid *Rhopalosiphum padi* as revealed by maternally inherited markers. *Molecular Ecology* 5, 659–670.

Martinez-Torres, D., Moya, A., Hebert, P.D.N. and Simon, J.C. (1997) Geographic distribution and seasonal variation of mitochondrial DNA haplotypes in the aphid *Rhopalosiphum padi* (Hemiptera: Aphididae). *Bulletin of Entomological Research* 87, 161–167.

Maynard Smith, J. and Haigh, J. (1974) The hitch-hiking effect of a favourable gene. *Genetical Research, Cambridge* 23, 23–35.

McPheron, B.A. (1990) Genetic structure of apple maggot fly (Diptera: Tephritidae) populations. *Annals of the Entomological Society of America* 83, 568–577.

Meglecz, E., Pecsenye, K., Varga, Z. and Solignac, M. (1998) Comparison of differentiation pattern at allozyme and microsatellite loci in *Parnassius mnemosyne* (Lepidoptera) populations. *Hereditas* 128, 95–103.

Meglecz, E., Neve, G., Pecsenye, K. and Varga, Z. (1999) Genetic variations in space and time in *Parnassius mnemosyne* (L.) (Lepidoptera) populations in north-east Hungary: implications for conservation. *Biological Conservation* 89, 251–259.

Mopper, S. (1998) Local adaptation and stochastic events in an oak leaf-miner population. In: Mopper, S. and Strauss, S.Y. (eds) *Genetic Structure and Local Adaptation in Natural Insect Populations*. Chapman & Hall, New York, pp. 139–155.

Mopper, S. and Strauss, S.Y. (eds) (1998) *Genetic Structure and Local Adaptation in Natural Insect Populations*. Chapman & Hall, New York.

Mun, J.H., Song, Y.H., Heong, K.L. and Roderick, G.K. (1999) Genetic variation among Asian populations of rice planthoppers, *Nilaparvata lugens* and *Sogatella furcifera* (Hemiptera: Delphacidae): mitochondrial DNA sequences. *Bulletin of Entomological Research* 89, 245–253.

Napolitano, M., Geiger, H. and Descimon, H. (1988) Structure démographique et génétique de quatre populations provençales de *Parnassius mnemosyne* (L.) (Lepidoptera, Papilionidae): isolement et polymorphisme dans des populations 'menacées'. *Génétique, Sélection, Evolution* 20, 51–62.

Napolitano, M. and Descimon, H. (1994) Genetic structure of French populations of the mountain butterfly *Parnassius mnemosyne* L. (Lepidoptera: Papilionidae). *Biological Journal of the Linnaean Society* 53, 325–341.

Nevo, E., Rashkovetsky, E., Pavlicek, T. and Korol, A. (1998) A complex adaptive syndrome in *Drosophila* caused by microclimatic contrasts. *Heredity* 80, 9–16.

Nuismer, S.L., Thompson, J.N. and Gomulkiewicz, R. (1999) Gene flow and geographically structured coevolution. *Proceedings of the Royal Society of London B* 266, 605–609.

Odonald, P. and Majerus, M.E.N. (1992) Non-random mating in *Adalia bipunctata* (the 2-spot ladybird). 3. New evidence of genetic preference. *Heredity* 69, 521–526.

Parkin, D.T. (1979) *An Introduction to Evolutionary Genetics*. Edward Arnold, London.

Partington, A. (ed.) (1992) *The Oxford Dictionary of Quotations*. Oxford University Press, Oxford.

Pashley, D.P. and Johnson, S.J. (1986) Genetic population structure of migratory moths: the velvetbean caterpillar (Lepidoptera: Noctuidae). *Annals of the Entomological Society of America* 79, 26–30.

Peterson, M.A. (1996) Long-distance gene flow in the sedentary butterfly, *Euphilotes enoptes* (Lepidoptera: Lycaenidae). *Evolution* 50, 1990–1999.

Peterson, M.A. and Denno, R.F. (1998a) The influence of dispersal and diet breadth on patterns of genetic isolation by distance in phytophagous insects. *American Naturalist* 152, 428–446.

Peterson, M.A. and Denno, R.F. (1998b) Life-history strategies and the genetic structure of phytophagous insect populations In: Mopper, S. and Strauss, S.Y. (eds) *Genetic Structure and Local Adaptation in Natural Insect Populations*. Chapman & Hall, New York, pp. 263–322.

Rand, D.M. (1994) Concerted evolution and RAPping in mitochondrial VNTRs and the molecular geography of cricket populations. *Molecular Ecology and Evolution, Series Experientia Supplementum (EXS)* 69, 227–245.

Reynolds, D.R., Riley, J.R., Armes, N.J., Cooter, R.J., Tucker, M.R. and Colvin, J. (1997) Techniques for quantifying insect migration. In: Dent, D.R. and Walton, M.R. (eds) *Methods in Ecological and Agricultural Entomology*. CAB International, Wallingford, Oxford, UK, pp. 111–145.

Richardson, B.J., Baverstock, P.R. and Adams, M. (1986) *Allozyme Electrophoresis. A Handbook for Animal Systematics and Population Studies*. Academic Press, London.

Riley, J.R., Reynolds, D.R., Mukhopadhyay, S., Ghosh, M.R. and Sarkar, T.K. (1995) Long-distance migration of aphids and other small insects in north-east India. *European Journal of Entomology* 92, 639–653.

Roderick, G.K. (1996) Geographic structure of insect populations: gene flow, phylogeography, and their uses. *Annual Review of Entomology* 41, 325–352.

Roskam, J.C. and Brakefield, P.M. (1999) Seasonal polyphenism in *Bicyclus* (Lepidoptera: Satyridae) butterflies: different climates need different cues. *Biological Journal of the Linnaean Society* 66, 345–356.

Rubinsztein, D.C., Amos, W. and Cooper, G. (1999) Microsatellite and trinucleotide-repeat evolution: evidence for mutational bias and different rates of evolution in different lineages. *Philosophical Transactions of the Royal Society B* 354, 1095–1099.

Saccheri, I., Kuussaari, M., Kankare, M., Vikman, P., Fortelius, W. and Hanski, I. (1998) Inbreeding and extinction in a butterfly metapopulation. *Nature* 392, 491–494.

Scriber, J.M., Bossart, J.L. and Snider, D. (1992) Diagnostic alleles from electrophoresis distinguish two noctuid pest species, *Hydraecia immanis* and *H. micacea* (Lepidoptera: Noctuidae). *The Great Lakes Entomologist* 25, 91–98.

Sheppard, W.S., Steck, G.S. and McPheron, B.A. (1992) Geographic populations of the medfly may be differentiated by mitochondrial DNA variation. *Experientia* 48, 1010–1013.

Sherlock, P.L., Bowden, J. and Digby, P.G.N. (1985) Studies of elemental composition as a biological marker in insects. 4. The influence of soil type and host-plant on elemental composition of *Agrotis segetum* (Denis and Schiffermüller) (Lepidoptera, Noctuidae). *Bulletin of Entomological Research* 75, 675–687.

Shufran, K.A., Black, W.C. IV and Margolies, D.C. (1991) DNA fingerprinting to study spatial and temporal distributions of an aphid, *Schizaphis graminum* (Homoptera: Aphididae). *Bulletin of Entomological Research* 81, 303–313.

Shufran, K.A., Margolies, D.C. and Black, W.C. IV (1992) Variation between biotype E clones of *Schizaphis graminum* (Homoptera: Aphididae). *Bulletin of Entomological Research* 82, 407–416.

Simon, J.-C., Carrel, E., Hebert, P.D.N., Dedryver, C.A., Bonhomme, J. and Le Gallic, J.-F. (1996a) Genetic diversity and mode of reproduction in French populations of the aphid *Rhopalosiphum padi* L. *Heredity* 76, 305–313.

Simon, J.-C., Martinez-Torres, D., Latorre, A., Moya, A. and Hebert, P.D.N. (1996b) Molecular characterization of cyclic and obligate parthenogens on the aphid *Rhopalosiphum padi* (L.). *Proceedings of the Royal Society of London B* 263, 481–486.

Simon, J.-C., Baumann, S., Sunnucks, P., Hebert, P.D.N., Pierre, J.-S., Gallic, J.-F. and Dedryver, C.-A. (1999) Reproductive mode and population genetic structure of the cereal aphid *Sitobion avenae* studied using phenotypic and microsatellite markers. *Molecular Ecology* 8, 531–545.

Slatkin, M. (1985a) Gene flow in natural populations. *Annual Review of Ecology and Systematics* 16, 393–430.

Slatkin, M. (1985b) Rare alleles as indicators of gene flow. *Evolution* 39, 53–65.

Slatkin, M. (1987) Gene flow and the geographic structure of natural populations. *Science* 236, 787–792.

Slatkin, M. (1993) Isolation by distance in equilibrium and non-equilibrium populations. *Evolution* 47, 264–279.

Slatkin, M. (1995) A measure of population subdivision based on microsatellite allele frequencies. *Genetics* 139, 457–462.

Smith, M.A.H. and MacKay, P.A. (1990) Latitudinal variation in the photoperiodic responses of populations of pea aphid (Homoptera: Aphididae). *Environmental Entomology* 19, 618–624.

Smith, A.D. and Riley, J.R. (1996) Signal processing in a novel radar system for monitoring insect migration. *Computers and Electronics in Agriculture* 15, 267–278.

Stock, M.W., Pitman, G.B. and Guenther, J.D. (1979) Genetic differences between Douglas-fir beetles (*Dendroctonus pseudotsugae*) from Idaho and coastal Oregon. *Annals of the Entomological Society of America* 72, 394–397.

Stone, G.N. and Sunnucks, P. (1993) Genetic consequences of an invasion through a patchy environment – the cynipid gallwasp *Andricus quercuscalicis* (Hymenoptera: Cynipidae). *Molecular Ecology* 2, 251–268.

Suckling, D.M., Hill, R.L., Gourlay, A.H. and Witzgall, P. (1999) Sex attractant-based monitoring of a biological control agent of gorse. *Biocontrol Science and Technology* 9, 99–104.

Sunnucks, P., De Barro, P.J., Lushai, G., Maclean, N. and Hales, D. (1997) Genetic structure of an aphid studied using microsatellites: cyclic parthenogenesis, differentiated lineages, and host specialisation. *Molecular Ecology* 6, 1059–1073.

Sutcliffe, O.L., Thomas, C.D., Yates, T.J. and Greatorex-Davies, J.N. (1997) Correlated extinctions, colonisations and population fluctuations in a highly connected ringlet butterfly population. *Oecologia* 109, 235–241.

Taberner, A., Dopazo, J. and Castañera, P. (1997) Genetic characterization of populations of a *de novo* arisen sugar beet pest, *Aubeonymus mariaefranciscae* (Coleoptera, Curculionidae), by RAPD analysis. *Journal of Molecular Evolution* 45, 24–31.

Taimr, L. and Kríz, J. (1978) Stratiform drift of the hop aphid (*Phorodon humuli* Schrank). *Zeitschrift für angewandte Entomologie* 86, 71–79.

Thomson, A.L. (1926) *The Problems of Bird Migration*. Witherby, London.

Thompson, J.N. (1994) *The Coevolutionary Process*. The University of Chicago Press, Chicago, Illinois.

Urquhart, F.A. and Urquhart, N.R. (1977) Overwintering areas and migratory routes of the Monarch butterfly (*Danaus p. plexippus*, Lepidoptera: Danaidae) in North America with special reference to the western population. *Canadian Entomologist* 109, 1583–1589.

Vawter, A.T. and Brussard, P.F. (1983) Allozyme variation in a colonizing species: the Cabbage butterfly *Pieris rapae* (Pieridae). *The Journal of Research on the Lepidoptera* 22, 204–216.

Waser, P.M. and Strobeck, C. (1998) Genetic signatures of interpopulation dispersal. *Trends in Ecology and Evolution* 13, 43–44.

Wallis, G.P. (1999) Do animal mitochondrial genomes recombine? *Trends in Ecology and Evolution* 14, 209–210.

Watt, W.B. (1977) Adaptation at specific loci. I. Natural selection on phosphoglucose isomerase of *Colias* butterflies: biochemical and population aspects. *Genetics* 87, 177–194.

Watt, W.B., Cassin, R.C. and Swan, M.S. (1983) Adaptation at specific loci. III. Field behaviour and survivorship differences among *Colias* PGI genotypes are predictable from *in vitro* biochemistry. *Genetics* 103, 725–739.

Westbrook, J.K., Esquivel, J.F., Lopez, J.D., Jones, G.D., Wolf, W.W. and Raulston, J.R. (1998) Validation of bollworm (Lepidoptera: Noctuidae) migration across south-central Texas in 1994–1996. *Southwestern Entomologist* 23, 209–219.

Wiernasz, D.C. (1989) Ecological and genetic correlates of range expansion in *Coenonympha tullia*. *Biological Journal of the Linnaean Society* 38, 197–214.

White, M.J.D. (1978) *Modes of Speciation*. W.H. Freeman & Company, San Francisco.

Wigglesworth, V.B. (1976) The evolution of insect flight. In: Rainey, R.C. (ed.) *Insect Flight*, Seventh Symposium of the Royal Entomological Society of London, Blackwells Scientific Publications, Oxford, pp. 255–269.

Woiwod, I.P., Tatchell, G.M., Dupuch. M.J., Macaulay, E.D.M., Parker, S.J., Riley, A.M. and Taylor, M.S. (1988) Rothamsted Insect Survey: Nineteenth Annual Summary: Suction Traps 1987. In: *Rothamsted Experimental Station Report for 1987*, Part 2, Rothamsted Experimental Station, Harpenden, pp. 195–229.

Wynne, I.R. (1997) Population studies on farmland insects using genetic markers. PhD thesis, University of Nottingham.

Wynne, I.R., Howard, J.J., Loxdale, H.D. and Brookes, C.P. (1994) Population genetic structure during aestivation in the sycamore aphid *Drepanosiphum platanoidis* (Hemiptera: Drepanosiphidae). *European Journal of Entomology* 91, 375–383.

Zera, A.J. and Denno, R.F. (1997) Physiology and ecology of dispersal polymorphism in insects. *Annual Review of Entomology* 42, 207–230.

Zhang, D.-X. and Hewitt, G.M. (1996) Nuclear integrations: challenge for mitochondrial markers. *Trends in Ecology and Evolution* 11, 247–251.

Zhang, D.-X. and Hewitt, G.M. (1997) Insect mitochondrial control region: a review of its structure, evolution and usefulness in evolutionary studies. *Biochemical Systematics and Ecology* 25, 99–120.

# Coping with Modern Times? Insect Movement and Climate Change

**Camille Parmesan**

*Integrative Biology, Patterson Labs Building, University of Texas, Austin, TX 78712, USA*

## Introduction

In the mid-1970s, a set of papers appeared examining responses to climate change in birds, butterflies and various other taxa. Prescience? Not exactly. 'Climate change' was not then the euphemism for global warming that it is today. The premise at that time was that 'climate is in a continual state of flux and, directly and indirectly, its fluctuations affect the status of plants, insects, birds and mammals' (Lamb, 1975). What relevance do such studies have to a symposium on insect movement? It turns out that the responses of insects to long-term climate trends include large geographic shifts of species ranges, a process resulting from movement of individuals, both random and systematic.

This chapter shows first how insect range shifts have tracked climate change over geological time, then over historical time, and finally over this past century. Much of the chapter is focused on how recent range shifts have been studied in the light of responses to current global warming trends. Evidence is presented that infrequent, single extreme weather and climate events, as opposed to mean climatic trends, may be principal drivers of such range movements.

Some of the processes underlying these range changes occur at the population level, with host, habitat and other local environmental changes affecting population persistence. Others occur at the individual level, with dispersal ability affecting the relative likelihoods that an individual suffering adverse climate conditions will either leave its habitat or remain *in situ* and risk mortality. The properties of individuals also affect the likelihood of finding suitable habitat, and hence of establishing new populations, if dispersal does occur. These complexities are rarely considered in models which attempt to predict biome shifts under various global warming scenarios, an issue which needs to be addressed as the mechanisms underlying responses to climate change become better understood.

## Non-anthropogenic Climate Change and Insect Movement

### Geological time

Insects have provided some of the best studies of species range movement. Coleopteran studies form a classic body of work on Pleistocene range shifts driven by the glacial/interglacial cycles. Several excellent reviews of this work can be found in Coope (1995), Ashworth (1997), Morgan (1997) and Ponel (1997), which are summarized here. Only a few localities have the necessary characteristics to have preserved long time-series of the beetle fossil record, but such sites are distributed throughout North America and Europe, with a few in South America also. Some sites, such as a few in the UK, have good resolution in recent times, giving nearly 100-year resolution over the past 14,000 years. Other sites, such as the Grande Pile in northern France, are poor for recent (Holocene) records, but the remaining records go back 140,000 years, thus spanning more than one glacial/interglacial cycle. The striking similarities among all of these studies is the rarity of species extinctions and the systematic shifts in community structure at each (northern hemisphere) site towards northern (cold-adapted) species at glacial maxima and towards southern (warm-adapted) species during interglacial periods. The morphology of beetles appears to have been fairly stable throughout this time and many species in the fossil records are still extant. Thus, the presence of a particular species at a given site in geological time can be compared to its current distribution. By this means, range shifts of hundreds of kilometres, and as much as 2000 km, have been documented for individual species associated with mean temperature changes of 4–8°C over as little as 500 years.

Beetles are particularly good dispersers and extinction or stasis may have been more common for less mobile insects (Williams, 1989), but little exploration of the fossil record has been undertaken to test this idea, probably because few other insects are as well preserved as beetles in the fossil record.

### Historical time

Some of the best studies of more recent species range movement are of the ebbs and flows of population densities and range boundaries of the Lepidoptera in northern European countries (Ford, 1945; Kaisila, 1962;

Heath, 1974; Burton, 1975; Heath et al., 1984). With some records dating back to the 18th century, these authors were able to track shifts to the north and east during periods of warming (often termed 'climatic amelioration'), and contractions to the south and west during periods of cooling ('climatic deterioration'). Thus, although the Little Ice Age technically ended in the early 1800s, records indicate that extended periods of wet summers and cool temperatures in these latitudes persisted until about 1890. Around 1920, there began a steep rise in mean temperatures. By 1940, dramatic northward expansions had occurred in dozens of butterflies and moths in the UK and Finland (Ford, 1945; Kaisila, 1962). Unfortunately for those new colonies, climate again 'deteriorated' (cooler and wetter summers) so that by 1970, most of those new range boundaries had contracted southward, almost to where they had been in the previous century (Kaisila, 1962; Heath et al., 1984; Mikkola, 1997). Thus, it is no surprise that, as another warming period began in the 1970s, and butterflies and moths again slowly colonized northward in Britain, Denmark, Sweden and Finland, this was initially interpreted as merely another tiny cycle within the greater shift from glacial to interglacial global conditions that had begun 12,000 years ago (Heath, 1974; Burton, 1975).

However, during the 1990s, a consensus was gradually building among climatologists that a systematic shift in climate was taking place. Then came the second report of the Intergovernmental Panel on Climate Change (IPCC) which concluded that there is a discernible human influence on the climatic trends in the present century (Houghton et al., 1996).

## Global Warming and Global Change: a Brief Overview

### Global warming patterns

Studies subsequent to the 1995 IPCC Report (Houghton et al., 1996) have not only confirmed a significant warming trend this century (Karl et al., 1996; Easterling et al., 1997), but have strengthened the causal link between increasing greenhouse gases and global warming trends. It is becoming increasingly apparent that other aspects of climate and weather events are changing alongside mean global temperature (reviewed by Easterling et al., 2000a,b; Meehl et al., 2000a,b).

For instance, precipitation patterns are changing (Easterling et al., 2000a,b). In general, total precipitation is increasing, but it is falling in fewer, more extreme, events (Karl et al., 1996; Groisman et al., 1999). Snowpack is increasing in the far north, while a few regions are becoming drier (e.g. southwestern United States, Sahara desert). El Niño events have increased in frequency and intensity, and the 'normal' state by 2050 is predicted to resemble 'El Niño-like conditions' (Meehl et al., 2000b).

The patterns in temperature are less clear, but even if the distribution remains stable, as the mean increases, so do the numbers of 'extreme' temperature events, i.e. those days falling outside some absolute temperature threshold set by basic physiological tolerance (Easterling et al., 2000a,b; Meehl et al., 2000a; Parmesan et al., 2000). For instance, at northern latitudes, the trend has been towards fewer days/nights below freezing in wintertime, and more days surpassing a given heat index in summer (Changnon et al., 2000). These events may in some cases simply reflect the same level of variation but around a new mean, but it is possible that the variance in temperatures might also increase.

Finally, the nature of current temperature rise bears the mark of human influence. The warming period of the 1930s and 1940s was qualitatively different from the current warming trend (Easterling et al., 1997). The earlier warming event is generally believed to have been a natural occurrence driven by solar forces and was characterized by an increase in maximum daytime temperature. The current warming trend comes primarily from an increase in minimum night-time temperatures, which are likely to be a consequence of increased cloudiness which, in turn, is linked with an increase in atmospheric $CO_2$ (Houghton et al., 1996; Beniston et al., 1998).

## Global warming embedded in global change

Climate change is part of a family of global changes driven by human activities – wildlife is beset by a bevy of non-climatic anthropogenic forces that affect population dynamics, community stability and species distributions. Of primary concern to terrestrial insects are land use change, pollution and invasive species, although increasing UV radiation may also be gaining importance. All of these forces act to alter, degrade or eliminate natural habitats, leading to the loss of species dependent on those habitats, and to range expansions of species that benefit from human activities.

Overt destruction of habitat due to urbanization, conversion to agriculture, clear-cut logging or creation of reservoirs is easily assessed. Habitat degradation is more subtle and often results from changes in land management, such as changes in grazing intensity/timing, changes in fire intensity/frequency, water diversion, or changes in forestry practices such as coppicing, logging techniques and reforestation practices. Loss of habitat by any of these means not only causes extinctions at that site, but endangers populations in surrounding good habitat patches by increased fragmentation. As good habitat patches become smaller and more isolated, the populations in those patches are more likely to go permanently extinct (Thomas, 1995; Hanski, 1999). All of these factors confound interpretation of observed changes in plant and animal distributions. It is also likely that their combined effects would produce responses that could not be predicted from the summation of single effects.

## Insects under Global Warming

Just 5 years ago, a previous Royal Entomological Society Symposium, dedicated to 'Insects in a Changing Environment', made a series of predictions concerning the potential effects of climate change on insects (Harrington and Stork, 1995). Predicted effects ranged from the simple expectations of poleward and upward shifts of species ranges (Harrington et al., 1995; Lawton, 1995; Porter, 1995; Rogers, 1995; Sutherst et al., 1995), to more subtle phenological shifts (Fleming and Tatchell, 1995), to very complex shifts in species' interactions caused by disruption of community composition at many trophic levels (Davis et al., 1995; Lawton, 1995; Sutherst et al., 1995). Now, just 5 years later, many of those predictions have been supported by observations of change in a diversity of taxa and geographic regions during the 20th century (reviewed by Christianson, 1999; Palevitz, 1999; Hughes, 2000; Wuethrich, 2000).

Of particular relevance to the topic of insect movement are the many studies documenting changes in range boundaries. Changes in distributions along segments of range edges show general expansion into traditionally cooler habitat for amphibians (Pounds et al., 1999), birds (Pounds et al., 1999; Thomas and Lennon, 1999), alpine flora (Grabherr et al., 1994) and the red fox (Hersteinsson and MacDonald, 1992). Shifts in entire species' ranges towards the poles and up mountainsides have been found in butterflies in both North America and Europe (Parmesan, 1996; Parmesan et al., 1999). These relatively straightforward patterns of change have their most likely explanation in recent climatic trends and have emerged despite the influences of a plethora of independent confounding factors which also impact the quality and spatial distribution of habitats.

### Migratory species and climate change

For very mobile or migratory animals, such as many birds, large mammals, pelagic fish and some insects, shifts of species' range occur via the process of individuals changing their patterns of movement or their migration destinations. Thus, these movements track yearly climatic fluctuations. Because of the resulting large year–year variance, several decades of annual data would be needed to discern a long-term trend in migratory behaviour. There is little documentation available to look for systematic long-term changes in location of overwintering or breeding grounds for migratory insects, but there is a substantial body of literature documenting the importance of weather and climate in shaping the timing and destinations of migrations.

The overwintering locations, the timing and destinations of northward migration, as well as the return southward in the Monarch butterfly (*Danaus plexippus* (L.)) all may be ultimately determined by local climate. Calvert and Brower (1986) showed that their overwintering ground in Mexico provides a narrow range of micro-climatic conditions which is cool and moist enough to prevent desiccation and allow individuals to survive the winter on stored reserves, yet warm enough to prevent freezing. This habitat is climatically specialized, with forest cover, slope and aspect interacting to satisfy the specific needs of this species. Malcolm *et al.* (1987) demonstrated that the northward spring migration of adult *D. plexippus* tracks optimum climatic conditions for breeding, indicating that either reproduction and/or larval development is restricted not only by extreme low temperatures, but also by extreme highs.

Drake and Farrow (1985) suggest a similar pattern for moths in Australia, where moths overwinter in coastal mountainous areas of moderate climate, then reinvade the more climatically erratic lowland territories and otherwise hostile inland regions during good spring climate conditions. Rapid, long-distance movement would be required to effect this tracking over a single generation. There are frequent suggestions in the literature that various moth species may travel hundreds of kilometres non-stop by riding high altitude winds associated with unusual short-term weather patterns (Drake and Farrow, 1985; Ryrholm and Kallander, 1987; Lindfors *et al.*, 1989), but these scenarios are typically extrapolated from a variety of indirect sources. However, an experimental study by Showers *et al.* (1989) using mark–recapture over a large geographic region found similar dynamics of moth movement. They documented movement of several hundreds of kilometres over just 2–4 days.

These examples clearly indicate that changes in range or destinations of migratory insects are likely under global warming. Sufficient records may exist for the more well documented species, such as migratory pests, that a synthesis of these could be analysed for long-term poleward movement.

## Sedentary species and climate change

In contrast to these very mobile examples, the bulk of wild species are sedentary, living their lives in a single spot, such as in a single field or wood, either because they cannot move (because of limited mobility), or because they lack the behavioural mechanisms which would cause them to disperse from their sites of birth. Most phases of plant life are sedentary, as well as many insects, amphibians, reptiles and small mammals. Rather than occurring by changes in the direction or distance of individual movements, range changes in sedentary species operate by the much slower process of population extinctions and colonizations. Most documented shifts in species distributions have been in relatively sedentary species because the continuous time series needed to track highly mobile or migratory species rarely exist for longer than the past 10–20 years. While changes during such a brief time period are important to note, they are difficult to analyse for trends and do not provide the temporal context necessary to link those changes with decadal climatic trends.

## Butterflies as Model Indicators

### Butterfly distributions

Butterflies have been put forth as model indicators of climate change (Dennis, 1993; Woiwod, 1997; Parmesan, 2000) because they provide a rare combination of documented climatic sensitivity and sufficiently accurate historical records. Casual observation, extensive field surveys and detailed experiments have established that the physiology, ecology and evolution of many lepidopteran species are dependent on climate (Dennis, 1993).

The distributions of butterfly species have received particular attention in Europe where biologists have become concerned over large declines and regional extinctions of many species. Because the primary concern has largely been with conservation, published studies have tended to focus on the overall status of a species. They have addressed whether the species is stable, in decline or in general expansion. When such a study encompasses an area that coincides with a section of a species' range boundary, it records changes in population density and expansion or retraction along that section of the boundary.

By this means, several studies of Lepidoptera have described expansion of northern boundaries situated in Finland (Martila *et al.*, 1990; Mikkola, 1997), Britain (Pollard, 1979; Warren, 1992; Pollard and Eversham, 1995; Hill *et al.*, 2000) and across Europe (Parmesan *et al.*, 1999). Depending on the study, some 30–75% of northern boundary sections were recorded as shifting north, a smaller portion (<20%) had contracted southward and the remainder were classified as stable. There are fewer studies mentioning altitudinal changes, but at least some species in North America (Parmesan, 1996) and France (Descimon, 1995) are showing a tendency for the lowest elevation populations to go extinct, effectively shifting their ranges up the mountains.

Such observations are very important to conservation, but are often difficult to interpret causally because they are rarely designed to tease apart responses to climate change from responses to other human influences, principally habitat change. In most parts of the world, both loss of traditional habitat and creation of 'weedy' habitat have had huge impacts on wild plant and animal distributions. Even areas with a long history of human activities have experienced recent dramatic shifts in the types of activity that have had rapid strong impacts on wildlife.

A well understood example comes from Europe, where the 5000–10,000-year history of human cultivation has resulted in the dependence of many flower and insect species on traditionally managed hay meadows and grazing pastures. Without periodic cutting or grazing, the meadows undergo succession first to shrubland (bramble, blackthorn, etc.) and then to forest. It is not known exactly from where most of these species came, and the extent to which they have evolved adaptations to traditional hay-meadow management since they arrived. Regardless of the initial pre-Holocene conditions, the current switch in agricultural policies towards fertilization (leading to loss of herbs and towards a grass monoculture), active planting, or conversion to crops has resulted in a massive loss of meadow habitat and meadow species. At the same time, the expansion of road building and the abandonment of pastures has increased the area of 'weedy' habitat which has been exploited by species tolerant of these conditions. Because many European butterfly species are denizens of meadows, or of ecotones between forest and meadow, many of the local extinctions and extensive distributional changes (boundary expansions and contractions) that have been documented could be due solely to land management changes.

## Focus on response to climate change: two case studies

In an attempt to tease apart these many influences on distributional change, Parmesan (1996) and Parmesan et al. (1999) designed studies which made intensive effort to exclude changes which were identifiable as due to habitat deterioration, loss or expansion. These two studies focused on current and historical population data from regions which had maintained stable, suitable butterfly habitat over the time period of the records (up to 100 years).

### Case 1: *Euphydryas editha* (Boisduval)

For *Euphydryas editha* (Edith's checkerspot) in western North America, habitat assessment was done by direct field census of a subset of historical population sites over a 5-year period (1992–1996). If the site was no longer suitable butterfly habitat (either through overt destruction or through loss of host plants or nectar sources), then it was excluded from the final analysis. If the site still contained the necessary vegetation in sufficient abundance to support a local butterfly population, then it was surveyed for the presence of adults, larvae or eggs. By this means, current population presence could be compared to the historical distribution (dating from 1904). Populations along the southern edge of the range in Baja California, Mexico, were four times as likely to have gone extinct as populations along the northern range boundary in Canada (80% extinct versus 20% extinct; Fig. 18.1). Populations from lower elevations (sea level to 2400 m) were more than three times as likely to have gone extinct than were those at the highest elevations (<15% had gone extinct above 2400 m; Fig. 18.1). These patterns of population extinction had effectively shifted the range of *E. editha* both northward and upward since the beginning of the century (Parmesan, 1996).

The magnitude of the range shift in *E. editha* matched the observed warming trend over the same region. The mean location of populations shifted 92 km northward and 124 m upward concurrent with mean yearly temperature isotherms shifting 105 km northward and 105 m upward (Karl et al., 1996; Parmesan, 1996). Further, the cline in frequency of population extinction had a breakpoint at 2400 m (fewer extinctions at the highest elevations). This breakpoint coincided exactly with that for snowpack depth and timing of snowmelt, which involved increased depth and later melt-date above 2400 m, decreased depth and earlier melt below 2400 m (Johnson, 1998). Patterns of surrounding habitat destruction in the vicinity of the target sites did not correlate with the natural extinction patterns (C. Parmesan, unpublished observations).

Unfortunately, it was not possible to evaluate whether range expansion was taking place at the cool margins of the species range, because these occur in rugged and remote terrain where historical records are far from complete. At the highest elevations, *E. editha* is restricted from further expansion by the range limit of its host plants – the highest known population is at 3400 m

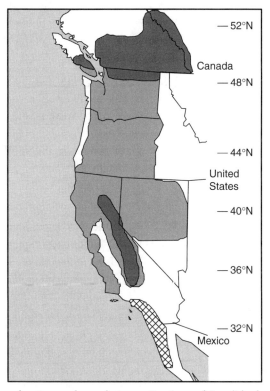

**Fig. 18.1.** Map of patterns of population extinctions of *E. editha* from 1904 to 1996 in western North American (modified from Parmesan, 2001). Each filled area represents averaged values from multiple populations. Shades of grey represent the proportion of populations extinct in a given area during the period 1992–1996 that were previously recorded as present during the period 1904–1983. Dark grey represents <20% extinctions, light grey represents 35–55% extinctions, cross-hatched represents >70% extinctions.

– and there are too few historical records to gauge whether the number of populations at high elevation has increased recently. The northernmost record of *E. editha* is from Jasper National Park in Canada. The territory north of this is poorly surveyed due to lack of roads coupled with scarcity of lepidopterists. New populations may have been colonized further north in recent decades, but it would be unlikely for these to have been discovered. In contrast, there is a clear shift along the southern range boundary, in Mexico. The most southern known population is currently 160 km further north than it was in 1970.

## Case 2: multiple European species

In Europe, because Parmesan *et al.* (1999) studied a suite of species, the filter for land use changes was generally less direct than with the Parmesan (1996) *E. editha* study. However, the approach remained the same: to isolate climatic impacts, as far as possible, species were eliminated from consideration if they were likely to have been severely affected by other, non-climatic, forces. As with *E. editha*, the principal direct factor believed to be affecting distributional changes of butterflies was land-use change. Therefore, we instituted criteria designed to restrict our study species to those that had not shown major impacts of land-use change on the position of either the northern or southern boundary.

A species was excluded from the study if: (i) it is extremely habitat restricted, e.g. requiring such a narrow combination of micro-climate, plant phenology and other characteristics that it is highly localized even within habitats containing its host; or (ii) it is known to be intolerant of even a modest level of human-mediated habitat modification. By using data from more than one country within a particular latitudinal range, we had replicated datasets for the same species across a single range boundary, but the countries differed in the type and degree of land-use change they had undergone during the study period (since the turn of the century). Therefore, for some species we excluded the data from a country if, within that country: (i) the range boundary lies in an area with so little potential habitat that its distribution could have changed dramatically via relatively small stochastic population extinctions coupled with a poor ability of the species to colonize highly dispersed habitat patches; or (ii) the species was known to have suffered severe habitat loss or degradation. Application of these criteria caused some species to be excluded from the study in some sections of their range boundaries and included in others. For example, *Argynnis paphia* was excluded from Britain, but retained in Finland.

Using these criteria, a few special clusters of species were eliminated from further analysis. For example, in some northern regions, specialists on dry meadows (*Cupido minimus* (Fuessly)), calcareous grasslands (*Lysandra bellargus* (Rottemburg) and *Lysandra coridon* (Poda)) and marshes

(*Euphydryas aurinia* (Rottenburg)) were excluded due to severe declines in these habitats in recent decades. All *Maculinea* species were excluded due to their extreme specialization. *Maculinea* are highly habitat specific and have complex host associations: early instars require a particular plant species and later instars require a particular ant species.

In northern Africa, a 7-year census (late 1980s to early 1990s) of historical butterfly collecting sites (1882–1932) was conducted by Tennent. Even though this area had suffered from large habitat destruction generally, there were many individual sites (often mountainous) which still supported high butterfly diversity and so could be included. As with *E. editha*, population extinctions at sites with good habitat were then overlaid on to the historical distributions to look for systematic shifts.

An obvious component of habitat suitability is the presence of host plants. There are some species of butterflies whose ranges reach the edge of distribution of their known hosts, and so they would be restricted in any response to climate change by the necessity for their host to first expand its own range. Thus, we also excluded species whose range is limited by their host-plant distribution.

Our final data filter was that we used only non-migratory species, thereby lowering the year-to-year variance in the geographic location of the range limit. This is an important point, because migratory species, by responding to yearly temperature fluctuations, will show rapid shifts in their range limits. In order to pick out an overall trend from the yearly variation, we would require very long time series of records with few gaps. Such continuous data exists for some species in the UK and Finland, but in all other countries there are gaps of 5–10 years in which particular areas were not well collected or censused. Because of the low dispersal rate of the species in our study, there is little likelihood of rapid, undocumented change in the intervening years.

These criteria thus focused the study on species in which responses to climate change were least likely to be confounded by other possible factors. They comprised species which were not habitat or host-plant limited, were non-migratory and were not extreme specialists. These were the only criteria used to exclude species; prior knowledge of boundary changes was not considered. The result was a replicated dataset of both single boundary and whole range distributions in a set of species chosen to be the most able to show effects of climate change uncontaminated by confounding factors.

We concluded that systematic northward shifts in both regional distributions and entire ranges had occurred over the course of this century. Of the 57 species for which we had data for at least one range boundary, 66% had shifted northward at a given boundary (by 35–240 km), and 3% (two species) had shifted southward (by less than 50 km). For 35 species for which we had data from both the northern and southern range limits, 63% had shifted their ranges to the north (by 35–240 km) and only 6% had shifted to the south (Parmesan *et al.*, 1999).

By studying many species simultaneously we were able to consider each species as a single data point. Among those species that showed range shifts, we asked whether the direction of these changes differed from random. That is, if range shifts were random, we would expect about 50% of the movements to be towards the north and about 50% to be towards the south (i.e. no significant difference between the numbers of southward and northward range shifts). The results showed a clear shift northwards, whether we examined data for single boundaries ($n = 92$ boundaries over 57 species, 44 moved north, three moved south, Binomial test, $P \ll 0.001$) or entire ranges ($n = 35$ species, 22 shifted north, two shifted south, Binomial test, $P \ll 0.001$).

## Comparisons with other studies

Parmesan (1996) and Parmesan et al. (1999) provide the only large-scale evidence of poleward shifts in entire species' ranges for any group of taxa. The results of these two studies are in accord with a suite of smaller-scale (regional) studies on butterflies, as well as on other insects, vertebrate and plant taxa. Most studies do not attempt to purge their datasets of confounding influences, yet provide similar indications of poleward shifting. In butterflies, for instance, Burton (1998) gives rather spartan summaries of European changes and found that 73% of 260 species had exhibited a 'north, northwest or westerly expansion of distribution'. It is unclear to which species or to which geographic areas within Europe this refers, but the summary figure is similar to that of Parmesan et al. (1999).

In Parmesan et al. (1999), when the British butterfly dataset was analysed prior to any 'filter' (i.e. including even those most severely affected by habitat loss), we found that 47% of all 38 non-migratory butterflies had shifted north at the northern boundary and only 8% had shifted south. The ratio of northward to southward movement of the northern boundary is significantly different from the 50:50 ratio expected by random ($P < 0.001$), again leading to the same conclusion of a general response to regional warming.

Close examination of the British dataset reveals that northward movement has, indeed, sometimes occurred in spite of general habitat-related decline for a species in recent years. The rather unexpectedly strong responses to the current warming trends appear to be swamping the counteractive forces of habitat loss, and may account for the similarities among studies of very different levels of detail, design and scale.

## Distinguishing Range Movements from Local Distributional Changes

A crucial part of the North American and European butterfly studies (Parmesan, 1996; Parmesan et al., 1999) was that both the northern and the southern range limits were analysed for changes. By this means, these studies could distinguish general range expansions or contractions from actual range shifts, and local or regional distributional changes from species-wide range movement.

Does this make a difference? At least in some cases, it does. For example, *Araschnia levana* (L.) (the map butterfly) has shown large population increases and rapid northerly colonization in the northern parts of its range, in northern France, the Netherlands and Fenno-scandia (Henriksen and Kreutzer, 1982; Radigue, 1994; van Swaay, 1995; Mikkola, 1997; N. Ryrholm, L. Kaila and J. Kullberg, personal communication). Since this dramatic expansion has accompanied a period of warm springs, superficially this might appear to be a climate-mediated northward shift, as predicted if it were responding to global warming. However, a closer look shows that it has expanded equally rapidly in the south. The first sighting of *A. levana* in Spain was in 1962 in the Pyrenees. Since then, *A. levana* has steadily expanded into Catalonia and is now common nearly to Barcelona (Viader, 1993; C. Stefanescu, personal communication). Thus, for unknown reasons, *A. levana* is expanding its range in all directions (Fig. 18.2). The southward extension is unlikely to have been climate mediated, and its existence cautions us against interpreting expansions at northern range boundaries as a northward shift of an entire species' range.

A second example comes from *Carterocephalus palaemon* (Pallas) (the chequered skipper). Over the past 30 years, *C. palaemon* has become extinct in the southern parts of both the UK and Finland. The most northern populations, near the northern coasts, have remained stable over the same time (Fig. 18.3). Examination of the southern boundary of the species, in Spain, shows that it has remained stable in recent decades (since the 1950s; Parmesan *et al.*, 1999). Species-wide interpretation of the regional patterns in northern countries would have led to a false conclusion of a poleward range shift when, in fact, the range of *C. palaemon* has remained stable over this century.

Such an interpretation was never suggested, for the biology of *C. palaemon* is well understood. Regional losses of habitat in the more disturbed, southern parts of Britain and Finland are the clear culprits for loss of this very sensitive butterfly (Thomas and Lewington, 1991; J. Kullberg and L. Kaila, personal communication). For many other parts of the world, however, such detailed biological knowledge is lacking and the raw changes in distribution may be the only available information.

## Focus on Mechanistic Links Between Climate Change and Movement – a Single-species Approach

Multi-species studies are powerful because they provide replication of a given phenomenon. However, they are often difficult to interpret in any detailed fashion and mechanistic explanations are typically lacking.

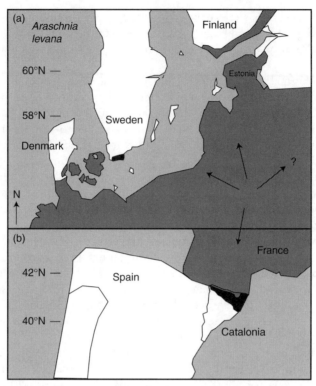

**Fig. 18.2.** Map of changes in distribution of *A. levana* over this century (modified from Parmesan, 2001). Black = areas of extension since 1970, medium grey = distribution from 1900 to 1969. (a) Northern range boundary: first seen in Sweden in 1970s, disappeared in cold summer of 1987, returned in 1992 with steady expansion ever since; first seen in Finland in 1973, established breeding by 1983; (b) southern range boundary: first seen in Pyrenees in 1962, steady expansion into Spain ever since.

Knowledge of precise causal factors underlying large-scale patterns of distributional change is usually unavailable, either because the basic biology is not well enough understood, or because observations were not made in sufficient detail during relevant periods of time. Though the field is still in its infancy, mechanistic links between climatic change and patterns of response in wildlife are beginning to emerge. There is increasing evidence that short-term extremes of weather and climate events may be driving many of the long-term, large-scale patterns of change being observed in natural systems (Easterling *et al.*, 2000b; Parmesan *et al.*, 2000).

One group which has been particularly well studied with respect to climate influences is the Lepidoptera. One species of lepidopteran, *E. editha*, has a 30-year history of basic research by more than a dozen researchers from which we can draw a population-based, mechanistic understanding of

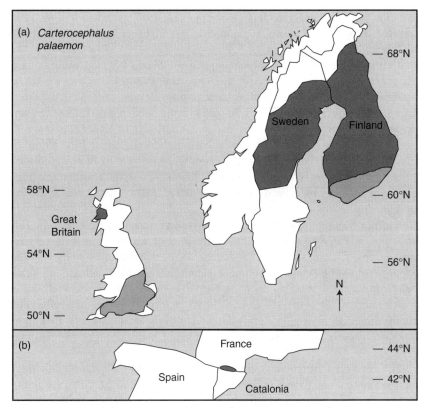

**Fig. 18.3.** Map of changes in distribution of *C. palaemon* over this century (modified from Parmesan *et al.*, 1999). Medium grey = areas of stability, i.e. no change in known population locations or densities. Light grey = areas of retraction, i.e. all known populations have gone extinct.

continental-wide distributional changes. In *E. editha*, single extreme weather events and single extreme climatic years appear to drive local population dynamics (including population extinctions and re-colonizations) which in turn drive changes to range limits.

The broad patterns of change in both climate and range of *E. editha* match, giving good support to climatic warming as a candidate for the cause of the range shift, but this tells us little about mechanism. A review of studies at the population level gives us that insight. Many extinctions of *E. editha* have been associated with particular climatic events (Singer and Ehrlich, 1979; Ehrlich *et al.*, 1980; Singer and Thomas, 1996). The 1975–1977 severe drought over California caused the extinction of five out of 21 surveyed populations (Singer and Ehrlich, 1979; Ehrlich *et al.*, 1980). Extremely wet years caused opposite responses in two subspecies. Following winters with 50–150% more precipitation than the average, *E. editha bayensis* crashed in the vicinity of San Francisco Bay (Dobkin et al., 1987), while *E. editha quino* exhibited population booms in northern Baja California, Mexico (Murphy and White, 1984).

Twenty years of studies at one site in the Sierra Nevada mountains of California have implicated three extreme weather events in carving a pathway to extinction of a whole set of *E. editha* populations at 2400 m elevation (Singer and Thomas, 1996). The first catastrophe occurred in 1989 when very low winter snowpack led to an early and unusually synchronous adult emergence in April, almost 2 months earlier than the usual June flight. So early, in fact, were the adults that flowers were not yet in bloom and most adults died, apparently from starvation. Just 1 year later another relatively light snowpack again caused adults to emerge early. Just when most had emerged, and many were sitting on the ground as mating pairs, a snowstorm that was normal for the season buried the insects under >15 cm of snow. The butterflies, adapted to summertime conditions of warmth and sun, suffered many deaths. Each of these events decreased the population size by an order of magnitude. The finale came but 2 years later in 1992 'when (unusually low) temperatures of $-5°C$ on June 16, without insulating snowfall, killed an estimated 97% of the *Collinsia* (host) plants . . . The butterflies had already finished flying and left behind young (caterpillars) that were not killed directly but starved in the absence of hosts' (Singer and Thomas, 1996).

The observed northward and upward range shift of *E. editha* during this century has occurred as a result of increased numbers of population extinctions at the southern range boundary and at lower elevations, with a symmetrical tendency towards population survival along the northern range boundary and at the highest elevations (Parmesan, 1996). Direct observations of population extinctions implicate influences of extreme weather events on the insect–host plant relationship. Thus, infrequent and severe climatic events appear to be driving a gradual range shift in this butterfly species, via short-term responses at the population level.

## Roles of Extreme Weather and Climate Events

There is increasing evidence that extreme weather and climate events have been increasing in magnitude and frequency (reviewed by Easterling et al., 2000a,b), and that such extreme events drive a slew of biotic processes (reviewed by Easterling et al., 2000b; Parmesan et al., 2000). Among these examples are several instances of long-term changes in species distributions or community structure which have been related to a few isolated extreme climatic events (Allen and Breshears, 1998; Sagarin et al., 1999). Thus, both climatologists and biologists are beginning to steer away from simple analyses of mean annual or seasonal temperature and precipitation trends to more complex analyses of monthly mean changes, suites of climate variables, and variability of climate itself and their subsequent impacts on natural biological systems.

## Interactions Among Ecological Parameters

Now that several of the early predictions have come to pass, it is increasingly clear that natural systems do respond to the small levels of climatic change they have recently experienced. With this foundation in hand, we can begin to look for more complex responses, including those involving changes in species' interactions. The basis for expecting changes in species' interactions and community structure are the dissimilarities among species in their physiological tolerances, life-history strategies, probabilities of population extinctions/colonizations and dispersal abilities. These individualistic traits shape a species' response to environmental change, including climate change. The speed and degree with which species respond to global warming through individual movements or range shifts is, thus, likely to be shaped by other ecological factors impacting individuals and populations.

### Ecological response vs. evolution

A factor often downplayed in discussions of responses to global warming is the propensity for evolution of a population *in situ* to the selective forces brought about by climate change. Few examples exist, but one study shows that even apparent 'movement' within a species may, in fact, be a result of local evolution. Fossil evidence from a beetle (*Helophorus aquaticus* (L.)) indicated that the cline between two morphological races has shifted location with the changing climate, tracking (presumably) the movement of the two races over the past 100,000 years (Angus, 1973). De Jong and Brakefield (1998) documented a clinal change over the past 20 years in a coccinellid, the two-spot ladybird (*Adalia bipunctata* (L.)). A previous study of this ladybird had shown that the extent of melanism in the elytra was genetically controlled and under strong temperature-driven selection for non-melanic forms (red with black spots) in southern locations and melanic forms (black with red spots) in northern locations (Brakefield, 1985; de Jong *et al.*, 1996). Along one such cline in the Netherlands, the gradient between the two morphs has become steeper, changing the extent and shape of the cline. This change has been brought about through changes in gene frequencies among populations. This population-level evolution can be understood entirely by the reduction of known selective pressures for melanism associated with a local warming trend during recent decades. The *in situ* evolutionary interpretation is supported by physiological experiments and energetic models (Brakefield, 1984a,b; Brakefield, 1985; de Jong *et al.*, 1996; de Jong, 1997): no movement of individuals is required. This example highlights one of the problems of assigning causation in climate change biology. In some cases, it may require fairly detailed biological knowledge to distinguish an evolutionary response from active movement of individuals.

## Phenology and range movement

Phenological shifts in response to climate change have been primarily studied with respect to single species. However, more crucial than any absolute change in timing of a single species is the potential disruption of coordination in timing between the life cycles of herbivorous insects and their host plants, parasitoids and their host insects, and insect pollinators with flowering plants. In the UK, *Anthocharis cardamines* (L.) seemed able to track the phenology of its host plant even when bud-formation came 2–3 weeks early (Sparks and Yates, 1997). Conversely, in *E. editha*, population crashes have been caused by changes in insect emergence time relative to both the senescence times of annual hosts (Singer, 1972; Weiss *et al.*, 1988) and the time of blooming of nectar sources (Singer and Thomas, 1996; Thomas *et al.*, 1996). As well as affecting population dynamics, a mismatch in phenology between adult flight time and host senescence was shown to hinder colonization of empty, otherwise suitable, habitat well within the dispersal range of nearby *E. editha* populations (Boughton, 1999). Ultimately, then, these phenological mismatches between the butterfly and its hosts and nectar plants were a likely part of the cause of the observed range shift in this species.

## Habitat change, microclimate and range movement

In northern Europe, some species that specialize in hot microclimates may actually be experiencing cooling trends due to changes in human land management (Thomas, 1993; Warren, 1995). For example, the Lycaenid butterfly, *L. bellargus*, requires extremely short turf where the soil and host plants are especially warm. In northern Europe, this requirement has long restricted the insects to chalk downs. However, vegetation height has recently increased in these habitats, resulting in local cooling and thereby necessitating active management of the few remaining butterfly populations. Northward range expansion of this species in the UK is impeded by the patchiness of the very restricted habitats it requires – short, south-facing turf (C.D. Thomas and J.A. Thomas, personal communication). Thus, for *L. bellargus*, the regional warming trend in Britain would have to be much greater than has yet been observed, perhaps as much as a further rise of 3°C, before it would compensate for the microclimatic cooling caused by land use changes. Similarly, the Nymphalid butterfly, *Argynnis paphia* (L.), requires open woodland so that sun may penetrate to the forest floor where its larvae feed on violets. Closing of the forest canopy due to changes in woodland management has shaded the host plants and cooled the microclimate in the UK (Thomas, 1993). *A. paphia* is only now showing signs of northward expansion, a response lag to the warming trend of more than 20 years (C.D. Thomas, personal communication).

### Dispersal ability and range movement

The above examples demonstrate in part how microclimatic restrictions can interplay with dispersal abilities to limit range movement. A highly dispersive species would be better able to exploit small, isolated patches of habitat, even if its requirements were fairly narrow. Further, we already have evidence that the rate of range movement is very idiosyncratic both within and among broad taxonomic groups of different mobility. In comparing butterflies to herbaceous plants, for instance, we find that, in butterflies, the mean change in geographic location of the range is of the same order as the regional mean change in temperature isotherms – about 100 km for both (Parmesan, 1996; Parmesan *et al.*, 1999). A study of alpine plants in Switzerland, however, found that the rate of colonization towards higher elevations was about half that expected from the local rise in temperature (Grabherr *et al.*, 1994). Among European butterflies, there was a (non-significant) trend towards relative stability of range in a family which in general has lower dispersal abilities (the Lycaenidae – Blues and Coppers), compared with one with typically stronger fliers (the Nymphalidae; Parmesan *et al.*, 1999). Out of 10 species of Lycaenidae, half showed stable distributions and half had moved northward in range, while amongst 15 species of Nymphalids, 80% had shifted their ranges northward ($n = 25$, log-likelihood ratio test, $G = 3.6$, $P = 0.17$).

Disparities of response, due either to differential lag times or insensitivity to climate change, have occurred with past major climate changes and led to disruption of species' associations and formation of novel ones (Davis and Zabinski, 1992; Ashworth, 1997; Morgan, 1997). It is likely that variation in dispersal abilities of modern species will also result in changes in species interactions as some species move out of a given habitat, some remain and others move in.

## Power in Correlational Studies

The above discussions highlight the complications inherent in any single study of biotic response to climate change. Gathering data over small portions of the distribution and then extrapolating to whole range effects can lead to errors of interpretation. Patterns of habitat loss, pollution or other forms of anthropogenic manipulation can directly affect the species being studied and confound signals of climate change. Further, the biology of species' interactions under climate change scenarios is poorly understood. Complex interspecies and community dynamics may not only be difficult to predict, but it may difficult to correctly interpret current observed patterns of biotic change with respect to a causal link to climate change.

In spite of all this real and potential complexity, the patterns that are, in fact, being documented in natural systems are surprisingly simple. Changes in species' distributions are providing some of the most powerful evidence that 20th century global warming has already impacted wildlife. This power comes not from perfect data from a single system, but from the cumulative synthesis of many studies. Distributional change in any individual species, taxon or geographic region may have a number of possible explanations. However, when we look for systematic trends over a multitude of species and studies, small aberrations in data (i.e. some species changing their distributions for other reasons) add to the overall variance (noise) but do not affect the averaged trend (the signal of biotic response).

The power of replication helps to interpret studies which do not have large geographic coverage. Thus, even though there are problems with interpreting individual studies which contain only data from a small portion of the species' distribution or range boundary, when many such studies are viewed together, they can be analysed for systematic patterns of change. Extending this idea, we can consider replication across taxa to look for systematic changes in natural systems in general. By this means, combining a suite of small- and large-scale correlational studies can provide a robust means of distinguishing climatic impacts from other known factors when analysing observed changes in natural systems.

## Implications for Future Work

In summary, the earliest (and simplest) predictions of responses to climate warming have been qualitatively borne out. Poleward and upward distributional movements over the 20th century have been documented across many taxa and multiple geographic regions. Such a systematic trend across all studies points to a single common causal factor, global warming being the most likely candidate. However, the examples discussed here show that it will not be easy to provide accurate predictions of biotic responses to the expected further warming (2–4°C by 2050). Other anthropogenic impacts, as well as complex interactions of those forces with a particular species' ecology, will ultimately shape the re-distribution of species and communities globally (Schneider and Root, 1996).

Future predictions should incorporate effects of habitat loss, land management changes and dispersal abilities. While a few single-species studies have attempted to address some of these issues, such complexities are ignored by the more comprehensive, biome-based, climate change scenarios. More research is needed to understand the mechanistic basis of these long-term responses to gradual mean warming trends. Given that evolution (climatic adaptation) can occur, how likely is this to play an important role in natural systems? What specific aspects of climate change are the most important drivers of biotic responses? How much do species/taxa differ in their responses?

Climate change is occurring simultaneously with rising atmospheric $CO_2$ levels. Given that the very nature of the overall warming trend is being shaped by the build-up of greenhouse gases (Easterling et al., 1997; Beniston et al., 1998), how much do biotic responses to the climatic change alter under variable $CO_2$ concentrations? There is a growing body of literature showing strong interactions between $CO_2$ levels and the effects of temperature and precipitation on plant morphology, growth, physiology, carbon:nitrogen ratio and secondary chemistry (reviewed by Cotrufo et al., 1998; Peñuelas and Estiarte, 1998). The resulting responses in plants have, in turn, often been shown to impact their insect herbivores (reviewed by Lincoln et al., 1993; Lindroth, 1996a,b; Bezemer and Jones, 1998). Thus, while it is not expected that insect dispersal itself would be directly impacted by increased $CO_2$, insect range movement may be secondarily affected by the combined influences of rising global $CO_2$ and temperature on host-plant distributions. Even more subtly, local host-plant populations may change in their acceptability or suitability for the local insect herbivores resulting in population-level insect booms or extinctions which, as with *E. editha*, could also ultimately affect range movement.

The qualitative differences between early and late-century global warming periods make it important to analyse changes in natural biota over both time periods. Response to recent, presumably anthropogenic, climate change is more likely to reflect future responses, but comparison with a 'natural' climate shift helps us to understand mechanistic drivers of such responses.

Finally, a major oversimplification inherent in the typical vegetation model for 2050 or 2090 climate change scenarios is the lack of reference to either barriers to movement (such as large urban areas or cropland), or to differential dispersal among species. Scenario models for animal movement, including insect movement, are sparse in the literature and hardly surface in past IPCC reports. Yet, predictions from these reports are one of the main guidelines used by governments around the world to shape their future policies with respect to global warming. The barrage of forces affecting natural systems, and the potential for novel complex responses to them, make it important to attempt more complex scenarios. Overall patterns indicate that a main impact of climate change is on wildlife movement. It is becoming clear that realistic models incorporating large-scale dispersal are required to underpin both the science and politics of biotic responses to global warming.

# References

Allen, C.D. and Breshears, D.D. (1998) Drought-induced shift of a forest-woodland ecotone: rapid landscape response to climate variation. *Proceedings of the National Academy of Sciences USA* 95, 14839–14842.

Angus, R.B. (1973) Pleistocene *Helophorus* from Borislav and Starunia in the Western Ukraine with a reinterpretation of M. Lomnicki's species. *Philosophical Transactions of the Royal Society of London* B 265, 299–326.

Ashworth, A.C. (1997) The response of beetles to Quaternary climate changes. In: Huntley, B., Cramer, W., Morgan, A.V., Prentice, H.C. and Allen, J.R.M. (eds) *Past and Future Rapid Environmental Changes: the Spatial and Evolutionary Responses of Terrestrial Biota*. Springer-Verlag, Berlin, pp. 119–127.

Beniston, M. *et al.* (1998) Europe. In: Watson, R.T., Zinyowera, M.C. and Moss, R.H. (eds) *The Regional Impacts of Climate Change: an Assessment of Vulnerability. A Special Report of Intergovernmental Panel on Climate Change Working Group II*. Cambridge University Press, Cambridge, pp. 149–185.

Bezemer, T.M. and Jones, T.H. (1998) Plant–insect herbivore interactions in elevated atmospheric $CO_2$: quantitative analyses and guild effects. *Oikos* 82, 212–222.

Boughton, D.A. (1999) Empirical evidence for source–sink dynamics in a butterfly: temporal barriers and alternative states. *Ecology* 80, 2727–2739.

Brakefield, P.M. (1984a) Ecological studies on the polymorphic ladybird *Adalia bipunctata* in the Netherlands. I. Population biology and geographical variation in melanism. *Journal of Animal Ecology* 53, 761–774.

Brakefield, P.M. (1984b) Ecological studies on the polymorphic ladybird *Adalia bipunctata* in the Netherlands. II. Population dynamics, differential timing of reproduction and thermal melanism. *Journal of Animal Ecology* 53, 775–790.

Brakefield, P.M. (1985) Differential winter mortality and seasonal selection in the polymorphic ladybird *Adalia bipunctata* (L.) in the Netherlands. *Biological Journal of the Linnaean Society* 24, 189–206.

Burton, J.F. (1975) The effects of recent climatic changes on British insects. *Bird Study* 22, 203–204.

Burton, J.F. (1998) The apparent effects of climatic changes since 1850 on European Lepidoptera. *Memoires de la Société Royale Belge d'Entomologie* 38, 125–144.

Calvert, W.H. and Brower, L.P. (1986) The location of monarch butterfly (*Danaus plexippus* L.) overwintering colonies in Mexico in relation to topography and climate. *Journal of the Lepidopterists' Society* 40, 164–187.

Changnon, S.A., Pielke, R.A. Jr, Changnon, D., Sylves, R.T. and Pulwarty, R. (2000) Human factors explain the increased losses from weather and climate extremes. *Bulletin of the American Meteorological Society* 81, 437–442.

Christianson, G.E. (1999) *Greenhouse: the 200-Year Story of Global Warming*. Greystone Books, Vancouver, Canada.

Coope, G.R. (1995) Insect faunas in ice age environments: why so little extinction? In: Lawton, J.H. and May, R.M. (eds) *Extinction Rates*. Oxford University Press, Oxford, pp. 55–74.

Cotrufo, M.F., Ineson, P. and Scott, A. (1998) Elevated $CO_2$ reduces the nitrogen concentration of plant tissues. *Global Change Biology* 4, 43–54.

Davis, A.J., Jenkinson, L.S., Lawton, J.H., Shorrocks, B. and Wood, S. (1995) Global warming, population dynamics and community structure in a model insect assemblage. In: Harrington, R. and Stork, N.E. (eds) *Insects in a Changing Environment*. Academic Press, London, pp. 431–439.

Davis, M.B. and Zabinski, C. (1992) Changes in geographical range resulting from greenhouse warming: effects on biodiversity in forests. In: Peters, R.L. and Lovejoy, T.E. (eds) *Global Warming and Biological Diversity*. Yale University Press, New Haven, Connecticut, pp. 297–308.

Descimon, H. (1995) La conservation des *Parnassius* en France: aspects zoogéographiques, écologiques, démomographiques et génétiques. *Rapports d'études de l'OPIE*, Vol. 1.

Dennis, R.L.H. (1993) *Butterflies and Climate Change*. Manchester University Press, Manchester.

Dobkin, D.S., Olivieri, I. and Ehrlich, P.R. (1987) Rainfall and the interaction of microclimate with larval resources in the population dynamics of checkerspot butterflies (*Euphydryas editha*) inhabiting serpentine grassland. *Oecologia* 71, 161–166.

Drake, V.A. and Farrow, R.A. (1985) A radar and aerial-trapping study of an early spring migration of moths (Lepidoptera) in inland New South Wales. *Australian Journal of Ecology* 10, 223–236.

Easterling, D.R., Horton, B., Jones, P.D., Peterson, T.D., Karl, T.R., Parker, D.E., Salinger, M.J., Razuvayev, V., Plummer, N., Jamason, P. and Folland, C.K. (1997) Maximum and minimum temperature trends for the globe. *Science* 277, 364–367.

Easterling, D.R., Evans, J.L., Groisman, P.Y., Karl, T.R., Kunkel, K.E. and Ambenje, P. (2000a) Observed variability and trends in extreme climate events: a brief review. *Bulletin of the American Meteorological Society* 81, 417–425.

Easterling, D.R., Meehl, G.A., Parmesan, C., Chagnon, S., Karl, T. and Mearns, L. (2000b) Climate extremes: observations, modeling, and impacts. *Science* 289, 2068–2074.

Ehrlich, P.R., Murphy, D.D., Singer, M.C., Sherwood, C.B., White, R.R. and Brown, I.L. (1980) Extinction, reduction, stability and increase: the responses of checkerspot butterfly (*Euphydryas editha*) populations to the California drought. *Oecologia* 46, 101–105.

Fleming, R.A. and Tatchell, G.M. (1995) Shifts in the flight periods of British aphids: a response to climate warming? In: Harrington, R. and Stork, N.E. (eds) *Insects in a Changing Environment*. Academic Press, London, pp. 505–508.

Ford, E.B. (1945) *Butterflies*. Collins, London.

Grabherr, G.M., Gottfried, M. and Pauli, H. (1994) Climate effects on mountain plants. *Nature* 369, 448.

Groisman, P.Y., Karl, T.R., Easterling, D.R., Knight, R.W., Jamason, P.F., Hennessy, K.J., Suppiah, R., Page, C.M., Wibig, J., Fortuniak, K., Razuvaev, V.N., Douglas, A., Førland, E. and Zhai, P. (1999) Changes in the probability of heavy precipitation: important indicators of climatic change. *Climatic Change* 42, 243–283.

Hanski, I. (1999) *Metapopulation Ecology*. Oxford University Press, New York.

Harrington, R. and Stork, N.E. (eds) (1995) *Insects in a Changing Environment, Symposia of the Royal Entomological Society of London no. 17*. Academic Press, London.

Harrington, R., Bale, J.S. and Tatchell, G.M. (1995) Aphids in a changing climate. In: Harrington, R. and Stork, N.E. (eds) *Insects in a Changing Environment*. Academic Press, London, pp. 125–155.

Heath, J. (1974) A century of changes in the Lepidoptera. In: Hawksworth, D.L. (ed.) *The Changing Flora and Fauna of Britain*. Academic Press, London, pp. 275–292.

Heath J., Pollard, E. and Thomas, J.A. (1984) *Atlas of Butterflies in Britain and Ireland*. Viking, Penguin Books, Harmondsworth.

Henriksen, H.J. and Kreutzer, I.B. (1982) *The Butterflies of Scandinavia in Nature*. Skandinavisk Bogforlag, Denmark.

Hersteinsson, P. and Macdonald, D.W. (1992) Interspecific competition and the geographical distribution of red and arctic foxes *Vulpes vulpes* and *Alopex lagopus*. *Oikos* 64, 505–515.

Hill, J.K., Thomas, C.D and Huntley, B. (2001) Modelling present and potential future ranges of European butterflies using climate response surfaces. In: Boggs, C.L., Watt, W.B. and Ehrlich, P.R. (eds) *Evolution and Ecology Taking Flight: Butterflies as Model Systems*. University of Chicago Press, Chicago, Illinois, in press.

Hughes, L. (2000) Biological consequences of global warming: is the signal already apparent? *Trends in Ecology and Evolution* 15, 56–61.

Houghton, J.T., Meira, L.G., Callender, B.A., Harris, N., Kattenberg, A. and Maskell, K. (eds) (1996) *Intergovernmental Panel on Climate Change Second Assessment Report: Climate Change 1995: the Science of Climate Change*. Cambridge University Press, Cambridge.

Johnson, T. (1998) Snowpack accumulation trends in California. M.S. thesis, Bren School of Environmental Sciences, University of California at Santa Barbara, Santa Barbara, California.

de Jong, P.W. (1997) Evolutionary genetics of the two-spot ladybird. PhD thesis, Leiden University, The Netherlands.

de Jong, P.W. and Brakefield, P.M. (1998) Climate and change in clines for melanism in the two-spot ladybird, *Adalia bipunctata* (Coleoptera: Coccinellidae). *Proceedings of the Royal Society London* 265, 39–43.

de Jong, P.W., Gussekloo, S.W.S. and Brakefield P.M. (1996) Differences in thermal balance, body temperature and activity between non-melanic and melanic two-spot ladybird beetles (*Adalia bipunctata*) under controlled conditions. *Journal of Experimental Biology* 199, 2655–2666.

Kaisila, J. (1962) *Immigration und Expansion der Lepidopteren in Finnland in den Jahren 1869–1960*. Acta Entomologica Fennica, Helsinki.

Karl, T.R., Knight, R.W., Easterling, D.R. and Quayle, R.G. (1996) Indices of climate change for the United States. *Bulletin of the American Meteorological Society* 77, 279–292.

Lamb, H.H. (1975) Our understanding of the global wind circulation and climatic variations. *Bird Study* 22, 121–141.

Lawton, J.H. (1995) The response of insects to environmental change. In: Harrington, R. and Stork, N.E. (eds) *Insects in a Changing Environment*. Academic Press, London, pp. 3–26.

Lincoln, D.E., Fajer, E.D. and Johnson, R.H. (1993) Plant–insect herbivore interactions in elevated $CO_2$ environments. *Trends in Ecology and Evolution* 8, 64–68.

Lindfors, C.E., Mikkola, K. and Ahti, K. (1989) First record of *Protexarnis squalida* new-record Lepidoptera Noctuidae from northern Europe with analysis of a long-range migration. *Notulae Entomologicae* 69, 5–12.

Lindroth, R.L. (1996a) $CO_2$-mediated changes in tree chemistry and tree-lepidoptera interactions. In: Koch, G.W. and Mooney, H.A. (eds) *Carbon Dioxide and Terrestrial Ecosystems*. Academic Press, San Diego, pp. 105–120.

Lindroth, R.L. (1996b) Consequences of elevated atmospheric $CO_2$ for forest insects. In: Körner, C. and Bazzaz, F.A. (eds) *Carbon Dioxide, Populations, and Communities*. Academic Press, San Diego, pp. 347–361.

Malcolm, S.B., Cockrell, B.J. and Brower, L.P. (1987) Monarch butterfly voltinism effects of temperature constraints at different latitudes. *Oikos* 49, 77–82.

Marttila, O., Haahtela, T., Aarnio, H. and Ojalainen, P. (1990) *Suomen Päiväperhoset*. Kirjayhtymä, Helsinki.

Meehl, G.A., Karl, T., Easterling, D.R., Changnon, S., Pielke, R. Jr, Changnon, D., Evans, J., Groisman, P.Y., Knutson, T.R., Kunkel, K.E., Mearns, L.O., Parmesan, C., Pulwarty, R., Root, T., Sylves, R.T., Whetton, P. and Zwiers, F. (2000a) An introduction to trends in extreme weather and climate events: observations, socioeconomic impacts, terrestrial ecological impacts, and model projections. *Bulletin of the American Meteorological Society* 81, 413–416.

Meehl, G.A., Zwiers, F., Evans, J., Knutson, T., Mearns, L. and Whetton, P. (2000b) Trends in extreme weather and climate events: issues related to modeling extremes in projection of future climate change. *Bulletin of the American Meteorological Society* 81, 427–436.

Mikkola, K. (1997) Population trends of Finnish Lepidoptera during 1961–1996. *Entomologica Fennica* 3, 121–143.

Morgan, A.V. (1997) Fossil Coleoptera assemblages in the Great Lakes region of North America: past changes and future prospects. In: Huntley, B., Cramer, W., Morgan, A.V., Prentice, H.C. and Allen, J.R.M. (eds) *Past and Future Rapid Environmental Changes: the Spatial and Evolutionary Responses of Terrestrial Biota*. Springer-Verlag, Berlin, pp. 129–142.

Murphy, D.D. and White, R.R. (1984) Rainfall, resources, and dispersal in southern populations of *Euphydryas editha* (Lepidoptera: Nymphalidae). *Pan-Pacific Entomology* 60, 350–354.

Palevitz, B.A. (1999) Global warming, organisms feel the heat. *The Scientist* 13, 1, 8.

Parmesan, C. (1996) Climate and species' range. *Nature* 382, 765–766.

Parmesan, C. (2001) Butterflies as bio-indicators of climate change impacts. In: Boggs, C.L., Watt, W.B. and Ehrlich, P.R. (eds) *Evolution and Ecology Taking Flight: Butterflies as Model Systems*. University of Chicago Press, Chicago, Illinois, in press.

Parmesan, C., Ryrholm, N., Stefanescu, C., Hill, J.K., Thomas, C.D., Descimon, H., Huntley, B., Kaila, L., Kullberg, J., Tammaru, T., Tennent, W.J., Thomas, J.A. and Warren, M. (1999) Poleward shifts in geographical ranges of butterfly species associated with regional warming. *Nature* 399, 579–583.

Parmesan, C., Root, T.L. and Willig, M.R. (2000) Impacts of extreme weather and climate on terrestrial biota. *Bulletin of the American Meteorological Society* 81, 443–450.

Peñuelas, J. and Estiarte, M. (1998) Can elevated $CO_2$ affect secondary metabolism and ecosystem function? *Trends in Ecology and Evolution* 13, 20–24.

Pollard, E. (1979) Population ecology and change in range of the white admiral butterfly *Ladoga camilla* L. in England. *Ecological Entomology* 4, 61–74.

Pollard, E. and Eversham, B.C. (1995) Butterfly monitoring 2 – interpreting the changes. In: Pullin, A.S. (ed.) *Ecology and Conservation of Butterflies*. Chapman & Hall, London, pp. 23–36.

Ponel, P. (1997) The response of Coleoptera to late-Quaternary climate changes: evidence from northeast France. In: Huntley, B., Cramer, W., Morgan, A.V., Prentice, H.C. and Allen, J.R.M. (eds) *Past and Future Rapid Environmental Changes: the Spatial and Evolutionary Responses of Terrestrial Biota*. Springer-Verlag, Berlin, pp. 143–151.

Pounds, J.A., Fogged, M.P.L. and Campbell, J.H. (1999) Biological response to climate change on a tropical mountain. *Nature* 398, 611–615.

Porter, J. (1995) The effects of climate change on the agricultural environment for crop insect pests with particular reference to the European corn borer and grain maize. In: Harrington, R. and Stork, N.E. (eds) *Insects in a Changing Environment*. Academic Press, London, pp. 93–123.

Radigue, F. (1994) Une invasion pacifique: la Carte géographique (*Araschnia levana* L.) dans l'Orne (1976–1992). *Alexanor* 18, 359–367.

Rogers, D.J. (1995) Remote sensing and the changing distribution of tsetse flies. In: Harrington, R. and Stork, N.E. (eds) *Insects in a Changing Environment*. Academic Press, London, pp. 177–193.

Ryrholm, N. and Kallander, C.F.R. (1987) The invasion of *Autographa mandarina*, Lepidoptera Noctuidae, in eastern Sweden in 1985. *Entomologisk Tidskrift* 108, 130–134.

Sagarin, R.D., Barry, J.P., Gilman, S.E. and Baxter, C.H. (1999) Climate-related change in an intertidal community over short and long time scales. *Ecological Monographs* 69, 465–490.

Schneider, S.H. and Root, T.L. (1996) Ecological implications of climate change will include surprises. *Biodiversity and Conservation* 5, 1109–1119.

Showers, W.B., Whitford, F., Smelser, R.B., Keaster, A.J., Robinson, J.F., Lopez, J.D. and Taylor, S.E. (1989) Direct evidence for meteorologically driven long-range dispersal of an economically important moth. *Ecology* 70, 987–992.

Singer, M.C. (1972) Complex components of habitat suitability within a butterfly colony. *Science* 176, 75–77.

Singer, M.C. and Ehrlich, P.R. (1979) Population dynamics of the checkerspot butterfly *Euphydryas editha*. *Fortschritte der Zoologie* 25, 53–60.

Singer, M.C. and Thomas, C.D. (1996) Evolutionary responses of a butterfly metapopulation to human and climate-caused environmental variation. *American Naturalist* 148, S9–S39.

Sparks, T.H. and Yates, T.J. (1997) The effect of spring temperature on the appearance dates of British butterflies 1883–1993. *Ecography* 20, 368–374.

Sutherst, R.W., Maywald, G.F. and Skarrat, D.B. (1995) Predicting insect distributional in a changed climate. In: Harrington, R. and Stork, N.E. (eds) *Insects in a Changing Environment*. Academic Press, London, pp. 59–91.

van Swaay, C.A.M. (1995) Measuring changes in butterfly abundance in The Netherlands. In: Pullin, A.S. (ed.) *Ecology and Conservation of Butterflies*. Chapman & Hall, London, pp. 230–247.

Thomas, C.D. (1995) Ecology and conservation of butterfly metapopulations in the fragmented British landscape. In: Pullin, A.S. (ed.) *Ecology and Conservation of Butterflies*. Chapman & Hall, London, pp. 46–64.

Thomas, C.D. and Lennon, J.J. (1999) Birds extend their ranges northwards. *Nature* 399, 213.

Thomas, C.D., Singer, M.C. and Boughton, D.A. (1996) Catastrophic extinction of population sources in a butterfly metapopulation. *American Naturalist* 148, 957–975.

Thomas, J.A. (1993) Holocene climate changes and warm man-made refugia may explain why a sixth of British butterflies possess unnatural early-successional habitats. *Ecography* 16, 278–284.

Thomas, J.A. and Lewington, R. (1991) *The Butterflies of Britain and Ireland.* Dorling Kindersley, London.

Viader, J. (1993) Papallones de Catalunya: *Araschnia levana* (Linnaeus, 1758). *Bulletin of the Society of Catalonian Lepidoptera* 71, 49–62.

Warren, M.S. (1992) The conservation of British butterflies. In: Dennis, R.L.H. (ed.) *The Ecology of Butterflies in Britain*. Oxford University Press, Oxford, pp. 246–274.

Warren, M.S. (1995) Managing local microclimates for the high brown fritillary, *Argynnis adippe*. In: Pullin, A.S. (ed.) *Ecology and Conservation of Butterflies*. Chapman & Hall, London, pp. 198–210.

Weiss, S.B., Murphy, D.D. and White, R.R. (1988) Sun, slope, and butterflies: topographic determinants of habitat quality for *Euphydras editha*. *Ecology* 69, 1486–1496.

Williams, N.E. (1989) Factors affecting the interpretation of caddisfly assemblages from quaternary sediments. *Journal of Paleolimnology* 1, 241–248.

Woiwod, I.P. (1997) Detecting the effects of climate change on Lepidoptera. *Journal of Insect Conservation* 1, 149–158.

Wuethrich, B. (2000) How climate change alters rhythms of the wild. *Science* 287, 793–795.

# Analysing and Modelling Range Changes in UK Butterflies

**19**

Jane K. Hill,[1]* Chris D. Thomas,[2] Richard Fox,[3] Dorian Moss[4] and Brian Huntley[1]

[1]*Environmental Research Centre, Department of Biological Sciences, University of Durham, Durham DH1 3LE, UK;* [2]*Centre for Biodiversity and Conservation, School of Biology, University of Leeds, Leeds LS2 9JT, UK;* [3]*Butterfly Conservation, Conservation Office, PO Box 444, Wareham, Dorset BH20 5YA, UK;* [4]*Centre for Ecology and Hydrology – Monks Wood, Abbots Ripton, Huntingdon, Cambridgeshire PE17 2LS, UK*

## Introduction

Over the past 200+ years, UK butterflies have undergone marked changes in their distributions and abundances (Robson, 1902; Tutt, 1905; South, 1941; Ford, 1945; Thomson, 1980; Heath *et al.*, 1984; Emmet and Heath, 1990). Although the general pattern during the 20th century was one of decline, at least 12 species (approximately 20% of the UK fauna) are currently expanding (Pollard and Eversham, 1995). Many of these expanding species are currently common and widespread throughout the UK, although a few rare and localized species are also expanding (Pollard and Eversham, 1995). In many cases, these expansions have been well documented (Pollard 1979; Jackson, 1980; Pollard *et al.*, 1997), but understanding why some species are expanding and others are not has received little attention.

The distributions of many butterfly species are constrained by climate (Pollard, 1979; Turner *et al.*, 1987; Dennis and Shreeve, 1991; Dennis, 1993), and recent range expansions and range shifts are consistent with warming 20th-century climates (Parmesan, 1996; Hill *et al.*, 1999; Parmesan *et al.*, 1999). By the end of the 21st century, future climate scenarios predict further increases in global temperatures of about 2°C, compared with <1°C during

---

* Present address: Department of Biology, University of York, PO Box 373, York YO10 5YW, UK.

the 20th century (Houghton et al., 1996), and the general consensus is that in north temperate regions, species' distributions will continue to shift northwards, tracking these changing climates (Coope, 1978). What is not clear, however, is which species will be able to track changing climates and which will not. The distribution of suitable habitat is likely to affect expansion rates, but recent anthropogenic destruction of habitat may mean that new, climatically suitable habitats are beyond the reach of migrants. This would prevent species' distributions from shifting northwards, and there is already some evidence in the UK that species are lagging behind current climates, possibly because of habitat fragmentation (Hill et al., 1999). The extent and frequency of these distribution–climate lags has not been studied.

In this paper, we compare recent range changes in three satyrid butterfly species, the speckled wood butterfly (*Pararge aegeria* (L.)), ringlet (*Aphantopus hyperantus* (L.)) and gatekeeper (*Pyronia tithonus* (L.)). This choice of species allows comparison of species with broadly similar ecologies and dispersal rates but different current and historical distributions, and different habitat specificities (and thus distribution of suitable habitat). We investigate evidence for lags between recent climate warming and range shifts in these three species, and test the prediction that any lags will be less for species where habitats are more widely available. To do this, we use a 'climate response surface' model to investigate the role of climate in determining butterfly range limits at a continental scale, and then investigate the importance of habitat availability, in addition to climate, in determining range limits at a regional scale in the UK. We investigate the extent of distribution–climate lags by comparing current UK distributions with those predicted by the model. We also analyse 20th-century range expansions in the UK with respect to climate and habitat and test the hypothesis that rates of recent range expansion differ among species in relation to habitat specificity. We also predict potential future UK butterfly distributions under a climate change scenario for the period 2070–2099, taking account of current habitat availability, and we use rates of recent range expansion to predict times to colonize suitable habitat which becomes climatically suitable in the future.

## Study Species

During the 19th century, speckled wood (*P. aegeria*), ringlet (*A. hyperantus*) and gatekeeper (*P. tithonus*) were all widespread in the UK, occurring as far north as central (*P. tithonus*, *P. aegeria*) and northern (*A. hyperantus*) Scotland (Thomson, 1980; Fig. 19.1). Towards the end of the 19th century, distributions of these species contracted, as did many other UK species (Burrows, 1916; Gibbs, 1916; Downes, 1948; Chalmers-Hunt and Owen 1952; Jackson, 1980) and between 1915 and 1939, *P. aegeria* was essentially restricted to southwest England and Wales, but with a refuge population persisting in western Scotland (Emmet and Heath, 1990). *P. tithonus* disappeared from Scotland and much of northern England, and distributions of *A. hyperantus* also contracted, although some populations persisted throughout the UK. From the 1940s onwards, distributions have been expanding (Fig. 19.1), although not all previously occupied areas have been re-colonized, especially for *P. aegeria* (Fig. 19.1a) and *P. tithonus* (Fig. 19.1c). Current European distributions of the three species differ; *P. aegeria* and *A. hyperantus* reach their northern range margins in central Fennoscandia and Scotland (Fig. 19.2a and b), whereas *P. tithonus* reaches its northern range margin further south in England and continental Europe (Fig. 19.2c). *P. aegeria* and *P. tithonus* reach their southern range margin in North Africa, whereas *A. hyperantus* reaches its southern limit in northern Spain. The three study species are all satyrids, and larvae feed on a variety of species of grass. Adults are a similar size (female wingspan range 42–56 mm; Emmet and Heath, 1990), and probably have broadly similar dispersal rates, although there is little specific information for any of the species. *A. hyperantus* and *P. tithonus* are univoltine, whereas *P. aegeria* can develop through between 1.5 and 2 generations per year, depending on location and temperature. These three satyrid species are ideal for this type of study because:

- their distributions have fluctuated markedly over the past 150+ years, probably in response to changing climates;
- their larval host plants (a range of grass species) are widely distributed and do not limit current distributions;
- there are good current and adequate historical distribution records;
- all three species can be recorded reliably from ground-based surveys, and are not easily confused with other species;
- their habitat requirements are quite well known and can be determined from remotely sensed (satellite) land cover data.

## Modelling the Role of Climate in Determining Butterfly Range Limits

### Generating the model at a European scale

We generated climate response surfaces for each of the three study species using methods described in detail elsewhere (Beerling *et al.*, 1995; Huntley *et al.*, 1995; Hill *et al.*, 1999). We will only briefly describe them here. Current continental European distributions were obtained from Tolman (1997) and current UK records were obtained from Emmet and Heath (1990) and Asher *et al.* (2001). Records were converted to presence/absence on a 50-km UTM (universal transvers Mercator) grid extending from the Azores east to 30°E longitude (records were not considered to be reliable for areas further east), and from the Mediterranean Sea (reliable records could not be obtained for North Africa) north to Svalbard (total of 2648 grid squares;

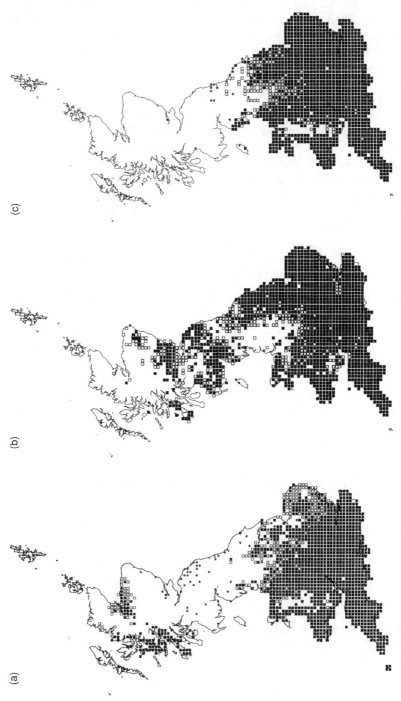

Fig. 19.1. Current (a) *P. aegeria*, (b) *A. hyperantus* and (c) *P. tithonus* distributions in the UK at a 10-km grid resolution. Black squares – species recorded 1940–1989, hollow squares – species first recent record 1990–1998, asterisks – historical pre-1915 records.

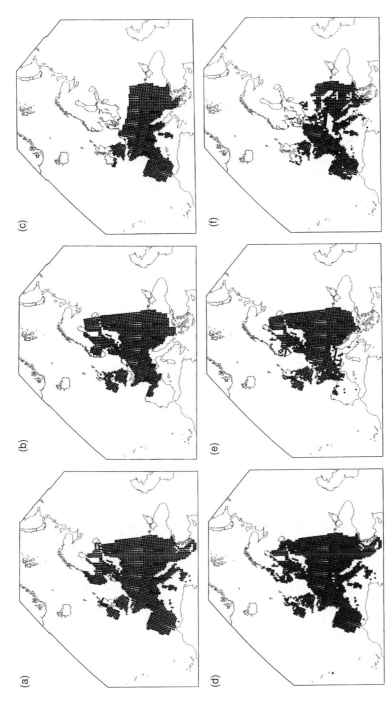

**Fig. 19.2.** Butterfly distributions for areas west of 30°E longitude (see text) on a 50-km UTM grid (a and d – *P. aegeria*, b and e – *A. hyperantus*, c and f – *P. tithonus*). (a–c) Current records, (d–f) simulated current distribution (*P. aegeria* probability of occurrence ≥0.50; *A. hyperantus* probability of occurrence ≥0.61; *P. tithonus* probability of occurrence ≥0.58).

Fig. 19.2a,b,c). We computed three bioclimatic variables that reflect principal limitations on butterfly growth and survival:

- annual temperature sum above 5°C (developmental threshold for larvae – Lees, 1962; Blakeley, 1997; GDD5),
- coldest month mean temperature (related to overwintering survival; MTCO),
- moisture availability (related to host plant quality and expressed as an estimate of the ratio of actual to potential evapotranspiration – Huntley et al., 1995; AET/PET).

We computed values for these three variables at the mean altitude of the grid square for the climate normal period of 1931–1960 (Leemans and Cramer, 1991), and fitted climate response surfaces describing European distributions of the three species in terms of these three variables. Previous studies have evaluated different combinations of several other variables in addition to those used here, but the three bioclimate variables listed above consistently produced the best models and so we only present results for models incorporating GDD5, MTCO and AET/PET (Hill et al., 1999).

## Simulating current butterfly distributions

We used the response surfaces to simulate distributions of the three species for the current climate and the closeness-of-fit between observed and simulated distributions was tested using the kappa statistic (Monserud and Leemans, 1992). For all three species, there was a good fit between current observed and simulated butterfly distributions (*P. aegeria*, kappa = 0.805 at a threshold probability of butterfly occurrence of 0.50, 2183 simulated occurrences versus 2153 observed occurrences, Fig. 19.2d; *A. hyperantus*, kappa = 0.788 at a threshold probability of butterfly occurrence of 0.61, 1601 simulated occurrences versus 1674 observed occurrences, Fig. 19.2e; *P. tithonus*, kappa = 0.752 at a threshold probability of butterfly occurrence of 0.58, 1310 simulated occurrences versus 1403 observed occurrences, Fig. 19.2f). Given, however, that all three species occupy a high proportion of grid cells (e.g. >80% of cells are currently occupied by *P. aegeria*), a more valuable test of the response surface is its ability to simulate accurately the species' range margins; visual inspection showed that this was also very good overall, although there were a number of mismatches between observed and simulated distributions:

1. *P. aegeria*. The model predicted occurrence in several areas beyond the species' current limits (Fig. 19.2d). In the UK, these areas included the Isle of Man and the Western Isles, localities which currently are probably too isolated to be occupied (Dennis and Shreeve, 1997). However, the model also predicted occurrence in areas along the east and west coasts of England and

Scotland where the species was historically present before its 19th-century range contraction, indicating that distributions are lagging behind current climates. The model predicted absence in some mountainous regions where *P. aegeria* is restricted to specific habitats (e.g. south-facing slopes and valley bottoms) below the mean elevation of the grid cell; such predicted absences are to be expected given that the model was fitted to bioclimate values for the mean elevation of the grid squares. An alternative response surface model fitted to climate variables calculated for the minimum elevation of grid squares successfully simulated occurrence in these mountainous regions; however, this model was less successful at simulating distributions at northern range margins and so we present data only for models fitted to mean elevation climate data (Hill *et al.*, 1999).

2. *A. hyperantus*. At the northern margin, observed and simulated distributions in Sweden and Finland were very similar (Fig. 19.2b and e), although the response surface did not simulate quite such extensive distributions as are currently observed in southern Scotland and Norway. At the southern margin, the response surface simulated a few occurrences in sites beyond its observed range in Spain and Italy, and did not simulate extensive distributions in northern Spain. As with *P. aegeria*, some areas where the response surface did not perform as well were mountainous regions where *A. hyperantus* occurs only in very specific habitats below the mean elevation of the grid cell.

3. *P. tithonus*. The response surface performed very well in the UK and continental Europe, but simulated a few occurrences in more northerly areas in Poland and southern Scandinavia (Fig. 19.2f). It also simulated a few occurrences in sites beyond its observed range in southern Italy and Greece. As with *P. aegeria* and *A. hyperantus*, occurrence was not predicted in some mountainous regions.

Although there were excellent fits between observed and simulated distributions, observed distributions were mainly taken from coarse-grained published maps and some of the mismatches may be due to lack of precision in published maps, rather than an inability of the bioclimatic variables to match existing distributions. To allow comparison among species, current distributions were determined from the same data sources, even though more accurate distribution data have been collated for *P. aegeria* for continental Europe (Hill *et al.*, 1999). In order to investigate the implications of using coarse-grained published maps on the performance of the response surfaces, we compared results for *P. aegeria* in this study (Fig. 19.2b) with those obtained using the more accurate data set (Fig. 2b in Hill *et al.*, 1999). Comparison of simulated distributions of *P. aegeria* from the two response surfaces generated using the different data sources showed that, although there were a few differences at range margins, differences were not that great. The response surface used in this study simulated a slightly more northerly distribution in England, Scotland and Norway than did the response surface generated using more accurate collated records. Thus both response surface models support the notion that *P. aegeria* distributions are lagging behind current climates. Given the similarity of results for *P. aegeria* generated from the different data sources, we are confident that results obtained for *A. hyperantus* and *P. tithonus* are unlikely to change greatly even if more accurate distribution data are used. For all species, the simulated distributions appear to be good representations of current distributions.

## Role of Climate and Habitat Availability in Determining Range Margins at a Regional Scale

We focused on UK distributions at a finer (10-km) resolution to investigate the importance of habitat availability, in combination with climate, in determining distributions (Hill *et al.*, 1999). We obtained 10-km resolution data for the current distributions of the three study species in the UK (Asher *et al.*, 2001). We derived values for the same three bioclimate variables, for locations at the mid-point and mean elevation of each 10-km cell (total of 2805 cells) using the same data sets and techniques as before (see above; Hill *et al.*, 1999). We then applied the climate response surfaces generated from the 50-km resolution data to these finer-scale climate data to simulate the probability of occurrence of each of the three butterfly species in the UK at a 10-km resolution, for the current climate. Distribution of potential habitat was measured for each species using data from the ITE Landcover Dataset, derived from satellite remote-sensed data (Fuller *et al.*, 1994). Availability of potential habitat for each species was calculated by summing the appropriate land cover types, chosen to best represent the known habitat of the species (Emmet and Heath 1990; Table 19.1). In previous analyses of *P. aegeria* distribution (Hill *et al.*, 1999), we used a slightly different method to determine availability of *P. aegeria* habitat; results and distribution maps presented here are thus slightly different from those in Hill *et al.* (1999). Figure 19.3 shows availability of habitat for the three species in the UK. Habitat availability throughout the UK differs among species in the order *P. tithonus* > *A. hyperantus* > *P. aegeria*.

We then used logistic regression to model the three species' observed UK distributions in relation to climate suitability and habitat cover. For all three species, both climate and habitat variables were significantly and positively related to butterfly presence (Table 19.2). The final models predicted >77% of butterfly presence/absence in grid cells correctly (probability of butterfly presence $\geq 0.5$; *P. aegeria*, model $X^2 = 1214.9$, 2 degrees of freedom (df), $P < 0.0001$; *A. hyperantus*, model $X^2 = 948.0$, 2 df, $P < 0.0001$; *P. tithonus*, model $X^2 = 2392.9$, 2 df, $P < 0.0001$; Fig. 19.4).

There was an excellent fit between current and predicted current *P. tithonus* distributions (model correctly assigned >90% of butterfly presence/absence in grid squares; Fig. 19.4c). There were a few areas where *P. tithonus*

**Table 19.1.** Remotely sensed land cover classification (derived from LANDSAT imagery; Fuller et al., 1994) used to determine habitat availability for the three satyrid butterflies.

| Land cover type | Species |
|---|---|
| Deciduous woodland | ag, hy, ti |
| Coniferous woodland | ag, hy, ti |
| Scrub/orchard | hy, ti |
| Rough grass/marsh grass | hy, ti |
| Grass heath | ti |
| Mown/grazed turf | ti |
| Meadow/verge/semi-natural | ti |
| Open shrub moor | ti |
| Dense shrub moor | ti |
| Bracken | ti |
| Dense shrub heath | ti |
| Ruderal weed | ti |
| Felled forest | ti |
| Open shrub heath | ti |
| Sea/estuary | |
| Inland water | |
| Beach and coastal bare | |
| Saltmarsh/intertidal vegetation | |
| Tilled ground | |
| Suburban/rural development | |
| Continuous urban | |
| Inland bare ground | |
| Moorland grass | |
| Upland bog | |
| Lowland bog | |
| Unclassified | |

Habitat availability (based on known habitat requirements, e.g. Emmet and Heath 1990) was calculated for each species by combining the appropriate land cover types. Species ag = *P. aegeria* (habitat combines two land cover types), hy = *A. hyperantus* (four types), ti = *P. tithonus* (14 types).

currently occurs but is not predicted to occur. As at a European scale, in most cases these are areas of high relief (e.g. the Welsh mountains, the Lake District in northwestern England) where the bioclimate variables used reflect the mean elevation whereas the butterfly occurs in specific habitats at low elevation. There were a few coastal areas in central and southern Scotland where *P. tithonus* is predicted to occur and which were occupied historically (Fig. 19.1c) but that it has not yet re-colonized. None the less, the overall pattern is that *P. tithonus* has more or less kept up with changing recent climates.

As with *P. tithonus*, the majority of the grid cells where *P. aegeria* and *A. hyperantus* currently occur but were not simulated are also grid cells of

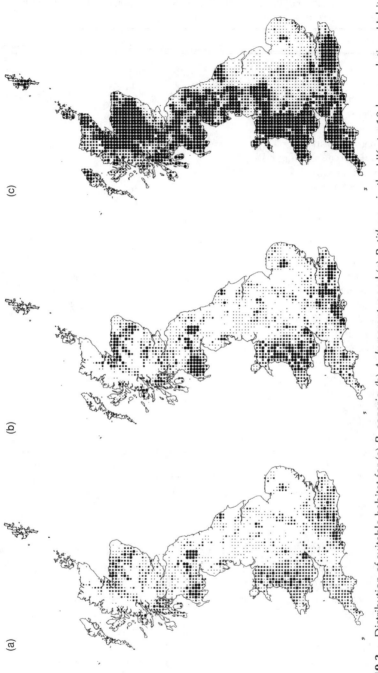

**Fig. 19.3.** Distribution of suitable habitat for (a) *P. aegeria*, (b) *A. hyperantus* and (c) *P. tithonus* in the UK at a 10-km resolution. Habitat availability is derived by summing appropriate land cover types (Table 19.1). (a and b) Small circles = 5–10% cover, medium = 11–20%, large >20%. (c) Small circles = 20–40% cover, medium = 41–60%, large >60%. Land cover data were not available for the Isle of Man.

**Table 19.2.** Results from logistic regression models predicting butterfly distributions in the UK (10-km resolution) in terms of climate (derived from response surfaces fitted to European 50-km distribution data) and habitat availability (land cover data – Table 19.1).

| Species | % Correct | Climate (SE) | Habitat (SE) | Constant (SE) |
|---|---|---|---|---|
| P. aegeria | 79.8 | 2.97 (0.11) | 8.64 (0.69) | −3.74 (0.15) |
| A. hyperantus | 76.7 | 2.38 (0.10) | 8.33 (0.64) | −2.18 (0.11) |
| P. tithonus | 90.0 | 5.56 (0.21) | 2.93 (0.25) | −4.64 (0.23) |

'% Correct' shows percentage of butterfly presence/absence in grid squares correctly assigned by the model; 'climate', 'habitat' and 'constant' are variables in the logistic equation.

high relief (e.g. the Welsh Mountains and western highlands of Scotland for *P. aegeria*; the Lake District and Cairngorm Mountains for *A. hyperantus*). However, both *P. aegeria* and *A. hyperantus* were predicted to occur in more areas currently beyond their range margins (e.g. northern England and southern Scotland for *P. aegeria*, and northwest England for *A. hyperantus*), areas which these species occupied in the 19th century (Fig. 19.1a and b). This pattern contrasts with that of *P. tithonus*, and indicates that both *P. aegeria* and *A. hyperantus* failed to keep up with the changing climates of the 20th century. As at a European scale, we compared results for *P. aegeria* in this study with those obtained from a response surface generated from more accurately collated data (Hill *et al.*, 1999). Again as at a European scale, there were few differences in the predicted distributions, regardless of which distribution datasets were used (Fig. 19.4a; cf. Fig. 1c in Hill *et al.*, 1999). Very similar predicted distributions obtained from the two different data sources may partly be due to accurate UK distribution records (from Asher *et al.*, 2001) being used in generating both response surfaces (see Methods). Because the same UK data source was also used for the other two species, we are confident that results for *A. hyperantus* and *P. tithonus* are likely to be robust and are unlikely to change greatly even if more accurate continental European distribution data are used to determine climate suitability.

## Role of Climate and Habitat Availability in Determining Recent UK Range Expansions

### Analysis of UK range changes

Since 1940, northern range margins have shifted northwards (difference in mean distance north of ten most northerly 10-km grid squares occupied in 1998 compared with 1939) by up to 180 km (*P. aegeria*, mean difference = 107 km northwards (England and Wales distributions only), *P. tithonus* =

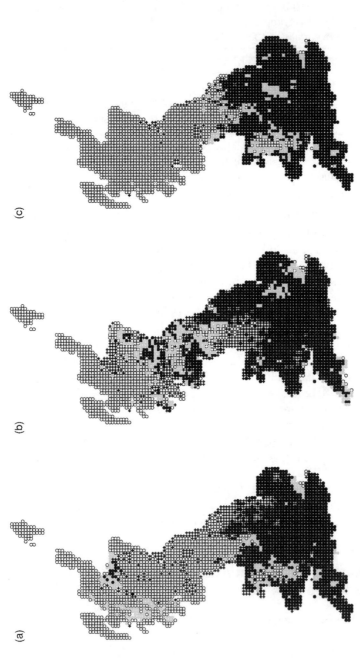

**Fig. 19.4.** Simulated current distribution of (a) *P. aegeria*, (b) *A. hyperantus* and (c) *P. tithonus* from logistic regression models incorporating climate suitability and habitat availability (threshold probability of occurrence ≥0.5); black circles – simulated occurrence coinciding with recorded presence, hollow circles – simulated absence coinciding with recorded absence, crosses – simulated occurrence coinciding with recorded absence, light grey circles – simulated absence coinciding with recorded presence. Land cover data were not available for the Isle of Man.

112 km northwards, *A. hyperantus* = 178 km northwards). This has corresponded with an approximate tenfold increase in number of 10-km grid squares occupied by each species during this period. Although increased recent sampling effort will have accounted for some of this increase (Asher et al., 2001), it should not invalidate the comparison of the species, nor should it alter the broad patterns of recorded distributions. Range expansions are unlikely to have been uniform since 1940, and temporary reversal and/or slowing up of range expansions occurred during relatively cool periods in the 1960s (Jackson, 1980) and 1980s (Hill et al., unpublished observations). The butterfly data sets used in this study were not at a sufficiently high temporal or spatial resolution to show these short-term retreats, but this is unlikely to invalidate comparisons of range expansions among species as all three species were probably affected similarly.

We used the area method (van den Bosch et al., 1990) to calculate the rate of areal expansion of each species in the UK since 1940. We plotted the area occupied each decade (square root of the area of 10-km squares with butterfly records) against year, and calculated the marginal velocity of range expansion ($E$) from the slope ($C$) of this line ($E = C/\sqrt{\pi}$; Lensink, 1997). There were no differences among species in either the slope or elevation of the relationship of area occupied over time (ANCOVA of √area occupied by species, with year as a covariate; species $F_{2,17} = 1.37$, $P = 0.3$; species × year interaction $F_{2,15} = 0.26$, $P = 0.8$; Fig. 19.5). This resulted in estimates of marginal velocity of range expansion ($E$) for *P. aegeria* of 2.82 km year$^{-1}$, *P. tithonus* = 2.76 km year$^{-1}$ and *A. hyperantus* = 2.98 km year$^{-1}$. However, differences in the historical distributions of each species make it difficult to correct for geometric distribution of potential new areas for colonization (Lensink, 1997). We therefore focused on six contiguous sub-areas of the UK. These areas correspond to six 100-km grid squares (Ordnance Survey grid squares SU, SP, SK, SE, SO, TL; Fig. 19.6). These six squares were chosen because they are the only grid squares which are within the current distributions of all three species, and which are almost entirely covered by land so that range expansions are unlikely to be constrained by coastlines. For each of the six 100-km grid squares, we plotted the area occupied by each species each decade (√area of 10-km squares with butterfly records from 1920 to the present) by year. In the subsequent analyses we only use data for a particular square and species which cover the linear phase of range expansion. For example, the most southerly (SU) grid square was fully occupied at a 10-km resolution by all three species by 1979, and so data from 1920 to 1979 only are included in analyses. However, in the most northerly (SE) square, butterflies did not start expanding until the 1940s and so only data from 1940 to the present are included. For any species or grid square, subsequent analyses are based on data covering a minimum of six decades. There were significant differences among species and among grids in expansion rates (ANCOVA of √area of 10-km squares with butterfly records each decade, with species and grid as factors, and year as a covariate;

species × year interaction, $F_{2,91} = 7.05$, $P = 0.001$; grid × year interaction, $F_{5,91} = 4.81$, $P = 0.001$).

## Effects of latitude on expansion rate

In order to investigate potential climatic impacts on expansion rates, we assigned each grid to one of four 100-km latitudinal categories, representing the distance north of the grid (1 = SU; 2 = SO, SP, TL; 3 = SK; 4 = SE). For each species in each grid we calculated the rate of areal expansion from the slope of the regression of area occupied against year (C; see above). There were significant differences among grids in rate of areal expansion in relation to latitude (two-way ANOVA of rate of areal expansion, with species and latitude as factors; latitude, $F_{3,11} = 16.01$, $P = 0.003$). The most southerly 100-km square (SU) had the fastest overall rate of expansion, and the most northerly square (SE) had the slowest rate (latitude 1, mean rate of expansion = 1.66, SD = 0.15; latitude 2, mean = 1.36, SD = 0.13; latitude 3, mean = 1.21, SD = 0.34; latitude 4, mean = 1.12, SD = 0.25). As before,

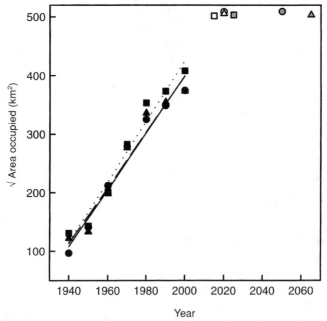

**Fig. 19.5.** Expansion of *P. aegeria* (circles and solid line), *A. hyperantus* (squares and dotted line) and *P. tithonus* (triangles and dashed line) in the UK from 1900 to the present (solid symbols) and predicted extent of climatically suitable habitat for the period 2070–2099 (hollow symbols). Extrapolations predict time to colonize new areas, and are based on expansion rates since 1900 (hollow symbols), and since 1970 (shaded hollow symbols).

there were also significant differences in expansion rates among species ($F_{2,11} = 16.29$, $P = 0.004$); *P. tithonus* had the fastest rate of areal expansion (mean expansion rate, $C = 1.51$, SD = 0.19, $n = 6$) and *P. aegeria* had the slowest rate (mean = 1.20, SD = 0.26), with *A. hyperantus* having intermediate values (mean = 1.33, SD = 0.20). The interaction term between species and latitude was not significant ($P > 0.1$). These values result in marginal velocities of range expansion ($E$) of 0.68, 0.75 and 0.85 km year$^{-1}$ for *P. aegeria*, *A. hyperantus* and *P. tithonus*, respectively. Taking into account that *P. aegeria* develops through at least 1.5 generations per year (Blakeley, 1997), whereas the other two species are univoltine, this results in an even lower per generation rate of expansion for *P. aegeria* compared with the other two species.

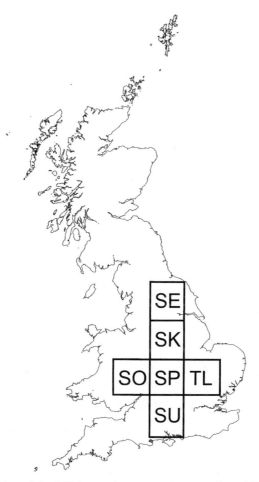

**Fig. 19.6.** Location of the 100-km study squares. Letters refer to UK Ordnance Survey grid system.

## Effects of habitat availability on expansion rate

We calculated the availability of habitat (percentage cover) in each of the six 100-km grid squares for each species by combining the appropriate land cover type classes (Table 19.1). There was a significant positive correlation of slope of areal increase (C; see above) with habitat availability for the whole data set ($r = 0.57$, $n = 18$, $P = 0.014$; Fig. 19.7), but within species, the correlation was not significant ($P > 0.1$ in all cases), indicating that at a 100-km resolution, habitat availability did not constrain expansion rates.

# Predicting Potential Future Distributions

## Predicting future European distributions

We predicted potential future European distributions of the three species for the end of the 21st century using methods described in Hill et al. (1999). We

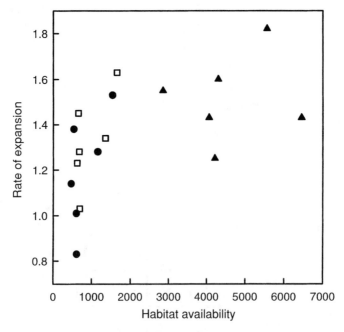

**Fig. 19.7.** Rate of range expansion (slope of √area of 10-km grid squares with butterfly records each decade since 1920 (see text), by year) in relation to habitat availability in six 100-km study squares. Habitat availability is calculated as the summed percentage cover of habitat of 10-km grid squares within each study square. *P. aegeria* – solid circles, *A. hyperantus* – hollow squares, *P. tithonus* – solid triangles.

obtained output from a transient climate change simulation, made using the HADCM2 general circulation model, for the period 2070–2099 to compute values for the three bioclimate variables (GDD5, MTCO and AET/PET) for the climate scenario for 2070–2099. These predicted future values were then used with the climate response surfaces to generate simulated potential distributions for the three butterfly species for the period 2070–2099. Table 19.3 shows mean differences for the three bioclimate variables between the climate normal period of 1931–1960 and future predicted climates. These data indicate that climates within the European study area will on average get warmer and drier by the end of the 21st century. Simulated future distributions for the three species are shown in Fig. 19.8 (using the same probability thresholds of occurrence as used for simulating current distributions, see above).

The model predicts considerable northward extension of the potential distribution of all three species, and indicates that they would have the potential to extend their ranges throughout most of the UK and Fennoscandia (Fig. 19.8). The southern range margin of *P. aegeria* (in North Africa) was not included in the current response surface making it impossible to predict future changes at the southern margin, and so areas south of 45°N latitude have been excluded from the simulated future distribution. The model predicts northward contraction of the southern range margin of *A. hyperantus*, with only isolated populations predicted to persist in central Europe (Fig. 19.8b), and it also predicts some contraction northwards of the southern range margin of *P. tithonus* in Iberia (Fig. 19.8c).

## Predicting future UK distributions, taking account of current habitat availability

In order to simulate potential butterfly distributions in the UK at a 10-km resolution, we computed values for the three bioclimate variables for the period 2070–2099 as before for the UK 10-km grid. We used the climate response surfaces generated from the 50-km grid European datasets to

Table 19.3. Mean values for three bioclimate variables for the 50-km UTM grid.

|  | GDD5 (degree days) | MTCO (°C) | AET/PET |
| --- | --- | --- | --- |
| 1931–1960 | 1915.5 (SD = 1079.0) | −1.36 (SD = 6.21) | 0.842 (SD = 0.288) |
| 2070–2099 | 2788.6 (SD = 1354.4) | 3.41 (SD = 5.17) | 0.837 (SD = 0.201) |
| Change | +873.1 (SD = 320.5) | +4.77 (SD = 1.32) | −0.006 (SD = 0.255) |

Data for the climate normal period 1931–1960 are from Leemans and Cramer (1991), values for the period 2070–2099 are derived from HADCM2 scenario. GDD5 = annual temperature sum >5°C, MTCO = mean temperature of the coldest month, AET/PET = actual/potential evapotranspiration.

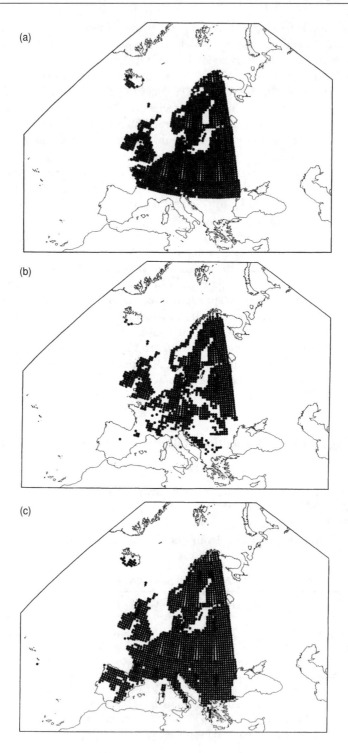

predict the probability of occurrence of each species in the UK under the changed climate scenario. We then used these values for probability of occurrence in the logistic regression equations (Table 19.2) to predict potential distributions of each species in the UK taking account of habitat availability (Table 19.1). The habitat availability values were those for the present because we have no basis for predicting how this might change over the next century. Predictions from the logistic regression model are for all three species to occur throughout the UK for the period 2070–2099, with the exception of the high mountain areas in central Scotland (Fig. 19.8). As at a European scale, we suspect that the species would find suitable habitats in sheltered valleys throughout most of these mountainous areas. Under a warmer climate, it appears that, at a 10-km resolution, sufficient habitat is present almost everywhere that it does not generally constrain potential future ranges for any of the three species. Under warmer climates it is also possible that species will be able to occupy poorer quality marginal habitats (Thomas et al., 1998), also indicating that habitat availability may not constrain potential future ranges.

## Predicting time-to-colonize newly available habitat

To predict the time required for the three study species to colonize newly available, climatically suitable habitats in the UK in the future, we used the regression equations describing the rate of areal expansion of each species in the UK since 1940 (see above; Fig. 19.5). Extrapolation of these rates of areal expansion for each species suggests that they could colonize all potentially suitable habitat by 2020. However, these estimates do not take account of the location of new habitat (Fig. 19.9), much of which is remote from current distributions. Thus these estimates are likely to be unrealistic. In addition, there was some indication from Fig. 19.5 that rates of expansion have been slower over the past 30 years; recalculating the marginal velocity of range expansion since 1970 results in reduced estimates of 1.16–1.70 km year$^{-1}$ for the three species. Extrapolation from these data suggests that the three species could colonize all potentially suitable habitat by 2060. As before, these estimates take no account of location of suitable habitat, and are also likely to be unrealistic. We also used expansion rates calculated from the six 100-km study squares to estimate time-to-colonize newly available habitat. These data predicted much slower colonization times, and predicted species

**Fig. 19.8** (opposite). Simulated future butterfly distribution for the period 2070–2099 for areas west of 30°E longitude on a 50-km UTM grid. (a) *P. aegeria*, (b) *A. hyperantus*, (c) *P. tithonus* (threshold probability of occurrence for each species are the same as in Fig. 19.1d–f). Because the southern range margin of *P. aegeria* was not included in the response surface model, only areas north of 45°N latitude are plotted in (a).

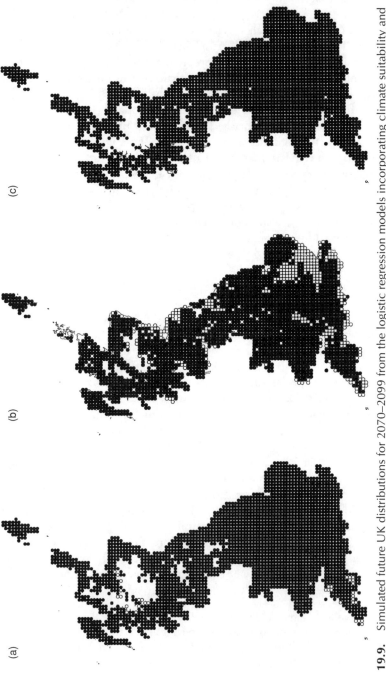

**Fig. 19.9.** Simulated future UK distributions for 2070–2099 from the logistic regression models incorporating climate suitability and current habitat availability (probability of occurrence ≥0.5; black circles). Hollow circles – sites where butterflies are currently recorded but not predicted in 2070–2099. (a) *P. aegeria*, (b) *A. hyperantus*, (c) *P. tithonus*. Land cover data were not available for the Isle of Man.

distributions to be in equilibrium with 2070–2099 climates within 150 years (*P. aegeria* = 143 years; *A. hyperantus* = 83 years; *P. tithonus* = 113 years).

## Discussion

Many butterfly species had more extensive distributions in the UK during the 19th century (Thomson, 1980), but towards the end of the 19th century, distributions contracted, and several previously common species disappeared from Scotland and large areas of northern England (Downes, 1948; Jackson, 1980). Of the three species considered here, distributions of *P. aegeria* became the most restricted during this time, but distributions of the other two species also contracted greatly, although a few *A. hyperantus* populations continued to persist in Scotland. There are no historical climate data available to test whether 19th-century contractions were due to climate change. However, historical climate and butterfly distribution data since 1900 indicate that, at least for *P. aegeria*, range expansions in the UK since 1940 are consistent with a general warming of the climate during this time (Hill *et al.*, 1999). All three species have yet to re-colonize all areas that they occupied historically.

### Performance of models

The high kappa statistics for the response surfaces fitted to 50-km resolution data indicate that climate is important in determining distributions of all three study species at a European scale. The three study species have broadly similar ecologies, and the same three bioclimate variables gave excellent fits to current butterfly distributions. These results indicate that this approach is likely to be successful for other UK species, although different combinations of bioclimate variables may be more appropriate for other species with contrasting life histories. For *P. aegeria*, the response surface predicted occurrence in some areas beyond its current northern margin in the UK, indicating that *P. aegeria* may be lagging behind current climates. The response surface also indicated that *A. hyperantus* is lagging behind current climates in parts of its range, but for *P. tithonus*, the fit between current recorded and simulated distributions in the UK was excellent. We currently have no information on whether these bioclimate variables determine range limits through their effects on local population dynamics; this clearly needs more study.

At the UK scale, both climate suitability and habitat availability were significantly and positively related to butterfly distributions using logistic regression models. These models predicted broadly similar results for distributions of the three species in the UK as were obtained from the European-scale model. For example, it predicted occurrence of *P. aegeria* in several regions beyond its current range margin (e.g. northern England and southern Scotland), in areas it occupied historically but has not yet re-colonized. It also predicted occurrence in areas that apparently have never been occupied by *P. aegeria* (e.g. Western Isles of Scotland, Isle of Man), and which may be too isolated to be colonized (Dennis and Shreeve, 1997). As at a European scale, the logistic model gave an excellent fit to *P. tithonus* distributions. There were a few scattered squares beyond its current range margin where *P. tithonus* was predicted to occur, but these areas may be too small and isolated to be currently occupied. Out of the three study species, the logistic regression model performed least well for *A. hyperantus*, but none the less still predicted >76% of butterfly presence/absences correctly. The model predicted more extensive distributions in northwest England and northeast Scotland, areas where *A. hyperantus* occurred historically. Of the three species, *A. hyperantus* has the most extensive distribution in Scotland, the area of the UK for which the current records are likely to be least reliable due to poor recording effort and this may contribute to the increased number of mismatches, compared with the other two study species.

## Impacts of habitat fragmentation

Availability of habitat in the UK varies among the three study species in the order *P. tithonus* > *A. hyperantus* > *P. aegeria* (Fig. 19.3; Emmet and Heath, 1990). This rank order is robust even if one argues about the exact land use categories appropriate for each species. For all species, habitat availability, in addition to climate suitability, was significantly related to butterfly distribution in logistic regression models. However, even after incorporation of habitat availability into the models, distributions of *P. aegeria* and *A. hyperantus* were still lagging behind current climates. The logistic regression models were generated for the whole of the UK, and any regional changes in habitat requirements among species were not included. For example, *P. aegeria* is much less restricted to woodland in the south and west of its UK range, and habitat requirements are often less specific in core areas compared with range margins (Thomas *et al.*, 1998). Thus the models may be predicting occurrence in areas at range margins with very little suitable habitat (e.g. northwest England for *A. hyperantus*, Vale of York for *P. aegeria*; Fig. 19.3a and b). Also, all areas included within the appropriate land cover types are assumed to be suitable, although this is unlikely to be the case. Thus for *P. aegeria* and *A. hyperantus*, the model may be overestimating the extent of current distributions and contributing to observed lag effects. However, for both these species, the models predict occurrence in several areas with high availability of habitat (e.g. southern Scotland for *P. aegeria*; northeast Scotland for *A. hyperantus*; Fig. 19.3a and b), indicating that at least some of these lag effects are real, and it is likely that habitat fragmentation at a finer scale than that considered here is affecting colonization rates in these two species. There was no evidence for any climate-distribution lags in *P. tithonus*, the species with the most widely available habitat, supporting the notion that habitat availability is affecting range expansion in these species. There are few data on dispersal rates for any of the three study species but some of the lag effects may also be due to failure of species to keep up with rapidly warming climates. However, given that *P. tithonus* has generally kept up with changing climates and probably does not have very different dispersal ability from the other two species, suggests that habitat availability rather than dispersal ability has had the greatest impact on species' ability to keep track of changing climates.

## Rates of range expansion

Range margins of the three species have shifted northwards in the UK by 107–178 km since 1940. The warming that has occurred in Europe this century has resulted in climatic isotherms shifting northwards by an average of 120 km (Beniston et al., 1998). Thus observed range shifts in these butterflies are in agreement with climatic changes. The slightly larger range shift of 178 km for *A. hyperantus* may be an inflated estimate because it is based on range changes in Scotland, and therefore includes areas where recording effort is poor, especially historically; the estimate may be inflated because historical range margins were actually further north than records suggest. This will be less of a problem for the other two species, where range shifts were calculated in England and Wales, where recording coverage is much better.

Comparison of rates of areal expansion in the six 100-km study grids since 1929 showed significant differences in expansion rates among species in the order *P. tithonus* > *A. hyperantus* > *P. aegeria*. This pattern mirrored the pattern of habitat availability among species and gives further support to the notion that habitat availability affects expansion rates. However, there was no effect of habitat availability in the within-species analysis, and as with the logistic regression models, it is likely that habitat availability averaged over a 100-km area is too coarse to detect habitat fragmentation at a scale relevant to dispersing butterflies, which is likely to be at a much finer scale than that considered here. The measures of habitat availability used here also do not take account of spatial distribution of habitat, which is likely to be a more important determinant of colonization rates than is availability of habitat *per se* (Thomas and Jones, 1993; Hill et al., 1996; Thomas and Hanski, 1997).

Extrapolation of expansion rates measured over the past 30 years indicate that the three species could colonize all suitable habitat by 2060. However, this estimate is unlikely to be realistic because it does not take account of location of suitable habitat, much of which is remote from current distributions (especially for *P. tithonus*). Expansion rates calculated over a smaller spatial scale (100-km grid squares) predict much slower colonization times, and predict equilibrium with late 21st-century climates within 150 years. These data also show that expansion rates vary among regions in the UK, and indicate that colonization of more isolated areas is likely to take much longer than estimates suggest. For the two species that are lagging behind current climates, these lags are likely to become even greater.

Maps of simulated potential future distributions (e.g. Fig. 19.8) do not represent forecasts of future distributions but do provide an indication of the magnitude of the potential impact of climate change on species' distributions. The actual outcomes will be difficult to predict given that responses of species to climate change are also likely to be influenced by factors resulting from associations with new predators/competitors, etc., in newly suitable areas; such factors may affect realized future distributions (Davis et al., 1998a,b). Results presented here show that among three species, rates of range expansion were related to availability of habitat, and that for the two species with the most restricted habitat requirements, there was evidence that distributions are lagging behind current climates. It is possible that species-specific dispersal rates, in addition to different habitat specificities, may account for some of these differences in range expansion rates, and more work is needed to determine the relative importance of habitat availability and dispersal on patterns and rates of range expansion. However, our results show that even moderately mobile species which are relative habitat generalists appear to be lagging behind current climates. Many other UK species are more sedentary with more restricted habitat requirements than the species considered here, and may therefore have much greater problems keeping track of changing climates. Whether such lags are widespread among the UK butterfly fauna remains to be seen.

## Summary

We investigate the distributions of three species of satyrid butterfly (*P. aegeria*, *A. hyperantus* and *P. tithonus*) that are currently expanding in the UK. These species were chosen because they have broadly similar ecologies and dispersal abilities but differ in their habitat requirements and thus in the availability of suitable breeding habitat. Since the 1940s, range margins of these species have shifted northwards in the UK in agreement with climatic changes. Rates of expansion differ among the species in the order *P. tithonus* > *A. hyperantus* > *P. aegeria*, mirroring patterns of habitat availability. We show that climate is important in determining their distributions at a European scale. At a UK scale, both climate suitability and habitat availability are significantly and positively related to butterfly distributions, but there is evidence that *P. aegeria* and *A. hyperantus*, the two species with the most restricted habitat requirements, are lagging behind current climates. All three species will have the potential to extend their ranges northwards under predicted future climate warming, but the distribution of habitat is likely to impact on their ability to track climate change. Many UK species have more restricted habitat requirements and are more sedentary than the species considered here; they are likely to have much greater problems tracking changing climates.

## Acknowledgements

We thank the huge number of volunteers whose records have contributed to the BRC and Butterfly Conservation data sets. We also thank Wolfgang Cramer (PIK Potsdam) for providing the spline surfaces used to interpolate both the present climate and the HADCM2 anomalies. The output from the HADCM2 transient simulation was supplied by the Climate Impacts LINK Project (Department of the Environment Contract EPG 1/1/16) on behalf of the Hadley Centre and the UK Meteorological Office. David Viner (CRU UEA) kindly facilitated access to these data. This study was funded by NERC grants GR9/3016 and GR3/12542.

## References

Asher, J., Warren, M., Fox, R., Harding, P., Jeffcoate, G. and Jeffcoate, S. (2001) *The Millennium Atlas of Butterflies in Britain and Ireland*. Oxford University Press, Oxford.

Beerling, D.J., Huntley, B. and Bailey, J.P. (1995) Climate and the distribution of *Fallopia japonica*: use of an introduced species to test the predictive capacity of response surfaces. *Journal of Vegetation Science* 6, 269–282.

Beniston, M. *et al.* (1998) In: Watson, R.T., Zinyowera, M.C. and Moss, R.H. (eds) *The Regional Impacts of Climate Change*. IPCC Working Group 2. Cambridge University Press, Cambridge, pp. 149–185.

Blakeley, D.S. (1997) Overwintering biology of *Pararge aegeria*. M.Phil. thesis, University of Leeds, Leeds.

van den Bosch, F., Metz, J.A.J. and Diekmann, O. (1990) The velocity of spatial population expansion. *Journal of Mathematical Biology* 28, 529–565.

Burrows, C.R.N. (1916) The disappearing *Pararge aegeria*. *The Entomologist's Record* 28, 112–114.

Chalmers-Hunt, J.M. and Owen, D.F. (1952) The history and status of *Pararge aegeria* (Lep. Satyridae) in Kent. *The Entomologist* 85, 145–154.

Coope, G.R. (1978) Constancy of insect species versus inconstancy of Quaternary environments. In: Mound, L.A. and Waloff, N. (eds) *Diversity of Insect Faunas*. Blackwell, Oxford, pp. 176–187.

Davis, A.J., Jenkinson, L.S., Lawton, J.H., Shorrocks, B. and Wood, S. (1998a) Making mistakes when predicting shifts in species range in response to global warming. *Nature* 391, 783–786.

Davis, A.J., Lawton, J.H., Shorrocks B. and Jenkinson, L.S. (1998b) Individualistic species responses invalidate simple physiological models of community dynamics under global environmental change. *Journal of Animal Ecology* 67, 600–612.

Dennis, R.L.H. (1993) *Butterflies and Climate Change*. Manchester University Press, Manchester.

Dennis, R.L.H. and Shreeve, T.G. (1991) Climatic change and the British butterfly fauna, opportunities and constraints. *Biological Conservation* 55, 1–16.

Dennis, R.L.H and Shreeve, T.G. (1997) Diversity of butterflies on British islands: ecological influences underlying the roles of area, isolation and the size of the faunal source. *Biological Journal of the Linnaean Society* 60, 257–275.

Downes, J.A. (1948) The history of the speckled wood butterfly (*Pararge aegeria*) in Scotland, with a discussion of the recent changes of range of other British butterflies. *Journal of Animal Ecology* 17, 131–138.

Emmet, A.M. and Heath, J. (1990) *The Butterflies of Great Britain and Ireland*. Harley Books, Colchester.

Ford, E.B. (1945) *Butterflies*. Collins, London.

Fuller, R.M., Groom, G.B. and Jones, A.R. (1994) The land cover map of Great Britain: an automated classification of Landsat thematic mapper data. *Photogrammetric Engineering and Remote Sensing* 60, 553–562.

Gibbs, A.E. (1916) The disappearing *Pararge aegeria*. *The Entomologist's Record* 28, 122–126.

Heath, J., Pollard, E. and Thomas, J.A. (1984) *Atlas of Butterflies in Britain and Ireland*. Viking, London.

Hill, J.K., Thomas, C.D. and Lewis, O.T. (1996) Effects of habitat patch size and isolation on dispersal by *Hesperia comma* butterflies: implications for metapopulation structure. *Journal of Animal Ecology* 65, 725–735.

Hill, J.K., Thomas, C.D. and Huntley, B. (1999) Climate and habitat availability determine 20th century changes in a butterfly's range margins. *Proceedings of the Royal Society B* 266, 1197–1206.

Houghton, J.T., Meira Filho, L.G., Callander, B.A., Harris, N., Kattenberg, A. and Maskell, K. (eds) (1996) *Climate Change 1995: the Science of Climate Change*. Intergovernmental Panel on Climate Change (IPCC). Cambridge University Press, Cambridge.

Huntley, B., Berry, P.M., Cramer, W. and McDonald, A. (1995) Modelling present and potential future ranges of some European higher plants using climate response surfaces. *Journal of Biogeography* 22, 967–1001.

Jackson, S.M. (1980) Changes since 1900 in the distribution of butterflies in Yorkshire and elsewhere in the north of England. *The Entomologist's Record* 105, 139–142.

Leemans, R. and Cramer, W. (1991) Research Report RR-91-18. International Institute for Applied Systems Analysis (IIASA), Laxenburg, Austria.

Lees, E. (1962) Factors determining the distribution of the speckled wood butterfly (*Pararge aegeria* (L.)) in Gt Britain. *Entomologist's Gazette* 13, 101–113.

Lensink, R. (1997) Range expansion of raptors in Britain and the Netherlands since the 1960s: testing an individual-based diffusion model. *Journal of Animal Ecology* 66, 811–826.

Monserud, R.A. and Leemans, R. (1992) Comparing global vegetation maps with the Kappa statistic. *Ecological Modelling* 62, 275–293.

Parmesan, C. (1996) Climate and species' range. *Nature* 382, 765–766.

Parmesan, C., Ryrholm, N., Stefanescu, C., Hill, J.K.,Thomas, C.D., Descimon, H., Huntley, B., Kaila, L., Kullberg, J., Tammaru, T., Tennant, J., Thomas, J.A. and Warren, M. (1999) Polewards shifts in geographical ranges of butterfly species associated with regional warming. *Nature* 399, 579–583.

Pollard, E. (1979) Population ecology and change in range of the White Admiral butterfly *Ladoga camilla* L. in England. *Ecological Entomology* 4, 61–74.

Pollard, E. and Eversham, B.C. (1995) Butterfly monitoring 2 – interpreting the changes. In: Pullin, A.S. (ed.) *Ecology and Conservation of Butterflies*. Chapman & Hall, London, pp. 23–36.

Pollard, E., Rothery, P. and Yates, T.J. (1997) Annual growth rates in newly established populations of the butterfly *Pararge aegeria*. *Ecological Entomology* 21, 365–369.

Robson, J.E. (1902) A catalogue of the Lepidoptera of Northumberland, Durham and Newcastle-upon-Tyne. *Natural History Transactions of Northumberland, Durham and Newcastle-upon-Tyne* 11, 1–152.

South, R. (1941) *The Butterflies of the British Isles*. F. Warne & Co, London.

Thomas, C.D and Hanski, I. (1997) Butterfly metapopulations. In: Hanski, I.A. and Gilpin, M.E. (eds) *Metapopulation Dynamics: Ecology, Genetics and Evolution*. Academic Press, San Diego, California, pp. 359–386.

Thomas, C.D. and Jones, T.M. (1993) Partial recovery of a skipper butterfly (*Hesperia comma*) from population refuges: lessons for conservation in a fragmented landscape. *Journal of Animal Ecology* 62, 472–481.

Thomas, C.D., Jordano, D., Lewis, O.T., Hill, J.K., Sutcliffe, O.L and Thomas, J.A. (1998) Butterfly distributional patterns, processes and conservation. In: Mace, G.M., Balmford, A. and Ginsberg, J.R. (eds) *Conservation in a Changing World*. Cambridge University Press, Cambridge, pp. 107–138.

Thomson, G. (1980) *The Butterflies of Scotland*. Croom-Helm, London.

Tolman, T. (1997) *Butterflies of Britain and Europe*. HarperCollins, London.

Turner, J.R.G., Gatehouse, C.M. and Corey, C.A. (1987) Does solar energy control organic diversity? Butterflies, moths and the British climate. *Oikos* 48, 195–205.

Tutt, J.W. (1905) *British Butterflies*. G. Gill & Sons, London.

# Index

Abiotic factors 114, 262, 361
*Acyrthosiphon pisum* (pea aphid) 26, 263, 264
*Adalia bipunctata* (two-spot ladybird) 264, 271, 403
Adaptation (adaptive response) 6, 43, 161, 174, 176, 184, 210, 212, 348, 369, 407
Adaptive performance 374
Adipokinetic hormones (AKHs) 66–77, 79–80
    age-related changes in responsiveness to AKHs 74, 79
    release of AKHs 66–68, 71, 72, 76
    sequences of peptides in 71, 72
Aerodynamic power *see* Power
Aerodynamics 2, 27–30, 43–61, 200
African armyworm moth 4, 11, 375
    *see also Spodoptera exempta*
Aggregation 4, 88, 97, 160, 209, 269
*Agrotis exclamationis* (heart and dart moth) 190
    *A. segetum* (turnip moth) 130, 152
Airspeed (= flight speed of insect) 6–7, 26, 28–30, 34, 50, 54–56, 90, 183, 185, 190, 191, 195–200, 212
    during host location by mosquitoes, tsetse 98, 99
    of foraging bees 136, 137, 139, 141, 143–148
    of high-flying migrants 210, 223, 224, 229
    of migrating butterflies 6–7, 193, 195–197, 200

    of migrating dragonflies 197, 200
Alanine 74, 75–77, 78
Allele, frequencies 207–208, 228, 370
Allometry 26–28
*Allonemobius* crickets 371
Allozymes 13, 345–346, 362–365, 366, 367, 369–376
Alternating selection 373
Ambrosia beetles 102
Amino acid 65, 66, 68, 71, 72, 73, 74, 77, 197
Amplified fragment length polymorphisms (AFLPs) 363
*Anartia fatima* 199
*Anax junius* 28
*Andricus quercuscalicis* (Knopper gall wasp) 366, 367, 372
Anemotaxis *see* Taxes
*Anopheles gambiae* 100, 101
Anteromotorism 22–24, 26
*Anthocharis cardamines* (orange tip) 322, 404
*Anticarsia gemmatalis* (velvetbean caterpillar) 375
Ants 189, 266, 281, 282, 322–323, 397
    army ants 1, 5–6, 9, 281–296, 299, 300, 309–315, 317
    desert ants 134
    *see also Cataglyphis*
    *see also Cataglyphis*; Dorylinae; *Dorylus* spp.; *Dorylus molestus*; *D. wilverthi*; *Eciton burchelli*; *E. rapax*; Ecitonninae; *Myrmica* spp.

Aphid 3, 7, 8, 87, 102, 112, 113, 115–118, 121, 122, 160–161, 167, 263–275, 363, 364, 366, 367, 369–370, 373
   see also Acyrthosiphon pisum; Aphis fabae; Diuraphis noxia; Drepanosiphum platanoidis; Myzus persicae; Phorodon humuli; Rhopalosiphum padi; Schizaphis graminum; Sitobion avenae; S. fragariae
Aphantopus hyperantus (ringlet butterfly) 416–431, 433–438
Aphidius 5, 118, 119
   A. ervi 115
   A. rhopalosiphi 118
   A. rosae 265
Aphis fabae (black bean aphid) 3, 160–161, 162, 166, 167
Aphrissa 186, 187, 195
   A. statira (sulphur butterfly) 135, 186, 195
Apis mellifera 130, 189, 199
   see also Honey bee
APLC see Australian Plague Locust Commission
Apoprotein 68
   apoprotein-III (Apo-III) 68–70, 80
Apple 352, 371
Aptery 1, 24, 25, 165, 166
Apterygota 20, 21, 30, 45, 50
Architecture see Host plant architecture
Araschnia levana (map butterfly) 399, 400
Arena see Migration, arena
Argynnis paphia (silver-washed fritillary) 396, 404–405
Armyworm moth 11, 164
   see also African armyworm moth; Oriental armyworm moth
Arrestant 119, 120
Artificial selection 162, 163, 164, 167, 168
Ascia monuste (great southern white) 194
Asclepias syriaca see Milkweed
Asteraceae 216, 217
Asynchronous flight muscle see Flight muscle, asynchronous
Atmospheric carbon dioxide 98, 390, 407
Atmospheric oxygen 30–31, 59
Attractant 119
Aubeonymus mariaefranciscae (sugar beet weevil) 367
Australia 7, 207–230
   semi-arid inland of 208, 217
Australian plague locust 207–230
   see also Chortoicetes terminifera
Australian Plague Locust Commission (APLC) 215–216, 229
AVHRR 216
Axillae 50, 53–55, 59, 392
Azimuth, of sun 5, 143–144, 185–188

Bark beetle 117, 166
   see also Ips
Barro Colorado Island 309–314
Bees 4, 5, 104, 189, 200, 364
   see also Apis mellifera; Bombus; Bumblebee; Honeybee
Beetles 3, 12, 13, 65, 103, 117, 148, 251, 356, 368, 369, 388, 403
   see also Coleoptera; Ambrosia beetle; Bark beetle; Carabid beetles; Colorado potato beetle; Onitine dung beetles; Pine beetle; Scarabaeine dung beetles
Behavioural-related patch fidelity 372
Bemisia tabaci (white fly) 366
Bimotorism 23–24
Biological control 119, 123
Biomass 19, 242
Biomechanics
   of army ant foraging 296
   of flight 1, 2, 19–36, 43–61
Biotypes 369
Birth 9, 301, 322–325, 328, 331, 333, 372, 392
Bivoltine 373
Black bean aphid see Aphis fabae
Blackheaded fireworm moth see Rhopobota naevana
Blattaria 24
Blattinopsidae 58–60
Blattinopsis augustai 58–59
Blueberry 371
Body angle (during flight) 29
Body length 30, 31, 33
   size 20, 21, 30–34, 36, 162, 169, 174, 287, 289
Bojophlebia prokopi 60
Bolitotherus cornutus (forked fungus beetle) 346
Bombus spp. 5
   B. terrestris 129, 130–134, 136–139, 141–149
Boundary layer see Flight boundary layer
Brachyptery 165, 166, 170, 171, 176, 236, 237, 240, 244, 246, 247, 250, 252
   see also Wing polymorphism
Broad bean 115, 116
Britain 367, 375, 388, 389, 393, 396, 397, 398, 399, 404, 405, 415–439
Bug see Heteroptera
Bumblebee 5, 29, 129, 130–134, 136, 139, 141–149
   see also Bombus spp., Bombus terrestris
Butterfly 6, 7, 11, 26, 51, 53, 129, 135–136, 183–200, 322, 323, 326, 327, 330, 346, 352, 367, 374, 393, 394, 397, 398, 404, 415–417, 420, 422, 431, 435, 436, 438
   blues and coppers see Lycaenidae
   heliconiine see Heliconiine butterflies

*Index*

monarch see *Danaus plexippus*
nymphalid see Nymphalidae
satyrid see *Coenonympha tullia*;
  *Pararge aegeria*; *Aphantopus hyperantus*; *Pyronia tithonus*
sulphur see *Aphrissa statira*; *Phoebis argante*; *Phoebis sennae*
swallowtail see Swallowtail butterflies; *Papilio glaucus*
see also *Anartia fatima*; *Araschnia levana*; *Argynnis adippe*; *A. paphia*; *Carterocephalus palaemon*; *Colias* spp.; *Cupido minimus*; *Cyanaris semiargus*; *Euphydryas editha*; *E. aurinia*; *Heliconius charithonia*; *H. erato*; *H. himera*; *H. peruviana*; *Maculinea* spp.; *Marpesia chiron*; *Pararge aegeria*; *Parnassius mnemosyne*; Pieridae; *Pieris brassicae*; *P. rapae*; *Plebejus argus*; *Urbanus proteus*

Cabbage root fly see *Delia radicum*
*Cadra cautella* (almond moth) 95, 96
California 248, 349, 375, 376, 401, 402
*Caloneura dawsoni* 58–59
Caloneurodea 58–60
Canada 183, 371, 372, 376, 394, 395, 396
Carabid beetles 12, 251, 368
Carbohydrate 2, 3, 66, 67, 69, 73, 77, 197
  mobilization and hypertrehalosaemia 73–74
  utilization 65, 66, 75
Carboniferous 19, 22, 23, 30, 35, 44, 57–61
*Cardiospermum corindum* (native balloon vine) 169, 170, 171
Caribbean 162, 169, 196
*Carterocephalus palaemon* (chequered skipper butterfly) 399, 401
*Cataglyphis* ants 134, 189
Cecidomyiidae 273
Cellular automata (CA) 8, 9, 285, 301, 306, 309, 311–312, 315, 317
Central America 169, 375
  see also Panama
*Ceratitis capitata* (Mediterranean fruit fly) 344, 376
Cereal aphid see *Sitobion avenae*; *Sitobion fragariae*
Chalk downs 404
*Chaoborus* larvae 271
Characterization (of a migration system) 207–230
*Chortoicetes terminifera* 7, 207–230
  see also Australian plague locust
Chromosomal markers 362–363
Chrysomelidae 72

Circadian rhythm 89, 100, 185
Circadian clock 185
Clap and fling see Flight mechanisms
Climate and *S. alterniflora* 242
Climate change 9, 11, 367, 387–394, 397, 399, 403, 404–407, 415, 423, 431, 435, 437–439
  change scenarios 407, 416, 431
  extremes 400, 402
  future 431, 438
  and lifecycle strategies 373
  response 387
  response surfaces 416, 417, 420, 422, 431
  see also Distribution, distribution-climate lags; Evolution and climate change; Glacial cycles; Global change; Global warming
Climatic gradient 373
Cline 345, 348–351, 354, 371, 372, 373, 395, 403
Clock-shift 186, 187, 199
Clone 116, 264, 268, 271, 370, 373
Clouded Apollo butterflies see *Parnassius mnemosyne*
Clouded yellow butterflies see *Colias*
Cloudless sulphur butterfly see *Phoebis sennae*
Clustering analysis 372
$CO_2$, carbon dioxide 98, 100–101, 390, 407
*Coccinella septempunctata* 273
Coccoidea 26, 251
  see also Scale insect
*Coenonympha tullia inornata* (inornate ringlet butterfly) 372
Coleoptera 20, 23, 33, 34, 35, 55, 65, 72, 75, 388
  see also *Adalia bipunctata*; *Aubeonymus mariaefranciscae*; Beetles; *Bolitotherus cornutus*; *Coccinella septempunctata*; *Cotinus mutabilis*; Chrysomelidae; *Decapotoma lunata*; *Dendroctonus pseudotsugae*; *Dorcus parallelopipedus*; *Gareta nitens*; *Geotrupes stercorosus*; *Helophorus aquaticus*; *Ips* spp.; *Leptinotarsa decemlineata*; *Melanophila acuminata*; Meloidae; *Melolontha melolontha*; *Onitis aygulus*; *O. pecuarius*; *Pachnoda sinuata*; *Onymacris plana*; *O. rugatipennis*; *Phalacrus substriatus*; Scarabaeoidea; *Scarabaeus deludens*; *S. rugosus*; *Tenebrio molitor*; Tenebrionidae

Coexistence, fugitive 299, 302–304, 316, 317
Coexistence, predator-mediated 299, 303–305, 309, 316, 317
*Colias* spp. (clouded yellow butterflies) 374
*Collinsia* 402
Colonization 8, 9, 237, 300, 302, 326, 331, 370, 371, 392, 401, 404
  by army ant colonies 310–311
  by Knoper gall wasp 366, 367
  in metapopulation models 300–304, 306, 307, 313, 329
  by planthoppers 238, 242, 244
  and range changes 389, 399, 405, 415–417, 423, 433–438
  and Slatkin's paradox 346–347, 375
Colorado potato beetle 102
  see also *Leptinotarsa decemlineata*
*Columba livia* (pigeon) 199
Compensation axis 324–325, 331–333
Competition 209, 268, 272, 302–304, 307, 316, 317, 323
Concerted evolution 364
Conditional dispersal strategy 262–263, 266, 274, 275
Conditioning 120, 121–122, 123, 371
Conservation 1, 11, 123, 194, 299–300, 315, 317, 321, 324, 325, 329, 355, 390, 393–394, 404
  see also Extinction; Habitat fragmentation; Habitat destruction
Corn 115, 116, 370
Corpora cardiaca 67, 71, 72, 76
Corridors, habitat 313–315, 316, 317
*Cotinus mutabilis* 148
Cotton 115, 116, 345
Course 12, 129, 135, 143, 144, 185, 190–193, 196, 200, 212
  see also Orientation; Heading
Cricket 161, 189, 329, 371
  see also *Allonemobius*; *Gryllus assimilis*; *Gryllus firmus*; *Gryllus rubens*
Crosswind drift 5, 135–136, 137, 143, 190–194, 196, 197, 200
  see also Flight, crosswind; Headwind; Optimal wind drift compensation; Tailwind
Cues, responses to 5, 87–91, 98, 99, 101–104, 111–121, 143, 160, 183, 189, 199, 210, 213–214, 223, 237, 267, 271–272
*Culex pipiens* 264, 265, 351
*Culiseta longiareolata* 265
*Cupido minimus* (small blue) 396
*Cyaniris semiargus* (mazarine blue) 323
Cyclic AMP (cAMP) 76

*Danaus* 187, 188, 198
  *D. plexippus* (monarch butterfly) 135, 136, 183, 187, 189, 199, 209, 374, 376, 391, 392
*Daphnia* 267, 271
Death 9, 301, 307, 313, 322–325, 328, 331, 333, 402
  see also Mortality
*Decapotoma lunata* 72, 73, 74, 75
  *Del*-CC in 72, 74
Decision-support system 11, 229
*Del*-CC (adipokinetic hormone) 72, 74
*Delia radicum* (cabbage root fly) 101, 104
Delphacidae 7, 236, 245–246, 248–252
  see also *Javesella simillima*; *Megamelus paleatus*; *Nesosydne* spp.; *Nilaparvata lugens*; Planthoppers; *Prokelisia dolus*; *Prokelisia marginata*; *Stobaera*; *Toya venilia*; *Tumidagena minuta*
Deme 354, 364, 370, 376, 377
*Dendroctonus pseudotsugae* 73
Departure direction 222, 224
Dermaptera 24, 55
Desert locust see Locust, desert; *Schistocerca gregaria*
Devonian 2, 19, 361
Diacylglycerol (fats) 66–69, 79
  mobilization and hyperlipaemia 68–69, 70, 72, 73
  transport 68, 69, 70
*Diaeretiella* sp. 5,
Diapause 3, 87, 112, 162, 166–167, 228–229
Diaphanopterodea 22, 60
Dictyoptera 55, 60
Diptera 23, 24–26, 31, 33, 34, 35, 36, 56, 65, 98, 122
  see also *Anopheles gambiae*; *Ceratitis capitata*; *Culex pipiens*; *Culiseta longiareolata*; *Delia radicum*; *Drosophila* spp.; *Glossina morsitans*; *Lutzomyia longipalpis*; *Phormia regina*; *Psila rosae*; *Rhagoletis mendax*; *R. pomonella*
Diptery 23, 25, 26
Dispersal 4, 7, 34, 160, 209, 322–323, 330, 362, 370, 372, 375, 404, 407, 416, 438
  ability 236–237, 263, 368–369, 373, 387, 396, 405, 437, 438
  definitions of 4, 88, 160, 261
  and gene flow 337–351, 354–355, 367, 368
  'natal' 160
  predation and the evolution of 261–275
  spatial modelling of 299–317
  see also Conditional dispersal strategy; Evolution of dispersal; Gene flow vs. dispersal;

Index

Predator-induced dispersal
(PID); Unconditional dispersal
strategy
Displacement 6, 34, 134–135, 137, 153, 183,
361, 365, 366, 373, 377
accidental 88
distance 195–199, 214, 222–224, 227,
236
*see also* Scale (of population
movements)
potential 214
Distribution changes 9, 269, 322–324, 327,
329, 388–389, 390–407, 415–439
distribution, future 406–407, 416,
430–435, 438
distribution-climate lags 416, 420, 421,
437
*Diuraphis noxia* (Russian wheat aphid) 269,
270
Diversity (faunal) 19–20, 22, 30–31, 57, 61,
299, 300, 309, 316–317, 397
DNA fingerprinting 363, 364
DNA-based markers 362–364
Dominant and co-dominant markers 363
*Dorcus parallelopipedus* (lesser stag beetle)
77
Dorylinae 281
*Dorylus* 6, 281–283, 287, 289, 292, 293, 295
*D. molestus* 294, 295
*D. wilverthi* 282, 287, 288, 289,
290–291, 295
Drag 27, 30, 46–48, 147, 198
Dragonflies 23, 29, 34, 60, 184, 189, 192–194,
197, 200
*see also* Odonata
*Drepanosiphum platanoidis* (sycamore aphid)
370
Drift *see* Crosswind drift; Genetic drift
*Drosophila* 28, 90, 352, 356, 375
*D. funebris* 90
*D. mauritiana* 353
*D. melanogaster* 89, 353
*D. pseudoobscura* 346–347, 349, 353,
354
*D. simulans* 349, 353
*D. viridis* 28
*D. willistoni* 353, 375
Drought 401
Duration of flight 6, 162–163, 164, 210, 213,
219, 223–224, 374
*see also* Flight capacity
*Dysdercus* spp. (cotton stainer bug) 166


East Africa 375
Ecdysteroids 164, 168
*Eciton* 6, 282, 283, 287, 289, 295
*E. burchelli* 281–295, 299, 300, 309, 317
*E. rapax* 284, 286
Ecitonninae 281

Ecotypes 369
Edith's checkerspot *see Euphydryas editha*
'Effective size' of a population ($N_e$) 338–340,
347, 365
El Niño 389
Electrophysiology 118
Elytrization 20, 23, 25
Emigration 211, 261–263, 265, 266, 268, 274,
275, 291–292, 295, 325, 328, 331–333
*Encarsia formosa* 28
Endocrine control 6, 65–80, 161, 164,
166–169, 172–173, 176, 237
Endothermic warm-up 77–79
Energetic costs 2, 29–30, 31, 34, 88, 134,
146–149, 164, 166, 167, 170, 185,
194–198, 252, 272, 289
Environmental heterogeneity 184
Enzyme phenotype 374
Enzymes (flight muscle) 69–70, 74, 166, 172,
175, 176
Ephemeroptera 21, 23, 26, 55, 60
*see also* Mayflies
*Epirrita dilutata* (November moth) 375
*Euphydryas aurinia* 397
*E. editha* (Edith's checkerspot
butterfly)
323, 346, 394, 395, 396,
400–402, 404
*editha quino* 402
*editha bayensis* 401
*Euphilotes enoptes* (lycaenid blue butterfly)
372
*Euproctis* spp. 120
Europe 199, 327, 352, 366, 367, 371, 388, 391,
393, 394, 396, 398, 404, 405, 417,
419, 420, 421, 425, 430–431, 432,
435, 436, 437, 438
Evolution 6, 91, 116, 121, 161–176, 184, 266,
338, 347, 355, 361, 364, 367
of army ant raids 281–296
and climate change 393, 403–404, 407
convergent 6, 281–282, 287, 295, 369
of dispersal/migration 6, 7, 8, 34,
159–177, 235–253, 261–275, 322
divergent 295
of flight 1, 2, 19–23, 26, 29, 30, 36,
43–45, 48, 50–51, 53, 54, 55,
57–61
performance 2, 34–35
of flightlessness 165–167, 170–172,
236–253
*see also* Aptery; Wing
dimorphism/polymorphism
of gigantism and miniaturization 2, 20,
30–33, 36, 59
of life history *see* Life history evolution.
Expansion (of hind wing) 51, 55, 60
Extinction
extinct taxa 30–31, 55, 60

extinction/colonization cycles and
    Slatkin's paradox 346
  local 8, 9, 273, 300–302, 309, 326, 329,
    330, 331, 372, 394, 401–402, 407
  modelling of 293, 299–317
  rate 273, 302, 330, 331, 368
  regional 393–395, 399, 402, 407

Fat 65, 66, 68, 69, 71, 197
Fat body 65–70, 73, 75, 76, 79
Fatty acids 66–67, 69, 102
  lipoprotein lipase 68–69
  synthesis 68
  utilization (oxidation) 68, 69, 70, 73, 172
Fecundity 6, 162–163, 165, 167, 169, 176,
    236, 237, 238, 272, 322, 323, 376
Finland 329, 372, 389, 393, 396, 397, 399,
    400, 421
Flapping 20–23, 26–29, 31, 36, 43, 46–50,
    53–54, 57, 60, 68, 198
Flexion lines 20, 54, 56, 59–60
Flight 1–3, 5, 19–36, 43–61, 129–154,
    160–167, 170–172, 175, 176, 265,
    281, 292, 361, 370, 374, 404
  boundary layer (fbl) 6–7, 183–200, 210, 212
  capacity 7, 34, 35, 211, 213, 214, 229
    see also Duration of flight;
      Migratory capacity
  crosswind 5, 7, 90, 96, 135, 137, 139,
    141–143, 146, 153, 212
    see also Crosswind drift;
      Headwind; Tailwind
  direction 132, 134–135, 183, 185, 186,
    188, 189, 199, 209, 210, 214,
    218–219, 223, 362, 377
  distance 34, 111, 131, 150, 151, 160,
    183, 196, 210, 213, 218,
    222–224, 227, 236, 323, 374, 377
    see also Displacement distance
  duration see Duration of flight
  endocrine control of 6, 65–80
  evolution see Evolution of flight;
    Evolution of flightlessness
  forward flight 26, 27, 28–30, 60
  foraging see Foraging flight
  fuels 2, 65–66, 67, 68, 73, 74–75, 79, 80,
    99, 175, 196, 197, 198, 361
  height see Height of flight
  high altitude migratory 210–214,
    210–214, 219, 222–229
  in planthoppers 236–241, 244–253
  low-altitude 6–7, 12, 129–154, 183–201, 211
  manoeuvrability 2, 22, 23, 27, 29,
    34–35, 36, 43, 54, 55, 60, 197
  manoeuvre see Roll; Yaw; Pitch
  mechanisms 2, 27–29, 46–61

  clap and fling 2, 55
  delayed stall 2, 53–54
  miniaturization of body size and
    see Miniaturization of body size
  rotational circulation 2, 28
  wake capture 2
    see also Aerodynamics;
      Biomechanics of flight;
      Kinematics; Reynolds number;
      Unsteady airflows
  muscle 6, 7, 13, 20, 29, 31–34, 35, 36,
    43, 53, 65–70, 73–80, 146, 148,
    164, 166–167, 170–172, 175,
    176, 272
    asynchronous 31–34, 35, 36
    physiology 65–70, 74–80, 172
    synchronous 31–32
  to odour sources 92–103, 118
  origin of 2, 20–22, 44–46, 50, 55, 57, 61, 361
  periodicity 89, 98–100, 159, 185, 374–375
  pheromone-following 91–97, 101, 102,
    103, 104, 113, 117–121, 151–154
  polymorphism see Polymorphism, flight
  ranging 89–90, 152
    see also Ranging movements
  radar or radio-tracking studies of
    see Radar; Radio-telemetry
  speed see Airspeed
  suppression by wind/temperature 211
  termination (of migratory flight) 213, 228
  tethered see Tethered flight
  track see Track, flight
  trajectory see Trajectory
    see also Migration; Optimal flight
      speed; Orientation flight
      (in bees)
Fluorescent in situ hybridization (FISH) 363
Foraging 4–6, 12, 88–104, 111–115, 118–119,
    121, 122, 123, 129–154, 159–160,
    187, 237, 251, 264, 265, 268–269,
    271, 281–295, 328, 362
  cue 5, 88–89, 90, 91, 98–104, 111–123
  flight 4–5, 89–104, 111, 113, 114,
    118–119, 123, 129–154,
    159–160, 187, 238, 251
  movements in ants 1, 5–6, 134, 189,
    281–295, 309–311
  range 130, 133–134
  strategy 12, 97–98, 101, 113, 130,
    133–134, 238
Forecasting 7, 10, 216, 219, 227, 228–230,
    313–314, 388, 389, 391, 406,
    415–416, 430–438

Fossil insects 2, 19, 21–23, 30–31, 35, 44, 55, 57–61, 282, 388, 403
Founders 346, 368, 371, 374
Fragmentation *see* Habitat fragmentation; Phase transition
France 269, 351, 367, 373, 388, 393, 399
Fruit fly 117
   *see also Ceratitis capitata*; *Drosophila*; Tephritid fruitflies
$F_{ST}$ 339–347, 365, 367, 368
Fuels *see* Flight fuels

Gambler's ruin 304, 313
*Gareta nitens* 72
   *Scd*-CC-I, *Scd*-CC-II in 72
Gatekeeper *see Pyronia tithonus*
Gene flow 10, 123, 337–356, 364–365, 367–377
   flow vs. dispersal 337
   transfer, horizontal 337
Genetics 6, 8, 10, 120–121, 123, 161, 162–163, 167–169, 173, 175, 176, 207–208, 214, 228, 235, 262, 337–356, 403
Genetic
   bottlenecks 364, 368, 376
   complex 207–208, 228
   correlations 6, 162, 163, 164, 167, 169
   distances 365, 367, 372, 373
   drift 10, 339–347, 368, 375
   heterogeneity 364, 367, 368, 370, 372, 374
   'hitchhiking' effect 364
   homogeneity 343, 346, 367, 368, 369, 370, 371, 375
   structure *see* Population genetic structure
Geographic barriers (genetic effects of) 367–368, 376
Geographical Information Systems (GIS) 10, 11, 216, 229, 417
Geometrids 375
*Geotrupes stercorosus* 72
   *Mem*-CC in 72
Gigantism 30–31, 59–60
Gill-plate theory (of origin of insect wings) *see* Wing origin
Gills 21, 44, 45, 46, 50
GIS *see* Geographical information system
Glacial cycles 10, 367, 388–389
Gliding 2, 22, 45–46, 48, 50, 59, 61
Global change 389, 390
   warming 387, 389–392, 394, 398, 399, 403–407, 415–416, 431, 433, 435, 437, 438
   *see also* Climate change; Glacial cycles; Oxygen concentration
Glycogen 66, 69, 75, 78
Glycolysis 66, 67, 172, 375
Global positioning system (GPS) 11

*Glossina* 75, 76, 79
   *G. morsitans* 75
   *see also* Tsetse fly
Goal-seeking 89, 183, 209
Goldenrain trees (*Koelreuteria elegans* and *K. paniculata*) 169–170, 174
*Grapholitha molesta* (oriental fruit moth) 95
Ground speed 6, 7, 132, 137, 144–145, 148, 150, 151, 191, 193, 195, 196, 211–212, 215, 224, 226–227
   *see also* Migration, speed of
*Gryllus*
   *G. assimilis* 166
   *G. firmus* 166–168
   *G. rubens* 167, 168

Habitat
   accessibility, poor 308–309, 313, 315, 317
   corridors 313, 315, 316, 317
   destruction/loss 299, 307, 311, 313, 317, 390, 393–397, 399, 416
   dimensionality 236, 239, 244–253
   fragmentation 9, 248, 299, 300, 308, 313, 317, 324, 326, 333, 368, 390, 416, 436, 437
   isolation 9, 236, 247–248, 250, 252, 329, 331, 390, 405, 420, 436, 438
   patchy 8, 171, 208, 219, 227, 248, 261, 263, 292, 299, 300–302, 306, 308, 311, 312, 316, 317, 322, 326–333, 390, 391, 396, 404, 405
   persistence 7, 8, 235, 236, 238–244, 248–253, 329, 368
   predictability of 208, 210, 227
   temporary 236, 238, 240, 241, 249, 250, 252, 253
      *sensu* Southwood 261, 262
   structure 7, 184, 236, 245, 248
   spatial variation in *see* Habitat, patchy
Habituation 154
Haematophagous insects 5, 20, 91, 97–101
   *see also* Mosquito; Sandfly; Tsetse
Haemolymph 43, 65–69, 71, 73–80
Haplotypes 353, 373, 376
Hardy-Weinberg equilibrium 365, 367, 374
Hawaii 245, 247, 248, 352, 376
Hawthorn 352, 371
Heading 5, 94, 135, 136, 137, 143–144, 185, 187, 190–194, 196, 197, 200, 211–212, 229
   *see also* Course; Orientation
Headwind 93–94, 146–148, 153, 195, 196, 210
Height of flight 6, 7, 129, 137, 142, 144–145, 148, 153, 154, 183, 190, 194, 210–211, 224, 392
   *see also* Migration, altitude of; Migration profile

Heliconiine butterflies 352
*Heliconius* 350, 356
   *H. charithonia* 353
   *H. erato* 350, 352
   *H. himera* 353
   *H. peruviana* 353
*Helicoverpa* (= *Heliothis*) *zea* 97
*Helicoverpa punctigera* (native budworm moth) 7, 207–230
*Heliothis virescens* 95, 96, 345
*Helophorus aquaticus* 403
Hemiptera 6, 25, 34, 56, 167, 173, 236, 292
   *see also* Heteroptera
Hermit beetle *see Osmoderma enemita*
*Hesperia comma* 326–329, 331, 332
Heterogeneity, spatial *see* Spatial heterogeneity; Habitat, patchy
Heteroptera 117, 251, 292
   cotton stainer *see Dysdercus* spp.
   fire bug *see Pyrrhocoris apterus*
   milkweed bug *see Oncopeltus fasciatus*
   soapberry bug *see Jadera hematoloma*
Heterosis 374
Heterozygote excess 374
Hind wing expansion *see* Expansion (of hind wing)
HOAD (b-hydroxyl CoA dehydrogenase) 73, 172
Holocycly 370
Homoptera 24, 31, 114, 117
Honeybee 130, 136, 143–146, 149, 187, 189–191, 199
   *see also Apis mellifera*
Horizontal gene transfer 337
Host location (by parasitoids) 4, 5, 89, 111–123, 134
Host plant architecture 236, 239, 244–247, 249
Hovering 27–29, 34, 35, 55–56, 57, 60, 149, 150
*Humulus lupulus* (hop) 374
Hybridization 337, 348, 350, 352–3, 356, 371
*Hydraecia* 371
   *H. immanis* (native hop vine borer) 370
   *H. micacea* (potato stem borer) 370
Hymenoptera 23, 26, 31, 33, 35, 56, 144, 363
   *see also* Ants; Bees; Parasitoids
Hyperlipaemia *see* Diacylglycerol
Hyperprolinaemia *see* Proline
Hypertrehalosaemia *see* Carbohydrate

Immigration 261, 265, 274, 275, 320, 323–325, 331, 333, 339, 347, 350
Indirect and direct orientation 5, 90–91
Inducible defence 266–267, 271
   *see also* Predator-induced defence
Inland Insect Migration Project (IIMP) 228, 210, 215, 216, 225, 228
Innate, two-way movements 376

Insect monitoring radars (IMRs) *see* Radar
Insecticide resistance 345, 347–348, 351, 355, 365, 374
Intergovernmental Panel on Climate Change (IPCC) 389, 407
Inter-host genetic patterns 371
*Ips* spp. (bark beetles) 166
Island model, of population structure 339–342
Isolation by distance model (IBD) 340, 342, 344–345, 368
Isoptera 23
   *see also* Termite
Italy 421

*Jadera haematoloma* (soapberry bug) 6, 161, 169–175, 176
*Javesella simillima* 239
Johnson, C.G. 6, 159, 161, 261, 377
Jumping 22, 46, 251
Juvenile hormone (JH) 6, 161, 164, 167, 168, 173, 175, 176
Juvenile hormone esterase 167, 168, 169, 175

Kairomone 117, 119, 120, 267, 271
Kennedy, J.S. 3, 6, 159–161, 176
Kestrel 199
Kinematics 2, 26–30, 44, 146, 200
Kinesis 5, 88, 89, 159
   klinokinesis 88
   orthokinesis 88
Kinetic parameters 374
Knopper gall wasp *see Andricus quercuscalicis*
*Koelreuteria elegans see* Goldenrain tree
   *K. paniculata see* Goldenrain tree

Ladybird beetle 263, 271–275, 403
   *see also* Coccinellidae
Lake Gatún 195–196, 197
Land use change 390, 394, 396, 404
Landmark orientation 88, 134, 183, 185, 190–194, 200
Leading edge vortex 27–28, 36, 54
Leaf miner 369
Learning 88, 112, 113, 120, 121, 122, 123, 134, 185, 187, 199, 287
   *see also* Orientation flight
Lemmatophoridae 57
Lepidoptera 7, 35, 56, 113, 114, 115, 116, 119, 120, 121, 184, 188, 251, 292, 352, 356, 369, 387–407, 415–439
   *see also* Butterfly; Moth
*Leptinotarsa decemlineata* (Colorado potato beetle) 72–76, 79, 102, 166
   *Pea*-CAH-I, *Pea*-CAH-II in 72, 76
Life history 7, 159, 162–163, 169–172, 175–177, 235–237, 251, 403, 435

Life cycle synchrony 372
Lift 2, 20, 22, 28, 46–51, 53–55, 59–61, 78, 199
Ligation of DNA 363
Light-traps see Traps, light-trap
Linkage disequilibrium 343–344, 350–351, 365
Linolenic acid 116
Lipophorin 68–69
    high density lipophorin (HDLp) 68, 70
    low density particles (LDLp particles) 68–69, 70, 80
Lipoprotein 68–69
Lipoprotein lipase see Fatty acids
Locust 3, 8, 11, 51, 53, 65, 66, 67, 68, 69–70, 71–72, 79, 80, 87, 129, 184, 190
    Australian plague see Chortoicetes terminifera
    desert 11, 87, 210
        see also Schistocerca gregaria
    migratory see Locusta migratoria
Locusta migratoria
    adipokinetic hormones in 69, 71
        Lom-AKH-I 69, 71, 80
        Lom-AKH-II 69, 71
        Lom-AKH-III 69, 71
Long-distance migration see Migration, long distance
Lutzomyia longipalpis (sandfly) 101
Lycaenidae 372, 404, 405
Lysandra bellargus 396, 404
    L. coridon 396

Macroptery 162–163, 165–174, 236–253
    see also Wing dimorphism/polymorphism
Maculinea spp. butterflies 397
    M. rebeli 322
Magnetic (or geomagnetic) compass orientation 88, 183, 188–190, 199, 200
Magnolia 369
'Mainland-island' systems 328–330, 339
Manduca 53
    M. sexta 27, 77
Mantodea 24
Mark–release–recapture (MRR) 9, 130, 149, 326, 331, 346, 347, 351, 362, 363, 366, 367, 375, 392
Marpesia chiron (many-banded daggerwing) 135
Mass movements (of army ants) 281, 291–292, 309–310
Massif Central 367
Maternal inheritance 349, 364
Matrix 261
Mayflies 23, 26, 33, 44–46, 50, 60
    see also Ephemeroptera
Mazonopterum wolfforum 58–59

Meadows 393, 394, 396
Mean-field approximation 301
Mecoptera 56
Medfly see Ceratitis capitata
Megaloptera 24, 56
Megamelus paleatus 239
Megasecoptera 22, 30, 58–60
Melanism 403
Melanophila acuminata 103
Melitaea cinxia (Glanville fritillary) 329, 372
Meloidae 72
Melolontha melolontha (cockchafer) 72
    Mem-CC in 72
Mem-CC (adipokinetic hormone) 72–74, 76
Mendelian markers 363
Metapopulation 8, 9, 14, 228, 235, 262, 300, 326–329, 331–333, 339, 341, 342, 368, 372
    spatially implicit 301–304, 307, 312, 339, 341, 342
Mexico 135, 162, 183, 375, 376, 391, 394, 395, 396, 402
Microclimate 404, 405
Microsatellites (single sequence repeats, SSRs) 363, 364, 365, 367, 369, 372, 373
Migration 3, 4, 6–8, 88, 135–136, 159–177, 183–200, 207–230, 235–244, 247–253, 261–265, 292, 361, 364, 366, 391, 392
    altitude of 3, 212, 215, 219, 222–224, 366, 392
        see also Height of flight
    'arena' 207–208, 210, 213, 215, 225, 227–228
    of butterflies 7, 129, 135, 183–201, 209, 376
    of Chortoicetes terminifera 211, 218–224
    deferring departure 222–223
    direction of 7, 135–136, 183, 185, 188, 189, 190, 199, 209, 210, 218, 222–223, 224, 226–227
        see also Track direction; Departure direction
    distance see Displacement distance; Flight distance
    of Helicoverpa punctigera 211, 218–219, 229
    initiation of 211–212
    intensity 224–225
    Kennedy's definition of 3, 88, 159–161, 176
    locust phase, and adipokinetic hormones 71–72
    long-distance 6–7, 34, 162, 183–186, 194, 199–200, 210, 211, 214, 219, 236, 263, 282, 347, 351, 362, 370, 375, 376, 392

Migration *continued*
   'pathway' 207–208, 225, 227, 228, 229
   of planthoppers 236–253
   and population genetic structure 399–341, 364, 367, 370, 374, 375, 376
   predation and the evolution of 262–263, 265–266, 268, 269, 273, 275
   profile 221
      *see also* Migration, altitude of
   rate 10, 355, 365
   seasonal variation of 224–225
   spatial variation in 227
   speed of 197–198, 224, 226–227, 377
      *see also* Ground speed
   stopover 194, 197, 198
   syndrome 4, 6, 159–176, 184, 207–208, 228
   system 7, 208–209, 213, 225, 228, 230
      characterization of *see* Characterization (of a migration system)
      functioning of 227
      persistence of 227–228
   *see also* Emigration; Evolution of migration; Immigration; Optimal migration strategy
Migration-in-progress 208, 215, 223–224
Migratory *see* Migration
   capacities 211, 214
      *see also* Flight capacity
   potential 214, 228
   transport 212
      *see also* Atmospheric transport
   *see also* Non-migratory populations; Non-migratory species
Milkweed bug *see* Oncopeltus fasciatus
Milkweed 162, 376
Miniaturization of body size 2, 20, 30, 31–34, 35, 36
Minisatellites 363, 364
Mitochondrial DNA (mtDNA) 363, 364
Mobility axis 326, 328, 331, 332
Model airplane 12
Models 8, 10
   aerodynamic 27, 29
   cellular automata (CA) 8, 9, 301, 306, 311–313, 315, 317
   climate response surface 416, 417, 420, 422, 431
   conceptual (of metapopulation) 8, 9
   conceptual (of migration systems) 7, 208
   of evolution of dispersal 262, 265, 268
   of foraging 130, 134, 265
   incidence function 8
   logistic regression 8
   optimality *see* Optimality models

patch occupancy 292
population 228
simulation 10, 229, 284, 415–439
'stepping stone' model 339
*see also* Island model, of population structure; Isolation by distance model; Metapopulation; Numerical weather forecasting; Optimal foraging models; Regression
Molecular ecology 8, 13, 366
   markers 9–10, 13, 338, 342, 345, 351, 353, 354, 355, 362–377
   techniques 8, 13, 14, 362–377
Monarch butterfly *see* Danaus plexippus
Morphologically cryptic species 371
Mortality 261, 263, 265, 268, 387
   differential 208
   during movement 164, 252, 265
   risks 164, 263, 268, 272, 274, 387
   *see also* Death
Mosquito 60, 89, 90, 98, 99, 100, 101, 264, 265, 351, 356
   *see also* Anopheles gambiae, Culex pipiens, Culiseta longiareolata
Moth 4, 5, 26, 29, 66, 80, 91, 94, 95–97, 103, 104, 120, 129, 151–154, 184, 188, 190, 192–193, 195, 200, 211, 214, 218, 219, 371, 389, 392
   *see also* Agrotis exclamationis; A. segetum; Cadra cautella; Epirrita dilutata; Euproctis spp.; Grapholitha molesta; Helicoverpa punctigera; H. zea; Heliothis virescens; Hydraecia immanis; H. micacea; Mythimna separata; Noctua pronuba; Pseudoletia unipunctata; Rhopobota naevana; Spodoptera exempta; S. exigua; Thaumetopoea pityocampa; Urania spp.; Zeiraphera diniana
Mouthparts (stylets) 6, 169, 170, 172, 174, 176
'Moving deme' hypothesis 364, 376
Multivoltine 112, 241
Mutation 341, 342, 365, 375
Mutation rate 363
*Myrmica* spp. 322–323
*Mythimna separata* (oriental armyworm moth) 210
*Myzus persicae* (peach-potato aphid) 364

Native balloon vine *see* Cardiospermum corindum
Native budworm 207–230
   *see also* Helicoverpa punctigera
Native hop vine borer *see* Hydraecia immanis

*Index* 453

Natural selection 10, 35, 36, 46, 116, 161–163, 165, 171, 175, 176, 194, 207, 210, 213, 236, 238, 250, 263, 286, 295, 322, 338, 341, 342, 345–351, 354, 355, 364–366, 369, 373, 375, 376, 403
Navigation 12, 88, 104, 129, 134, 149, 150, 151, 154, 183–185, 194, 199, 200, 292, 361
NDVI *see* Normalized difference vegetation index
$N_e$ *see* 'effective size' of population
Neighbourhood population size 340, 342, 344–345, 355, 372
*Nesosydne* spp. 247
Netherlands 399, 403
Neuroptera 56
Niche patchiness 372
*Nicotiana* plant 115
*Nilaparvata lugens* (brown planthopper) 167, 239, 240, 250, 372
NOAA 216
*Noctua pronuba* (large yellow underwing) 190
'Noise' 5, 88, 91, 97, 98, 102, 104, 114
Non-migratory populations 162, 163, 176
Non-migratory (sedentary) species 368, 372, 392, 397, 398, 438, 439
Non-molecular markers 362
Normalized difference vegetation index (NDVI) 216–217
North America 31, 33, 162, 169, 240, 241, 247, 352, 368, 369, 370, 375, 376, 388, 391, 393, 394, 395, 398
Nuclear DNA (nDNA) 363, 365
Nuclear organizing regions (NOR) 362–363
Numerical weather forecasting 216, 219, 221–222, 227, 229
Nymphalidae 193, 356, 405


Octopamine 66–67
Odonata 7, 21, 29, 51, 53, 55–56, 60
    *see also* Dragonflies; *Pantala*
Odour plumes 5, 88–104
Olfactometer 102, 118
*Ona*-CC (adipokinetic hormone) 72–74
*Oncopeltus fasciatus* 6, 161–164, 166, 167, 176, 177
Onitine dung beetles 75
*Onitis aygulus* 72, 74
    *Ona*-CC in 72, 73
*Onitis pecuarius* 72, 73, 74
    *Ona*-CC in 72, 73
*Onymacris plana* 72
    *Tem*-Hr-TH in 72
*Onymacris rugatipennis* 72
    *Tem*-Hr-TH in 72
*Ooencyrtus pityocampae* 120

Oogenesis-flight syndrome 6, 161, 166
Optical flow 5, 137, 143–145, 148
Optimal flight speed 29, 195–196
    *see also* Flight speed
Optimal foraging models 265
Optimal fuel loading 197, 198
Optimal migration strategy 185, 194–197, 200
Optimal wind drift compensation 194, 200
Optimality models 185, 200
Optomotor anemotaxis *see* Taxes
Oriental armyworm moth 210
    *see also Mythimna separata*
Orientation behaviour, mechanisms 6, 12, 89–92, 104, 183–191
    direction 186–187, 190–192, 195, 210, 211, 213, 215
    *see also* Course; Heading;
    orientation of migrating butterflies 183–191
    orientation to pheromone source 94–97, 119, 153
    *see also* Flight, pheromone following
    orientation without the sun 189, 190
    *see also* Indirect orientation;
        Landmark orientation;
        Magnetic compass orientation;
        Sun compass
Orientation flight (= learning flight of bees) 149–151
Orthoptera 20, 24, 55, 60, 251, 266
*Osmoderma enemita* (hermit beetle) 13
Overwintering 135, 161, 162, 183, 199, 241–244, 250, 351, 376, 391, 392, 402, 420
Oviposition preference 369, 371
Oxygen *see* Atmospheric oxygen


*Pachnoda* 73, 74, 75, 79
    *marginata* 72
        *Mem*-CC in 72–74
    *sinuata* 72, 73, 77
Palaeodictyoptera 22, 23, 30, 57–60
Palaeozoic 30, 31, 36
Panama 185–188, 193, 196, 197
Panama Canal 188, 193, 309, 310
Panmixis 376
*Pantala* (dragonflies) 193, 197
    *flavescens* 197
    *hymenaea* 197
*Papilio glaucus* (tiger swallowtail butterfly) 369
Papilionidae (swallowtail butterflies) 352
*Parabuteo unicinctus* (Harris hawk) 199
Parachuting 2, 46, 61
*Pararge aegeria* (speckled wood butterfly) 416–439

Parasitoids 4, 5, 31, 89, 102, 111–123, 134,
        262, 265, 269, 404
    of aphids 5, 115, 116–119, 265
    of Lepidoptera 114, 119–120
    see also *Aphidius ervi*; *A. rhopalosiphi*;
        *A. rosae*; *Encarsia formosa*;
        *Ooencyrtus pityocampae*;
        *Pauesia unilachni*; *Praon* spp.;
        *Telenomus euproctidis*;
        *Tetrastichus servadeii*;
        *Trichogramma evanescens*;
        *T. sibericum*
*Parnassius mnemosyne* (clouded Apollo
    butterflies) 367
'Pathway' see Migration, 'pathway'
Patchiness (of populations) 8–9, 219–220,
    225, 268–269, 294, 300–308,
    311–313, 321–333, 390, 396, 404
    see also Habitat, patchy;
        Metapopulation
*Pauesia unilachni* 265
PCR (polymerase chain reaction) 363
Pea aphid 263, 264, 271, 273
    see also *Acyrthosiphon pisum*
*Pea*-CAH-I and II 72, 76
Pedunculate oak 366
P-elements 353
Percolation 307, 308, 317
*Periplaneta americana* 72
    *Pea*-CAH-I, *Pea*-CAH-II in 72
Permian 23, 30, 35, 57, 59
Permothemistida 23, 60
Pest management 1, 7, 11, 119, 120, 123, 152,
    215, 230, 321, 351
Pesticide 112, 351
Pesticide resistance see Insecticide
    resistance
*Phalacrus substriatus* 346
Phantom midge larvae 267
    see also *Chaoborus*
Phase polymorphism
    gregarious (crowded) locusts 71, 72
    solitary locusts 71, 72
Phase transition (to fragmentation) 309, 313,
    317
Phasmatodea 24, 251
Phenology 184, 219, 323, 391, 396, 404
Phenotypic polymorphisms 362
Pheromone 5, 91, 92–93, 94–97, 101–104,
    113, 117–121, 130, 134, 151–154
    trails of army ants 284–285, 286, 287,
        295
    trap see Traps, pheromone
    see also Flight, pheromone-following
Phoebis 186, 187
    *P. argante* (apricot sulphur
        butterfly) 186, 188
    *P. sennae* (cloudless sulphur) 196
Phoresy 3, 120

*Phormia regina* 75
*Phorodon humuli* (damson-hop aphid) 374
Photoperiod 160, 173, 211, 373
Phylogenetic non-independence 241, 252
Phytophagous insects 5, 79, 91, 97, 99,
    101–104, 236, 368
Phytophagy 20, 34–36
Pieridae 193, 196
*Pieris brassicae* 199
    *P. rapae* 375
Pine beetles 102
Pine processionary moth see *Thaumetopoea
    pityocampa*
Pitch (flight manoeuvre) 29
Pityolure 120
Plant
    defence 114, 115, 116
    signalling 114, 115, 116
    volatile 5, 101, 102, 103, 113, 114, 115,
        116
Plant community transition zone (ecotone)
    370
Planthopper 7, 167, 236–253, 263 see also
    Delphacidae
*Plebejus argus* (silver-studded blue butterfly)
    330
Plecoptera (stoneflies) 46, 56, 263
*Plecotus auritus* (long-eared bat) 199
Plume see Odour plume
Polarized light 104, 144, 189, 200
Pollination 20, 34–36, 91, 104, 404
Polymorphism 341, 353, 362, 363, 371, 373
    in ant workers 6, 282, 287, 289, 295
    phase see Phase polymorphism
    wing see Wing dimorphism/
        polymorphism
    see also Phenotypic polymorphisms
Polyphenism 165
Population
    closed 367
    density and dispersal 266, 269
    density and wing-morph 173–174
    effective population density 340, 342
    effective population size see 'Effective
        size' of a population
    distributions see Population patterns
    dynamics 5, 8, 122, 269, 282, 291, 292,
        322–333, 401, 404, 435
    genetic structure 9–10, 123, 235,
        337–356, 362–377
    persistence 228, 229, 329–331, 387, 416,
        435
    spatial structure 8–9, 299, 301, 321–333
    see also 'Effective size' of a population;
        Island model; Metapopulation;
        Model, population;
        Neighbourhood population
        size; Patchiness (of
        populations); 'Pathway'

(population movements through space and time); Subpopulations; Scale (and population structure)
Potential displacement 214
Posteromotorism 23–26
Power (requirements for flight) 27, 29–30, 31, 34, 146–148, 191, 195, 197, 198, 200
   parasite power 30, 147
   profile power 30
   the power curve 29, 30, 34, 146–147, 195, 197, 198
*Praon* 5, 118, 119
   *P. volucre* 118
Predation 7, 8, 20, 22, 34–35, 60, 102, 134, 160, 244, 261–275, 281–296, 303–305, 307, 309, 316, 317
   risk 88, 98, 134, 197, 198, 261–275
Predator avoidance 117, 262, 265, 267, 268, 274
Predator-induced defence 266, 267, 271, 272, 274, 275
   conditions for evolution of 267
Predator-induced dispersal (PID) 261–275
Predator-mediated coexistence *see* Coexistence, predator-mediated
Pre-mating barriers 371
Pre-reproductive period 164, 211, 213, 214
*Prokelisia* spp. 7
   *P. dolus* 237, 241–245, 248
   *P. marginata* 239–245
Proline 3, 65, 74, 79, 80
   hyperprolinaemia 74
   synthesis 75–77
   utilization, oxidation 74, 75, 78
Protein markers 362
Protodonata 30
Protorthoptera 23, 58–59
*Prunus* spp. 374
   *P. serotina* 369
*Pseudaletia unipunctata* (armyworm) 164
Pseudosink 323, 324
*Psila rosae* (carrot fly) 103
*Pya*-AKH 80
Pyrenees 367, 399, 400
*Pyronia tithonus* (gatekeeper butterfly) 416–438
*Pyrrhocoris apterus* (fire bug) 79, 80
   *Pya*-AKH (adipokinetic hormone) in 80
   walking activity in 79

Radar
   harmonic 5, 12, 129–154, 190
   insect monitoring (IMR) 7, 12, 207–230, 362
   X-band scanning 3, 12, 129
Radio-labelling 362, 377

Radio-telemetry (radio-tracking) 12, 13
Rainfall variability 184, 208, 210, 215, 217, 226–227
Rainforest 295
   neotropical 184, 300, 309, 317
Random amplified polymorphic DNA (RAPDs) 363, 364, 367
Range, change 9, 11, 387–407, 415–439
   and climate 388–402, 406, 415, 428
   *see also* Global warming
   contraction 327, 389, 393, 394, 415, 416, 417, 421, 427, 431, 435
   expansion 327, 366, 367, 373, 389, 391, 393, 399–400, 405, 415, 416, 417, 425, 427–430, 435, 437, 438
   future 430–433
   rate of 427–430, 433, 437
   and habitat 327, 404–405, 406, 425, 430, 437
   shift 9, 387–406, 425, 437
Ranging movements 4, 89, 90, 104, 152, 153, 160, 171
Rare allele method 342, 365
Recombination rate 350, 351, 364
Rectification 211
Reflexes (in migration) 161, 162
Refuelling 197, 198
Regression (analysis) 196, 242, 290, 372, 422, 425, 428, 433, 435, 436
Remote sensing 7, 10, 11, 422–423
   *see also* Satellite remote sensing
Reproductive isolation 338, 355
Resource finding 4, 5, 87–104, 123
   *see also* Vegetative movement
Restriction fragment length polymorphisms (RFLPs) 363
Restriction of DNA 363
Reynolds number 27–28, 36
*Rhagoletis* 371
   *R. mendax* 371
   *R. pomonella* (apple maggot fly) 104, 352, 371
*Rhopalosiphum padi* (bird cherry-oat aphid) 373
*Rhopobota naevana* (blackheaded fireworm moth) 120
Ringlet butterfly *see Aphantopus hyperantus*
Risk of predation *see* Predation risk
Roll (flight manoeuvre) 29
Route fidelity 130, 134
Rules of thumb (for army ant behaviour) 284–287
Russian wheat aphid *see Diuraphis noxia*

Sandfly *see Lutzomyia longipalpis*
*Sapindus saponaria* (soapberry tree) 170
Satellite remote sensing 11, 215–217, 229, 417, 422
Satyridae 372, 416

Scale (and population structure) 321–330, 333
Scale (of population movements) 215, 227
    see also Displacement distance
Scale insect 87, 117, 369
    crawlers 3
    see also Coccoidea
Scarabaeine dung beetles 75
Scarabaeoidea 72, 74
*Scarabaeus deludens* 72
    *Scd*-CC-I, *Scd*-CC-II in 72
*Scarabaeus rugosus* 73, 74
*Schistocerca gregaria* 13, 71, 210
    see also Desert locust
*Schizaphis graminum* (green bug) 370
Sedentary species 368, 372, 392
    see also Non-migratory species
Self-organization 6, 282, 283, 286, 295
Sequencing 363
Settling response 3, 160, 176
Sexual selection 34–36
Shuttle system for lipoproteins 68, 70
Sink, demographic 315, 317, 322–326, 333
Site fidelity 130
*Sitobion avenae* (cereal aphid) 118, 367, 373
    *S. fragariae* (cereal aphid) 370
Slatkin's paradox (of gene flow) 345–349, 351
Snowpack 389, 395, 402
Soapberry (tree) see *Sapindus saponaria*
Soapberry bug see *Jadera hematoloma*
Sources and sinks 323–326
South America 350, 375
Southwood, T.R.E. 159, 261, 262
Spain 323, 367, 399, 417, 421
*Spartina alterniflora* (cordgrass) 241–242
Spatial autocorrelation 345
Spatial heterogeneity/variability 208, 216, 262, 267–271, 292, 299–317, 321–333, 391
    see also Habitat, patchy;
Speciation 355, 371, 372
Species richness 19, 28, 31, 33
Speckled wood butterfly see *Pararge aegeria*
*Sphecoptera brongniarti* 58–59
Spider mites 263
    see also *Tetranychus urticae*
*Spodoptera exempta* (African armyworm) 4, 375
    *S. exigua* 116
Stabilizing selection 375, 377
Station-keeping 3, 4, 88, 159, 160, 161
*Stenodictya pigmaea* 58–59
'Stepping stone' model 339, 342, 372
*Stobaera* spp. 250
Stoneflies 263
    see also Plecoptera
Straightness (of migration or air path) 213, 222, 227
    see also Straightness factor

Straightness factor 213–214, 222–223
    see also Straightness (of migration or air path)
Strepsiptera 23, 33
Subpopulations 123, 339, 345, 346, 368, 375
Sun compass 88, 144, 185–190, 199
Sun's azimuth see Azimuth of sun
Surface-skimming 46, 48, 59
Survey (of populations, habitats) 7, 215–216, 229–230, 393, 401, 417
Swallowtail butterflies (Papilionidae) 352
    see also *Papilio glaucus*
Sycamore 370
Sycamore aphid see *Drepanosiphum platanoidis*
Synchronous flight muscle 31–32
Syndromes see Migration syndrome; Oogenesis-flight syndrome

Tailwind 147–148, 195–197, 200
Take-off 46, 16–161, 211, 223
Taxes 5, 88–89, 90, 159
    anemotaxis 89, 93, 99, 103
        optomotor anemotaxis 94, 95, 100, 153
    chemotaxis 89, 92, 95
    klinotaxis 88, 93
    phonotaxis 89
    phototaxis 89
    tropotaxis 88, 93
Taylor, L.R. 159
Teams, super-efficient 282, 287, 289, 295
Tegminization 20, 23–26
*Telenomus euproctidis* 120
*Tem*-Hr-TH (adipokinetic hormone) 72–74
Temperature
    allele for flight at lower temperatures 374
    associated with range shifts 388, 394, 402, 405, 415
    daily activity cycles and 89
    dispersal and 266
    and global climate change 389, 390, 407, 415
        see also Global warming
    klinokinesis/orthokinesis 88
    migratory flight capacity and 214
    remote sensing of 11, 216
    seasonal temperatures and migration patterns 184, 208, 227
    and selection for melanism in *Adalia* 403
    thoracic 77, 78
        see also Endothermic warm-up
    threshold for larval development 371, 392, 420

thresholds for initiating and sustaining migratory flight 211, 212, 222, 228
physiological tolerance 390, 420
Tempo (of worker ants) 282, 289, 291, 293, 295
Temporary habitats *see* Habitat, temporary
*Tenebrio molitor* 72, 73, 74
  *Tem*-Hr-TH in 72, 73, 74
Tenebrionidae 72
Tephritid fruitflies 371
  *see also Rhagoletis*
Termination (of migratory flight) *see* Flight, termination
Termite 60, 281, 282
  *see also* Isoptera
Terrestriality 21
Tethered flight 78, 145, 162–164
*Tetranychus urticae* (spider mite) 263
Tetraptery 25
*Tetrastichus servadeii* 120
*Thaumetopoea pityocampa* (pine processionary moth) 119, 120
Thoracic segments 22
Thrust 27, 28, 46–50, 78
Thysanoptera 35
Time compensation 186–189, 199
Time-compensated sun compass 186–188
*Toya venilia* 248, 252
Tracheal system 21, 30–31, 43
Track
  direction 5, 135, 141–144, 146, 183, 190–193, 196, 211, 212
    *see also* Migration, direction of flight tracks
      during 'ranging' 90
      of foraging bees 5, 131–150
      of insects orienting to odour source 95, 96–97
      of migrants 7, 135
        *see also* Trajectory
      of turnip moths 153
Tracking *see* Radio-telemetry; Radar; Video
Trade-offs 7, 165, 166, 369
Trajectory (of insect or air) 213–214, 219, 221–223
Traps
  light-trap 104, 215, 218, 220
  pheromone trap 118, 119, 120, 152–154, 229, 362, 366
  suction trap 5, 366, 370, 374
  water trap 118, 119
Transit (phase of dispersal) 261
Transit (mass transit systems of ants) 281, 295
Transoceanic flights 213
Transponder 12, 130, 132–133, 149, 150, 152, 153

Transport, atmospheric 211–212, 216, 219, 222–223, 225, 227, 229
Transposons 363
Trehalose 66, 67, 68, 69, 73
Triacylglycerol 66, 79, 80
Triassic 23
*Trichogramma* 119
  *T. evanescens* 120
  *T. sibericum* 120
Trichoptera 56
Trivial flight 66, 370
Trivial movement *see* Vegetative movement
Tsetse fly 87, 89, 98, 99, 100
  *see also Glossina*
*Tumidagena minuta* 239
Turkey oak 366
Turnip moth 130, 152–154
  *see also Agrotis segetum*
Two-spot ladybird *see Adalia bipunctata*


Unconditional dispersal strategy 262, 263
United States 169, 187, 346, 366, 371, 372, 373, 375, 389, 395, 402
Univoltine 112, 371, 373, 417, 429
Unsteady airflows 2, 27, 28, 29, 36, 51, 53–54
*Urania* 184, 188, 193–195, 200
  *U. fulgens* 184, 194, 195
*Urbanus proteus* 135


Vagility 368, 369, 370, 372
Vector analysis 136, 137, 191, 192, 196, 212
Vegetative movement 3, 4, 12, 160
Vein fusion 56
Veins 50–51, 54–57, 59
Video 11, 13, 90
Vision 103
Volicitin 116


Warm-up flight *see* Endothermic warm-up
Weather 135, 200, 207, 211, 266, 330, 387, 389, 391, 400, 401, 402
  disturbed 210–211
West Indies 375
Wheat 118
Whitefly *see Bemisia tabaci*
Wigglesworth, V.B. 2, 44, 45
Wind 3, 5, 34, 89, 90, 98, 135, 183, 185, 190–198, 200, 210, 212, 222, 228, 252, 366, 392
  compensation 5, 7, 129, 134, 136, 137, 143, 144, 190–197, 200
  convergence 4, 210
  direction 7, 89, 90, 92, 94, 99, 101, 132, 136, 141, 192, 193, 195–198, 200, 210, 221–222
  flight suppression by 211

vs. height (near the ground) 136, 140
speed 6, 98–100, 132, 136, 139, 140, 146, 183, 192–197, 200, 210, 211, 213, 221–223
see also Crosswind drift; Headwind; Tailwind
Wind tunnel 45, 95, 114, 115, 118, 145, 153, 374
Wing 20–36, 43–61
    camber 50–54, 59–60
    corrugation 50, 54, 57
    differentiation 20, 22–26, 30, 34, 55, 60
    dimorphism/polymorphism 7, 164, 165, 167, 170, 173, 174, 236, 239, 240, 241, 245–252, 263
    flexion 20, 54, 56, 59–60
    folding 55, 57, 60
    origin 2, 21–22, 36, 44–48, 50, 59, 61
        'gill-plate theory' 2, 21, 44–46, 361
        'paranotal lobes' theory' 2, 21
    rigidity 50, 54
Wing *Continued*
    torsion 51–57, 59–60
    twisting 48–53, 55–56, 60
    *see also* Anteromotorism; Aptery; Diptery; Posteromotorism, Tetrapty

wing length 6, 7, 162–163, 165–167, 169, 172–174
    *see also* Brachytery; Macroptery
Wingbeat frequency 12, 20, 22, 26, 31, 33, 36, 46, 199, 215
Wingless females 111, 282
Winglets 21, 22, 45–46, 48, 50, 57–59
*Wolbachia* 349
Woodland 75, 375, 404, 436

X-chromosome 364

Yaw 29

*Zeiraphera diniana* (larch budworm moth) 353, 371
*Zophobas rugipes* 72
    *Tem*-Hr-TH in 72
Zoraptera 32